Asymmetric Hydrogenation and Transfer Hydrogenation

Asymmetric Hydrogenation and Transfer Hydrogenation

Edited by

Virginie Ratovelomanana-Vidal
Phannarath Phansavath

WILEY-VCH

Editors

Dr. Virginie Ratovelomanana-Vidal
Chimie ParisTech
i-CLeHS, CNRS
11, rue P. & M. Curie
75005 Paris
France

Dr. Phannarath Phansavath
Chimie ParisTech
i-CLeHS, CNRS
11, rue P. & M. Curie
75005 Paris
France

Library of Congress Card No.:
applied for

British Library Cataloguing-in-Publication Data
A catalogue record for this book is available from the British Library.

Bibliographic information published by the Deutsche Nationalbibliothek
The Deutsche Nationalbibliothek lists this publication in the Deutsche Nationalbibliografie; detailed bibliographic data are available on the Internet at <http://dnb.d-nb.de>.

Print ISBN: 978-3-527-34610-3
ePDF ISBN: 978-3-527-82231-7
ePub ISBN: 978-3-527-82230-0
oBook ISBN: 978-3-527-82229-4

Typesetting SPi Global, Chennai, India
Printing and Binding CPI Group (UK) Ltd, Croydon, CR0 4YY

Printed on acid-free paper

C093158_190321

Contents

Foreword

It is a tremendous honor and a sheer delight to have been asked to write an introduction for this important and timely tome entitled "Asymmetric Hydrogenation and Transfer Hydrogenation," edited by my dear friends Dr. Virginie Vidal and Dr. Phannarath (Poki) Phansavath, themselves notable contributors to the field. This book will certainly stand as a landmark contribution that not only summarizes the state of the current art for this useful reaction genre but also sets the stage for practitioners to explore vistas of future importance.

As many readers will recognize, (transfer) hydrogenation has a very long and distinguished history in organic chemistry, playing a key role in the field and indeed perhaps serving as a bellwether for the development of synthetic methods in the discipline as a whole. Paul Sabatier (Nobel Prize, 1912) has been credited with the first advances in alkene hydrogenation in the late 1890s, and it was Knoevenagel who reported the first transfer hydrogenation reaction in 1903 – essentially a dimerization process involving disproportionation between identical donor and acceptor units promoted by palladium black in a *heterogeneous* transformation. Two decades later (1925) came the famous Meerwein–Ponndorf–Verley reaction, a redox coupling between a different donor and acceptor. Several more decades passed (1960s) before *homogeneous* transition metal catalysts were developed for both hydrogenation (e.g. Wilkinson's catalyst) and transfer hydrogenation, where Henbest, Mitchell, and coworkers hydrogenated carbonyls to alcohols using isopropanol as the hydride transfer reagent. This was subsequently followed by transition metal-catalyzed *asymmetric* processes for carbonyl, imine, alkene, and related reductions, in which the seminal work of Henri Kagan culminated in the award of the Nobel Prize in Chemistry to Noyori and Knowles for their contribution to the field in 2001. Thus began the explosive flurry of activity, leading to the current status of the field – a cluster of reactions that occupies an exalted status within the discipline owing to the fact that they are being used in the fragrance, flavors, pharmaceutical, and agrochemical industries worldwide on ton scale. Flow techniques, biocatalysis, and organocatalytic methods round out more recent additions to the armamentarium available to practitioners of the art of organic synthesis, pointing to exciting opportunities for further exploration.

Homogeneous catalytic asymmetric (transfer) hydrogenation transformations thus quite rightfully deserve the attention they are receiving in this volume. As pointed out throughout, these reactions represent perhaps the most efficacious and facile means to access enantiomerically enriched molecules, principally because they serve as supreme examples of atom-economical, environmentally sustainable reactions. Virginie and Poki have assembled a star-studded group of individuals to share their deep wisdom and rich experience on this topic with members of the community. Each contributor is a highly qualified expert in their field, and the result is a magnificently crafted contribution that will be of value for years to come. Congratulations to the authors Virginie and Poki for this marvelous effort!

Philadelphia, PA, USA *Gary Molander*
May 6, 2020

Preface

The homogeneous asymmetric hydrogenation of prochiral unsaturated compounds catalyzed by transition metal complexes is undoubtedly one of the most effective and straightforward methods to access enantiomerically enriched molecules. Moreover, catalytic asymmetric hydrogenation is the perfect example of an atom-economical reaction and constitutes an ideal reaction in the context of greener chemistry, warranting its use in both academic and industrial contexts. Another efficient method to approach these chiral compounds relies on asymmetric transfer hydrogenation. In this case, no hydrogen gas is needed, and hydrogen donor sources are used. The operational simplicity of this method makes it a highly useful complement to hydrogenation. The year 2001 marked the awarding of the Nobel Prize in Chemistry to Professors R. Noyori, W. S. Knowles, and K. B. Sharpless for their contributions to the development of asymmetric catalysis. As in any other scientific fields, these outstanding chemists relied on the efforts of many prior generations of chemists who laid the foundation for their exceptional discoveries. In this context, a plethora of scientists made major contributions in advancing the field of asymmetric hydrogenation and transfer hydrogenation reactions. Nowadays, these methods are widely used in academic laboratories and for process and manufacturing facilities worldwide in the pharmaceutical, agrochemical, and fine chemical industry. This book covers various aspects of homogeneous asymmetric (transfer) hydrogenation, and the different chapters are not intended to provide a comprehensive review but rather to present recent progress in the field of homogeneous asymmetric hydrogenation and transfer hydrogenation.

In the first chapter entitled "The Historical Development of Asymmetric Hydrogenation," J. M. Brown discusses the early developments that enabled the current levels of success in catalytic asymmetric synthesis, focusing on alkene hydrogenation.

In the following chapter, C. Bruneau presents the "Asymmetric (Transfer) Hydrogenation of Functionalized Alkenes During the Past Decade" and details a selection of relevant results obtained with transition metal catalysts during the past decade using combinations of new families of optically pure ligands with group 8 (Ru), 9 (Rh, Ir), and 10 (Pd) metals constituting the basis of the main recent advances in this field.

N. Arai and T. Ohkuma describe in the chapter entitled "Asymmetric (Transfer) Hydrogenation of Functionalized Ketones" the recent progress that includes the asymmetric hydrogenation of alkyl ketones, α,β-unsaturated ketones, α-aminoketones, α-hydroxyketones, and α-oxophosphonates. Highly enantioselective hydrogenations of these ketones are introduced with a main focus on the homogeneous catalytic systems.

J.-H. Xie and Q.-L. Zhou review in the fourth chapter the "Asymmetric (Transfer) Hydrogenation of Aryl and Heteroaryl Ketones," covering the recent progress of these effective and atom-economical methods to provide enantiomerically enriched alcohols and the development of transition metal catalysts containing chiral ligands.

The fifth chapter entitled "Asymmetric (Transfer) Hydrogenation of Substituted Ketones Through Dynamic Kinetic Resolution" written by P.-G. Echeverria, T. Ayad and ourselves, P. Phansavath, and V. Ratovelomanana-Vidal depicts the preparation of valuable building blocks bearing up to three stereogenic centers in one single step, showing that transition metal-promoted asymmetric hydrogenation and transfer hydrogenation are particularly efficient tools especially when combined with a dynamic kinetic resolution process, highlighting recent advances achieved in this field.

X. Zhang and P.-L. Shao, in the chapter entitled "Industrial Applications of Asymmetric (Transfer) Hydrogenation," outline the intensive use of asymmetric hydrogenation in the pharmaceutical, agrochemical, fragrance, and flavor industries, based on the privileges of high atom economy, high reactivity, mild reaction conditions, high cost efficiency, and sound environmental factors. Numerous unsaturated compounds, such as enamines, ketones, olefins, and imines, can be hydrogenated in excellent enantioselectivities.

In the chapter entitled "Tethered Ruthenium(II) Catalysts in Asymmetric Transfer Hydrogenation," V. K. Vyas, R. C. Knighton, and M. Wills recount the discovery and development of Ru(II)-tethered catalysts for the asymmetric transfer hydrogenation of ketones and imines, covering aspects of catalyst design and activity as well as applications of the catalysts, notably to industrially relevant targets. This class of catalyst has enjoyed a significant amount of interest from researchers worldwide and offers exciting prospects as a practical reagent in both industrial and academic laboratories.

Q.-H. Fan, Y.-M. He, F.-J. Li describe in the eighth chapter the "Homogeneous Asymmetric Hydrogenation of Heteroaromatic Compounds Catalyzed by Transition Metal Complexes" as a straightforward and atom-economical way to a wide range of chiral heterocyclic compounds. This chapter sums up the major developments and provides representative examples, giving an overall perspective for the homogeneous catalytic asymmetric hydrogenation of heteroaromatics utilizing hydrogen gas as reducing agent.

The chapter entitled "Asymmetric (Transfer) Hydrogenation of Imines" written by I. Peñafiel, J. Mangas-Sánchez, and C. Claver deals with the asymmetric reduction of imines via catalytic hydrogenation and transfer hydrogenation mediated by noble transition metal complexes as well as through biocatalysis. Particularly, because of intense research in enzyme discovery and directed evolution, biocatalytic imine reduction is envisaged to expand its substrate scope in the near future.

The chapter "Asymmetric Hydrogenation in Continuous-Flow Conditions," written by G. Farkas, J. Madarász, and J. Bakos, summarizes the development in the field of continuous-flow asymmetric hydrogenation reactions based on different types of catalysts such as chirally modified metal surfaces, chiral organometallic and organocatalysts, and unmodified transition metal surfaces used in chiral auxiliary-aided synthesis. Homogeneous or heterogeneous systems, as well as transfer hydrogenation reactions, are discussed.

In the last chapter entitled "Organocatalytic Asymmetric Transfer Hydrogenation Reactions," S. Das, V. N. Wakchaure, and B. List give an overview of the various types of organocatalytic asymmetric transfer hydrogenation methods using different types of acids, including chiral amines, thioureas, phosphoric acids, and disulfonimides using biomimetic Hantzsch esters as the reductants, demonstrating that various functional groups such as C=C, C=O, and C=N could be reduced to afford useful building blocks.

We are grateful to each of the outstanding authors for their invaluable contributions to this project. We deeply thank them for the professionalism and expertise they have displayed in making this book possible. We also thank the staff at Wiley-VCH for their crucial help in assembling the final product.

Chimie ParisTech-CNRS, Paris, France

Phannarath Phansavath and Virginie Vidal

1

The Historical Development of Asymmetric Hydrogenation

John M. Brown

Chemistry Research Laboratory, University of Oxford, 12 Mansfield Road, Oxford OX1 3TA, UK

1.1 Introduction

How did chemists gain the current levels of knowledge and expertise for controlling molecular chirality through hydrogenation or otherwise? The desirability of asymmetric synthesis was recognized in the 1880s by Emil Fischer and others, but practical solutions only arose more than 80 years later. The key reasons are explored here. This brief review has five main Sections 1.2–1.6, covering first the development of ideas underpinning our understanding of asymmetry, then the initial applications to asymmetric synthesis, and also the development of asymmetric heterogeneous hydrogenation of alkenes. The final sections on asymmetric homogeneous hydrogenation of alkenes are limited to work published in or before the early 1980s, in advance of extensive developments, and thus excluding the important inputs of iridium catalysts and more recently early transition metals.

1.2 Early Work on the Recognition of Molecular Asymmetry

Chemistry was an emerging science by the beginning of the nineteenth century with many opportunities for fundamental discovery. At that time scientists crossed disciplines easily; optics and mineralogy played important roles because of the ready accessibility and verifiable purity of solid substances. Malus had invented the first polarimeter in 1808, enabling measurement of both the sense and magnitude of rotation of plane-polarized light [4]. Following this, work by Arago and others on the interaction of polarized light with minerals intensified in the following decade [5]. Haüy had earlier concluded that each type of crystal has a fundamental primitive, nucleus or "integrant molecule" of a particular shape, that could not be broken further without destroying both the physical and chemical nature of the crystal. He had accidentally dropped and shattered a crystal of calcite that enabled him to make the deduction [6]. Biot observed the striking phenomenon that samples of plane sections

Asymmetric Hydrogenation and Transfer Hydrogenation, First Edition.
Edited by Virginie Ratovelomanana-Vidal and Phannarath Phansavath.

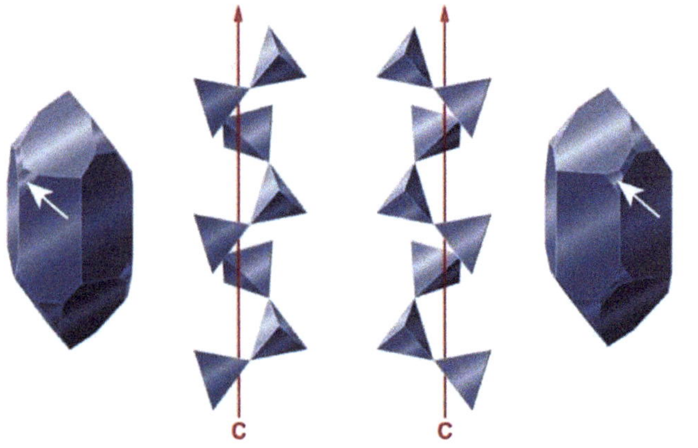

physica status solidi (b), Volume: 249, Issue: 11, Pages: 2057-2088, First published: 20 September 2012, DOI: (10.1002/pssb.201248188)

Figure 1.1 α-quartz crystals with object and mirror image hemihedral faces; inner part shows packing of sub-units. Created by K-H Ernst and reproduced with permission of Wiley-VCH.

of rock crystal (α-quartz) rotated the plane of plane-polarized light. Furthermore, some quartz crystals rotated polarized light to the right, and others to the left. The observations obtained by Biot for liquids in his designed polarimeter showed that diverse natural substances, either as liquids or in solution, showed the same phenomenon of optical activity with consistent rotations to the left or to the right for a given substance, which he quantified through Biot's Law. The vapor of oil of turpentine also demonstrated optical activity. By contrast, water, alcohol, and sulfuric acid were inactive [7, 8]. Biot deduced that the response to polarized light was a property exhibited by the individual molecules of the analyte, making a link to Haüy's proposal. The English scientist Herschel was aware of this work and was able to correlate the direction of rotation for α-quartz crystals with the structure of the crystal. He made it clear that the mirror-image pair differed by virtue of the hemihedral faces that were themselves object and mirror image (Figure 1.1) [1, 2]. Well over a 100 years then elapsed before the absolute configuration of an α-quartz crystal was determined by De Vries DeVries, using Bijvoet's recently developed anomalous dispersion method. The laevorotatory form is on the left of Figure 1.1a [3].

This laid the groundwork for explaining a puzzling observation. Cream of Tartar is a crystalline product isolated as a by-product from winemaking and was widely used in baking and otherwise. It was known to be the dipotassium salt of tartaric acid and showed an optical rotation as expected. A winemaker in the Vosges had isolated a second crystalline product at the same stage of production, and the ensuing isolated acid (racemic acid) had similar properties to tartaric acid but lacked optical activity. This became interesting to the prominent Swedish chemist Berzelius in the late 1820s [9]. He characterized his "paratartaric acid" thoroughly. It was identical to tartaric acid in analytical composition, had the same chemical composition and the same physical properties, and same distinct melting points. His student Mitscherlich [10], by then working in Berlin, discovered that an aqueous solution of paratartaric acid was "indifferent" to polarized light in contrast to the known optical activity of tartaric acid and its salts in solution, although his isolated crystal was active. From the crystal structures of the acids and their sodium ammonium double salts, he concluded that: “the nature and number of atoms, their arrangement and their distance from one another are the same in both bodies.” This sparked Pasteur’s interest, as he himself acknowledged; he repeated the crystallization of both sodium ammonium salts. Both sets of crystals were hemihedral with a key proviso: the hemihedral faces of the tartrate crystals all had the same sense, while those of paratartaric acid lay either to the left or to the right. Pasteur was able to separate the right-handed from the left-handed crystals on a 5-g scale and examine them separately in solution by polarimetry. Both were optically active but in near equal and opposite directions! This Eureka moment needed rechecking under the critical gaze of his mentor and senior colleague Biot. With that test successfully achieved, his reputation was then secured in 1848 when 26 years old. Later on, he extended this seminal sequence of experiments by resolving paratartaric acid into its enantiomers with the alkaloids quinidine or cinchonidine (CD) [11–13].

All this was carried out without a proper understanding of the molecular structure at the atomic level. The first insights into that came many years later, with the publications of Van’t-Hoff and Le Bel in late 1874 [14–16]. Prior to that aliphatic organic compounds were typically drawn (and presumably visualized) as a linear formula string or in-plane with dotted connections (e.g. typical Berichte papers, 1874). Van’t Hoff’s publication solved the existing problem of isomerism in saturated organic compounds at a stroke – “The theory is brought into accord with the facts if we consider the affinities of the carbon atom directed toward the corners of a tetrahedron of which the carbon atom itself occupies the center.” He also explained the existence of enantiomers in the case when a carbon atom had affinities to four different substituent groups. Van’t Hoff had claimed to be inspired by the work of Wislicenus on lactic acid, $CH_3CHOHCO_2H$, stereoisomerism [11]. Le Bel made similar conclusions and was specifically concerned with explaining the link between tetravalent carbon and optical activity. These analyses proved to be the foundation stone of modern organic chemistry.

The problem of visualizing representations with multiple stereocenters was first solved by Emil Fischer’s in-plane notation for sugars, compared with more modern representations in Figure 1.2, [17]. He had determined the relative configurations by chemical means. In order to specify the then unknown absolute configurations, the penultimate carbon in the chain of the dextrorotatory isomer was written with OH to the right, with the aldehyde (or equivalent group) placed at the top of the chain. His guess at the absolute configuration was proved correct, much later, through

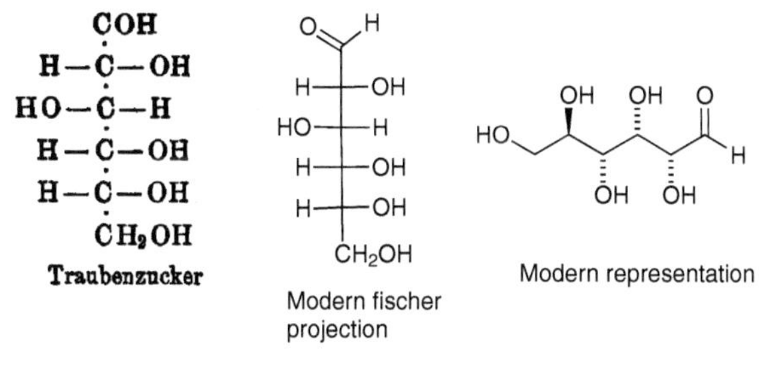

Figure 1.2 Fischer’s representation of the open-chain form of glucose and its modern variants. Source: Based on Lichtenthaler [17].

stereochemical correlation with Bijvoet's X-ray analysis of the sodium *rubidium* salt of tartaric acid, using zirconium Kα radiation that specifically excited Rb [18].

1.3 Origins and Early Development of Asymmetric Synthesis [19]

The contrast between optical activity in Nature and its absence in synthesized organic compounds led to the suggestion of a "vital force" that intervened to create natural chirality. This provided a challenge to several late nineteenth century organic chemists, with Emil Fischer prominent. He first suggested that chlorophyll was responsible for the conversion of CO_2 into single enantiomers of sugars by plants. Although this was not correct, it made a link between chirality and asymmetric synthesis. He discovered how to convert hexoses into their higher homologues by successive Kiliani hydrocyanation, nitrile hydrolysis to give a lactone, and Na amalgam reduction. He observed significant excesses of one diastereomer of product, formally an asymmetric induction, but he was aware of the possibility of the interconversion of isomers in the course of a multistage reaction, however. To avert this he studied the hydrocyanation of the tetraacetyl salicylaldehyde glucoside, helicin, the glucose moiety acting as a chiral auxiliary. The ensuing cyanohydrin was hydrolyzed and oxidized to the corresponding carboxylic acid with concurrent glycoside cleavage, Figure 1.3a. The product exhibited a very small rotation that implied some asymmetric induction had occurred but insufficient to make a secure claim of success [20]. It then remained for Marckwald to provide a more robust example in the decarboxylation of methylethylmalonic acid in the presence of 1 equiv of the alkaloid brucine (Figure 1.3b). After workup and complete removal of brucine, the monoacid had an optical rotation equivalent to 10% excess of one enantiomer [21]. The observation was carefully repeated and confirmed by Erlenmeyer and Landsberger [22]. From then on it was accepted that purely chemical asymmetric syntheses could be attained. But how could this be achieved in practice? The test of a synthetic method lies in its application, and by that criterion asymmetric synthesis remained a self-contained field for the next 60 years. Asymmetric synthesis played only a marginal part in the many impressive total syntheses that were achieved in that period. To reach an asymmetric target, the chirality tended to come from a natural source [23].

Some of the reasons for the slow development of practical and efficient asymmetric syntheses are easy to understand. The first half of the twentieth century experienced two major wars, with inevitable disruption of the progression of basic science. The early examples of asymmetric synthesis were based on empirical observation and carried out at a time before the theories that underpin experimental design – the nature of the chemical bond, electronic theory, and the process of bond making and bond breaking – were properly described. The development of organic reaction mechanisms coincided with an increased appreciation of organic stereochemistry and structure, and so by the 1950s serious attempts to build a rational foundation for asymmetric synthesis were in place. One of the most conspicuous of these was the extensive systematic study of asymmetric induction

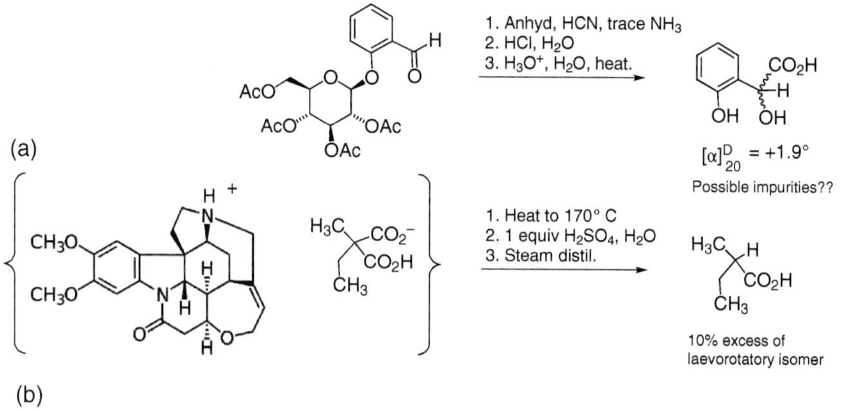

Figure 1.3 Early attempts at asymmetric synthesis: (a) Source: Based on McKenzie [20]. (b) Source: Marckwald [21] and Erlenmeyer et al. [22].

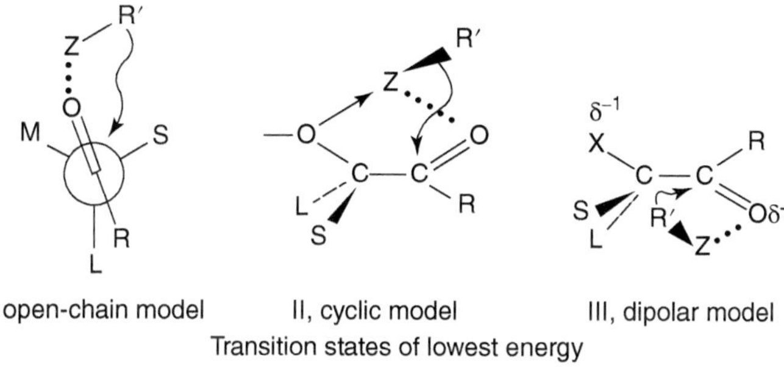

Figure 1.4 Cram's original model for 1,2-asymmetric induction. Source: Reprinted with permission from Cram and Wilson [24]. © 1963 American Chemical Society.

by Cram and coworkers. He was largely concerned with the relative configuration of diastereomers, generated through addition of nucleophilic reagents to ketones, but also recognized base-catalyzed interconversion when control reactions with pure enantiomers were performed [25, 26]. The sequence of papers led to the original Cram model for 1,2-asymmetric induction, later modified and refined [24] (Figure 1.4).

The contemporary work of Prelog was likewise directed to a precise understanding and extending of asymmetric synthesis. His analyses of mechanism and stereochemistry provided models for later workers. Bredig and Minaeff had discovered alkaloid-catalyzed aldehyde hydrocyanations, probably the first genuine example of asymmetric catalysis [28]. Prelog revisited this to elucidate the mechanism and was able to construct models that could explain the ≥25% ee involved in the cinchona alkaloid-catalyzed hydrocyanation of cinnamaldehyde [27, 29] (Figure 1.5).

All the early work suffers from a significant handicap in that polarimetry was the only rigorous tool available for estimation of enantiomeric purity – essentially so throughout the period covered by this article. If specific rotations were small, or the product was difficult to purify completely, this placed severe demands on accuracy. The optical yields quoted in the present text are taken directly from

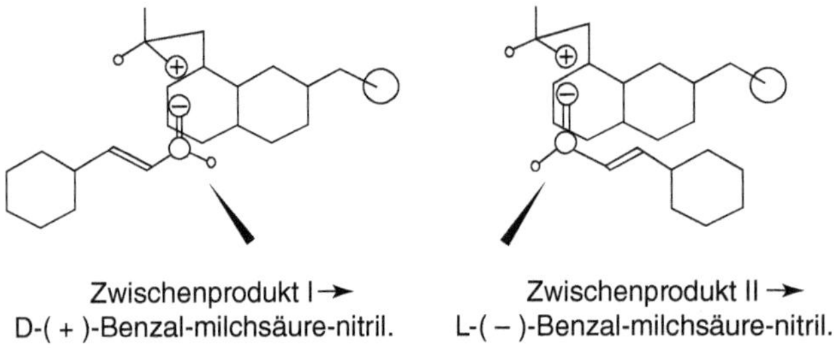

Figure 1.5 Prelog's 1954 models for alkaloid-catalyzed enantioselective HCN addition to cinnamaldehyde. Source: Reprinted with permission from Prelog and Wilhelm. © 1954, John Wiley & Sons [27].

the original literature references and may suffer from these shortcomings. The advent of chiral chromatography and NMR techniques changed perspectives and facilitated the upsurge in activity in asymmetric synthesis from the mid-late1970s onward. Enantiomer separation by gas chromatograph (GC) with chiral columns was demonstrated in Gilav's 1966 publication [30] but did not enter widespread use until commercially prepared columns were accessible several years later. Chiral HPLC, currently (2019) the most widely used technique, was developed a decade later and its convenience and accuracy prevailed [31]. Increasing use of NMR made it the primary tool for characterization of organic compounds by the mid-1960s. The invention of chiral derivatizing agents, notably Mosher's acid, allowed measurement of enantiomer excess by NMR [32]. Chiral europium shift reagents enabling direct NMR determination on a reaction product followed soon afterward [32]. These advances were critical in moving asymmetric catalysis, including hydrogenation, to a central role in synthesis. Catalytic reactions need careful optimization and without access to rapid analytical techniques that would have been a strong deterrent.

It makes a useful exercise to make comparisons of the first discoveries in asymmetric hydrogenation (1968–1972) with the contemporary state of the art in asymmetric synthesis [33]. The examples selected are intended to illustrate the breadth of effort in and shortly before that time, rather than provide a comprehensive survey.

Hydroboration of Alkenes afforded the first opportunity for a practical and general laboratory asymmetric synthesis from 1961 onward. The principle is simple in that the chiral entity lies in a natural product-derived enantiomerically enriched borohydride, and after hydroboration of the desired alkene, the borane reagent is removed stoichiometrically by oxidation [34, 35]. After further developments, a versatile synthesis of secondary alcohols derived from *cis*- or cyclic alkenes with 48–91% optical yield became available (Figure 1.6).

Asymmetric Diels–Alder Reactions are potentially of two types in that one reactant may bear an auxiliary resolved chiral group leading to asymmetric induction,

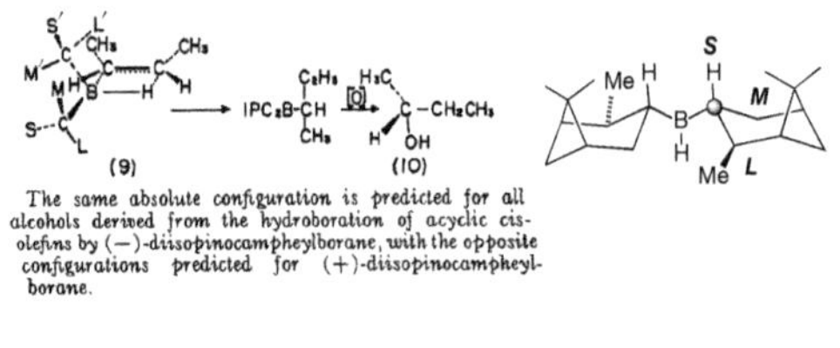

Figure 1.6 H. C. Brown's model for the asymmetric hydroboration of alkenes. Source: Based on Brown and Zweifel [34].

or alternatively the cycloaddition reaction may be driven by asymmetric catalysis, normally through chiral Lewis acid attachment to a basic group in the dienophile side chain. Early work was exclusively in the former category [36]. Walborsky's 1963 paper gave interesting insights; very low optical yields were obtained in the reaction of dimenthyl fumarate with simple dienes unless an achiral Lewis acid catalyst ($TiCl_4$ and $SnCl_4$) was present to activate the dienophile when values up to 78% could be obtained [37]. Further work by Farmer and by Sauer extended these results with varied reactants and Lewis acids [38, 39].

Asymmetric Ketone Reductions. Stereospecific NADH-promoted reductions of carbonyl compounds play a central role in metabolism and had challenged synthetic imitation. The challenge was met through Corey's need for a diastereomerically pure single secondary alcohol intermediate in his prostaglandin synthesis. He achieved this through preparing an enantiomerically pure borohydride reagent from limonene with careful optimization. Further improvement effected through precise changes to the adjacent ester group gave 92 : 8 selectivity in the desired step [40, 41] (Figure 1.7).

Asymmetric Additions to Ketenes. In a series of papers beginning in 1960, Pracejus made a robust kinetic and mechanistic study of a simple asymmetric reaction – the catalytic addition of nucleophiles to prostereogenic ketenes RR′C=C=O [42]. Addition of MeOH was effectively catalyzed by alkaloid bases, but only acetylquinidine gave optical yields ≥50%, and that only at temperatures below −80 °C, with an optimum of 74% at below −100 °C; Figure 1.8.

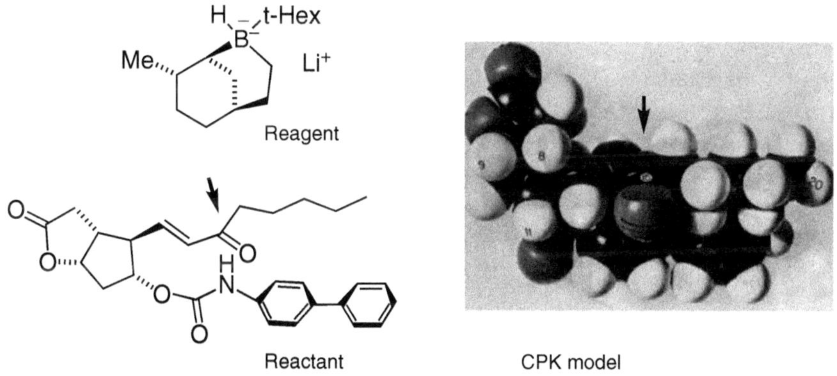

Figure 1.7 Chiral borohydride reduction of a prostaglandin precursor: CPK model. Source: Corey et al. [40], with permission from the American Chemical Society.

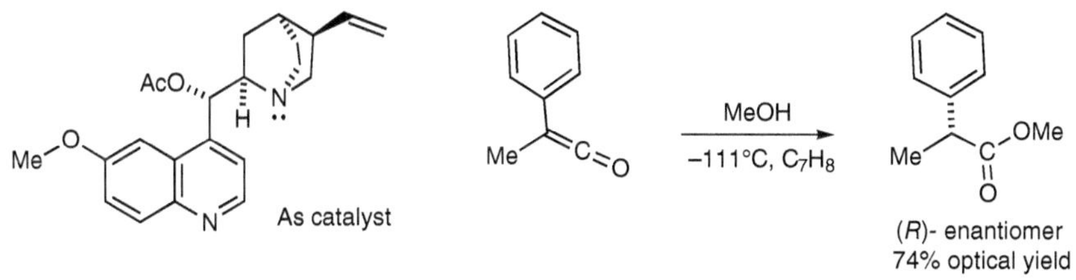

Figure 1.8 Catalytic asymmetric synthesis of α-chiral esters according to Pracejus.

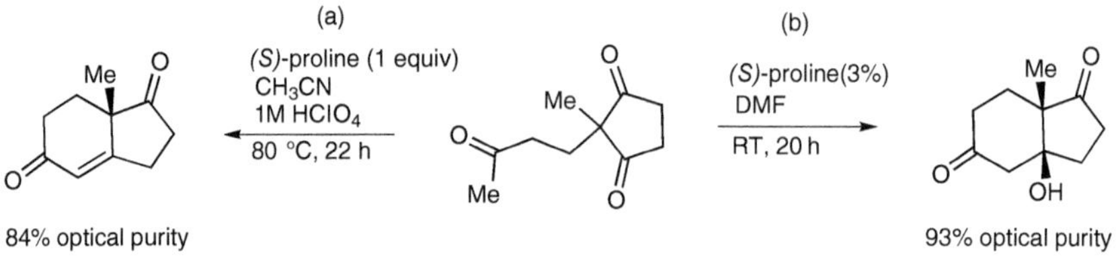

Figure 1.9 Early examples of the asymmetric aldol reaction: (a) Source: Based on Eder et al. [43]. (b) Source: Based on Hajos and Parrish [44].

Catalytic Asymmetric Aldol Condensations. The Robinson–Michael addition reaction was a standard method for preparing ring-fused cyclohexenones; the ring-closure reaction generates a new asymmetric center, and it was utilized in the preparation of the CD ring moiety in steroids. In 1971, two patents were published within weeks from separate pharmaceutical companies that demonstrated a practical method for the catalytic asymmetric synthesis of indenones (i.e. ring CD of steroids) and octalindiones in high enantiomer excess, with a simple amino acid as catalyst, preferably proline [45, 46] (Figure 1.9). The follow-up papers appeared in 1971 and 1974 [43, 44]; hence, the procedure is often called the Hajos–Parrish–Eder–Sauer–Wiechert reaction to acknowledge the dual discovery; the first two authors were based at Hoffman la Roche (New Jersey) and the other three at Schering (Berlin). Of all these early results in asymmetric synthesis, this has had the most lasting impact through the much later development of the acyclic asymmetric aldol reaction – and the (re)birth of organocatalysis [47].

1.4 Early Developments in the Asymmetric Heterogeneous Hydrogenation of Alkenes

Catalytic hydrogenation was known in the late nineteenth century. James Boyce (USA) converted plant oils into reduced edible oils for commercial use by hydrogenation over a nickel catalyst, but it was his French contemporary Paul Sabatier who undertook a systematic study of the reaction, introducing the general use of hydrogenation into organic synthesis. He demonstrated that finely divided metals could catalyze the hydrogenation of double, triple, and aromatic bonds [48].

When a C=C double bond is unsymmetrically disubstituted either at one or both termini, hydrogenation gives rise to enantiomers, depending on the face of H_2 addition, with the potential for catalytic control. This was recognized far earlier than any practical realization of asymmetric synthesis. So how could an asymmetric heterogeneous hydrogenation catalyst be created? Klabunovskii's approach involved depositing metals on single-handed α-quartz powder. The hydrogenation of ethyl 2-phenylcinnamate proceeded at 135 °C to a product that had a defined but very low optical rotation [50, 51]. Asymmetric hydrogenation reactions of higher selectivity were carried out by Akabori. He demonstrated that $PdCl_2$, deposited on silk fibroin and prereduced, catalyzed an imine–amine hydrogenation to a phenylalanine precursor in c. 25% optical yield. A similar approach was used with an alkene precursor

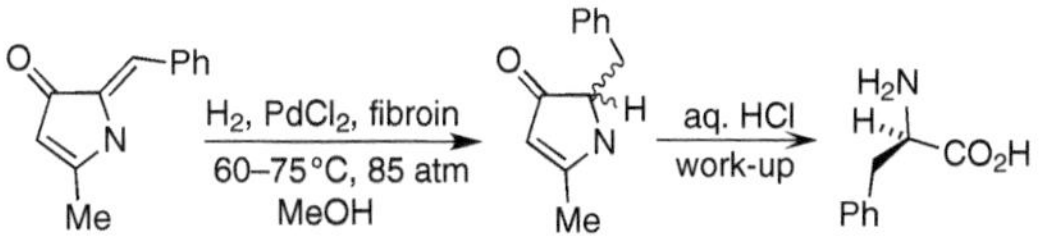

Optical yield (D)-phenylalanine, %	
Silk fibroin	36
Acetylated silk fibroin, wild	32
Acetylated silk fibroin, cultured	66

Figure 1.10 Asymmetric alkene reduction effected by chirality transfer from a metal supported on a natural protein. Source: Data from Izumi et al [53], and see also Izumi's 1971 review [54].

of phenylalanine, giving the desired product in 36% optical yield. This was claimed to be the first synthetic asymmetric catalyst, albeit using the natural chirality of fibroin. By employing acetylated fibroin from cultivated silkworms, an optical yield of 66% was achieved [52–54] (Figure 1.10). The current successful examples of heterogeneous asymmetric hydrogenation, arising from Orito's work on Pt/cinchona catalysts, continue the emphasis on C=N and C=O asymmetric reductions but with less encouraging results for C=C reductions [49].

Asymmetric induction in the heterogeneous hydrogenation of chiral alkenes where a stereogenic center is in proximity to the alkene had been observed in several distinct cases using unmodified metal catalysts. A very early example due to Bergmann and Tietzmann showed that enantiomerically enriched phenylalanine was formed on hydrogenation, and then hydrolysis, of the mixed diketopiperazine from (L)-proline and dehydrophenylalanine; the specific rotation of the crude product indicates high diastereoselectivity arising from asymmetric induction. Remarkably, the reversed product configuration was observed for hydrogenation of the ring-opened amination product from the bicyclic substrate [55] (Figure 1.11). In subsequent publications, Schmidt and coworkers endorsed and refined the basic

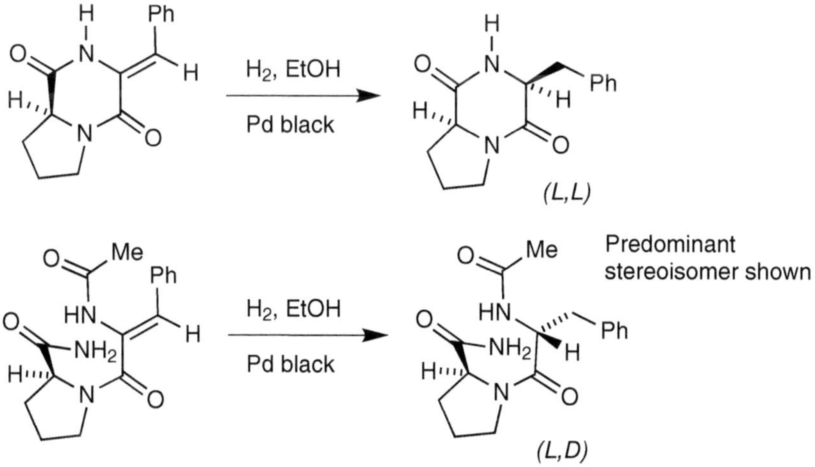

Figure 1.11 Hydrogenation of prolyl-dehydrophenylalanines according to Bergman and Tietzmann.

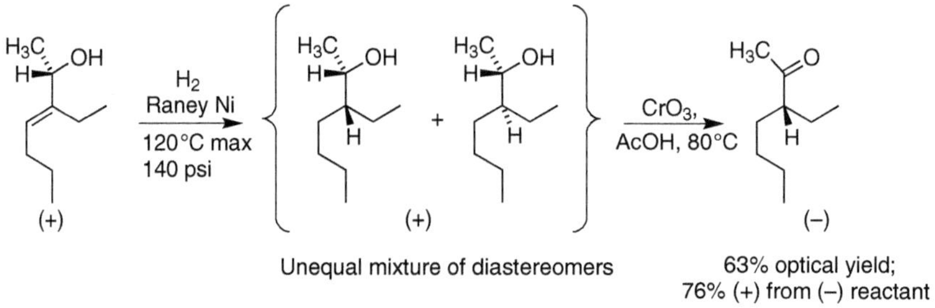

Figure 1.12 Asymmetric induction in hydrogenation; the (*R*–) configurational correlation was defined in later work by Mori et al. [58].

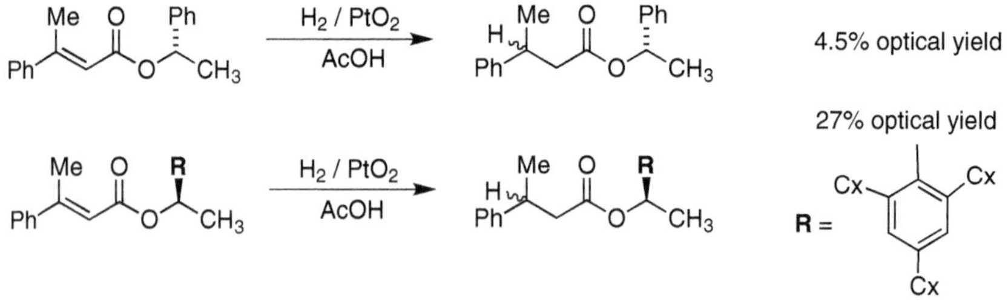

Figure 1.13 Hydrogenation of enantiomerically pure β-methylcinnamic acid esters. Source: Based on Arcus et al. [60].

elements of this work; proline was shown to be a necessary component for high diastereoselectivity [56, 57].

In a sequence of studies with simple acyclic reactants, Arcus and coworkers discovered very early examples of alkene hydrogenation directed by a hydroxyl group. He resolved his allylic alcohol reactant (Figure 1.12) and hydrogenated each hand separately to produce an unequal mixture of diastereomers. The level of stereoselectivity was revealed by oxidation of the secondary alcohol while retaining the new asymmetric center created in the hydrogenation reaction [59, 60].

Prelog systematically investigated steric effects on diastereoselectivity in hydrogenation, varying the bulk of the substituents at the stereogenic center remote from the alkene, in the further examples shown in Figure 1.13 [61].

In summary and despite much interesting work, there was little prospect of a synthetically useful heterogeneous asymmetric catalyst *for alkene reduction* by the early 1970s [62].

1.5 The Development of Rhodium Asymmetric Homogeneous Hydrogenation of Alkenes

Initial Work. After 1945, the practical use of metal catalysts was extensively explored. One of the first major applications in industry was the hydroformylation of terminal alkenes based on Roelen's pioneering work [63]. It was realized that hydroformylation of alkenes using $HCo(CO)_4$, the preferred homogeneous catalyst, could be accompanied by hydrogenation of the product aldehyde as a side reaction, and sometimes hydrogenation of the reacting alkene was observed (e.g. with dienes) [64]. There was an early demonstration of alkene hydrogenation as the preferred reaction pathway; high-temperature reaction of methyl acrylate with CO/H_2 with $Fe(CO)_5$ as catalyst in benzene leads to predominant reduction to methyl propionate as the main product at low CO pressure [65]. Halpern then achieved homogeneous hydrogenation under ambient conditions. In aqueous 3 M HCl, the coordination com plex $(NH_4)_2RuCl_6$ catalyzes the hydrogenation of simple unsaturated acids, and an alkene–Ru complex may be observed spectroscopically [66]. At about that time, the catalytic hydrogenation of alkenes by added tributylborane under forcing conditions was demonstrated [67].

Rhodium–Phosphine Hydrogenation Catalysts. In the following years, many further examples of homogeneous hydrogenation were recorded, although none held the promise of a general synthetic method. This goal was realized with the introduction by Wilkinson and coworkers of the low oxidation state PPh_3 complexes of rhodium and ruthenium that are simply prepared and bench stable. They showed that $ClRh(PPh_3)_3$ is a general homogeneous catalyst for alkene reduction, allowing extensive physicochemical studies [68–70], while $HRu(PPh_3)_3Cl$ is a selective catalyst for hydrogenation of terminal alkenes [71, 72]. The efficacy of "Wilkinson's catalyst" for a range of homogeneous alkene hydrogenations brought rhodium catalysis into the mainstream of organic chemistry. The advantages in selectivity over heterogeneous processes were manifestly demonstrated by selective monohydrogenation of dienes [73], selective deuteration [74], and face-selective hydrogenations in steroidal cycloalkenes [75] (Figure 1.14). Horner made a detailed study of reactivity with *in situ* generated P_3RhCl complexes and demonstrated that electron-releasing aryl groups enhanced reactivity. This work suggested that the catalytically active species was the dihydrido complex ClH_2RhPS, where S was a labile solvent molecule displaceable by the alkene [76].

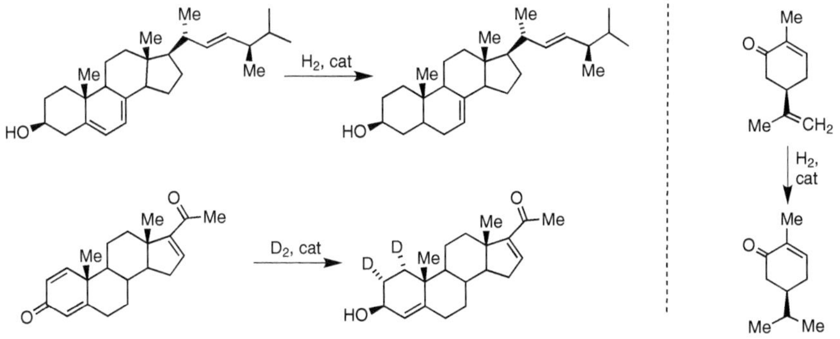

Figure 1.14 Early examples of selectivity in homogeneous hydrogenation with Wilkinson's catalyst.

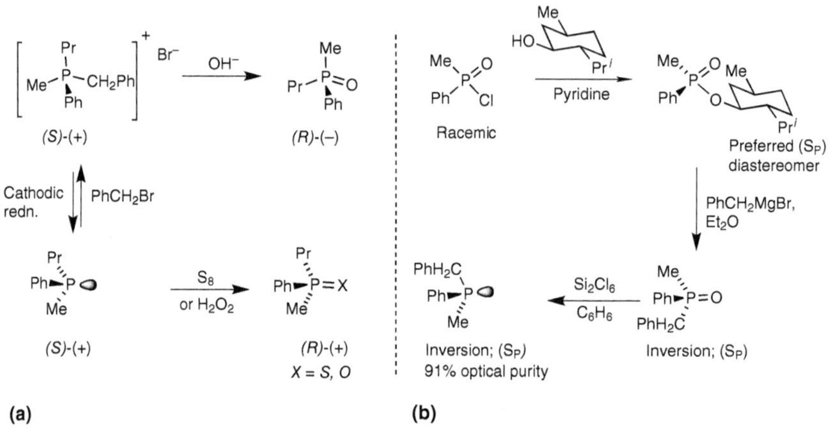

Figure 1.15 (a) Enantiomerically pure phosphines; (a) Horner's stereochemical correlation set; (b) Mislow's method for asymmetric synthesis of tertiary phosphines.

Phosphorus chirality. By the early 1960s it was known that enantiomerically pure benzylphosphonium salts underwent a stereospecific transformation to the corresponding phosphine oxide in aqueous base through debenzylation. Horner extended this chemistry, showing that electrochemical reduction of resolved phosphonium salts led to fragmentation of the most labile substituent with stereospecific formation of the phosphine, typically through loss of a benzyl group. The phosphines produced racemized with half-lives of a few hours at 130 °C [77, 78]. The absolute configuration of (*S*)-(+)-MePrPhP was established by the chemical correlation with the benzylphosphonium salt of the established absolute configuration [79]. In a later work, the direct reduction of phosphine oxides to the corresponding phosphines using $HSiCl_3$ was shown to occur with only a small loss of enantiomeric purity [80]. Mislow's work was initially directed to quantifying the slow pyramidal inversion in compounds of third-row elements, especially compounds of sulfur and phosphorus [81]. This enabled the development of a versatile synthesis of enantiomerically pure phosphine oxides using menthol as chiral auxiliary [82, 83]. Furthermore, he improved Horner's silane reduction procedure, using Si_2Cl_6 in place of $HSiCl_3$ [84]. These advances laid the groundwork for the rapid utilization of phosphine ligands in asymmetric hydrogenation (Figure 1.15).

First Rh-catalyzed asymmetric hydrogenations: Catalytic asymmetric hydrogenation began in 1968. A prior insight was given by Horner's comment in a paper on the effect of phosphorus ligands on alkene isomerization vs. hydrogenation: "*Aliphatische und grossvolumige Substituenten verlangsamen die Hydrierung. Diese Fakten müssen bei der von uns geplanten* ***stereospezifischen Hydrierung mit optisch aktiven tertiären Phosphinen*** *berücksichtigt werden.*" [76]. Shortly thereafter, however, Knowles published the first experimental demonstration using an optically impure tertiary phosphine prepared according to Mislow's procedure. The Rh(III) complex hydrogenated unsaturated carboxylic acids at 60 °C in up to 15% optical purity [85] (Figure 1.16). The paper ended on a prophetic note "*The inherent generality of this method offers almost unlimited opportunities for matching*

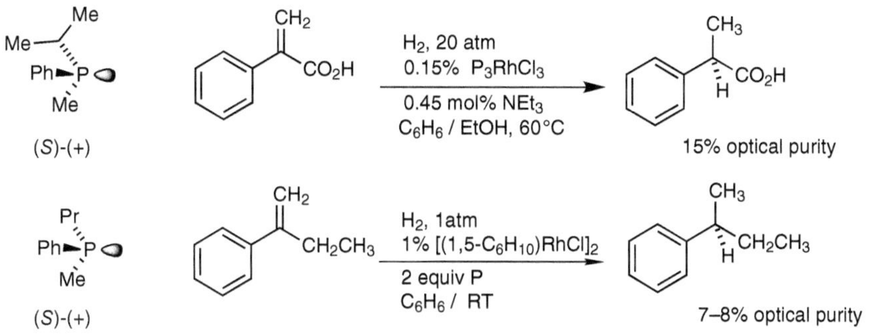

Figure 1.16 Asymmetric hydrogenations presented in the original Horner and Knowles papers.

substrates with catalysts in a rational manner and we are hopeful that our current effort will result in real progress towards complete stereospecificity." His emphasis on the need for catalyst optimization in asymmetric catalysis has later been proved to be correct. Possibly this publication forced Horner's hand, since it was quickly followed by their asymmetric synthesis of 2-phenylbutane in 7–8% optical yield through a similar asymmetric hydrogenation [86]. These papers were not widely publicized at the time; perhaps, the untapped potential had not yet been widely recognized.

Taking a distinct approach, McQuillin showed that a partially reduced Rh(III) complex incorporating an enantiomerically pure formamide catalyzed the hydrogenation of an unsaturated ester with up to 60% optical purity [88]. Crabtree indicated that the application of homogeneity tests to hydrogenations with this type of catalyst indicated that the catalytic species was colloidal or nanoparticulate rather than homogeneous. This was assumed to apply to McQuillin's work, although the homogeneity of his actual catalytic system was not specifically tested [89]. Morrison's contribution offered comparable levels of selectivity. He reasoned that chirality in the organic backbone of a tertiary phosphine might be easier to achieve than chirality at phosphorus. The synthesis of neomenthyldiphenyl phosphine involved difficult purification but was rewarded by the hydrogenation of a series of unsaturated acids including both α- and β-cinnamic acids in up to 61% optical yield, albeit under fairly forcing conditions [87] (Figure 1.17).

In a lecture published in late 1970, Knowles's work was extended to other enantiomerically enriched dialkylphenylphosphines, but optical yields ≤30% were obtained. But from the question session at the end of the lecture: *"**DR. ATKINSON**: I have another question based on the phosphorous rhodium bonds, a preferred conformation, that, of course, would not be the only conformation. Did you consider the possibility of using a bi-phosphine, optically active? **DR. KNOWLES**: We've considered that very strongly. The main problem is that of synthetically making it"* [91]. That consideration became reality with Kagan's first paper on asymmetric hydrogenation in early 1971. Using tartaric acid as the chiral scaffold, the ensuing DIOP ligand was the first effective chelating biphosphine for asymmetric catalysis. The favored reactants were dehydroamino acids, and optical yields of up to 72% were obtained

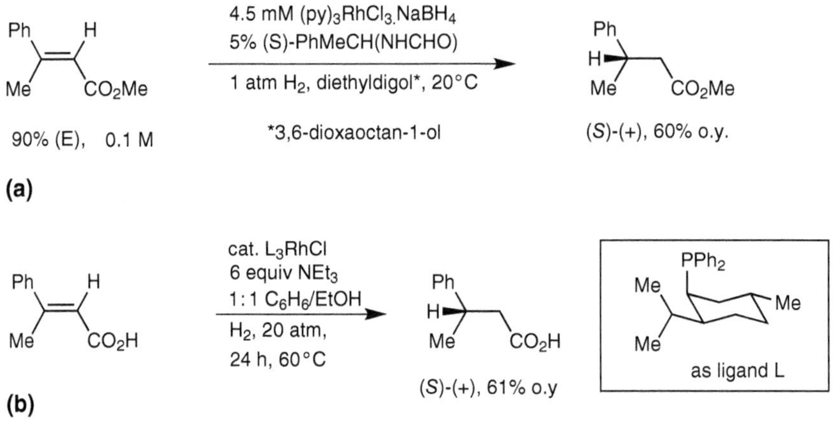

Figure 1.17 (a) McQuillin's chiral amide-catalyzed asymmetric hydrogenation. (b) Morrison's application of the NMDPP ligand in asymmetric hydrogenation. Source: Based on Morrison et al. [87].

under gentle conditions [92, 93]. This was improved to 80%, for the synthesis of *N*-acetyltyrosine, in the ensuing full paper [94]. In a later work with DIOP rhodium catalysts, building on the development of cationic dialkene bisphosphine complexes by Schrock and Osborn to provide the precursor [95], enantiomer excesses up to 92% were recorded with a simple enamide [90] (Figure 1.18). First Knowles, and then Kagan, had filed patents on asymmetric hydrogenation with 1970 priority dates, recognizing the potential for commercial as well as academic applications [96, 97].

Practical catalytic asymmetric hydrogenation. A further paper from Knowles and the Monsanto group revealed that the amino-acid L-Dopa, the standard treatment for alleviation of the symptoms of Parkinson's disease, was accessible to synthesis by asymmetric hydrogenation [98]. Their initial breakthrough came using (*R*)-(o-anisyl)PPhMe, (*R*)-PAMP, previously prepared by Mislow [84], that enabled hydrogenation of the protected (L)-Dopa precursor in 58% optical yield, starting with ligand of c. 95% optical purity. From that point, systematic optimization through ligand synthesis provided their preferred ligand (o-anisyl)PCxMe ((*R*)-CAMP) and gave the desired product with up to 90% optical purity. There is a strong positive nonlinear effect favoring (*R*,R**) over (*R,S*) for the bis-phosphine rhodium solvate intermediate derived from (*R*)-PAMP observed by NMR in solution, and in this case the effect could have diminished any contribution from (*S*)-impurity [99] (Figure 1.19).

This was almost but not quite good enough to provide the basis of a commercial process. Further synthesis, surely guided by Kagan's effective use of a chelating biphosphine [93], led to the oxidative dimerization of the P-oxide of (*R*)-CAMP, followed by stereospecific reduction to give (*R,R*)-DIPAMP [100, 101]. In this "double asymmetric induction" dimerization, the 95% enantiomeric purity of the monophosphine is amplified to c. 99% [102]. This synthesis formed the basis of the Monsanto (L)-DOPA process used therapeutically for over 30 years [103] (Figure 1.20). The

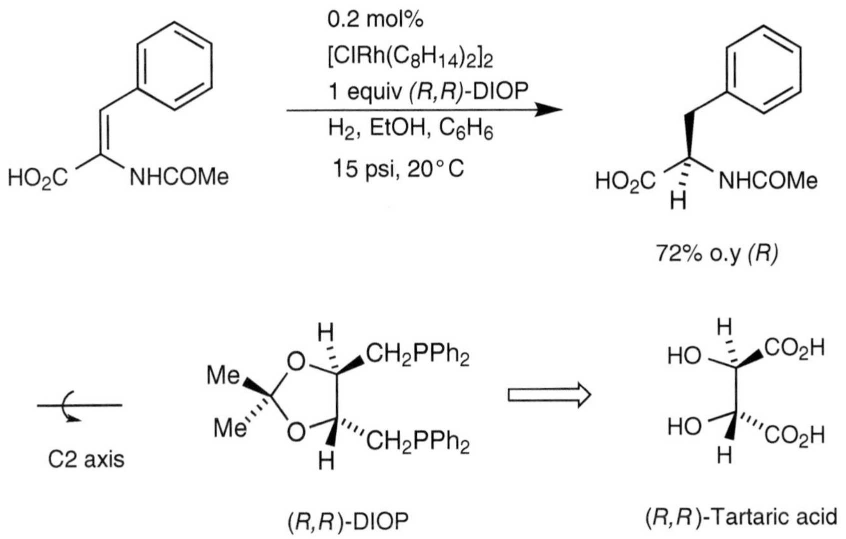

Figure 1.18 (*R,R*)-DIOP, derived from natural tartaric acid, forms an efficient and stereoselective rhodium catalyst for asymmetric hydrogenation. Source: Based on Sinou and Kagan [90].

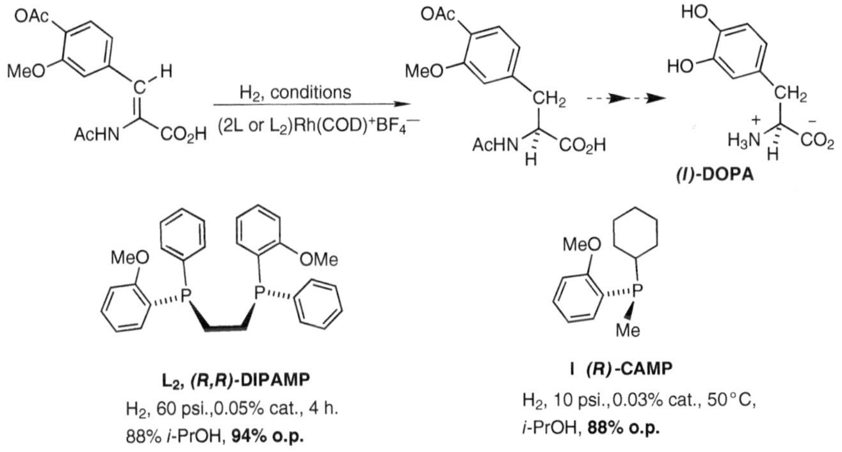

Figure 1.19 The asymmetric hydrogenation of an L-DOPA precursor; comparison of the bidentate (*R,R*)-DIPAMP and monodentate (*R*)-CAMP ligands.

Knowles and Kagan papers in particular led to an upsurge of interest in broadening the scope of asymmetric hydrogenation and in asymmetric catalysis more generally. In 2001, William Knowles shared a Nobel Prize with Ryoji Noyori (Ru in asymmetric hydrogenation and transfer hydrogenation, 1985 onward) and Barry Sharpless (Ti in asymmetric epoxidation, 1980 onward). The powerful influence of the early catalytic asymmetric hydrogenations on the direction of organic chemical research thereafter

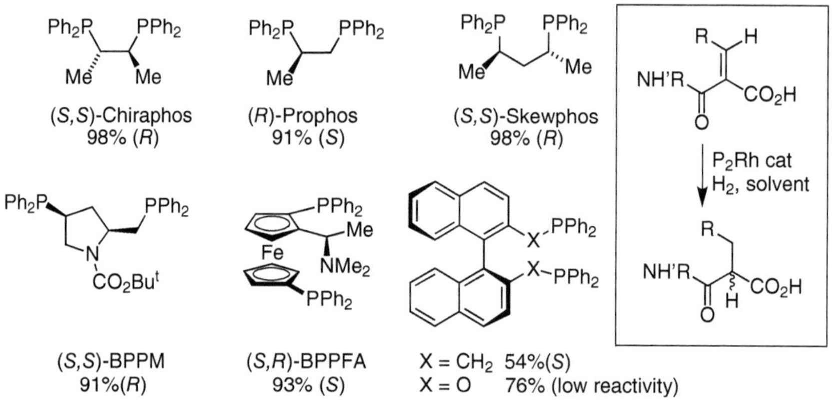

Figure 1.20 The efficiency of early examples of chelating biphosphines in asymmetric hydrogenation. Optical yield data is the highest value obtained with that ligand.

is abundantly clear. Contemporary organic chemistry widely engages catalytic reactions, and the rapid development of asymmetric rhodium hydrogenation up to 1975 should be regarded as a true "paradigm shift" [104].

With these early results, rhodium asymmetric hydrogenation opened up new challenges, both to extend the scope and to broaden the basis of ligand design. The second of these provided a purely synthetic challenge that was widely adopted, following the principles first demonstrated by Kagan: a moderately rigid scaffold, supporting two phenylphosphino or arylphosphino groups that could chelate to rhodium. An initial success was achieved by Bosnich using scaffolds based on enantiomerically pure analogues of diphenylphosphino-ethane or diphenylphosphino-propane [105]. 4-Hydroxyproline [106], or a 1,2-disubstituted ferrocene [107], provided alternative approaches. The potential of a biaryl ligand with asymmetry based on hindered rotation between atropisomers was recognized; both Kumada and Hayashi, and Grubbs, used a binaphthyl scaffold with spacers to the phosphine links, but only moderate enantiomer excesses were achieved in asymmetric hydrogenation [108, 109] (Figure 1.20). Early reviews could point to the broadening interest in the field, and also note the limitation to functional alkenes capable of chelation to rhodium in catalysis [110].

1.6 The Development of Ruthenium Asymmetric Homogeneous Hydrogenation of Alkenes

Ruthenium phosphine complexes had played a role in the development of effective homogeneous hydrogenation, which encouraged the development of asymmetric catalysts. The first success came from James's work in which the isolated complex Ru_2Cl_4[DIOP]$_3$ catalyzed the hydrogenation of *N*-acetyldehydroalanine in 60% optical yield, similar to the result when HRh[DIOP]$_2$ was employed and to the same preferred enantiomer [111, 112]. Later studies used HRuCl[DIOP]$_2$ or carbonyl ruthenium clusters such as $H_4Ru_4(CO)_8$[DIOP] for catalysis of hydrogenation [113]. This early work had demonstrated the potential of ruthenium complexes in asymmetric hydrogenation but did not lead to practical synthetic outcomes [114]. At about the same time, Noyori's efforts were directed toward the initially difficult synthesis and resolution of BINAP (resolved enantiomers of 2,2′-bis(diphenylphosphino)-1,1′-binaphthyl), as described in a retrospective review [111]. The first indication of success lay in a rhodium complex-catalyzed asymmetric synthesis of amino-acid derivatives that rivaled the optical purities achieved by Knowles or Bosnich in their best cases [115]. Optimization required fine-tuning with regard to the nature of the catalyst and lower substrate concentrations, however, and significantly higher optical yields were obtained with *N*-benzoyl enamides than with *N*-acetyl enamides. Subsequently, other research groups reported lower enantioselectivity in asymmetric hydrogenations with BINAP under standard conditions [116]. The utility of BINAP in rhodium catalysis was underscored by the demonstration of an industrially useful stereospecific alkene isomerization, however. This work linked catalyzed isoprene dimerization with fragrances related to monoterpenes [117].

Ruthenium BINAP hydrogenation chemistry was initiated by Ikariya [118]. He was able to synthesize a reactive dimeric Ru complex Ru_2Cl_4[BINAP]$_2$ directly from the corresponding cycloocta-[4, 8]-diene precursor and to show that it was a good hydrogenation catalyst for enamide substrates under mild conditions, giving the opposite enantiomer of product to BINAP–Rh hydrogenation for all (*Z*)-dehydroamino acid derivatives, albeit the same enantiomer as BINAP–Rh for the one example of an (*E*)-isomeric substrate tested; Figure 1.21.

At this stage it was far from obvious that using Ru complexes might confer any advantage over Rh in asymmetric hydrogenation. Further progress depended on the identification of a problem that had unsatisfactory solutions with existing methods. The monoterpene citronellol, existing in Nature as both (*R*)- and (*S*)-enantiomers, is commercially important for its flavor and fragrance properties. In principle, it is accessible by the asymmetric hydrogenation of geraniol or its (*Z*)-isomer nerol (Figure 1.22). Using ClRh(COD)((*R*)(+)BINAP) as catalyst at ambient temperature and elevated H_2 pressure, reaction occurred exclusively at the allylic double bond

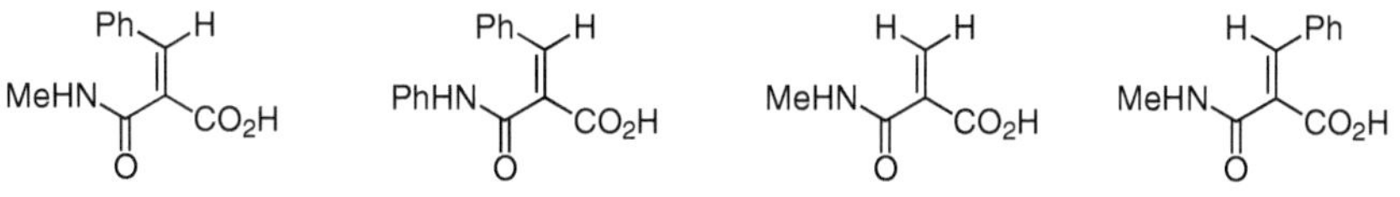

Ru catalysis: Ru_2Cl_2[(*S*)-BINAP]$_2$, NEt_3, 2 atm. H_2, 35°C, (EtOH, THF 1 : 1)

86% (S) **92% (S)** **76% (S)** **65% (S)**

Rh catalysis: Rh([(*S*)-BINAP](Solv)$_2$])$^+$ ClO_4^-, 3–4 atm. H_2, RT; (Solv = EtOH or THF)

84% (*R*) **96% (*R*)** **67% (*R*)** **87% (*S*)**

Figure 1.21 Comparative data for optical yields in the asymmetric hydrogenation of dehydroamino acids. Source: Data from Ikariya et al. [118].

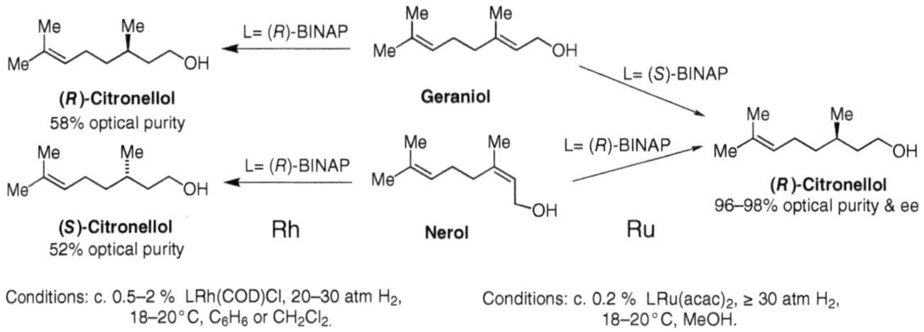

Figure 1.22 Comparison of rhodium and ruthenium BINAP catalysts in the asymmetric synthesis of citronellol.

giving (*R*)-citronellol in 58% ee (by hplc). Conversely, nerol gave the (*S*)-enantiomer in 52% ee [119]. Later work described ruthenium hydrogenation, notably using the catalyst precursor $(CH_3CO_2)_2Ru$(BINAP) or its trifluoroacetate analogue, described in patents [120]. The reaction occurred in high enantioselectivity for both geraniol and nerol under accessible reaction conditions and was further applied to both allylic and homoallylic alcohols [121].

It was clear that asymmetric hydrogenation by Ru complexes still required a chelating functional group in proximity to the reacting alkene but with wider substrate tolerance than for rhodium. A sequence of papers demonstrated that principle, further applied to unsaturated acids [122], benzomorphans and morphinans (*N*-formylenamides) [123], and isoquinoline alkaloids (*N*-acyl enamides) [124]. A major driving force for optimization of ruthenium asymmetric hydrogenation was the industrial potential, providing a parallel to rhodium catalysis for the synthesis of (L)-DOPA. In this particular case, Takasago Perfumery Co. synthesized citronellal from geraniol with BINAP as the Ru ligand, both for its direct use and also for application as an intermediate in Vitamin E synthesis. Their efficient BINAP–Ru-catalyzed process worked to provide 350 tons per annum of the desired product [125]. Rhodium and ruthenium asymmetric hydrogenations of prochiral alkenes continue to provide an important methodology in modern pharmaceutical process chemistry [126].

1.7 Conclusions

The original intent was to understand why there was a time lapse of nearly 70 years between identifying the aim of using synthetic chemistry to achieve effective asymmetric synthesis and its realization. Some of the key factors are clarified, especially the introduction of new methods for the measurement of enantiomer excess, that are not dependent on assaying by polarimetry. These advances were both instrumental (GC, LC, and NMR) and chemical (Mosher reagent and lanthanide shift reagents). With better methodology, the field advanced rapidly. The clearest advance came with the recognition that Wilkinson's catalyst provides a potential pathway for catalytic asymmetric hydrogenation and the diverse approaches of Knowles and Kagan. But there are other factors in play, deserving more detailed scrutiny. Early workers had no design template for an effective asymmetric synthesis and insufficient guidance on the variations in reactant structure that are presently taken for granted in systematic optimization procedures. The "gold rush" era of asymmetric synthesis was closely linked to the timing of general advances in organic synthesis based on a better understanding of structure and mechanism, the discovery of new stoichiometric and catalytic reactions, and a better appreciation of enzymology.

References

1 Herschel, J.W.F. (1822). *Trans. Cambridge Philos. Soc.* 1: 43–60. (accessible online; ex. Hathi Trust).

2 **Hemihedral**: Having only half the plane faces needed for the highest degree of symmetry in its system. See also the obsolete **Plagihedral** [OED]: Having or designating certain symmetrical crystal faces whose axes are each at the same oblique angle to the vertical, occurring especially in quartz, where the orientation of these faces indicates whether the crystal is laevorotatory or dextrorotatory.

3 De Vries, A. (1958). *Nature* 181: 1193.

4 Lyle, R.E. and Lyle, G.G. (1964). *J. Chem. Educ.* 41: 308–313.

5 Azzam, R.M.A. (2011). *Thin Solid Films* 519: 2584–2588.

6 (a) Authier, A. (2014). *Early Days of X-Ray Crystallography*. Chapter 12, 318. Oxford, UK: Oxford University Press. (b) For a brief accessible biography of Haüy, see *René Just Haüy – Wikipedia*; NB the quotation cited there (Ref. 2).

7 Mauskopf, S.H. (1976). *Trans. Am. Philos. Soc.* 66 (Pt. 3): 5; This excellent review "Crystals and Compounds: Molecular Structure and Composition in Nineteenth-Century French Science" is published separately in book form: Trans. Amer. Philos. Soc., New Series, Vol. 66, pt 3, American Philosophical Society, Philadelphia, PA, USA.

8 Mauskopf, S.F. (2011). *A history of chemistry*" in: Chapter 1. In: *Chiral Analysis* (eds. K.W. Busch and M.A. Busch). Amsterdam, The Netherlands: Elsevier Science.

9 Winderlich, R. (1948). Jöns Jakob Berzelius. *J. Chem. Educ.* 25: 500–505.

10 Winderlich, R. (1949). "Eilhard Mitscherlich" 1794-1869. *J. Chem. Educ.* 26: 358–361.

11 Richardson, G.M. (ed.) (1902). *The Foundations of Stereo-Chemistry; Lectures by Pasteur, van't Hoff, Le Bel and Wislicenus*. New York: American Book Company.

12 Bernal, J.D. (2006). *Science and Industry in the Nineteenth Century*. Chapter VII, 181. London: Routledge (reprint of his 1953 work).

13 Flack, H.D. (2009). *Acta Crystallogr. Sect. A* 65: 371–389.

14 Van't Hoff, J.H. (1874). *Arch. Neerl. Sci. Exact. Nat.* 9: 445–454. "A suggestion looking to the extension into space of the structural formulas at present used

in chemistry, and a note upon the relation between the optical activity and the chemical constitution of organic compounds".
15 LeBel, J.A. (1874). *Bull. Soc. Chim. Fr.* 22: 337–347. "On the relations which exist between the atomic formulas of organic compounds and the rotatory power of their solutions".
16 For Refs 11, 12 in translation, see: "Classic Papers" may be accessed from the main menu. https://www.chemteam.info
17 (a) Fischer, E. (1891). *Ber. Dtsch. Chem. Ges.* 24: 2683–2687. (b) For a detailed discussion and appreciation of Fischer's contributions, see:Lichtenthaler, F.W. (1992). *Angew. Chem. Int. Ed.* 104: 1577–1593.
18 Bijvoet, J.M., Peerdeman, A.F., and Vanbommel, A.J. (1951). *Nature* 168 (4268): 271–272.
19 For an interesting article on the early history of asymmetric synthesis, see: Kagan, H.B. and Gopalaiah, K. (2011). *New. J. Chem.* 35: 1933–1937.
20 (a)See the overview by: McKenzie, A. (1904). *J. Chem. Soc., Trans.* 85: 1249–1262. (b) The original reference for helicin work: Fischer, E. and Slimmer, M. (1903). *Ber. Dtsch. Chem. Ges.* 36: 2575–2587.
21 Marckwald, W. (1904). *Ber. Dtsch. Chem. Ges.* 37: 349.
22 Erlenmeyer, E. and Landsberger, F. (1914). *Biochem. Z.* 64: 366–381.
23 Nicolaou, K.C. and Sorensen, E.J. (eds.) (1996). *Classics in Total Synthesis: Targets, Strategies, Methods*, 1e. New York: Wiley.
24 Cram, D.J. and Wilson, D.R. (1963). *J. Am. Chem. Soc.* 85: 1245–1249.
25 Cram, D.J. and Elhafez, F.A.A. (1952). *J. Am. Chem. Soc.* 74: 5828–5835.
26 Cram, D.J. and Kopecky, K.R. (1959). *J. Am. Chem. Soc.* 81: 2748–2755.
27 Prelog, V. and Wilhelm, M. (1954). *Helv. Chim. Acta* 37: 1634–1660.
28 Bredig, G. and Minaeff, M. (1932). *Biochem. Z.* 249: 241–244.
29 For parallel work on the catalysed hydrocyanation of benzaldehyde, see: Albers, H. and Albers, E. (1954). *Z. Naturforsch.* 9b: 122–133, and 133–145.
30 Gilav, E., Feibush, B., and Charless, R. (1966). *Tetrahedron Lett.* 8: 1009–1012.
31 Mikes, F., Boshart, G., and Gilav, E. (1976). *J. Chromatogr.* 122: 205–221.
32 (a) Dale, J.A., Dull, D.L., and Mosher, H.S. (1969). *J. Org. Chem.* 34: 2543–2549. (b) Goering, H.L., Eikenberry, J.N., and Koermer, G.S. (1971). *J. Am. Chem. Soc.* 93: 5913–5914. (c) Whitesides, G.M. and Lewis, D.W. (1971). *J. Am. Chem. Soc.* 93: 5914–5916.
33 For a review of early asymmetric syntheses, see: Boyd, D.R. and McKervey, M.A. (1968). *Quart. Rev. Chem. Soc.* 22: 95–122.
34 Brown, H.C. and Zweifel, G. (1961). *J. Am. Chem. Soc.* 83: 486–487.
35 Brown, H.C., Ayyangar, N.R., and Zweifel, G. (1964). *J. Am. Chem. Soc.* 86: 397–403.
36 For very early examples of asymmetric Diels–Alder reactions using chiral maleate or fumarate esters w/o added Lewis acid, see: Korolev, A. and Mur, V. (1948). *Dokl. Akad. Nauk SSSR* 59: 251–253.
37 Walborsky, H.M., Davis, T.C., and Barash, L. (1963). *Tetrahedron* 19: 2333–2351.
38 Farmer, R.F. and Hamer, J. (1966). *J. Org. Chem.* 31: 2418–2419.
39 Sauer, J. and Kredel, J. (1966). *Tetrahedron Lett.* 8: 6359–6362.

40 Corey, E.J., Albonico, S.M., Koelliker, U. et al. (1971). *J. Am. Chem. Soc.* 93: 1491.

41 Corey, E.J., Varma, R.K., and Becker, K.B. (1972). *J. Am. Chem. Soc.* 94: 8616–8618.

42 Pracejus, H. (1960). *Ann. Chem.* 634: 9–22.

43 Eder, U., Sauer, G., and Weichert, R. (1971). *Angew. Chem. Int. Ed.* 10: 496–497.

44 Hajos, Z.G. and Parrish, D.R. (1974). *J. Org. Chem.* 39: 1615–1621.

45 Hajos, Z.G. and Parrish, D.R. (1970). Asymmetric synthesis of optically active polycyclic organic compounds. DE2012623A, 1971, priority date 21 January 1970.

46 Eder, U., Wiechert, R., and Sauer, G. (1970). Optically active 1,5-indandione and 1,6-naphthalenedione derivatives. DE2014757A, 1971, priority date 20 March 1970.

47 List, B., Lerner, R.A., and Barbas, C.F. (2000). *J. Am. Chem. Soc.* 122: 2395–2396.

48 For a brief biography of Paul Sabatier, see: Kagan, H.B. (2012). *Ang. Chem., Int. Ed.* 51: 7376–7382.

49 Orito, Y., Imai, S., Niwa, S., and Nguyengiahung (1979). *J. Synth. Org. Chem. Jpn.* 37: 173–174.

50 Terent'ev, A.P., Klabunovskii, E.I., and Patrikeev, V.V. (1950). *Dokl. Akad. Nauk SSSR* 74: 947–950.

51 Klabunovskii, E., Smith, G.V., and Zsigmund, A. (2006). *Heterogeneous Enantioselective Hydrogenation: Theory and Practice*. Berlin: Springer.

52 Akabori, S., Sakurai, S., Izumi, Y., and Fujii, Y. (1956). *Nature* 178: 323–324.

53 For variation of selectivity based on variation of the silk fibroin source and pre-acylation, see: Izumi, Y., Fujii, Y., and Akabori, S. (1957). *Nippon Kagaku Zasshi* 78: 886–888; data retrieved from the Scifinder Abstract of this paper.

54 For caution on catalyst stability and reproducibility for Refs 52, 53 see: Izumi, Y. (1971). *Ang. Chem., Int. Ed.* 10: 871–881.

55 Bergman, M. and Tietzmann, J.E. (1944). *J. Biol. Chem.* 155: 535.

56 Poisel, H. and Schmidt, U. (1973). *Chem. Ber.* 106: 3408–3420.

57 Schmidt, U., Kumpf, S., and Neumann, K. (1994). *J. Chem. Soc., Chem. Commun.*: 1915–1916.

58 Mori, K., Kiyota, H., and Rochat, D. (1993). *Liebigs Ann. Chem.*: 865–870.

59 Arcus, C.L. and Smyth, D.G. (1955). *J. Chem. Soc.*: 34–40; and refs. therein describing earlier attempts at asymmetric heterogeneous hydrogenation.

60 Arcus, C.L., Reid, J.A., and Page, J.M.J. (1963). *J. Chem. Soc.*: 1213, and intervening papers in *J. Chem. Soc.*

61 Prelog, V. and Scherrer, H. (1959). *Helv. Chim. Acta* 42: 2227–2232.

62 Klabunovskii, E. (1970). *Russ. Chem. Rev.* 35: 1035–1049; in English translation through the Institute of Physics on open access.

63 Cornils, B., Herrmann, A., and Rasch, M. (1994). *Ang. Chem. Int. Ed.* 33: 2144–2163.

64 (a) Wender, I., Orchin, M., and Storch, H.H. (1950). *J. Am. Chem. Soc.* 72: 4842–4843. (b) Adkins, H. and Williams, J.L.R. (1952). *J. Org. Chem.* 17: 980–987.

65 Uchida, H. and Bando, K. (1956). *Bull. Chem. Soc. Jpn.* 29: 953–956.

66 Halpern, J., Harrod, J.F., and James, B.R. (1961). *J. Am. Chem. Soc.* 83: 753.

67 Dewitt, E.J., Trapasso, L.E., and Ramp, F.L. (1961). *J. Am. Chem. Soc.* 83: 4672.

68 Young, J.F., Osborn, J.A., Jardine, F.H., and Wilkinson, G. (1965). *Chem. Commun.*: 131–132.

69 Osborn, J.A., Jardine, F.H., Young, J.F., and Wilkinson, G. (1966). *J. Chem. Soc. A*: 1711.

70 In parallel work, ICI Heavy Chemicals and Organics Division filed a patent on $ClRh(PPh_3)_3$ hydrogenations (Coffey, R.S. British Patent 1,121,642 (applied 18.2.65)); see the review by Candlin, J.P. and Oldham, A.R. (1968). *Disc. Faraday Soc.* 46: 60–71.

71 Hallman, P.S., Evans, D., Osborn, J.A., and Wilkinson, G. (1967). *Chem. Commun.*: 305–306.

72 Hallman, P.S., McGarvey, B.R., and Wilkinson, G. (1968). *J. Chem. Soc. A*: 3143–3150.

73 Ireland, R.E. and Bey, P. (1973). *Org. Synth.* 53: 63–65; see also Ref. 74 below.

74 Djerassi, C. and Gutzwill, J. (1966). *J. Am. Chem. Soc.* 88: 4537.

75 Birch, A.J. and Walker, K.A.M. (1966). *J. Chem. Soc. C*: 1894–1896.

76 Horner, L., Buthe, H., and Siegel, H. (1968). *Tetrahedron Lett.* 37: 4023–4026.

77 Horner, L., Winkler, H., Rapp, A. et al. (1961). *Tetrahedron Lett.* 5: 161–166.

78 Horner, L. (1964). *Pure Appl. Chem.* 9: 225–244.

79 Horner, L. and Winkler, H. (1965). *Ann. Chem.* 685: 1–10.

80 Horner, L. and Balzer, W.D. (1965). *Tetrahedron Lett.* 6: 1157–1162.

81 Mislow, K. (1971). *Pure Appl. Chem.* 25: 549–562.

82 Korpiun, O. and Mislow, K. (1967). *J. Am. Chem. Soc.* 89: 4784–4786.

83 Korpiun, O., Lewis, R.A., Chickos, J., and Mislow, K. (1968). *J. Am. Chem. Soc.* 90: 4842–4846.

84 Naumann, K., Zon, G., and Mislow, K. (1969). *J. Am. Chem. Soc.* 91: 7012–7022.

85 Knowles, W.S. and Sabacky, M.J. (1968). *Chem. Commun.*: 1445–1446.

86 Horner, L., Siegel, H., and Buethe, H. (1968). *Ang. Chem. Int. Ed.* 7: 942.

87 Morrison, J.D., Burnett, R.E., Aguiar, A.M. et al. (1971). *J. Am. Chem. Soc.* 93: 1301–1303.

88 (a) Abley, P. and McQuillin, F.J. (1969). *Chem. Commun.*: 477–478. (b)Full paper: Abley, P. and McQuillin, F.J. (1971). *J. Chem. Soc. C*: 844.

89 Anton, D.R. and Crabtree, R.H. (1983). *Organometallics* 2: 855–859.

90 Sinou, D. and Kagan, H.B. (1976). *J. Organomet. Chem.* 114: 325–337.

91 Knowles, W.S., Sabacky, M.J., and Vineyard, B.D. (1970). *Ann. N. Y. Acad. Sci.* 172: 232–237.

92 Dang, T.P. and Kagan, H.B. (1971). *Chem. Commun.*: 481.

93 Gillespie, H.B., Snyder, H.R., Herbst, R.M., and Shemin, D. (1939). *Org. Synth.* 19: 67–69, for an earlier example of heterogeneous hydrogenation of dehydroamino acids to racemic products.

94 Kagan, H.B. and Dang Tuan, P. (1972). *J. Am. Chem. Soc.* 94: 6429–6433.

95 Osborn, J.A. and Schrock, R.R. (1971). *J. Am. Chem. Soc.* 93: 2397–2407.

96 (a) Knowles, W.S. and Sabacky, M.J. (1970). Catalytic asymmetric hydrogenation of β-substituted α-(acylamido) acrylic acids and/or their salts. DE2123063A, 1971; to Monsanto, priority date 11 May 1970; (b) First patent referring to catalytic L-DOPA synthesis: 3-(3,4-Dihydroxyphenyl)-L-alanine. DE2210938A1, 1972; to Monsanto, priority date 8 March 1971.

97 Kagan, H. and Dang, T.P. (1970). Bidendate asymmetric diarylphosphine ligands for rhodium complexes as catalysts for asymmetric hydrogenation. DE2161200A, 1971; to Institut Français du Petrole, priority date 10 December 1970.

98 Knowles, W.S., Sabacky, M.J., and Vineyard, B.D. (1972). *J. Chem. Soc., Chem. Commun.*: 10–11.

99 Brown, J.M., Chaloner, P.A., and Nicholson, P.N. (1978). *J. Chem. Soc., Chem. Commun.*: 646–647.

100 Knowles, W.S., Sabacky, M.J., and Vineyard, B.D. (1973). Asymmetric catalysis. DE2456937A1, 1975. to Monsanto, priority date 3 December 1973.

101 Knowles, W.S., Sabacky, M.J., Vineyard, B.D., and Weinkauff, D.J. (1975). *J. Am. Chem. Soc.* 97: 2567–2568.

102 Harned, A.M. (2018). *Tetrahedron* 74: 3797–3841. (A review of Horeau's work on double asymmetric induction).

103 (a) Knowles, W.S. (2004). *Asymmetric hydrogenations – The Monsanto L-dopa process.* In: *Asymmetric Catalysis on an Industrial Scale.* Chapter 1, 23–38. Weinheim: Wiley-VCH. (b) See also Selke, R. *The other L-DOPA process.* In: *Asymmetric Catalysis on an Industrial Scale.* Chapter 2., 39–52. Weinheim: Wiley-VCH.

104 Kuhn, T.S. (1970). *The Structure of Scientific Revolutions.* Chapter V., 2e, 43–54. Chicago/London: Chicago University Press.

105 (a) Fryzuk, M.D. and Bosnich, B. (1977). *J. Am. Chem. Soc.* 99: 6262–6267. (b) Fryzuk, M.D. and Bosnich, B. (1978). *J. Am. Chem. Soc.* 100: 5491–5494. (c) MacNeil, P.A., Roberts, N.K., and Bosnich, B. (1981). *J. Am. Chem. Soc.* 103: 2273–2280.

106 Achiwa, K. (1976). *J. Am. Chem. Soc.* 98: 8265–8266.

107 Hayashi, T., Mise, T., Mitachi, S. et al. (1976). *Tetrahedron Lett.* 17: 1133–1134.

108 Tamao, K., Yamamoto, H., Matsumoto, H. et al. (1977). *Tetrahedron Lett.* 18: 1389–1392.

109 Grubbs, R.H. and DeVries, R.A. (1977). *Tetrahedron Lett.* 18: 1879–1880.

110 Knowles, W.S. (1983). *Acc. Chem. Res.* 16: 106–112.

111 (a) Noyori, R. (2013). *Angew. Chem., Int. Ed.* 52: 79–92. (b) Anon (1980). Asymmetric hydrogenation catalyst. Ajinomoto Co., Inc., Japan Patent JP55061937A, 1980.

112 (a) Tani, K., Yamagata, T., Otsuka, S. et al. (1982). *J. Chem. Soc., Chem. Commun.*: 600–601. (b) Tani, K., Yamagata, T., Tatsuno, Y. et al. (1985). *Angew. Chem.* 97: 232–234.

113 (a) Botteghi, C., Gladiali, S., Bianchi, M. et al. (1977). *J. Organomet. Chem.* 140: 221–228. (b) James, B.R. and Wang, D.K.W. (1980). *Can. J. Chem.* 58: 245–250.
114 Matteoli, U., Frediani, P., Bianchi, M. et al. (1981). *J. Mol. Catal.* 12: 265–319.
115 Miyashita, A., Yasuda, A., Takaya, H. et al. (1980). *J. Am. Chem. Soc.* 102: 7932–7934.
116 (a) Hopkins, J.M., Dalrymple, S.A., Parvez, M., and Keay, B.A. (2005). *Org. Lett.* 7: 3765–3768. (b) Alame, M., Pestre, N., and de Bellefon, C. (2008). *Adv. Synth. Catal.* 350: 898–908, and earlier papers.
117 Tani, K., Yamagata, T., Akutagawa, S. et al. (1984). *J. Am. Chem. Soc.* 106: 5208–5217.
118 Ikariya, T., Ishii, Y., Kawano, H. et al. (1985). *J. Chem. Soc., Chem. Commun.*: 922–924.
119 Inoue, S., Osada, M., Koyano, K. et al. (1985). *Chem. Lett.*: 1007–1008.
120 Takaya, H., Ohta, T., Noyori, R. et al. (1986). Ruthenium-phosphine complexes useful as hydrogenation catalysts. EP245959A2, 1987; original JP1986-108888, Appl 13 May 1986.
121 Takaya, H., Ohta, T., Sayo, N. et al. (1987). *J. Am. Chem. Soc.* 109: 1596–1597.
122 Ohta, T., Takaya, H., Kitamura, M. et al. (1987). *J. Org. Chem.* 52: 3174–3176.
123 Kitamura, M., Hsiao, Y., Noyori, R., and Takaya, H. (1987). *Tetrahedron Lett.* 28: 4829–4832.
124 Noyori, R., Ohta, M., Hsiao, Y. et al. (1986). *J. Am. Chem. Soc.* 108: 7171–7119.
125 Blaser, H.U., Spindler, F., and Studer, M. (2001). Enantioselective catalysis in fine chemicals production. *Appl. Catal., A* 221: 119–143.
126 (a) Beliaev, A. (2016). *Org. Process Res. Dev.* 20: 724–732. (b) Karlsson, S., Sorensen, H., Andersen, S.M. et al. (2016). *Org. Process Res. Dev.* 20: 262–269.

2

Asymmetric (Transfer) Hydrogenation of Functionalized Alkenes During the Past Decade

Christian Bruneau

Univ Rennes, CNRS, The Institute of Chemical Sciences – ISCR – UMR 6226, Rennes Cedex 35042, France

2.1 Introduction

Since the early steps of asymmetric hydrogenation of functionalized alkenes with chiral mono- and diphosphine ligands in the 1970s [1], tremendous developments of transition metal-catalyzed achievements have been carried out. This represents about 50 years of research on a very hot topic, which encountered many challenges due to the huge demand of industry for the production of optically enriched target molecules with biological activities. For this reason many scientific data focusing on asymmetric hydrogenation [2] and transfer hydrogenation [3] and applications in review articles, book chapters, and books have been published. This chapter essentially covers in a nonexhaustive manner the results that have been obtained during the past decade with rhodium, iridium, and ruthenium catalysts, which constitute the major actors in asymmetric (transfer) hydrogenation, but does not include mechanistic considerations. The most popular functionalized alkenes that have been used as model compounds for catalytic system evaluation are essentially acrylic and α- and β-amino acid derivatives.

2.2 Asymmetric Hydrogenation with Rhodium Catalysts

A wide range of new ligands have been prepared and used either to prepare rhodium complexes or generate catalysts *in situ* by association of the ligand with a rhodium(I) precursor.

2.2.1 Chiral Bisphosphine Ligands

Thirty years after the discovery of Dipamp (Scheme 2.1) by Knowles and coworkers, associated with the industrial production of L-DOPA [4], modification of the *o*-anisyl group by addition of some other substituents on the phenyl ring has led

Asymmetric Hydrogenation and Transfer Hydrogenation, First Edition.
Edited by Virginie Ratovelomanana-Vidal and Phannarath Phansavath.

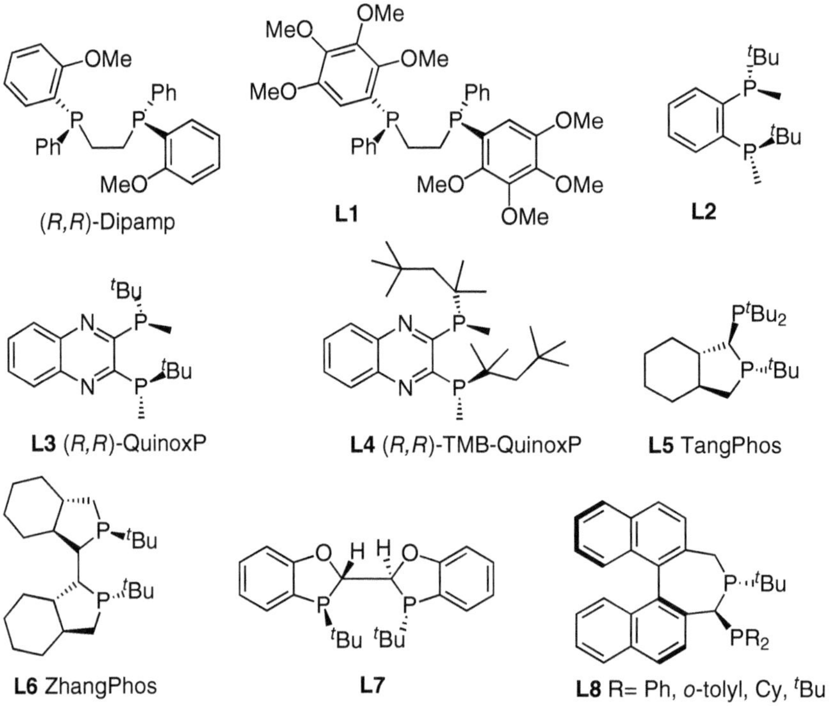

Scheme 2.1 Selected bisphosphine ligands.

to the production of new efficient catalytic systems. In particular, the use of ligand **L1** (Scheme 2.1) led to full conversion and higher enantioselectivities than the original Dipamp for the enantioselective hydrogenation of α- and β-dehydroamido esters **1**, **2**, dialkyl itaconate **3**, and atropic acid **4** (R^1 = H, R^2 = Ph) (Scheme 2.2) under similar conditions [5]. The 1,2-bis(*tert*-butylmethylphosphino)benzene ligand **L2** (Scheme 2.1) developed by Imamoto has shown excellent properties for the enantioselective hydrogenation of a variety of functionalized alkenes such as **1**, **2**, and **5** (Scheme 2.2). In methanol, at room temperature under 3 bar of hydrogen pressure, enantioselectivities higher than 99% were obtained with very high turnover frequencies (up to 10 000 h^{-1}) [6]. The introduction of the quinoxaline core as tether connecting the two phosphino groups provided the ligand **L3** [6], which also exhibited very high enantioinduction with substrates **1**, **2**, and **7**.

Ligand **L4** featuring more bulky alkyl groups at the phosphorus atom was prepared as an air-stable crystalline solid, and its association with $[Rh(cod)_2]X$ or $[Rh(nbd)_2]X$ (X = BF_4 or SbF_6) made possible the asymmetric hydrogenation of methyl α-acetamidocinnamate in 99.9% ee [7]. Acyclic β-acetylamino acrylonitriles were fully hydrogenated in 77–93% ee at 40 °C in methanol under 100 bar of hydrogen with [Rh(cod)(**L3**)]BF_4 as catalyst (Scheme 2.3) [8]. The (*R,R*)-QuinoxP (**L3**) ligand led to the major hydrogenated product with (−) specific rotation.

Zhang made a comparative study of Tangphos (**L5**) and Zhangphos (**L6**) ligands in the asymmetric hydrogenation of α-arylene acetamides, α-aryl enol acetates, α- and

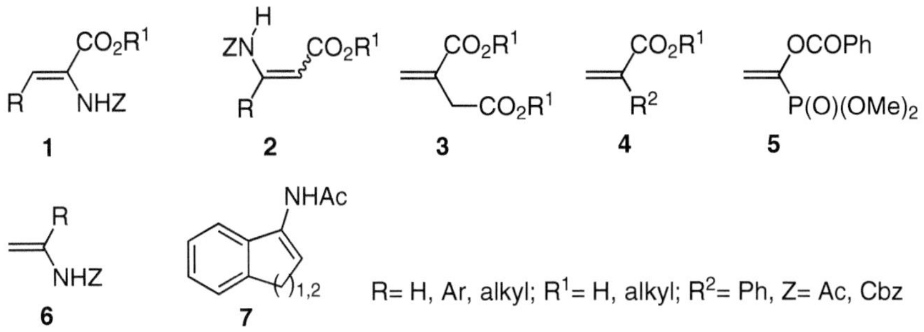

Scheme 2.2 Typical model substrates in asymmetric hydrogenation.

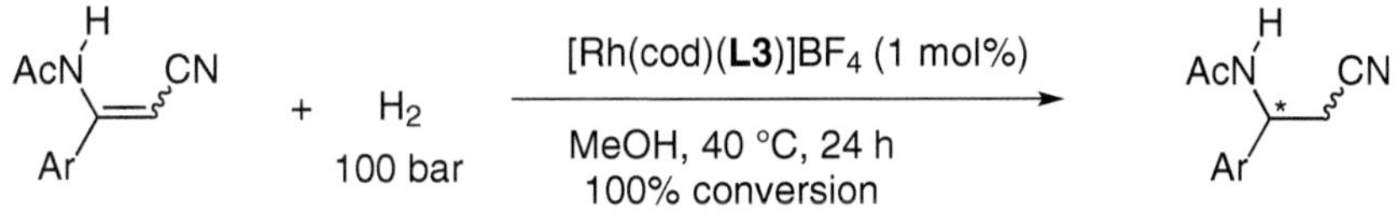

Ar= Ph (93% ee), *p*-MeC_6H_4 (89% ee), *p*-$MeOC_6H_4$ (89% ee), *p*-FC_6H_4 (84% ee)
p-$CF_3C_6H_4$ (77% ee), *m*-MeC_6H_4 (90% ee), *m*-ClC_6H_4 (90% ee)
2-thiophenyl (92% ee), 2-naphthyl (84% ee)

Scheme 2.3 Rhodium-catalyzed asymmetric hydrogenation of β-aryl β-acetylamino acrylonitriles. Source: Based on Ma et al. [8].

β-dehydroamino esters, and itaconic acid derivatives in methanol at room temperature under 1.5 bar with 1 mol% of rhodium catalyst. The high efficiency of **L5** and **L6** and the possibility of using the later even at high temperature (50 °C in trifluoroethanol, TFE) was presented as a promising advantage (Scheme 2.4) [9]. The structurally related ligand **L7** also exhibited excellent catalytic activity for the hydrogenation of the same substrates under mild conditions, e.g. at 0 °C under 7 bar of hydrogen for 12 hours [10].

Finally, the rigid bisphosphine ligand **L8** with C_1-symmetry constructed on the binepine core [11] could achieve good to excellent enantioselectivities with substrates of type **1** and **2** but did not compete favorably with the previous ones due to higher substrate dependence [12].

2.2.2 Chiral Ferrocenyl Bisphosphine Ligands

After the early success of rhodium (Josiphos) catalysts in asymmetric hydrogenation reactions reported in 1994 [13], variations on this organometallic diphosphine ligand have been undertaken (Scheme 2.5). The [Rh(cod)(**L9**)]BF_4 complex presented very low selectivity, leading to modest enantioselectivities (below 56% ee) in hydrogenation of substrates **1** and **3**. On the other hand, satisfactory enantioselectivities were obtained during the asymmetric hydrogenation of α-substituted cinnamic acid ammonium salts with this catalyst, whereas the use of the related **L10** ligand provided only 65–69% ee for the hydrogenation of the same cinnamic acid ammonium

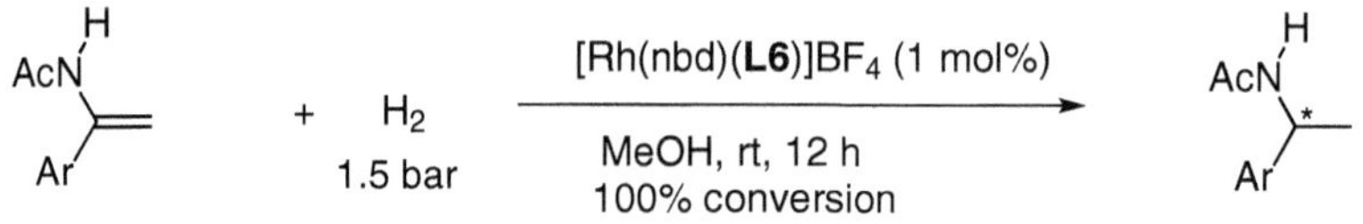

Ar= Ph, *m*-MeC_6H_4, *m*-$MeOC_6H_4$, *p*-MeC_6H_4, *p*-$MeOC_6H_4$, *m*-MeC_6H_4 (>99% ee)

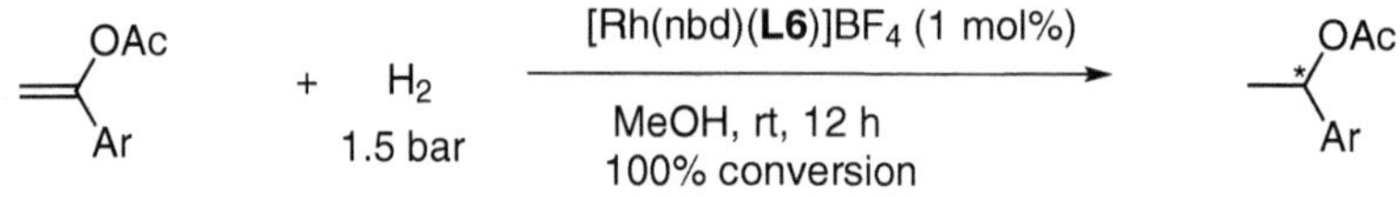

Ar= Ph (97% ee), *p*-FC_6H_4 (98% ee), *p*-ClC_6H_4 (97% ee), *p*-$NO_2C_6H_4$ (>99% ee)
2-naphthyl (99% ee)

Scheme 2.4 Rhodium-catalyzed asymmetric hydrogenation of α-arylene acetamides and enol esters.

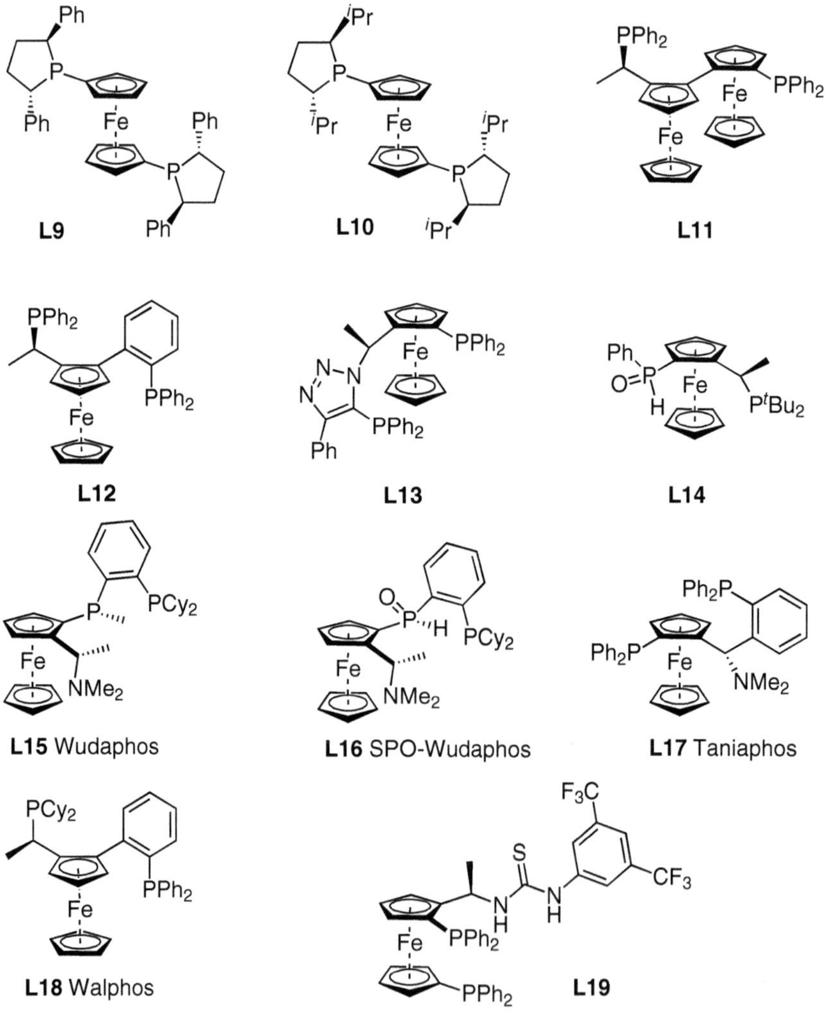

Scheme 2.5 Selected ferrocenyl bisphosphine ligands.

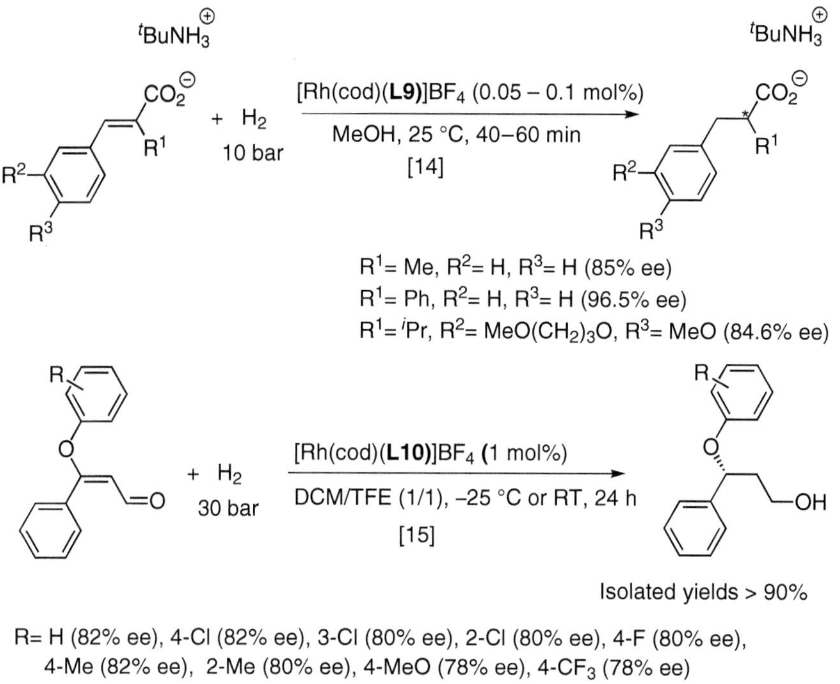

Scheme 2.6 Asymmetric hydrogenation with Rh(**L9**) and Rh(**L10**) catalysts.

salts (Scheme 2.6) [14]. However, [Rh(cod)(**L10**)]BF_4 made possible the sequential asymmetric hydrogenation of β-aryl-β-aryloxy acroleins at low temperature in excellent yields with enantioselectivities up to 82% (Scheme 2.6) [15].

At 20 °C under 1 bar of hydrogen pressure in methanol, the catalytic activity of the catalyst generated *in situ* from $[Rh(nbd)_2]BF_4$ and **L11** exhibited moderate enantioselectivity in the hydrogenation of substrates of type **1** and **3** (ee < 80%) as compared to the catalyst generated from the monoferrocenyl ligand **L12** (Walphos), which gave 97% ee for **1** and 87% for **3**. **L11** proved to be more efficient than **L12** only in the hydrogenation of α-methyl cinnamic acid, leading to 92% ee [16]. The ligand **L13** based on a triazoleferrocene backbone allowed the asymmetric hydrogenation of various α-acetamido cinnamates of type **1** at room temperature under 1 bar of hydrogen in more than 97% ee but was less efficient in the hydrogenation of α-acetamidostyrene affording only 84% ee [17]. Upon coordination to rhodium, the phosphine-secondary phosphine oxide **L14** provides an *in situ* generated rhodium(phosphine–hydroxyphosphine) complex that has revealed excellent performances for the asymmetric hydrogenation of model substrates **1–3** at room temperature under 1 bar of hydrogen with a catalyst loading of 1 mol%, leading to full conversion in less than two hours and enantioselectivity values in the range 90–99% [18]. Other ferrocenyl bisphosphine ligands have been used for the preparation of less-classical substrates. Thus, acrylic acids have been hydrogenated with excellent enantioselectivities with the Wudaphos ligand **L15** (Scheme 2.7) [19]. This ligand was also able to promote the asymmetric hydrogenation of α-aryl

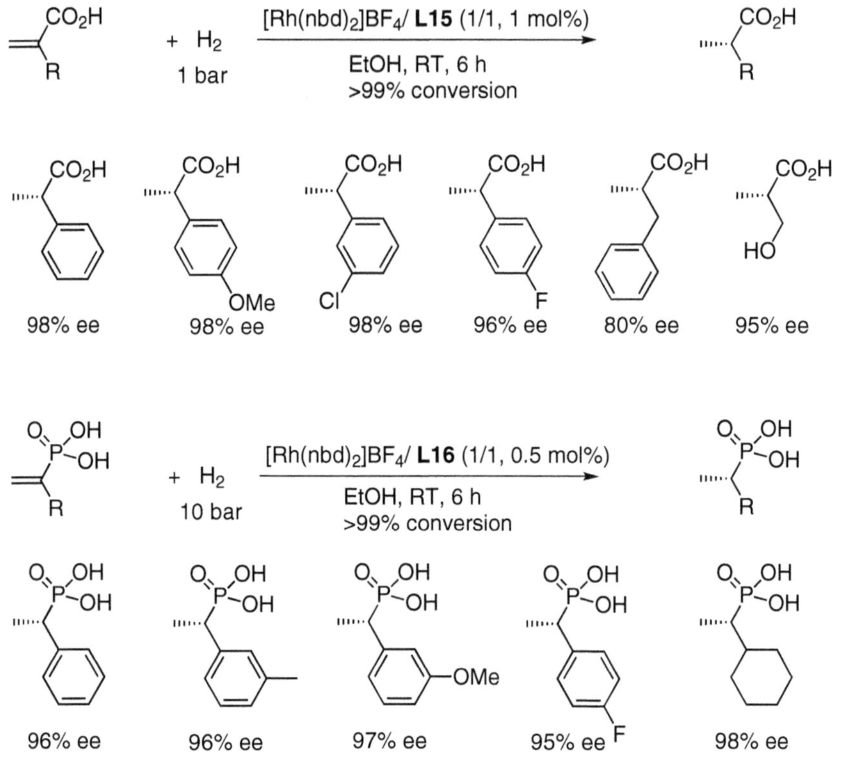

Scheme 2.7 Asymmetric hydrogenation of acrylic acids with a rhodium (Wudaphos) catalyst.

ethenylphosphonic acids in more than 90% ee, but better results were obtained with the utilization of the phosphine–phosphine oxide ligand **L16**, which gave more than 94% ee in most cases (Scheme 2.7) [20].

As illustrated in Scheme 2.8, ethyl 3-aryl-2-(phosphonomethyl)propanoates have been hydrogenated with very good enantioselectivities in the presence of (*S,S*)-Taniaphos ligand **L17** at room temperature under 10 bar of hydrogen [21]. Under closely related conditions, chiral 4-hydroxymethyl-3-fluoropiperidine has been obtained with very high diastereoselectivity and enantioselectivity upon hydrogenation of the corresponding 3,4-dehydropiperidinium chloride with a rhodium–Walphos (**L18**) system [22].

Ligand **L19** combines a ferrocenyl bisphosphine group and a thiourea motif able to interact with a substrate via hydrogen bonding. Thus, metal complexes equipped with this type of ligand may act both as metal catalyst and organocatalyst. This concept proved to be very efficient for the asymmetric hydrogenation of β,β-disubstituted nitroolefins such as β-nitro-α-methylstyrene derivatives [23] and β-amidonitroalkenes [24] (Scheme 2.9).

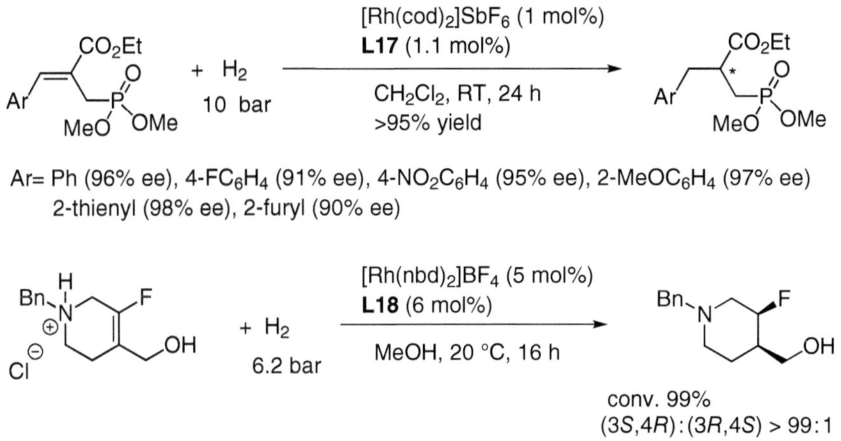

Scheme 2.8 Asymmetric hydrogenation of specific substrates with rhodium precursors and ferrocenyl bisphosphine ligands.

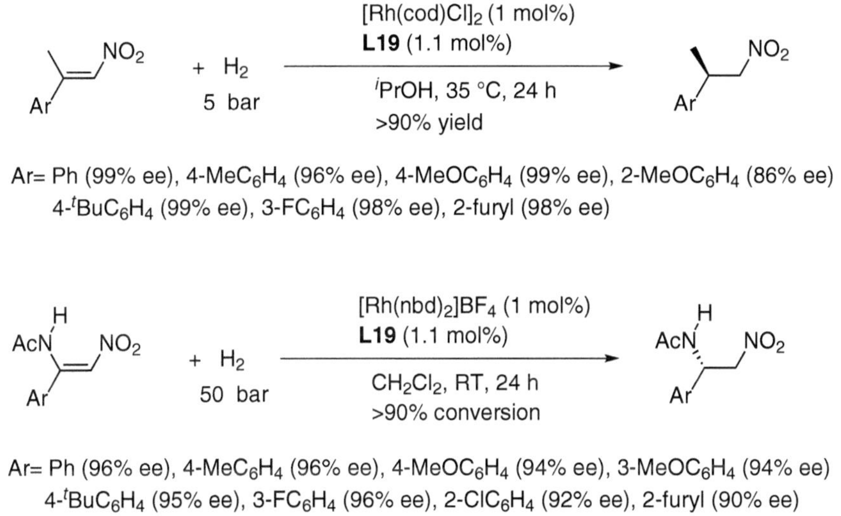

Scheme 2.9 Asymmetric hydrogenation of nitroalkenes with rhodium/bis(phosphine)–thiourea **L19**.

2.2.3 Chiral Phosphine–Phosphoramidite and Phosphine–Phosphite Ligands

Phosphine–phosphoramidite ligands **L20–L22** (Scheme 2.10) containing the optically pure atropoisomeric 1,1′-bi(2-binaphthol) have been prepared and evaluated in the asymmetric hydrogenation of functionalized olefins. Catalysts prepared from $[Rh(cod)_2]BF_4$ and **L20**, **L21**, or **L22** generated productive hydrogenation catalysts,

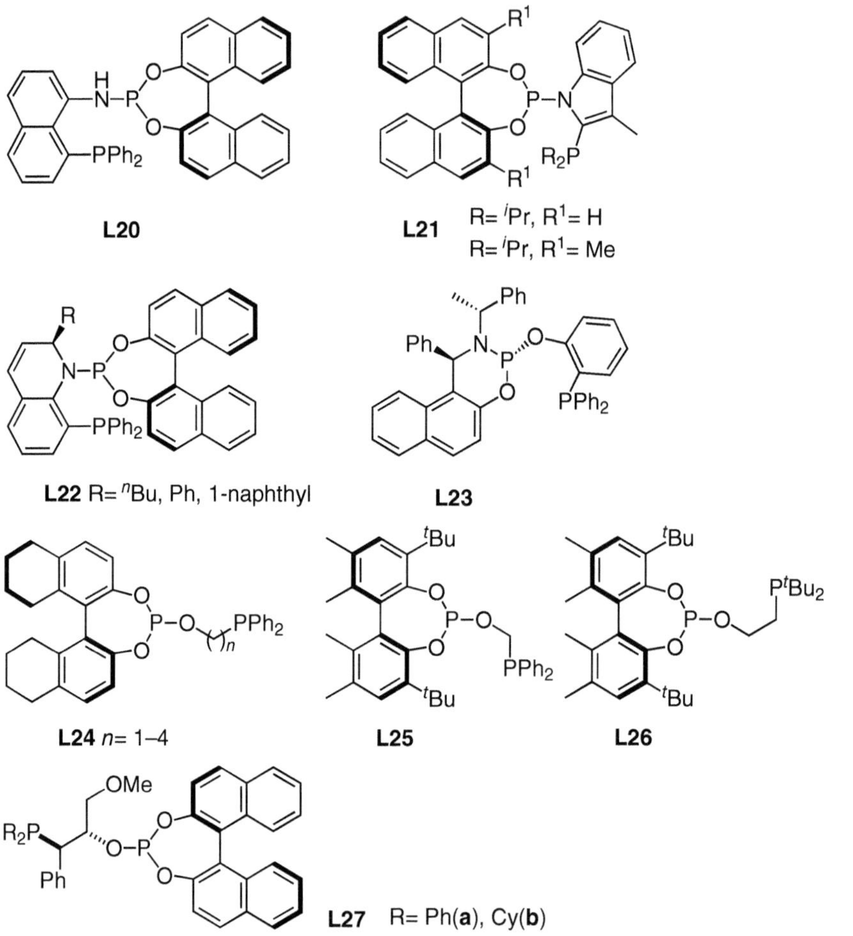

Scheme 2.10 Selected phosphine–phosphoramidite and phosphine–phosphite ligands.

leading to excellent enantioselectivities for α-acetamido cinnamates **1** (95–99% ee), enamides (94–99% ee), and α-enol ester phosphonates (87–99% ee) [25–27]. The ligand **L23** is not based on binaphthol but includes a chiral phosphorus atom at the phosphoramidite moiety. The [Rh(cod)(**L23**)]BF_4 complex used at 0.1 mol% of catalyst loading promoted the complete and highly enantioselective hydrogenation of selected substrates within three hours at room temperature under 15 bar of hydrogen (Scheme 2.11) [28].

At this stage, it is interesting to note that symmetrical bis(phosphoramidite) ligands where the two nitrogen atoms are part of a chiral 1,2-diamine tether are much less efficient in terms of reactivity and enantioselectivity than the phosphine–phosphoramidite ligands [29].

Phosphine–phosphite bidentate ligands **L24** [30], **L25** [31], and **L27** [32] have also been recently investigated in rhodium-catalyzed asymmetric hydrogenation. The best results are gathered in Scheme 2.11. It appears that these ligands also have

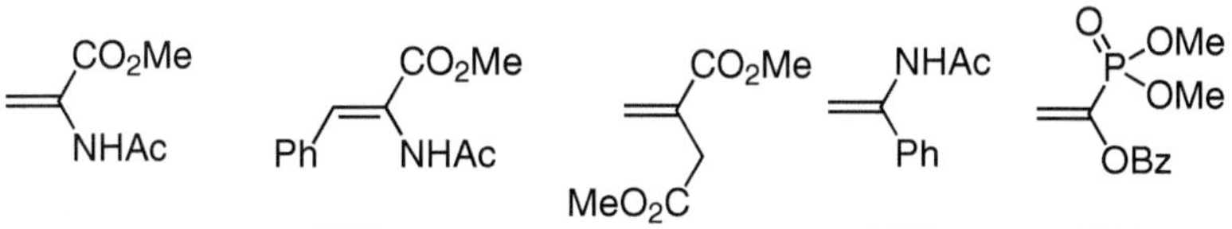

L20:		99% ee		96% ee	99% ee
L21:	97% ee	97% ee	98% ee	94% ee	87% ee
L22:	99% ee	>99% ee	>99% ee	>99% ee	
L23:	91% ee	92% ee	97% ee	96% ee	91% ee
L24:			97.5% ee		
L25:	91% cc	96% ee	89% ee		
L27a:	98% ee	99% ee	99% ee	98% ee	92% ee
L27b:	99% ee	98% ee	97% ee		98% ee

Scheme 2.11 Asymmetric hydrogenation with Rh/phosphine–phosphoramidite systems.

interesting properties for the asymmetric hydrogenation of the classical model substrates. [Rh(nbd)(**L26**)]BF_4 has been used with success in hydrogenation of a large scope of trisubstituted enol esters to produce optically active esters with chirality at the α-carbon [33].

2.2.4 Self-assembled Diphosphine Ligands

After the emergence of monodentate phosphine, phosphoramide, and phosphite ligands at the turn of the 2000s [34], a way to close the gap between mono- and bidentate ligands by self-assembling was rapidly investigated. This was done by the association in the same ligand of a chiral phosphorus-containing moiety with a motif able to generate hydrogen-bonding interactions, which stabilize the formation of chelated forms, "pseudobidentate ligands," around the rhodium center. It was achieved by introducing amides [35], sulfonamides [36], ureas [37] (Scheme 2.12, **L28**[Rh] and **L29**[Rh]), and even two different functionalized phosphorus-containing moieties [38]. In most case, the chiral information was dictated by the chiral phosphorus group, which was a phosphine [35, 37], a phosphonite [35], a phosphoramidite [36, 38], or a phosphite [39]. Excellent enantioselectivities were obtained with some assembled systems for the asymmetric hydrogenation of α-dehydroamino esters [35, 36, 39], itaconic esters [35], and α-hydroxymethyl acrylates, cyclic enamides, and methyl α-acetamidomethyl cinnamate (Scheme 2.13).

The supramolecular interaction of a phosphine–phosphoramidite equipped with pyridine groups with zinc–salphen (or a porphyrin template) generated a catalytic system able to hydrogenate dimethyl itaconate at 25 °C under 10 bar of hydrogen in more than 99% ee, whereas the free ligand provided very low conversion (2%) under similar conditions (Scheme 2.12, ligand **L30**) [40].

2.2.5 Monodentate Phosphorus Ligands

Monodentate phosphorus ligands have become popular in the 2000s because of their easy preparation, stability, and efficiency in asymmetric hydrogenation associated

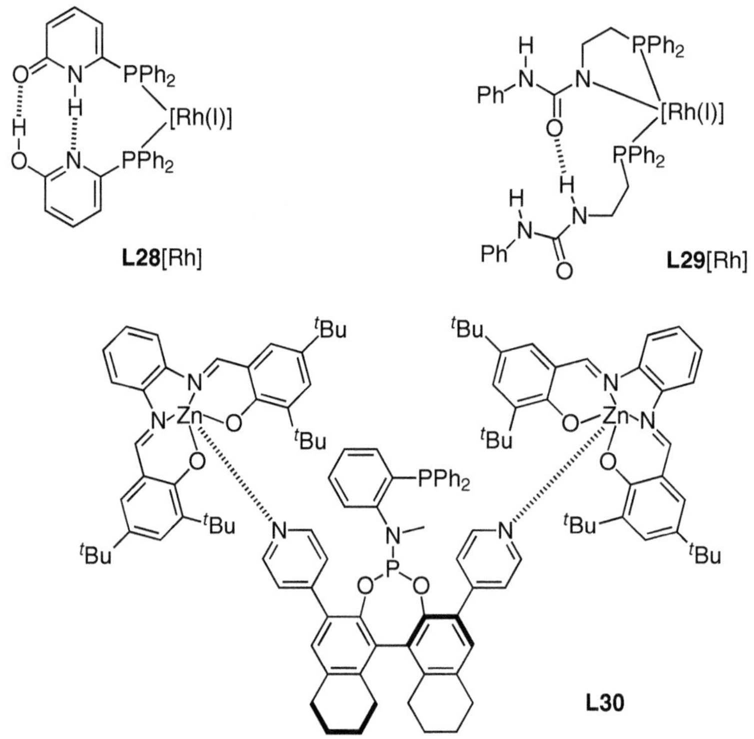

Scheme 2.12 Examples of self-assembling systems for asymmetric hydrogenation.

with rhodium precursors [34]. However, in the field of hydrogenation of functionalized olefins, very few new ligands have been studied during the past decade. Some novel structures featuring bulky carboranes **L32**, **L33**, or chiral moieties derived from pinene **L31** and amino acids **L34** are presented in Scheme 2.14. Rhodium species derived from these ligands were active in hydrogenation of α-dehydroamino esters, but the enantioselectivities were not improved as compared to previous results [41–43].

2.2.6 Asymmetric Transfer Hydrogenation with Rhodium Catalysts

The field of asymmetric transfer hydrogenation with rhodium catalysts was pioneered by Brunner and Leitner at the end of the 1980s [44]. The catalytic system and further examples were based on a rhodium(I) precursor associated with a chiral bisphosphine ligand such as **L35** [44] and **L36** [45] or a monodentate ligand such as **L37** [46] with ammonium formate as hydrogen source (Scheme 2.15). The transfer hydrogenation of the model substrates of type **1** and **3** took place but with moderate to acceptable enantiomeric excess (up to 97% in the case of itaconic acid with **L37**). Performing the reaction in aqueous medium with the water-soluble ligand **L38** did not improve the results [47]. On the other hand, asymmetric hydrogenation of β-methyl-β-aryl-nitroalkenes was efficiently performed using sodium

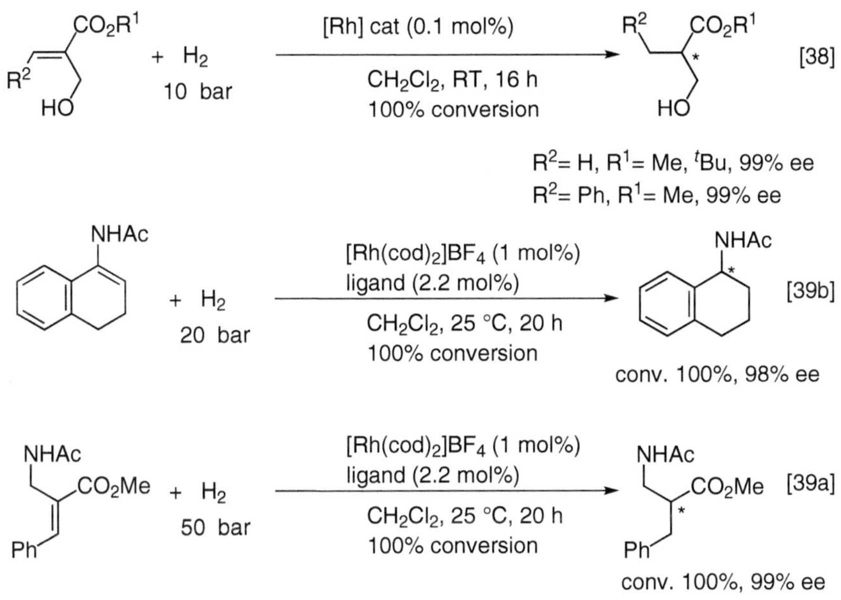

Scheme 2.13 Selected efficient asymmetric hydrogenation with self-assembled rhodium catalysts.

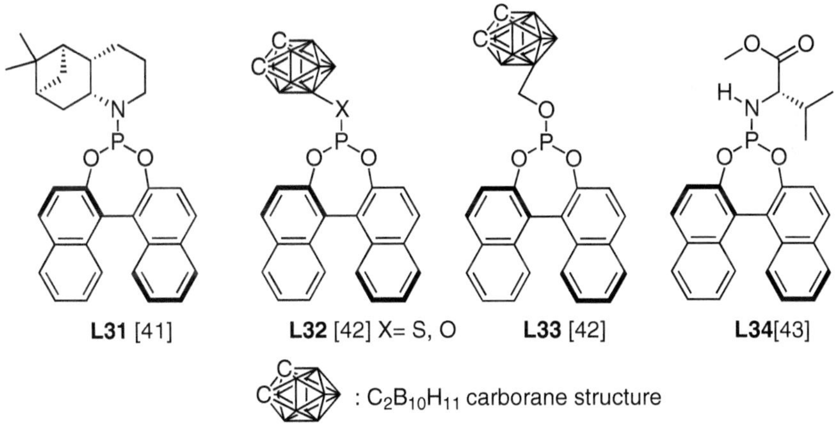

Scheme 2.14 Some monodentate phosphorus ligands.

formate/formic acid as hydrogen source and $[RhCp^*Cl_2]_2$ (1 mol%) in the presence of the diamine ligand **L39** at 28 °C in water leading to enantioselectivities located in the range 78–90%, depending on the nature of the aryl group [48]. It can be noted that these values are slightly lower than those obtained in hydrogenation with the ferrocenyl ligand **L19** (Scheme 2.9) [23].

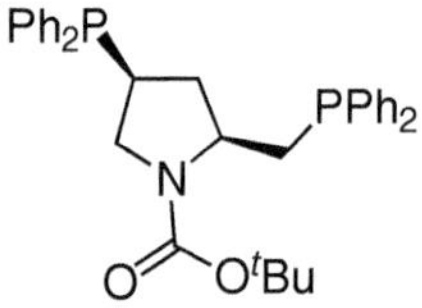

L35 BPPM

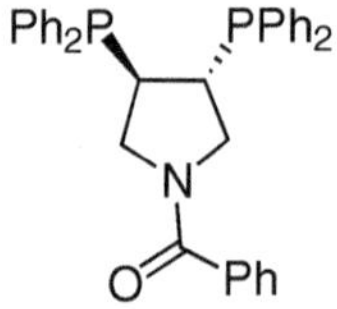

L36 DEGUPHOS

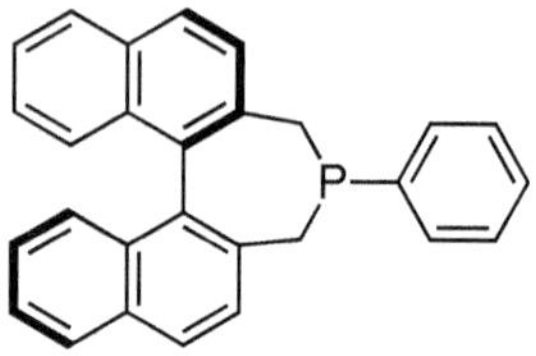

L37 Ph-BINEPINE

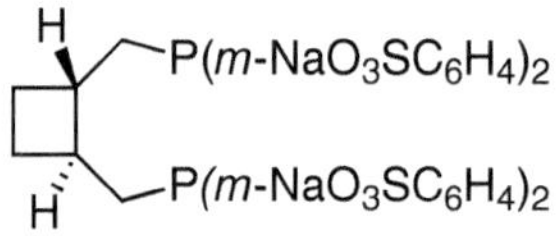

L38

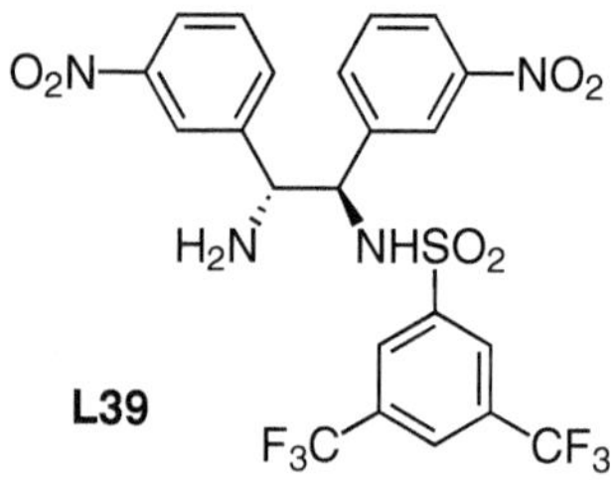

L39

Scheme 2.15 Some chiral ligands used in asymmetric transfer hydrogenation with rhodium catalysts.

2.3 Asymmetric Hydrogenation with Iridium Catalysts

During the past 20 years, iridium-catalyzed asymmetric hydrogenation of alkenes has known a tremendous development [2]. Besides bisphosphine ligands, a variety of bidentate ligands associated to a cationic iridium center have been introduced in iridium-catalyzed hydrogenation. They mainly consist in phosphine–(thio)oxazoline, phosphine–imidazole, phosphonite–oxazoline, phosphite–oxazoline, phosphine–carboxylate, phosphine–carbene, and oxazoline–*N*-heterocyclic carbene.

2.3.1 Chiral Bidentate Ferrocenyl Ligands

Ligand **L19** (Scheme 2.5) combining a ferrocenyl bisphosphine group and a thiourea group, which has been applied for asymmetric hydrogenation of β,β-disubstituted nitroolefins with a rhodium catalyst (Scheme 2.9), has also found useful application in the iridium-catalyzed hydrogenation of α-fluoro-β-enamino esters, leading to full conversion with excellent enantioselectivities and diastereoselectivities (>25 : 1 dr) (Scheme 2.17) [49]. The ferrocenyl ligand **L40** based on a spiranic chiral moiety (Scheme 2.16) provided remarkably high enantioselectivities (in most cases >99%) in the hydrogenation of β-acetamidonitroalkenes both with aryl and alkyl substituents in β-position (Scheme 2.17) [50]. The iridium complex containing the P,N-ligand **L41** (Scheme 2.16) promoted the enantioselective hydrogenation of α- and β-substituted acrylic esters and amides with enantioselectivities up to 92% at room temperature, but a long reaction time of 72 hours and 62 bar of hydrogen pressure were required to reach high conversions [51].

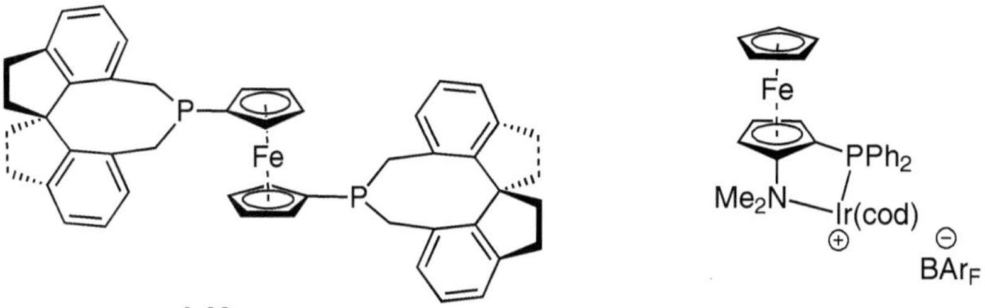

L40 L41

Scheme 2.16 Some ferrocenyl ligands used in asymmetric iridium-catalyzed hydrogenation.

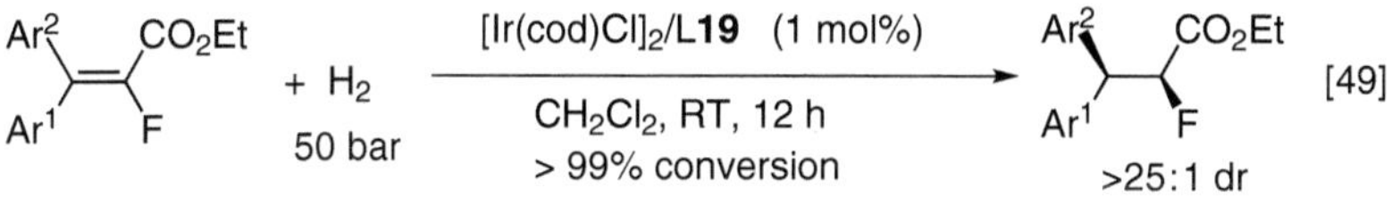

$Ar^1 = C_6H_5$; $Ar^2 = C_6H_5$ (98 ee), 2-FC_6H_4 (>99% ee), 2- and 4-$MeOC_6H_4$ (>99% ee), 3-MeC_6H_4 (98% ee)

$Ar^2 = C_6H_5$; $Ar^1 =$ 3-$MeOC_6H_4$ (95% ee), 4-$MeOC_6H_4$ (99% ee), 4-ClC_6H_4 (98% ee)

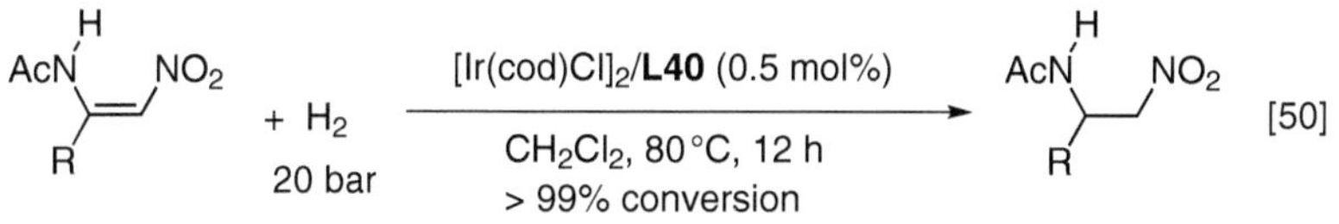

R= C_6H_5, 3-FC_6H_4, 4-FC_6H_4,3-MeC_6H_4, 4-MeC_6H_4, 3-$MeOC_6H_4$, 2-naphthyl : (99% ee)
4-ClC_6H_4, 4-BrC_6H_4 : > 99.9% ee; 4-$CF_3C_6H_4$ (99.6% ee)
isopropyl (96% ee); cyclohexyl (98% ee)

Scheme 2.17 Asymmetric hydrogenation with iridium–ferrocenyl ligands.

2.3.2 Other Chiral Bidentate P,N-ligands

Most of the phosphine–oxazoline and other ligands reported in Schemes 2.18 and 2.19 have been evaluated in the asymmetric hydrogenation of less-common substrates. For instance, apart from the ligands of type **L51**, which gave satisfactory (up to 98% ee) and moderate (19–83% ee) enantioselectivities with α-dehydroamido acid esters and aliphatic β-dehydroamido acid esters, respectively, these types of ligands have been scarcely studied [52]. On the other hand, α- and β-substituted cinnamates have been successfully hydrogenated, leading to good enantioselectivities with various ligands as shown in the selection gathered in Scheme 2.20.

A variety of acrylic acids have also been enantioselectively hydrogenated (ee > 95% in most cases) under low hydrogen pressure (3–15 bar) in methanol with iridium catalysts equipped with a bidentate spirophosphine–oxazoline ligand such as **L46** [58]. The neutral Ir(cod)(**L52**) containing the related anionic phosphine–carboxy ligand **L52** not only provided enantioselectivities in the range 94–99.4% in hydrogenation of acrylic acids but also allowed the enantioselective hydrogenation of challenging

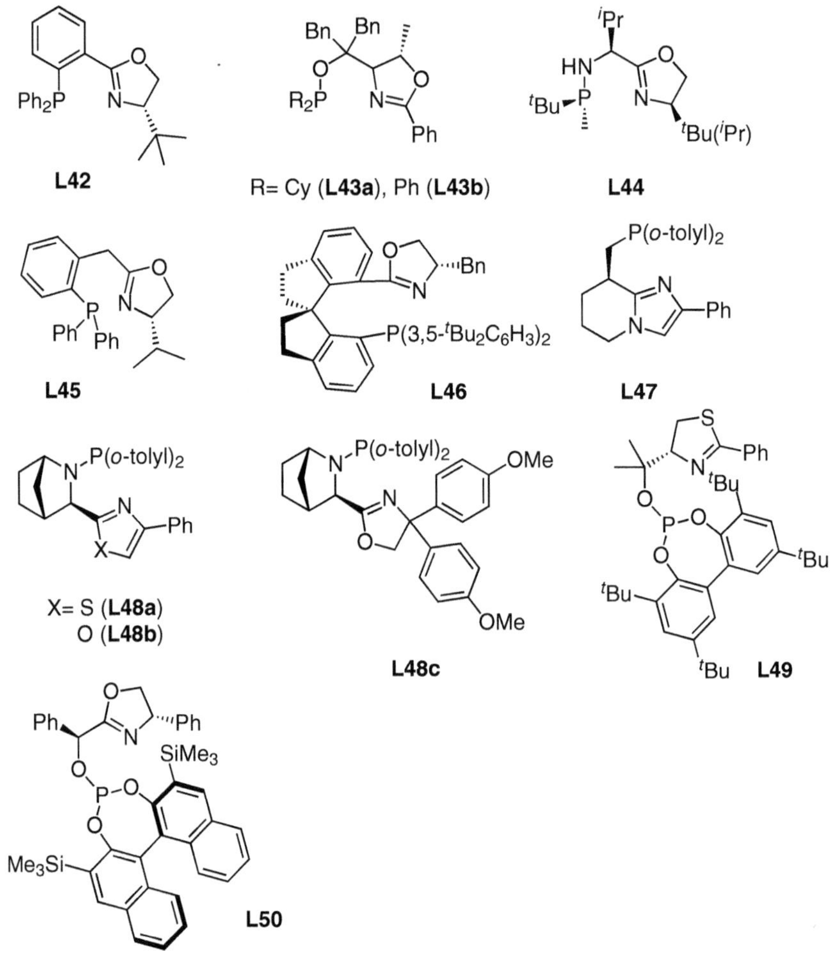

Scheme 2.18 Phosphine–oxazoline ligands used in asymmetric hydrogenation with iridium.

3-alkyl-3-methylenepropionic acids, where the double bond and the acid group are in remote positions [59].

Enamines having a terminal or trisubstituted double bond corresponding to α-styrene derivatives including cyclic substrates have been hydrogenated in dichloromethane at −20 °C to 25 °C under 10–50 bar of hydrogen pressure in the presence of iridium catalysts equipped with ligands **L42** [60] and **L43a** [61]. Excellent conversions were obtained with enantiomeric excesses, depending on the association of ligand/substrate but reaching more than 80% in most cases (up to 92% ee).

The iridium complex [Ir(cod)(**L43b**)]BAR_F revealed extremely high activity and enantioselectivity for the formation of optically active nitriles from α,β-unsaturated nitriles of (*E*)-configuration in methanol at 0 °C under 100 bar of hydrogen pressure (Scheme 2.21) [61].

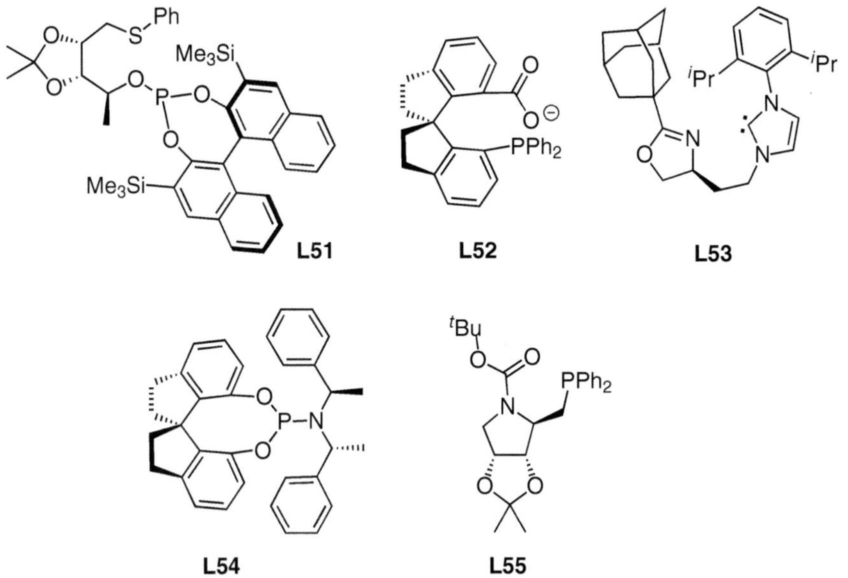

Scheme 2.19 Other mono- and bidentate ligands used in asymmetric hydrogenation with iridium.

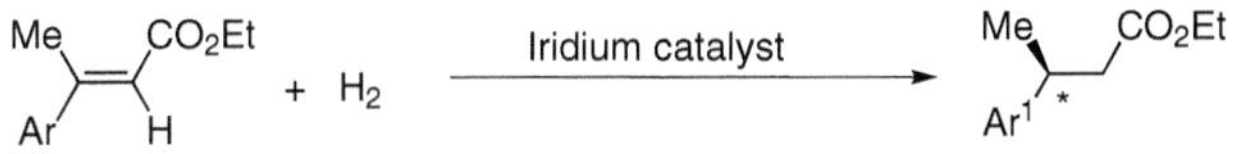

Ar	Catalyst	H_2 (bar)	Solvent	Temp.	Time (h)	Conv. (%)	ee (%)	References
Ph	[Ir(cod)(**L45**)]BAR$_F$	50	DCM	rt	24	>99	90	[53]
p-MeOC$_6$H$_4$	[Ir(cod)(**L45**)]BAR$_F$	50	DCM	rt	24	>99	92	[53]
Ph	[Ir(cod)(**L47**)]BAR$_F$	50	DCM	rt	17	>99	93	[54a]
Ph	[Ir(cod)(**L48a**)]BAR$_F$	50	DCM	rt	17	99	98	[55a]
Ph	[Ir(cod)(**L49**)]BAR$_F$	50	DCM	rt	2	100	96	[56a]
p-MeOC$_6$H$_4$	[Ir(cod)(**L49**)]BAR$_F$	50	DCM	rt	2	100	96	[56a]
p-ClC$_6$H$_4$	[Ir(cod)(**L49**)]BAR$_F$	50	DCM	rt	2	100	95	[56a]
Ph	[Ir(cod)(**L50**)]BAR$_F$	50	DCM	rt	4	100	94	[56b]
p-MeC$_6$H$_4$	[Ir(cod)$_2$]BAR$_F$ + **L55**	50	DCM	rt	4	100	97	[57]
p-MeOC$_6$H$_4$	[Ir(cod)$_2$]BAR$_F$ + **L55**	50	DCM	rt	4	98	96	[57]
p-FC$_6$H$_4$	[Ir(cod)$_2$]BAR$_F$ + **L55**	50	DCM	rt	4	94	95	[57]

Scheme 2.20 Asymmetric hydrogenation of ethyl β-methyl cinnamates with iridium catalysts.

The iridium complexes incorporating the bidentate oxazoline–aminophosphine **L44** or oxazoline–phosphonite **L50** ligands are among the best catalysts for the challenging enantioselective hydrogenation of cyclic enamides (Scheme 2.22) [56, 62].

Andersson has developed a series of aminophosphine–(thio)oxazoline iridium catalysts such as **L48a,b** with chiral ligands based on a bicyclic core. These catalysts

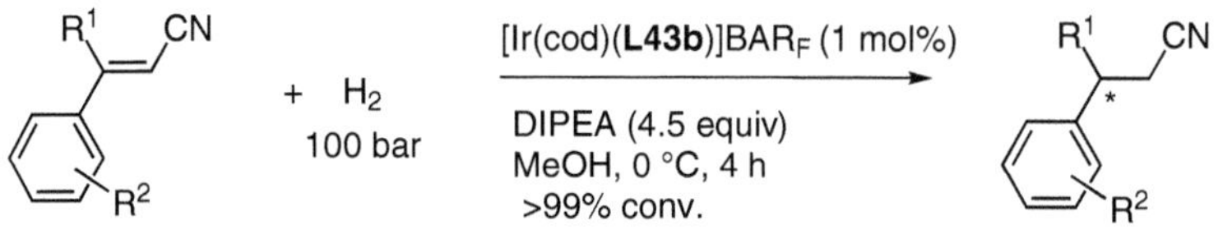

R^1= Me;

R^2= H (98% ee), *o*-Me (82% ee), *m*-Me (97% ee), *p*-Me (97% ee), *m*-MeO (98% ee) *m*-Cl (96% ee), *p*-Cl (>99% ee), *p*-F (96% ee), *p*-AcNH (98% ee), *m*-NO_2 (93% ee)

R^2= H;

R^1= Et (99% ee), Me_2CHCH_2 (97% ee)

Scheme 2.21 Asymmetric hydrogenation of acrylic nitriles with an iridium catalyst. Source: Based on Müller and Pfaltz [61].

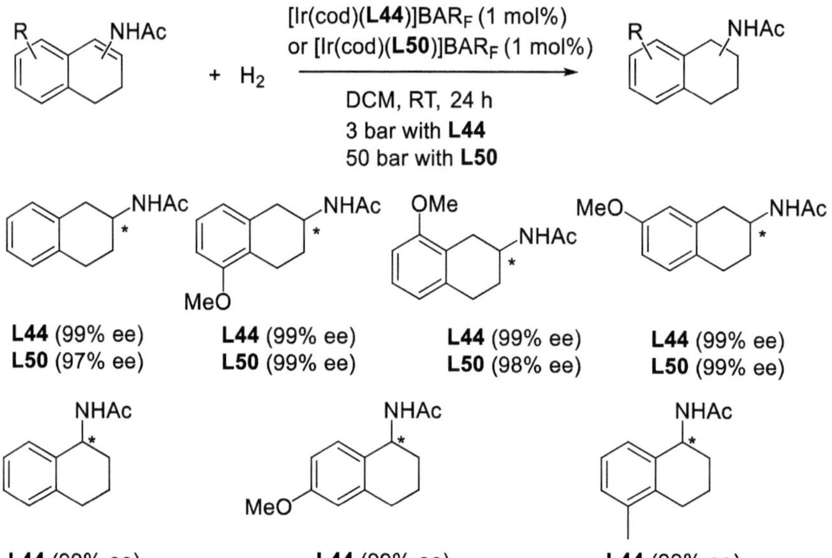

Scheme 2.22 Asymmetric hydrogenation of cyclic enamides with iridium catalysts. Source: Salomo et al. [62] and Biosca et al. [56].

showed excellent properties for the enantioselective hydrogenation of a variety of unsaturated sulfones including (*E*)- and (*Z*)-vinylic, (*E*)-allylic and homoallylic linear sulfones [55] (Scheme 2.23), and cyclic sulfones (Scheme 2.24) [55]. In most cases, full hydrogenation of the unsaturated substrates under 50 bar of hydrogen pressure at room temperature in dichloromethane and high enantiomeric excesses were obtained.

In addition, the complex based on **L48c** generated a highly efficient catalyst for the asymmetric hydrogenation of tetrasubstituted olefins, especially those derived from difunctionalized maleates [55]. As shown in Scheme 2.25, tetrasubstituted (*Z*)-maleate derivatives were hydrogenated in high yield with perfect diastereoselectivity and excellent enantioselectivity. The high enantioselectivities obtained

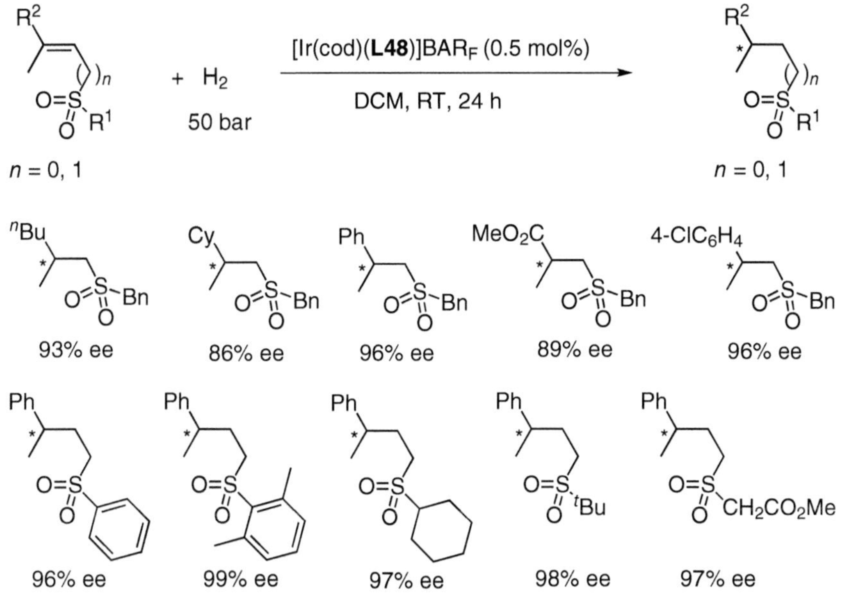

Scheme 2.23 Asymmetric hydrogenation of vinyl and (*E*)-allyl sulfones with iridium catalysts. Source: Based on Peters et al. [55].

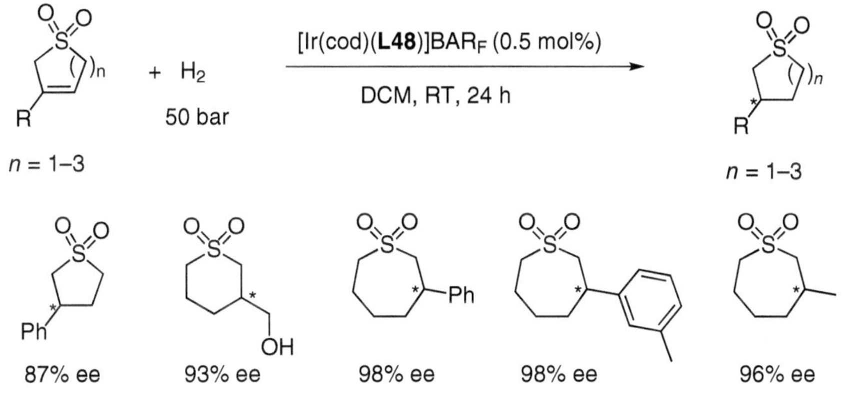

Scheme 2.24 Asymmetric hydrogenation of cyclic sulfones with iridium catalysts. Source: Based on Zhou et al. [55].

for these challenging substrates were attributed to the substitution of the oxazoline ring of the chiral ligand by two electron-rich *p*-methoxyphenyl substituents.

Allylic alcohols represent another class of functionalized olefins that have been hydrogenated in the presence of iridium catalysts. Satisfactory enantiomeric excesses (>90%) have been obtained from 3,3-disubstituted allylic alcohols with the phosphine–imidazole ligand **L47** [56] and phosphine–thiaoxazolines [55]. The iridium complex [Ir(cod)(**L53**]BAR_F featuring an *N*-heterocyclic carbene–oxazoline bidentate ligand has shown remarkable activity for the asymmetric hydrogenation

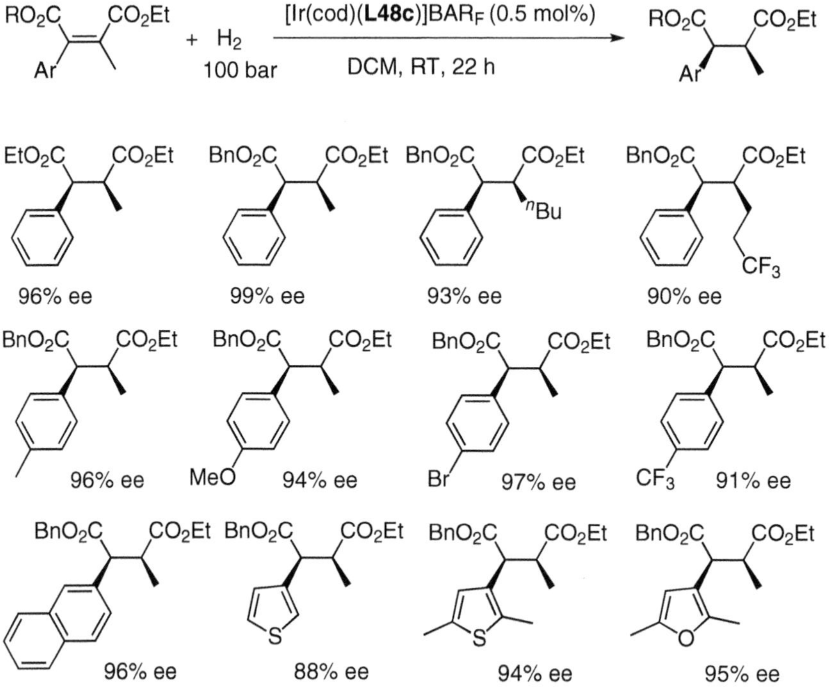

Scheme 2.25 Chiral succinates from asymmetric hydrogenation of tetrasubstituted maleates.

of chiral trisubstituted olefinic substrates with high stereo- and enantioselectivities. A variety of unprotected and protected allylic alcohols have thus been hydrogenated with high diastereo- and enantioselectivities resulting from matching of the chiral catalysts with substrates (Scheme 2.26) [63]. It is noteworthy that the same catalyst made possible the enantioselective access to Homo-Roche ester derivatives upon enantioselective hydrogenation of a trisubstituted acrylic ester surrounded by an allylic hydroxy group (Scheme 2.26) [63].

Finally, it is worth noting that monodentate phosphoramidite ligands based on the binaphthol core have revealed poor efficiency in asymmetric hydrogenation of β-dehydroamino esters with iridium complexes [64]. On the other hand, iridium complexes based on a spirophosphoramidite ligand such as **L54** are highly efficient for the synthesis of tetrahydroisoquinolines by asymmetric hydrogenation of the corresponding enamines featuring an exocyclic alkylidene moiety under 1 bar of hydrogen at room temperature in the presence of 5 mol% of I_2 as additive [65].

2.3.3 Asymmetric Transfer Hydrogenation with Iridium Catalysts

As far as asymmetric transfer hydrogenation of functionalized olefins with iridium catalysts is concerned, there are very few examples. A recent article reports on transfer hydrogenation of chalcone with a chiral IrCl(cod)(NHC) catalyst with

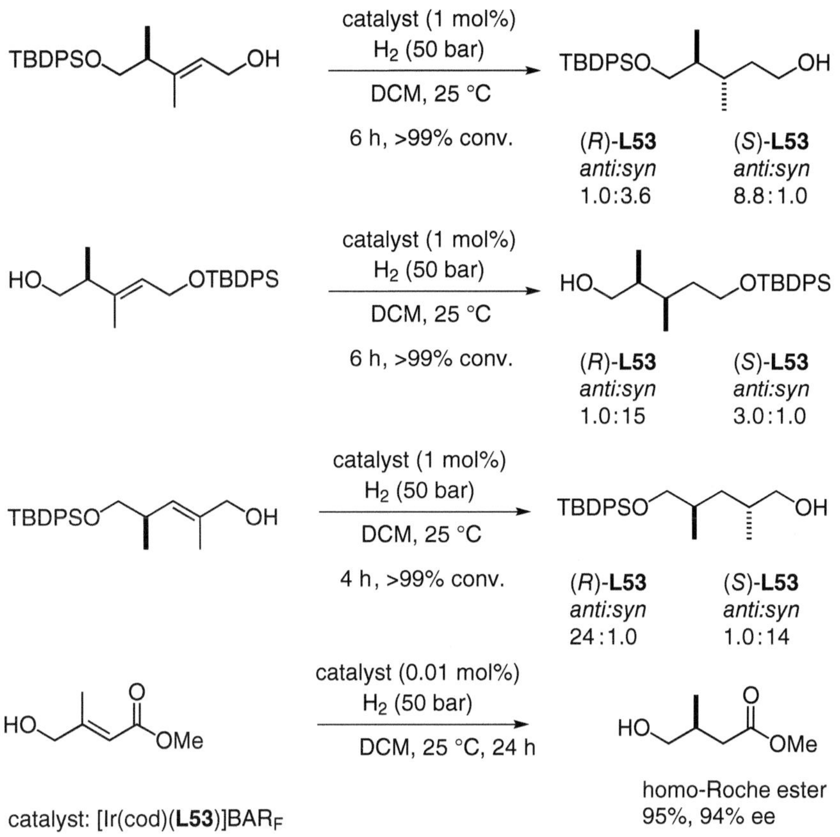

Scheme 2.26 Asymmetric hydrogenation of trisubstituted allylic derivatives.

isopropanol as hydrogen donor under basic conditions, but there is no chemoselectivity, both C=C and C=O bonds being reduced with very low enantioselectivities [66]. Much better results were obtained for the reduction of nitroolefins with aqueous formic acid at room temperature in the presence of the iridium complex based on the diamine **L56** (Scheme 2.27) [67].

2.4 Asymmetric Hydrogenation with Other Transition Metal Catalysts

2.4.1 Asymmetric Hydrogenation with Ruthenium Catalysts

Asymmetric hydrogenation of functionalized olefins such as **1–4** (Scheme 2.2) has been initially studied in depth with the help of ruthenium catalysts featuring an optically pure diphosphine ligand such as those derived from binaphthyl (**L57**) and biphenyl (**L58, L60–L63**) structures and other skeletons such as Duphos (**L60**) (Scheme 2.28) [2, 68]. During the past decade, much less results have appeared on this topic.

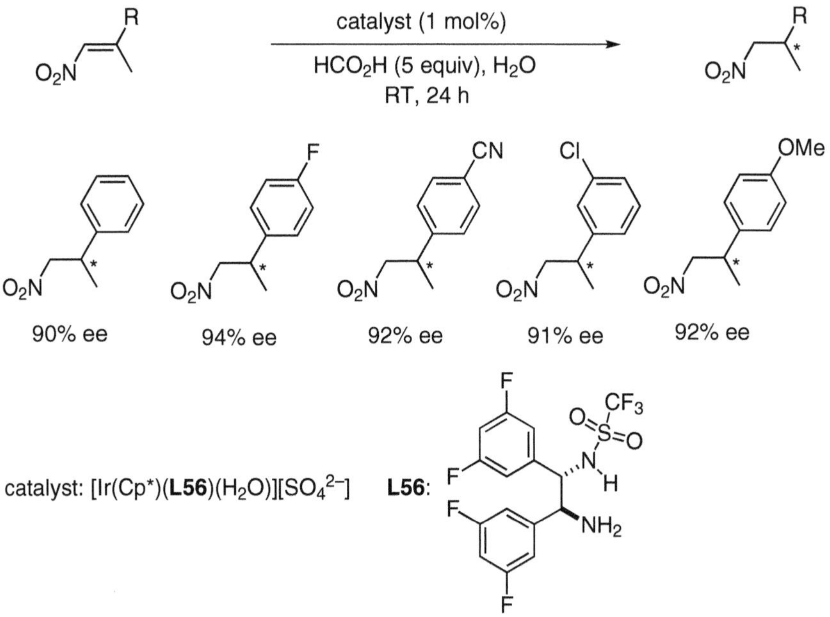

Scheme 2.27 Iridium-catalyzed asymmetric transfer hydrogenation of nitroolefins. Source: Based on Soltani et al. [67].

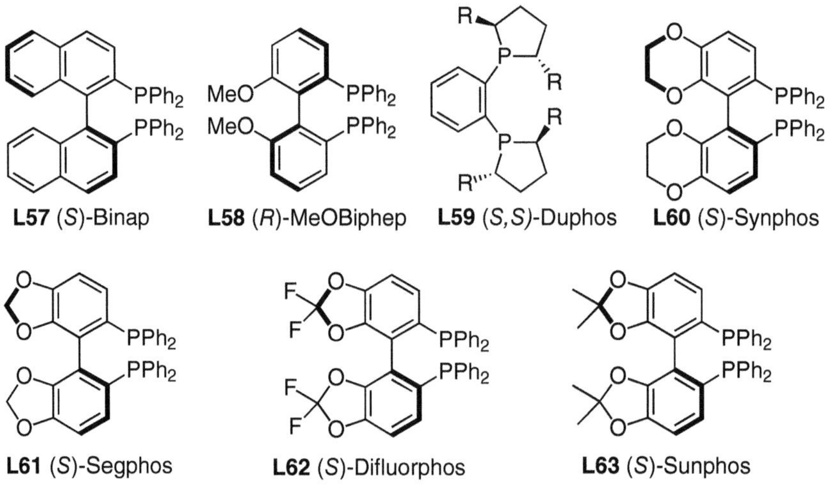

Scheme 2.28 Some typically atropoisomeric ligands used in ruthenium-catalyzed hydrogenation. Source: Ref. [2], Genêt [2], and Kitamura et al. [68].

A direct access to Roche ester derivatives in high yields and enantioselectivities has been obtained from the *in situ* generated catalytic system initially developed by Genêt et al. [69] based on the association of Ru(cod)(methylallyl)$_2$, an optically pure biphenyl ligand, and HBF_4 in methanol under 5 bar of hydrogen at 20 °C for 24 hours (Scheme 2.29) [70].

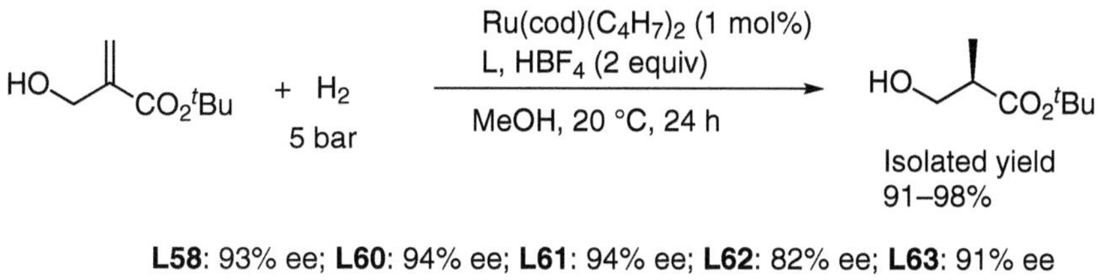

Scheme 2.29 Ruthenium-catalyzed enantioselective hydrogenation of *tert*-butyl 2-hydroxymethylpropenoate. Source: Based on Pautigny et al. [70].

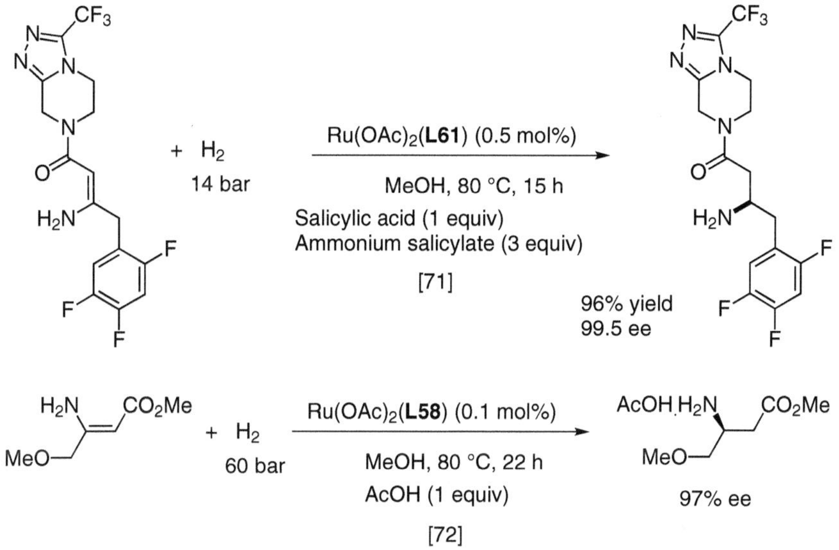

Scheme 2.30 Preparation of optically active amines via ruthenium catalysis. Source: Steinhuebel et al. [71] and Mattei et al. [72].

A few examples of asymmetric hydrogenation of β-enamine amides and esters featuring an unprotected NH_2 group leading to biologically active products of industrial interest have been successfully hydrogenated with ruthenium catalysts equipped with the same type of ligand (Scheme 2.30) [71, 72].

Following initial works based on ruthenium–Binap (**L57**) [73], trisubstituted enamides derived from 2-tetralones and 2-chromanones, where the carbon–carbon double bond is embedded in a six-membered ring, have been more efficiently hydrogenated (ee > 94%) with ruthenium catalysts involving the Synphos ligand **L60** [74, 75].

Asymmetric transfer hydrogenation of prochiral functionalized alkenes has been scarcely studied with ruthenium catalysts. The first examples dealing mainly with acrylic acid derivatives appeared back to the 1990s with secondary alcohols or formic acid as hydrogen donor and chiral Binap or diamine derivatives as ligands [76]. Nevertheless, efficient catalytic systems based on ruthenium have been developed in asymmetric hydrogenation of ketones, enones, and imine substrates [3].

2.4.2 Asymmetric Hydrogenation with Palladium Catalysts

Asymmetric hydrogenation of carbon–carbon double bonds with palladium catalysts has been less explored than with the previously reported metals [77]. During the past decade, enesulfonamides have been investigated with $Pd(O_2CCF_3)_2$ as metal precursor associated with chiral diphosphine ligands. In most cases, the reactions were performed in TFE as solvent under relatively high hydrogen pressure (14–48 bar). JosiPhos (**L64**) appeared to be the most appropriate ligand for the formation of six-membered cyclic sulfonamides in more than 95% enantioselectivity corresponding to enantioselective hydrogenation of an endocyclic double bond (Scheme 2.31) [78], whereas the hydrogenation of an exocyclic double bond was more favorably promoted at room temperature by the binaphane ligand **L65** (Scheme 2.32) [78].

Finally, the tetrasubstituted double bond of cyclic β-(arylsulfonamido)acrylates was carried out under similar conditions with the DuanPhos ligand **L66** bringing the chiral information, and enantioselectivities ranging from 87% to 93% were obtained in the presence of trifluoroacetic acid as additive [79]. Mechanistic studies, mainly deuterium labeling, revealed that during these hydrogenations (Schemes 2.31–2.33), *N*-sulfonylimines, tautomers of enesulfonamides, were formed as intermediates and that the presence of acidic proton played a crucial role.

Recently, using Binap (**L57**) as chiral ligand, hydantoin derivatives featuring an exocyclic double bond have been hydrogenated with moderate enantioselectivities

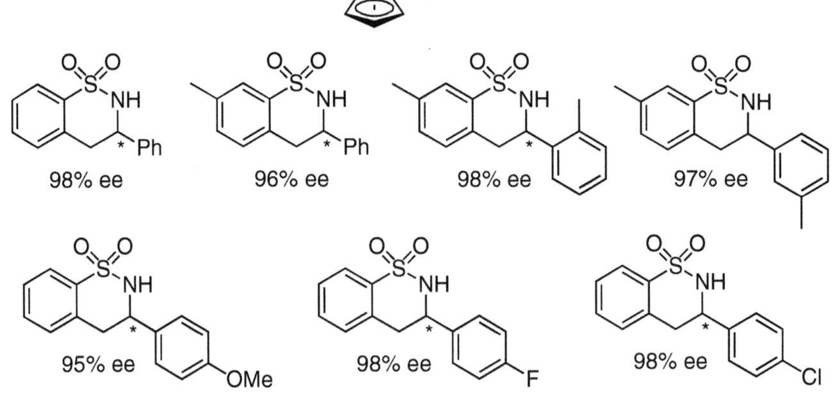

Scheme 2.31 Palladium-catalyzed enantioselective hydrogenation of cyclic enesulfonamides.

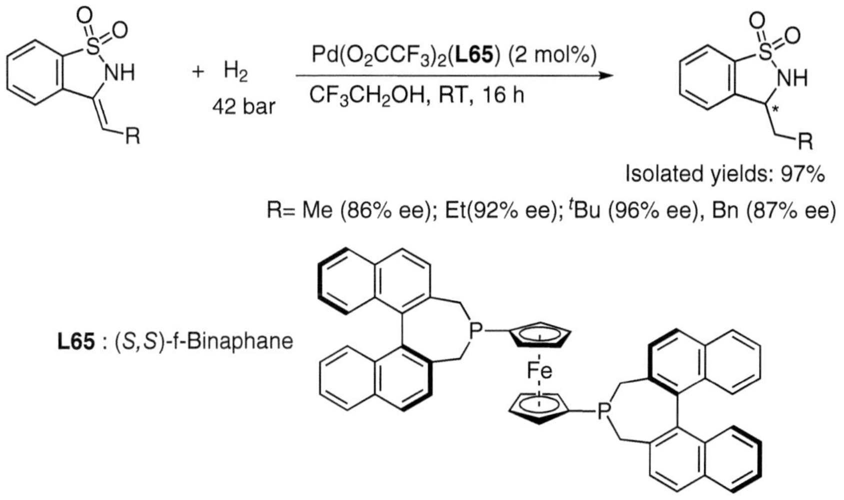

Scheme 2.32 Palladium-catalyzed enantioselective hydrogenation of exocyclic enesulfonamides. Source: Based on Yu et al. [78].

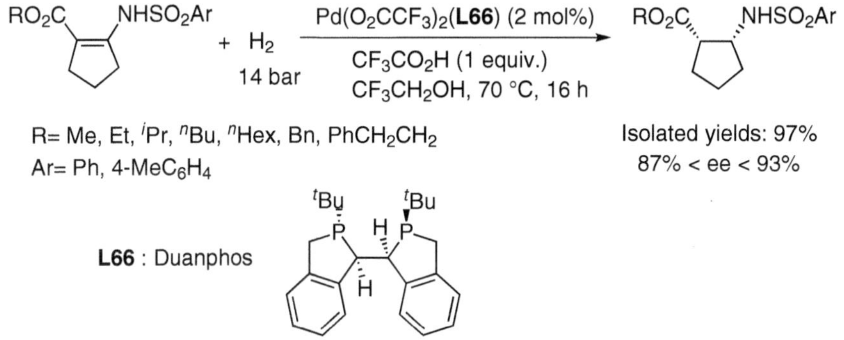

Scheme 2.33 Palladium-catalyzed enantioselective hydrogenation of cyclic tetrasubstituted β-enamide esters.

except in the case of 2-methoxy- and 2-methylphenyl derivatives, which gave 88% and 90% ee, respectively (Scheme 2.34) [80].

It can be noted that optically pure mono- and biscarbene ligands have also been involved in asymmetric hydrogenation of allylic alcohols and acrylic esters with some success but with limited scope [81].

If palladium-catalyzed hydrogen-transfer reactions from alcohols to unsaturated systems are known, there are not many asymmetric versions, especially in reduction of functionalized carbon–carbon double bonds [82]. Among them, the chemoselective hydrogenation of 3-methylcyclohexenone into 3-methylcyclohexanone in 40% yield with an enantiomeric excess of 30% was obtained using 3 mol% of $Pd(OAc)_2$(Me-Duphos) (**L59** with R = Me) as catalyst and formic acid/triethylamine as hydrogen donor [83]. Later on, the binuclear palladium catalysts, $[Pd(OH)(Binap)]_2[OTf]_2$ or $[Pd(OH_2)_2(Binap)][OTf]$, were

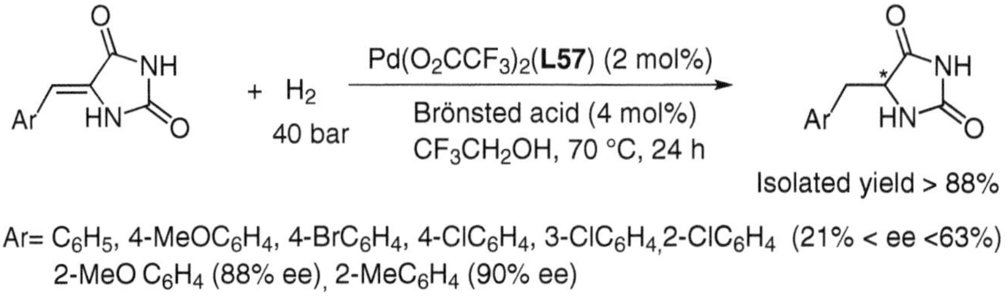

Scheme 2.34 Palladium-catalyzed enantioselective hydrogenation of prochiral hydantoins. Source: Based on Ma et al. [80].

also found to be efficient for the chemoselective asymmetric hydrogenation of conjugated enones into chiral ketones in ethanol as hydrogen donor at room temperature [84].

2.5 Asymmetric (Transfer) Hydrogenation with First-row Transition Metal Catalysts

Recently, earth-abundant first-row transition metals have found applications in hydrogen-transfer reaction. The asymmetric (transfer) hydrogenation of ketones, imines, and alkenes with these metals has been recently reviewed [85]. For the enantioselective hydrogenation of functionalized alkenes such as enamides, promising systems have been obtained with cobalt ($CoCl_2$ (10 mol%) or $Co(CH_2SiMe_3)_2$ (1 mol%)) associated to a diphosphine ligand such as Quinox (**L3**) or Duphos (**L59**) operating at room temperature under 35 bar of hydrogen pressure [86]. In transfer hydrogenation, nickel catalysts equipped with a diphosphine ligand such as (*S*)-Binapine (**L67**, Scheme 2.35) or DuPhos (**L59**) have led to excellent enantioselectivities with β-dehydroamino esters (94–99% ee), α-dehydroamino esters (86–88% ee), and β,β-disubstituted acrylic esters (54–94% ee) [87].

L67: (*S*)-Binapine

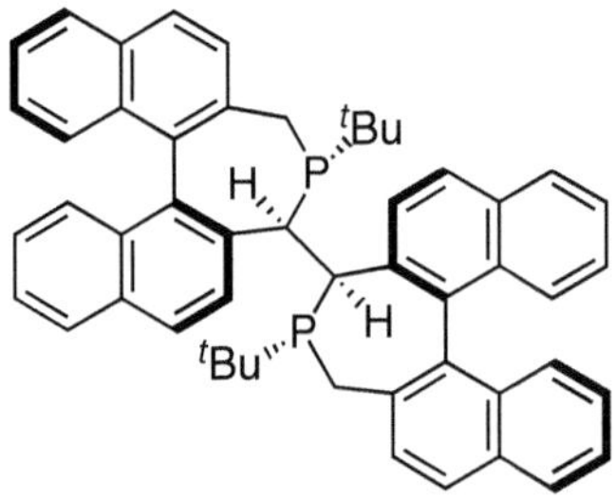

Scheme 2.35 (*S*)-Binapine ligand.

2.6 Conclusion

Asymmetric hydrogenation of prochiral functionalized olefins has been the subject of numerous studies initially carried out with rhodium and ruthenium catalysts.

Many catalytic systems involving a huge variety of ligands have led to efficient systems for the enantioselective hydrogenation of α- and β-dehydroamino acid derivatives and acrylic acids and esters, enamides as model substrates. During the past decade, rhodium systems have still been explored intensively, and new efficient iridium-based catalysts have been developed, especially with phosphine–oxazoline ligands. At the same time, new families of prochiral substrates such as enesulfonamides and nitroalkenes have been investigated with specific biologically active molecules targets as final objectives. It must also be noted that first-row transition metals are emerging in the field.

Abbreviations

BAR_F	tetrakis(3,5-bis(trifluoromethyl)phenyl)borate
cod	cycloocta-1,5-diene
Cp*	pentamethylcyclopentadienyl
L-DOPA	(L)-3,4-dihydroxyphenylalanine
NHC	*N*-heterocyclic carbene
nbd	norbornadiene.

References

1 (a) Knowles, W.S. and Sabacky, M.J. (1968). *Chem. Commun.*: 1445–1446. (b) Horner, L., Siegel, H., and Büthe, H. (1968). *Angew. Chem. Int. Ed. Engl.* 7: 942. (c) Dang, T.P. and Kagan, H.B. (1971). *Chem. Commun.*: 481.

2 (a) Chi, Y., Tang, W., and Zhang, X. (2005). *Modern Rhodium-Catalyzed Organic Reactions* (ed. P.A. Evans), 1–31. Weinheim, Germany: Wiley-VCH. (b) de Vries, J.G. and Elsevier, C.J. (eds.) (2007). *Handbook of Homogeneous Hydrogenation*. Weinheim Germany: Wiley-VCH. (c) Genêt, J.-P. (2008). *Modern Reduction Methods* (eds. P.G. Andersson and I.J. Munslow), 3–38. Weinheim, Germany: Wiley-VCH. (d) Diesen, J.S. and Andersson, P.G. (2008). *Modern Reduction Methods* (eds. P.G. Andersson and I.J. Munslow), 39–64. Weinheim, Germany: Wiley-VCH. (e) Church, T.L. and Andersson, P.G. (2008). *Coord. Chem. Rev.* 252: 513–531. (f) Bruneau, C., Renaud, J.-L., and Jerphagnon, T. (2008). *Coord. Chem. Rev.* 252: 532–544. (g) Xie, J.H., Zhu, S.-F., and Zhou, Q.-L. (2011). *Chem. Rev.* 111: 1713–1760. (h) Zhou, Q.-L. and Xie, J.-H. (2011). *Top. Organomet. Chem.* 36: 1–28. (i) Lam, F.L., Kwong, F.Y., and Chan, A.S.C. (2011). *Top. Organomet. Chem.* 36: 29–66. (j) Xie, J.-H., Zhu, S.-F., and Zhou, Q.-L. (2012). *Chem. Soc. Rev.* 41: 4126–4139. (k) Blaser, H.-U., Pugin, B., and Spindler, F. (2012). *Top. Organomet. Chem.* 42: 65–102. (l) Ager, D.J., de Vries, A.H.M., and de Vries, J.G. (2012). *Chem. Soc. Rev.* 41: 3340–3380. (m) Etayo, P. and Vidal-Ferran, A. (2013). *Chem. Soc. Rev.* 42: 728–754. (n) Verendel, J.J., Pamies, O., Dieguez, M., and Andersson, P.G. (2014). *Chem. Rev.* 114: 2130–2169. (o) Ayad, T., Phansavath, P., and Ratovelomanana-Vidal, V. (2016). *Chem. Rec.* 16: 2754–2771. (p) Seo, C.S.G. and Morris, R.H. (2019). *Organometallics* 38: 47–65.

3 (a) Zassinovich, G. and Mestroni, G. (1992). *Chem. Rev.* 92: 1051–1089. (b) Wang, D. and Astruc, D. (2015). *Chem. Rev.* 115: 6621–6686. (c) Barath, E. (2018). *Catalysts* 8: 671. https://doi.org/10.3390/catal8120671.

4 Vineyard, B.D., Knowles, W.S., Sabacky, M.J. et al. (1977). *J. Am. Chem. Soc.* 99: 5946–5952.

5 Zupancic, B., Mohar, B., and Stephan, M. (2008). *Adv. Synth. Catal.* 350: 2024–2032.

6 (a) Tamura, K., Sugiya, M., Yoshida, K. et al. (2010). *Org. Lett.* 12: 4400–4403. (b) Zhang, Z., Tamura, K., Mayama, D. et al. (2012). *J. Org. Chem.* 77: 4184–4188. (c) Imamoto, T., Tamura, K., Zhang, Z. et al. (2012). *J. Am. Chem. Soc.* 134: 1754–1769.

7 Imamoto, T., Horiuchi, Y., Hamanishi, E. et al. (2015). *Tetrahedron* 71: 6471–6480.

8 Ma, M., Hou, G., Wang, J., and Zhang, X. (2011). *Tetrahedron: Asymmetry* 22: 506–511.

9 (a) Zhang, X., Huang, K., Hou, G. et al. (2010). *Angew. Chem. Int. Ed.* 49: 6421–6424. (b) Huang, K., Zhang, X., Emge, T.J. et al. (2010). *Chem. Commun.* 46: 8555–8557.

10 Tang, W., Qu, B., Capacci, A.G. et al. (2010). *Org. Lett.* 12: 176–179.

11 Gladiali, S., Dore, A., Fabbri, D. et al. (1994). *Tetrahedron: Asymmetry* 5: 511–514.

12 Dai, Q., Li, W., and Zhang, X. (2008). *Tetrahedron* 64: 6943–6948.

13 Togni, A., Breutel, C., Schnyder, A. et al. (1994). *J. Am. Chem. Soc.* 116: 4062–4066.

14 Fox, M.E., Jackson, M., Lennon, I.C. et al. (2008). *J. Org. Chem.* 73: 775–784.

15 Liu, Y., Chen, J., Zhang, Z. et al. (2016). *Org. Biomol. Chem.* 14: 7099–7102.

16 Zirakzadeh, A., Gro, M.A., Wang, Y. et al. (2014). *Organometallics* 33: 1945–1952.

17 Oki, H., Oura, I., Nakamura, T. et al. (2009). *Tetrahedron: Asymmetry* 20: 2185–2191.

18 Landert, H., Spindler, F., Wyss, A. et al. (2010). *Angew. Chem. Int. Ed.* 49: 6873–6876.

19 Chen, C., Wang, H., Zhang, Z. et al. (2016). *Chem. Sci.* 7: 6669–6673.

20 Yin, X., Chen, C., Li, X. et al. (2017). *Org. Lett.* 19: 4375–4378.

21 Luo, L.-B., Wang, D.-Y., Zhou, X.-M. et al. (2011). *Tetrahedron: Asymmetry* 22: 2117–2123.

22 Krska, S.W., Mitten, J.V., Dormer, P.G. et al. (2009). *Tetrahedron* 65: 8987–8994.

23 Zhao, Q., Li, S., Huang, K. et al. (2013). *Org. Lett.* 15: 4014–4017.

24 Li, P., Zhou, M., Zhao, Q. et al. (2016). *Org. Lett.* 18: 40–43.

25 Yu, S.-B., Huang, J.-D., Wang, D.-Y. et al. (2008). *Tetrahedron: Asymmetry* 19: 1862–1866.

26 Wassenaar, J., Kuil, M., Lutz, M. et al. (2010). *Chem. Eur. J.* 16: 6509–6517.

27 Pullmann, T., Engendahl, B., Zhang, Z. et al. (2010). *Chem. Eur. J.* 16: 7517–7526.

28 Schmitz, C., Holthusen, K., Leitner, W., and Francio, G. (2017). *Eur. J. Org. Chem.*: 4111–4116.

29 (a) Drommi, D., Micalizzi, G., and Arena, C.G. (2014). *Appl. Organomet. Chem.* 28: 614–619. (b) Drommi, D. and Arena, C.G. (2017). *Appl. Organomet. Chem.* 31: e3837.
30 Farkas, G., Balogh, S., Madarasz, J. et al. (2012). *Dalton Trans.* 41: 9493–9502.
31 Kleman, P., Vaquero, M., Arribas, I. et al. (2014). *Tetrahedron: Asymmetry* 25: 744–749.
32 (a) Fernandez-Perez, H., Donald, S.M.A., Munslow, I.J. et al. (2010). *Chem. Eur. J.* 16: 6495–6508. (b) Fernandez-Perez, H., Pericas, M.A., and Vidal-Ferran, A. (2008). *Adv. Synth. Catal.* 350: 1984–1990. (c) Etayo, P., Nunez-Rico, J.L., and Vidal-Ferran, A. (2011). *Organometallics* 30: 6718–6725.
33 Leon, F., Gonzalez-Liste, P.J., Garcia-Garrido, S.E. et al. (2017). *J. Org. Chem.* 82: 5852–5867.
34 (a) Komarov, I.V. and Börner, A. (2001). *Angew. Chem. Int. Ed.* 40: 1197–1200. (b) Jerphagnon, T., Renaud, J.-L., and Bruneau, C. (2004). *Tetrahedron: Asymmetry* 15: 2101–2111.
35 (a) Schäffner, B., Holz, J., Verevkin, S.P., and Börner, A. (2008). *Tetrahedron Lett.* 49: 768–771. (b) Wieland, J. and Breit, B. (2010). *Nat. Chem.* 2: 832–837.
36 Patureau, F.W., Kuil, M., Sandee, A.J., and Reek, J.N.H. (2008). *Angew. Chem. Int. Ed.* 47: 3180–3183.
37 Meeuwissen, J., Detz, R., Sandee, A.J. et al. (2010). *Eur. J. Inorg. Chem.*: 2992–2997.
38 Daubignard, J., Detz, R.J., Jans, A.C.H. et al. (2017). *Angew. Chem. Int. Ed.* 56: 13056–13060.
39 (a) Pignataro, L., Boghi, M., Civera, M. et al. (2012). *Chem. Eur. J.* 18: 1383–1400. (b) Pignataro, L., Bovio, C., Civera, M. et al. (2012). *Chem. Eur. J.* 18: 10368–10381.
40 Bellini, R. and Reek, J.N.H. (2012). *Eur. J. Inorg. Chem.* 2012: 4684–4693.
41 Schmitz, C., Leitner, W., and Francio, G. (2015). *Eur. J. Org. Chem.* 13: 2889–2901.
42 (a) Lyubimov, S.E., Davankov, V.A., Petrovskii, P.V. et al. (2008). *J. Organomet. Chem.* 693: 3689–3691. (b) Lyubimov, S.E., Kalinin, V.N., Tyutyunov, A.A. et al. (2009). *Chirality* 21: 2–5.
43 Breuil, P.-A.R. and Reek, J.N.H. (2009). *Eur. J. Org. Chem.* 35: 6225–6230.
44 (a) Brunner, H. and Leitner, W. (1988). *Angew. Chem. Int. Ed. Engl.* 27: 1180–1181. (b) Brunner, H. and Leitner, W. (1990). *J. Organomet. Chem.* 387: 209–217. (c) Leitner, W., Brown, J.M., and Brunner, H. (1993). *J. Am. Chem. Soc.* 115: 152–159. (d) Lange, S. and Leitner, W. (2002). *J. Chem. Soc., Dalton Trans.*: 752–758.
45 Rocha Gonsalves, A.M.d.'A., Bayon, J.C., Pereira, M.M. et al. (1998). *J. Organomet. Chem.* 553: 199–204.
46 Alberico, E., Nieddu, I., Taras, R., and Gladiali, S. (2006). *Helv. Chim. Acta* 89: 1716–1729.
47 Sinou, D., Safi, M., Claver, C., and Masdeu, A. (1991). *J. Mol. Catal.* 68: L9–L12.
48 Tang, Y., Xiang, J., Cun, L. et al. (2010). *Tetrahedron: Asymmetry* 21: 1900–1905.
49 Han, Z., Guan, Y.-Q., Liu, G. et al. (2018). *Org. Lett.* 20: 6349–6353.

50 Yan, Q., Liu, M., Kong, D. et al. (2014). *Chem. Commun.* 50: 12870–12872.
51 Metallinos, C. and Van Belle, L. (2011). *J. Organomet. Chem.* 696: 141–149.
52 Margalef, J., Borras, C., Alegre, S. et al. (2019). *ChemCatChem* 11: 2142–2168.
53 Lu, W.-J., Chen, Y.-W., and Hou, X.-L. (2010). *Adv. Synth. Catal.* 352: 103–107.
54 (a) Kaukoranta, P., Engman, M., Hedberg, C. et al. (2008). *Adv. Synth. Catal.* 350: 1168–1176. (b) Tolstoy, P., Engman, M., Paptchikhine, A. et al. (2009). *J. Am. Chem. Soc.* 131: 8855–8860.
55 (a) Li, J.-Q., Paptchikhine, A., Govender, T., and Andersson, P.G. (2010). *Tetrahedron: Asymmetry* 21: 1328–1333. (b) Zhou, T., Peters, B., Maldonado, M.F. et al. (2012). *J. Am. Chem. Soc.* 134: 13592–13595. (c) Peters, B.K., Zhou, T., Rujirawanich, J. et al. (2014). *J. Am. Chem. Soc.* 136: 16557–16562. (d) Cheruku, P., Paptchikhine, A., Ali, M. et al. (2008). *Org. Biomol. Chem.* 6: 366–373. (e) Kerdphon, S., Ponra, S., Yang, J. et al. (2019). *ACS Catal.* 9: 6169–6176.
56 (a) Mazuela, J., Pamies, O., and Dieguez, M. (2013). *ChemCatChem* 5: 2410–2417. (b) Biosca, M., Magre, M., Pamies, O., and Dieguez, M. (2018). *ACS Catal.* 8: 10316–10320.
57 Elias-Rodriguez, P., Borras, C., Carmona, A.T. et al. (2018). *ChemCatChem* 10: 5414–5424.
58 (a) Li, M.-L., Yang, S., Su, X.-C. et al. (2017). *J. Am. Chem. Soc.* 139: 541–547. (b) Zhu, S.-F. and Zhou, Q.-L. (2017). *Acc. Chem. Res.* 50: 988–1001.
59 Yang, S., Che, W., Wu, H.-L. et al. (2017). *Chem. Sci.* 8: 1977–1980.
60 Baeza, A. and Pfaltz, A. (2009). *Chem. Eur. J.* 15: 2266–2269.
61 Müller, M.-A. and Pfaltz, A. (2014). *Angew. Chem. Int. Ed.* 53: 8668–8671.
62 Salomo, E., Orgué, S., Riera, A., and Verdaguer, X. (2016). *Angew. Chem. Int. Ed.* 55: 7988–7992.
63 (a) Zhu, Y. and Burgess, K. (2008). *J. Am. Chem. Soc.* 130: 8894–8895. (b) Zhao, J. and Burgess, K. (2009). *J. Am. Chem. Soc.* 131: 13236–13237. (c) Zhu, Y. and Burgess, K. (2013). *Adv. Synth. Catal.* 355: 107–115. (d) Khumsubdee, S., Zhou, H., and Burgess, K. (2013). *J. Org. Chem.* 78: 11948–11955.
64 Mrsic, N., Panella, L., Minnard, A.J. et al. (2011). *Tetrahedron: Asymmetry* 22: 36–39.
65 Yan, P.-C., Xie, J.-H., Hou, G.-H. et al. (2009). *Adv. Synth. Catal.* 351: 3243–3250.
66 Ramasamy, B., Prakasham, A.P., Gangwar, M.K., and Ghosh, P. (2019). *ChemistrySelect* 4: 357–365.
67 Soltani, O., Ariger, M.A., and Carreira, E.M. (2009). *Org. Lett.* 11: 4196–4198.
68 Kitamura, M., Tsukamoto, M., Bessho, Y. et al. (2002). *J. Am. Chem. Soc.* 124: 6649–6667.
69 Genêt, J.-P., Pinel, C., Ratovelomanana-Vidal, V. et al. (1994). *Tetrahedron: Asymmetry* 5: 665–674.
70 Pautigny, C., Jeulin, S., Ayad, T. et al. (2008). *Adv. Synth. Catal.* 350: 2525–2532.
71 Steinhuebel, D., Sun, Y., Matsumura, K. et al. (2009). *J. Am. Chem. Soc.* 131: 11316–11317.
72 Mattei, P., Moine, G., Püntener, K., and Schmid, R. (2011). *Org. Process Res. Dev.* 15: 353–359.
73 Renaud, J.-L., Dupau, P., Hay, A.-E. et al. (2003). *Adv. Synth. Catal.* 345: 230–238.

74 Pautigny, C., Debouit, C., Vayron, P. et al. (2010). *Tetrahedron: Asymmetry* 21: 1382–1388.
75 Wu, Z., Ayad, T., and Ratovelomanana-Vidal, V. (2011). *Org. Lett.* 13: 3782–3785.
76 (a) Brown, J.M., Brunner, H., Leitner, W., and Rose, M. (1991). *Tetrahedron: Asymmetry* 2: 331–334. (b) Saburi, M., Ohnuki, M., Ogasawara, M. et al. (1992). *Tetrahedron Lett.* 33: 5783–5786. (c) Liu, W., Cui, X., Cun, L. et al. (2005). *Tetrahedron : Asymmetry* 16: 2525–2530. (d) Chen, Y.-C., Xue, D., Deng, J.-G. et al. (2004). *Tetrahedron Lett.* 45: 1555–1558. (e) Xue, D., Chen, Y.-C., Deng et al. (2005). *J. Org. Chem.* 70: 3584–3591.
77 Chen, Q.-A., Ye, Z.-S., Duan, Y., and Zhou, Y.-G. (2013). *Chem. Soc. Rev.* 42: 497–511.
78 Yu, C.-B., Gao, K., Wang, D.-S. et al. (2011). *Chem. Commun.* 47: 5052–5054.
79 Yu, C.-B., Gao, K., Chen, Q.-A. et al. (2012). *Tetrahedron Lett.* 53: 2560–2563.
80 Ma, B.-D., Du, S.-H., Wang, Y. et al. (2017). *Tetrahedron: Asymmetry* 28: 47–53.
81 (a) Boronat, M., Corma, A., Gonzalez-Arellano, C. et al. (2010). *Organometallics* 29: 134–141. (b) Arnanz, A., Gonzalez-Arellano, C., Juan, A. et al. (2010). *Chem. Commun.* 46: 3001–3003. (c) Dasgupta, A., Ramkumar, V., and Sankararaman, S. (2015). *RSC Adv.* 5: 21558–21561.
82 Muzart, J. (2015). *Eur. J. Org. Chem.* 26: 5693–5707.
83 Drago, D. and Pregosin, P.S. (2002). *Organometallics* 21: 1208–1215.
84 (a) Tsuchiya, Y., Hamashima, Y., and Sodeoka, M. (2006). *Org. Lett.* 8: 4851–4854. (b) Monguchi, D., Beemelmanns, C., Hashizume, D. et al. (2008). *J. Organomet. Chem.* 693: 867–873.
85 Zhang, Z., Butt, N.A., Zhou, M. et al. (2018). *Chin. J. Chem.* 36: 443–454.
86 Friedfeld, M.R., Shevlin, M., Hoyt, J.M. et al. (2013). *Science* 342: 1076–1080.
87 (a) Yang, P., Xu, H., and Zhou, J. (2014). *Angew. Chem. Int. Ed.* 53: 12210–12213. (b) Guo, S., Yang, P., and Zhou, J. (2015). *Chem. Commun.* 51: 12115–12117. (c) Guo, S. and Zhou, J. (2016). *Org. Lett.* 18: 5344–5347.

3

Asymmetric (Transfer) Hydrogenation of Functionalized Ketones

Noriyoshi Arai and Takeshi Ohkuma

Hokkaido University, Division of Applied Chemistry and Frontier Chemistry Center, Faculty of Engineering, N13-W8, Kita-ku, Sapporo, Japan

3.1 Introduction

The asymmetric hydrogenation of ketones is one of the powerful methods that provides synthetically useful secondary alcohols in optically active form [1–3]. In comparison to asymmetric reductions using stoichiometric metal–hydride reagents modified or catalyzed by a chiral compound, or those by biochemical methods with enzymes, the asymmetric hydrogenation has several advantages. (i) The most distinct advantage over the other methods from a synthetic point of view is the cleanliness and effectiveness of the reaction. When an appropriate catalyst is selected, the reaction proceeds almost quantitatively in a highly enantioselective manner without formation of undesirable by-products. The atom efficiency of the reaction is 100%. (ii) The reducing agent in the reaction is molecular hydrogen, H_2, which is a low-cost and readily available reagent. (iii) Catalyst loading required in the reaction could be quite low. It can be reduced as low as millionth of the substrate in the best cases. These advantages feature the asymmetric hydrogenation as an economical and environmentally benign synthetic method. Thus, the asymmetric hydrogenation is regarded as indispensable core technology for preparing chiral nonracemic molecules in modern organic synthesis, especially in those of large scale. Transfer hydrogenation with a simple organic compound such as formic acid (formate) and 2-propanol as a hydride source is an alternative reaction. This process does not require pressure of hydrogen, although a dehydrogenated by-product (carbon dioxide, acetone, etc.) is formed. For readers' convenience, frequently used abbreviations of chiral diphosphines and diamines and their structures are shown in Figures 3.1 and 3.2.

Asymmetric Hydrogenation and Transfer Hydrogenation, First Edition.
Edited by Virginie Ratovelomanana-Vidal and Phannarath Phansavath.

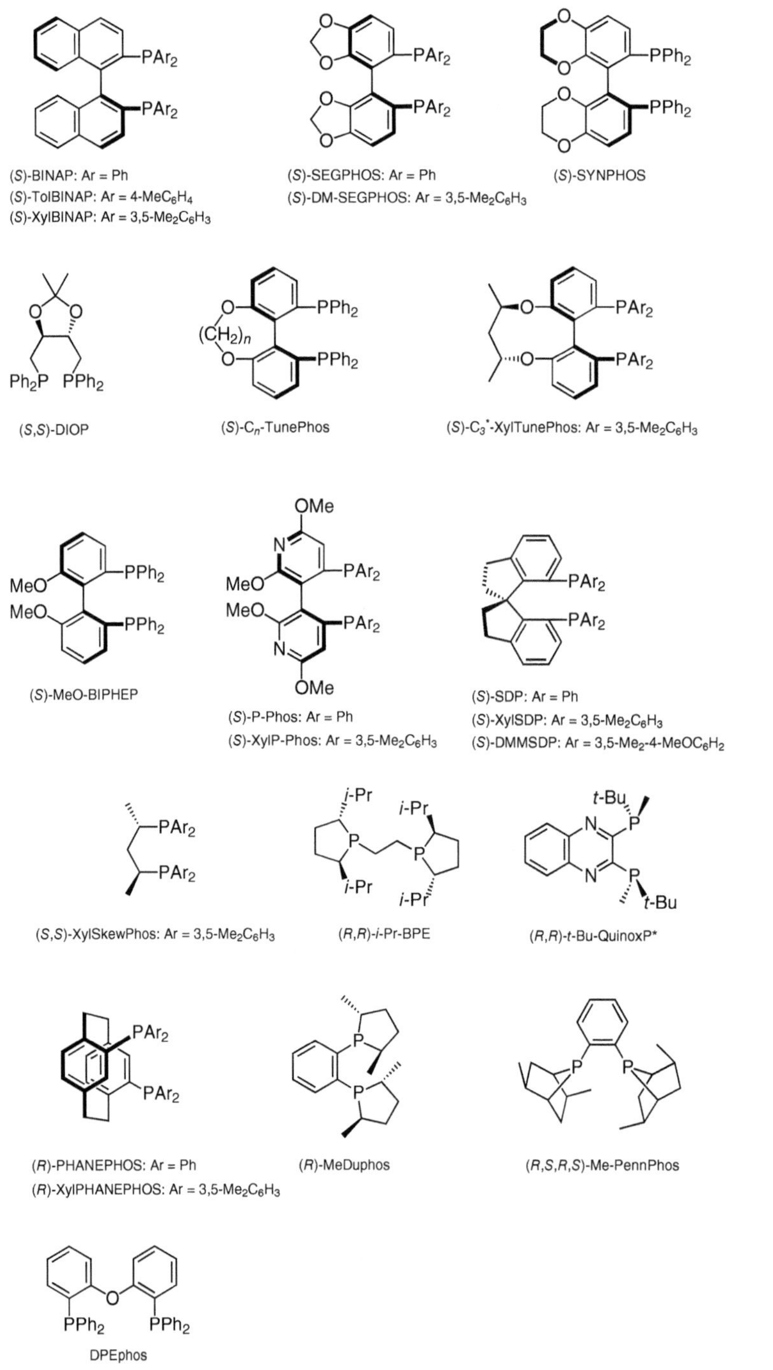

Figure 3.1 Abbreviation of representative phosphines with their structures.

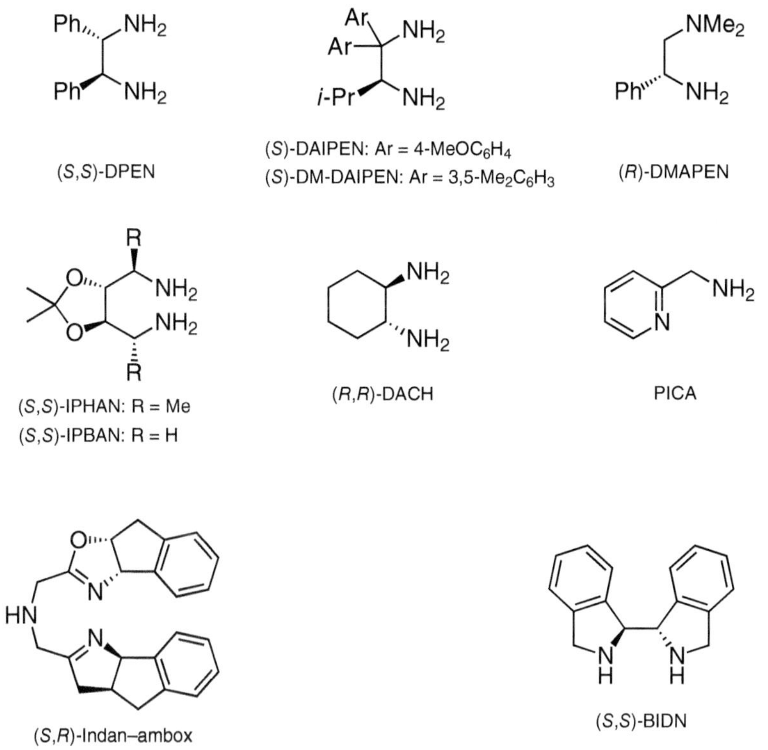

Figure 3.2 Abbreviation of representative amines with their structures.

3.2 Asymmetric (Transfer) Hydrogenation of Alkyl Ketones

Strict discrimination between two alkyl (sp^3-carbon) groups under artificial chiral environment is a longstanding problem to be solved in the area of organic synthesis, because no general chemical principles for this differentiation have been reported. Many studies on asymmetric hydrogenation of simple alkyl ketones are found in the literatures; however, reports of better than 80% ee are quite few, and most of them are hydrogenation of alkyl methyl ketones. For example, hydrogenation of cyclohexyl methyl ketone with the Ru–XylBINAP/DAIPEN system, which gives excellent selectivity in the asymmetric hydrogenation of aromatic ketones (briefly discussed later) [4], afforded the corresponding alcohol in 85% ee, while the reaction of 2-nonanone gave the nearly racemic alcohol (Scheme 3.1) [5].

In an early investigation, several 2-alkanones were hydrogenated over tartaric acid–NaBr-modified Raney nickel in moderate selectivity (63–80% ee) (Scheme 3.2) [6].

Hydrogenations on a polymer-supported metal catalyst were also reported to give moderate selectivity in the case of 2-hexanone; however, the scope of the reactions was not fully investigated (Scheme 3.3) [7].

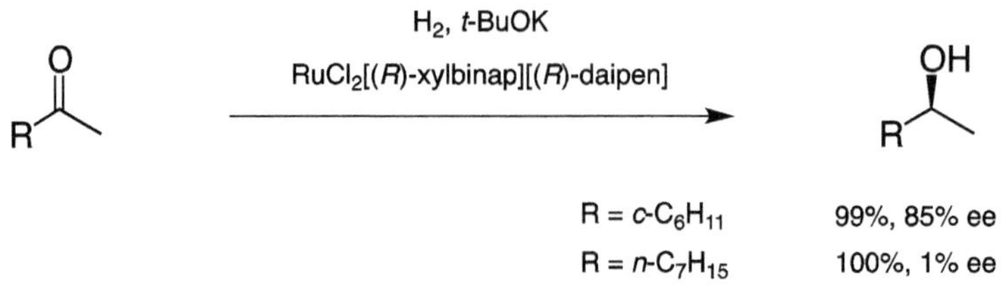

Scheme 3.1 Asymmetric hydrogenation of alkyl methyl ketones. Source: Based on Noyori and Ohkuma [5].

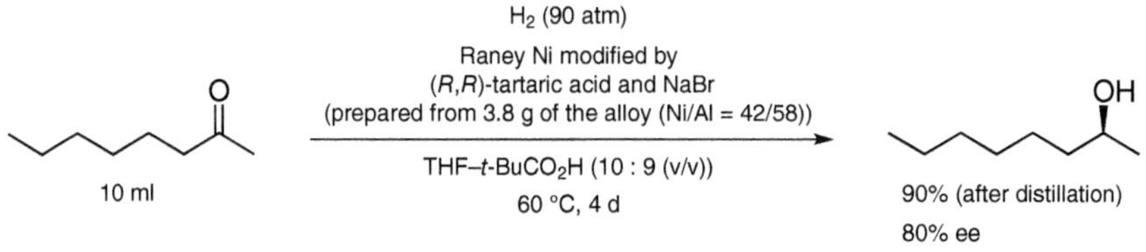

Scheme 3.2 Asymmetric hydrogenation of 2-octanone. Source: Based on Osawa [6].

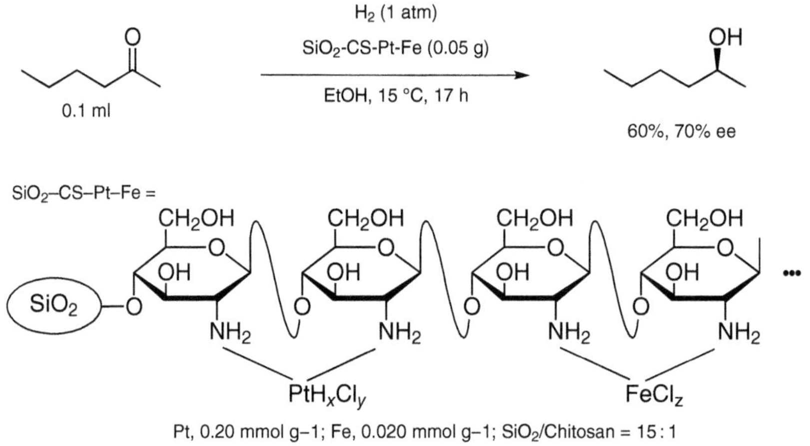

Scheme 3.3 Asymmetric hydrogenation of 2-hexanone with polymer-supported metal catalyst. Source: Modified from Wei et al. [7].

Among the reported methods for asymmetric hydrogenation of dialkyl ketones to date, the reactions with the Rh–PennPhos catalytic system [8] and the Ru–BINAP/picolylamine (PICA) system [9] seemed to have relatively wide substrate scope.

The Rh–PennPhos system showed medium to high enantioselectivity in the hydrogenation of various types of alkyl methyl ketones, though the relatively high catalyst loadings and the long reaction time were required to achieve the maximum conversion (Table 3.1) [8]. Addition of 2,6-lutidine and KBr was crucial to obtain high conversion and stereoselectivity. It was suggested that the substitution of chloride by bromide on the Rh complex changed the electronic nature of the catalyst,

Table 3.1 Asymmetric hydrogenation of alkyl methyl ketones by Rh–PennPhos system.

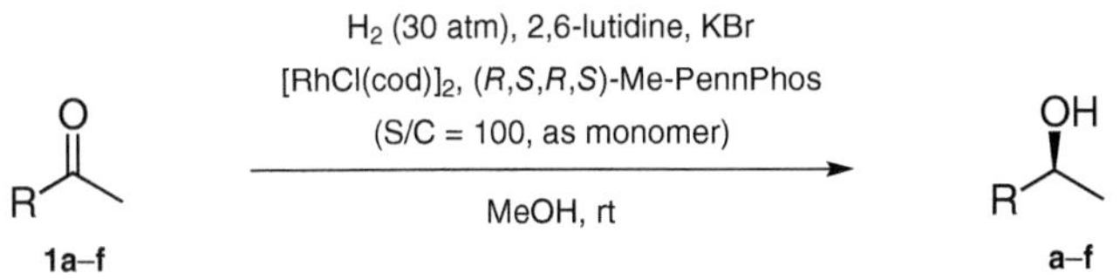

Entry	Substrate	R	Time (h)	Yield (%)	ee (%)
1	**1a**	*t*-Bu	96	51	94
2	**1b**	*c*-C_6H_{11}	106	90	92
3	**1c**	*i*-Pr	94	99	84
4	**1d**	*i*-Bu	75	66	85
5	**1e**	$Ph(CH_2)_2$	56	99	73
6	**1f**	*n*-Bu	48	96	75

Source: Based on Jiang et al. [8].

resulting in the improvement of the catalytic efficiency. The use of noncoordinating and weaker bases such as 2,6-lutidine accelerated the reductive elimination of the intermediate without formation of undesired resting species. High enantioselectivity was obtained for relatively bulky ketones **1a** and **1b** (Entries 1 and 2). The stereoselectivity decreased in the reaction of ketones with smaller alkyl groups. However, hydrogenation of the ketones with unbranched alkyl groups **1e** and **1f** gave the corresponding secondary alcohols **2e** and **2f** in still moderate enantioselectivities of 73% (Entry 5) and 75% (Entry 6).

Chiral RuXY(binap)(1,2-diamine) complexes (X, Y = Cl, Cl, or H, η^1-BH_4) are known to promote rapid enantioselective hydrogenation of various simple nonfunctionalized ketones in 2-propanol with [4, 5, 9] or without [10, 11] a strong base. A proposed catalytic cycle for the hydrogenation of acetophenone with *trans*-RuH(η^1-BH_4)[(*S*)-xylbinap][(*S,S*)-dpen] in 2-propanol based on kinetic and spectroscopic experiments is shown in Scheme 3.4 [10].

The excellent catalytic activity of this catalyst is rationalized by a nonclassical "*metal–ligand cooperative mechanism*" using the NH functionality. The precatalyst **A** is converted to the cationic species **B** with the loss of BH_4^- moiety in an alcoholic solvent. Then, **B** is transformed to the active RuH_2 species **D** through H_2-coordinated cationic intermediate **C**. This process is promoted by a base. Then, the ketone is reduced by **D**, resulting in the alcoholic product and the 16-electron Ru–amide complex **E**. This species is readily protonated in an alcoholic solvent to regenerate the cationic amino complex **B**. Complex **E** partially returns to **D** by reaction with H_2. The ketone is hydrogenated in the outer coordination sphere of **D**, where neither ketone/Ru nor alkoxy/Ru interaction is involved, through the six-membered pericyclic-type transition state (TS) **F**. The reaction with $RuCl_2$(diphosphine)(diamine) (**A′**)/alkaline base system is also suggested to proceed via basically the same catalytic cycle.

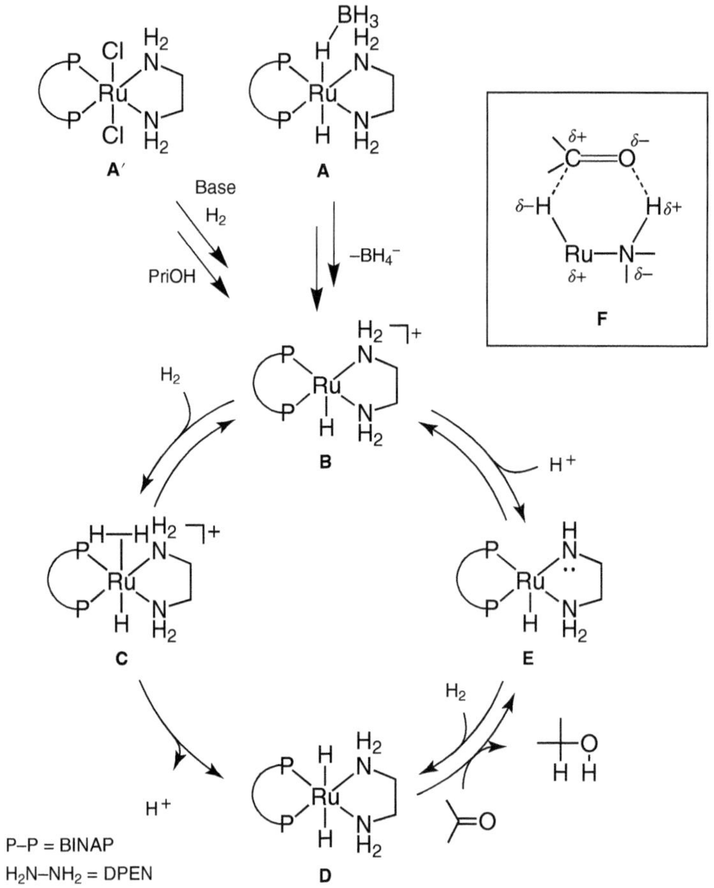

Scheme 3.4 Mechanism of the asymmetric hydrogenation with BINAP/DPEN–Ru(II) catalyst. Source: Based on Sandoval et al. [10].

TS models in hydrogenation of acetophenone with the (*S*)-TolBINAP/(*S,S*)-DPEN–RuH_2 catalyst are depicted in Figure 3.3 [10]. (*S*)-TolBINAP and (*S,S*)-DPEN bind to the Ru center in a same plane constructing the C_2-symmetric RuH_2 complex corresponding to **D** in Scheme 3.4. The axially directed hydrogens (H_{ax}) are more reactive than the equatorial ones, because the $H^{\delta-}$–$Ru^{\delta+}$–$N^{\delta-}$–$H_{ax}{}^{\delta+}$ moiety with a smaller dihedral angle suitably interacts with the $C^{\delta+}{=}O^{\delta-}$ group. Therefore, the hydride on Ru and the amine proton are concomitantly transferred onto the carbonyl carbon and the oxygen, respectively. A ketone approaches the difunctional reaction site in a manner that minimizes steric repulsion and maximizes electronic attractive interaction. The $\mathbf{G}_{Si}$ is a preferable TS to the diastereomeric $\mathbf{G}_{Re}$, because the *Re*-TS suffers serious nonbonded repulsion between the aromatic groups of TolBINAP and the larger substituent of the ketone. The secondary attractive interaction between the NH_{eq} and the phenyl ring could further stabilize the $\mathbf{G}_{Si}$ in the case of aromatic ketones.

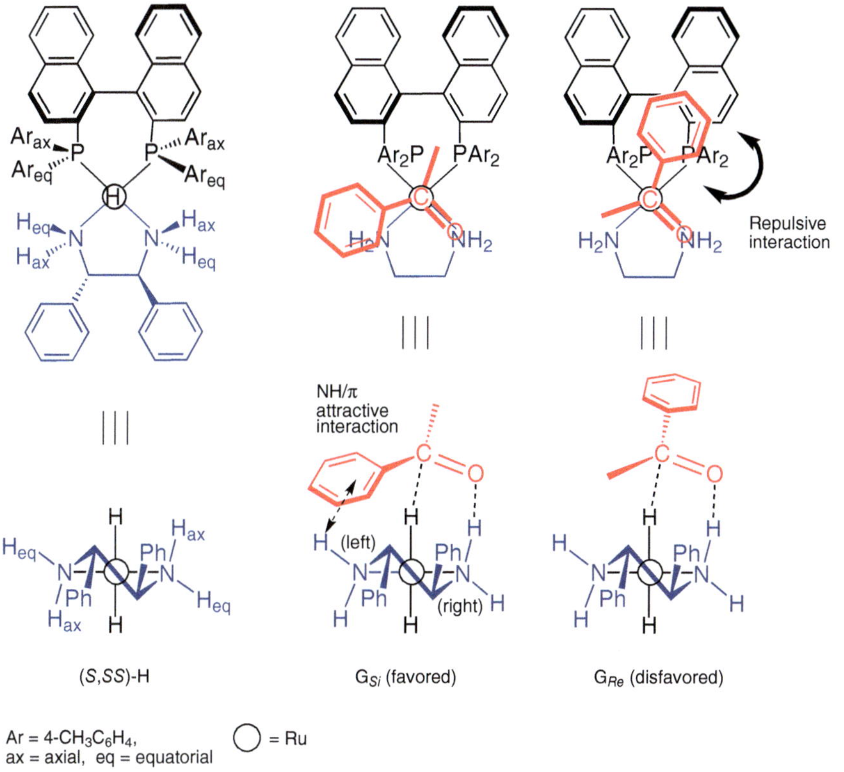

Figure 3.3 Transition state models for hydrogenation of ketone catalyzed by Ru(II)–(*S*)-tolBINAP/(*S*,*S*)-DPEN system. Source: Sandoval et al. [10]. © 2003 American Chemical Society.

In contrast to the case of less-hindered ketones, *tert*-alkyl ketones resisted this hydrogenation obviously by a steric reason. For example, $RuCl_2$[(*S*)-tolbinap][(*S*,*S*)-dpen] with a base hydrogenated *tert*-butyl methyl ketone (**1a**) in 2-propanol at 9 atm H_2 to give (*S*)-**2a** in only 20% yield and 14% ee [12]. This problem was overcome by replacement of the symmetrical 1,2-diamine ligand with an unsymmetrical NH_2/pyridine hybrid ligand, α-PICA. In this reaction, ethanol was a suitable solvent instead of 2-propanol. The new BINAP/PICA-based Ru catalyst system allowed practical asymmetric hydrogenation of a wide range of *tert*-alkyl ketones (Table 3.2) [12]. The reaction was able to be carried out under an atmospheric pressure of hydrogen (Entry 2). When R was an isopropyl group (**3b**, *tert*-alkyl *sec*-alkyl ketone), the hydrogenation hardly proceeded probably due to the steric hindrance (Entry 5).

It was reported that asymmetric hydrogenation of cyclohexyl methyl ketone (**1b**) catalyzed by an Ru(II)–indan–ambox complex gave alcohol (*R*)-**2b** in a highly stereoselective manner (95% ee) (Scheme 3.5) [13]. Unfortunately, this catalytic system afforded unsatisfactory selectivity in the hydrogenation of pinacolone (**1a**) and 3-methyl-2-butanone (**3b**) (42% ee (*R*) and 65% ee (*R*), each).

Table 3.2 Asymmetric hydrogenation of *tert*-alkyl ketones with the Ru–BINAP/PICA system.[a)]

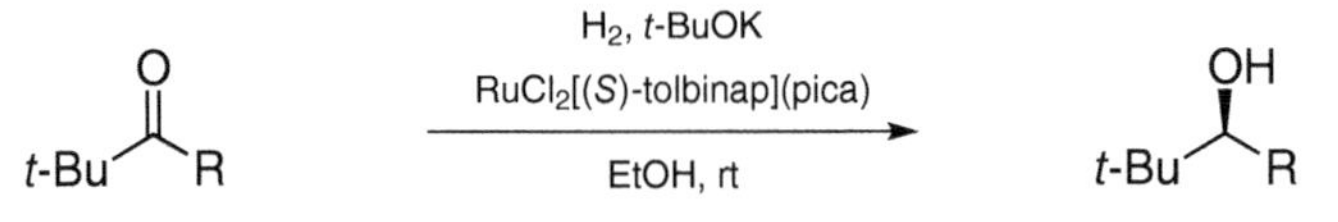

1a, 3a,b → **2a, 4a, b**

Entry	Substrate	R	S/C	H_2 (atm)	Time (h)	Yield (%)	ee (%)
1	**1a**	Me	2000	5	5	100	97
2	**1a**	Me	2000	1	9	100	98
3	**1a**	Me	100 000	20	24	100	88
4	**3a**	n-C_8H_{17}	2300	5	5	100	97
5	**3b**	i-Pr	2020	8	24	<5	nd[b)]

a) For simplicity, the table lists results with the Ru(II)–(*S*)-TolBINAP complex, though some experiments were actually performed with the (*R*)-complex, in which the configuration of the product was also opposite to that shown in the table.
b) Not determined.
Source: Based on Ohkuma et al. [12].

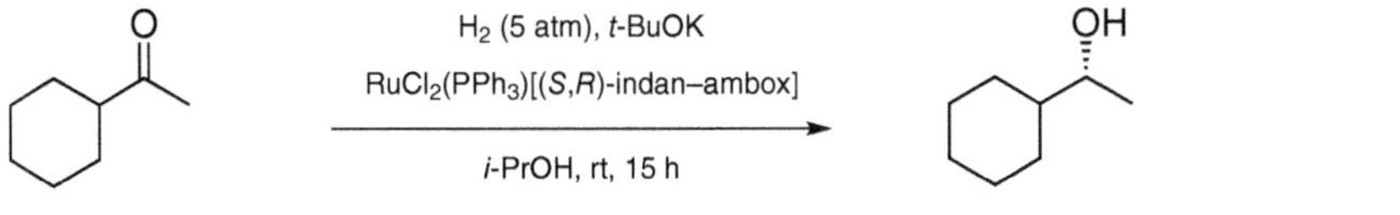

Scheme 3.5 Asymmetric hydrogenation of cyclohexyl methyl ketone. Source: Based on Li et al. [13].

As another recent example, chiral ruthenabicyclic complex **5** was applied to the asymmetric hydrogenation of ketones, which are difficult substrates to reduce with high reactivity and enantioselectivity using original Ru(II)–diphosphine/diamine systems (Scheme 3.6) [14]. It is noteworthy that quite a large turnover number (TON) was achieved in this catalytic system, though the enantioselectivity was moderate.

More recently, nonnoble metal catalysts for the asymmetric hydrogenation of aliphatic ketones with high selectivity have been developed. For example, hydrogenation of **1b** with an Mn complex **6** gave (*R*)-**2b** in 84% ee (Scheme 3.7) [15]. It is considered that the asymmetric induction derives from the approach of the ketone group perpendicular to the N—H and Mn—H groups in the *cis* position with either the *Re* or *Si* enantioface (Figure 3.4). The different steric interaction between the ketone substituents and methyl groups of the chiral phosphorus ligand in *R* and *S* configurations is suggested to be the origin of asymmetric induction.

There are few reports of asymmetric transfer hydrogenation (ATH) that gives high enantioselectivity with respect to dialkyl ketones. Transfer hydrogenation of aliphatic ketones catalyzed by Ru complexes linked to the secondary face of

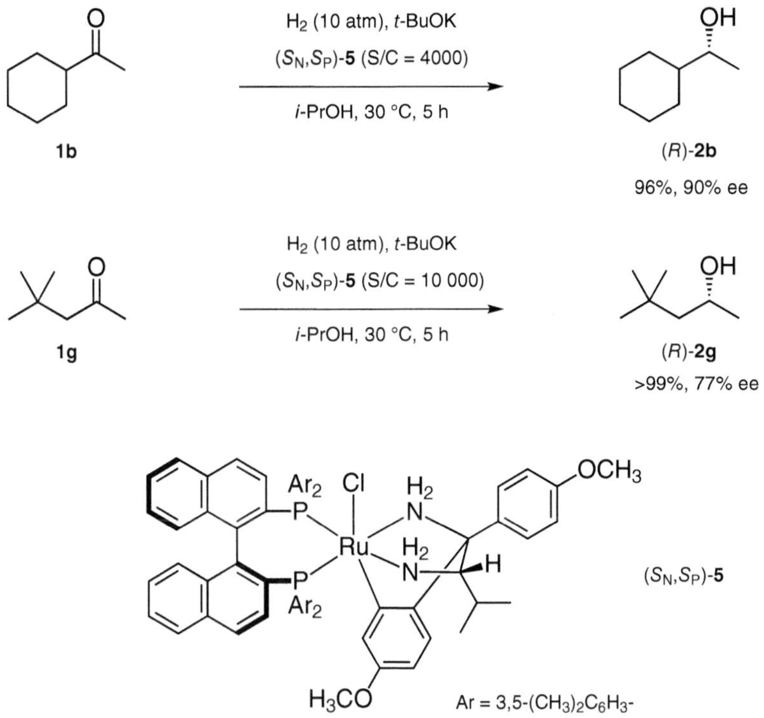

Scheme 3.6 Asymmetric hydrogenation with chiral ruthenabicyclic complex. Source: Based on Matsumura et al. [14].

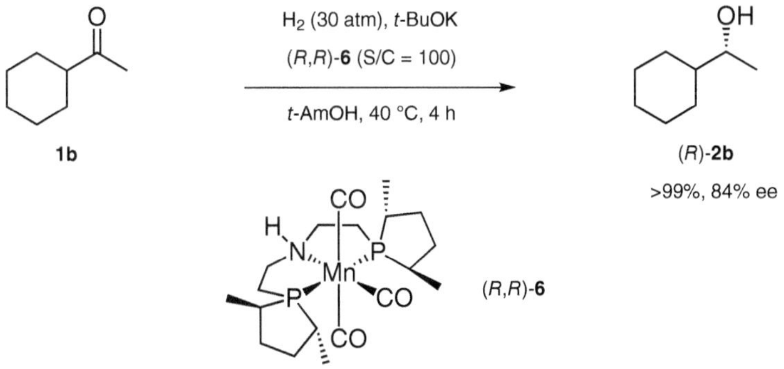

Scheme 3.7 Asymmetric hydrogenation of cyclohexyl methyl ketone with Mn(I) catalyst. Source: Based on Garbe et al. [15].

β-cyclodextrin is an outstanding example [16]. Woggon and Schlatter envisaged that the ruthenium arene complexes Ru-**7** (Figures 3.5 and 3.6) would simultaneously bind lipophilic molecules and participate in orientation and activation of the substrates, being useful for the enantioselective reduction of nonconjugated ketones by ruthenium complex-catalyzed transfer hydrogenations [17, 18]. The catalyst was prepared *in situ* prior to the catalytic reaction by mixing the CD ligand **7**

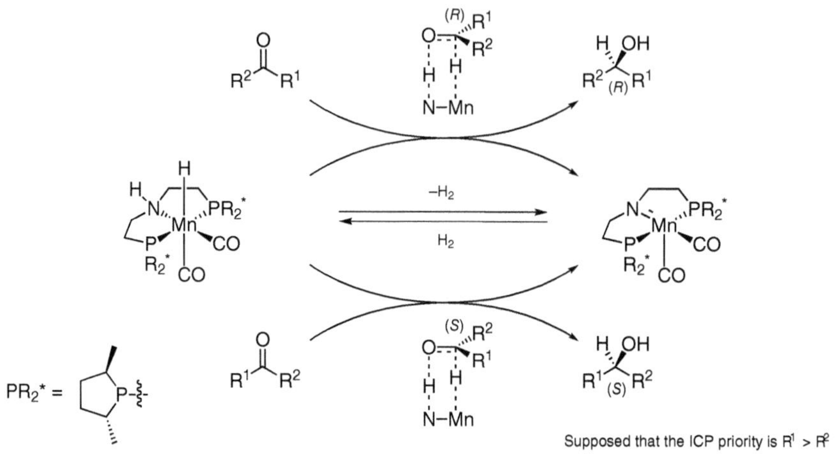

Figure 3.4 Proposed mechanism for hydrogenation of ketones by Mn(I)–chiral PNP pincer ligand system.

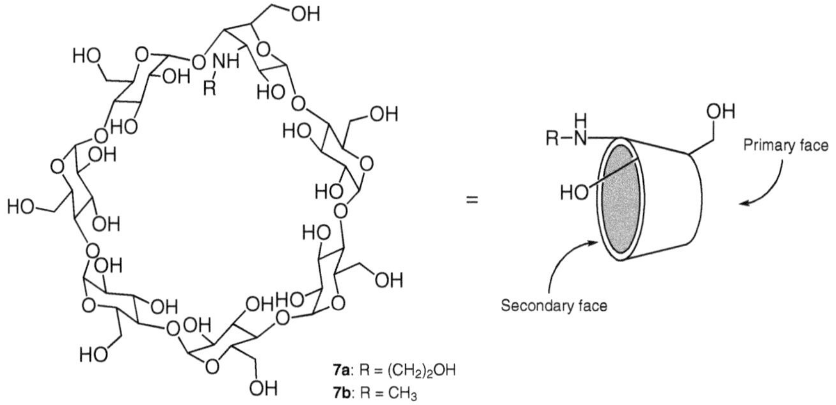

Figure 3.5 Structure of β-cyclodextrin modified by an amino group **7**.

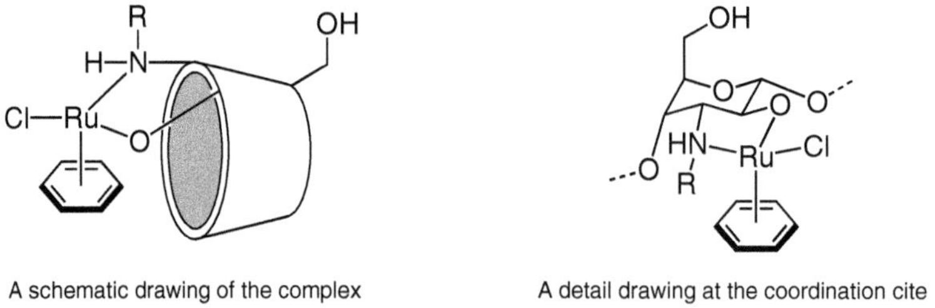

Figure 3.6 Presumed structure of Ru-**7** complex.

Table 3.3 Asymmetric transfer hydrogenation of dialkyl ketones by Ru-**7** catalytic system.

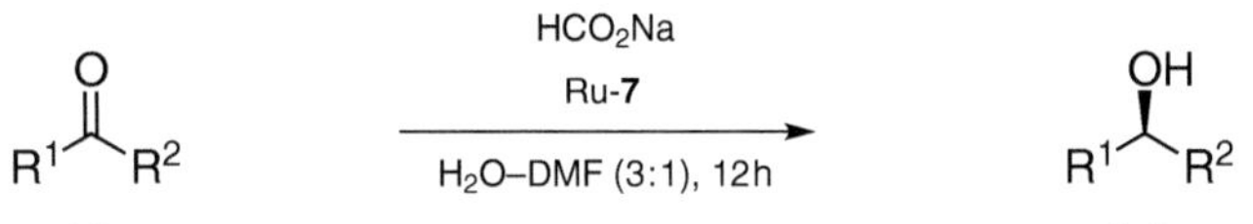

1,8 cat. : subst. : HCO_2Na = 1:10:100 **2,9**

Entry	Substrate	Catalyst	*R*[1]	*R*[2]	Temp. (°C)	Yield (%)	ee (%)
1	**1g**	Ru-**7a**	(*E*)-$(CH_3)_2C{=}CH(CH_2)_2C(CH_3){=}CH(CH_2)_2$	CH_3	50	35	98
2	**1g**	Ru-**7b**	(*E*)-$(CH_3)_2C{=}CH(CH_2)_2C(CH_3){=}CH(CH_2)_2$	CH_3	25	76	93
3	**1h**	Ru-**7a**	$(CH_3)_2C{=}CH\,(CH_2)_2$	CH_3	50	95	90
4		Ru-**7b**			25	99	86
5	**1i**	Ru-**7a**	$(CH_3)_2CH\,(CH_2)_3$	CH_3	50	60	89
6	**1f**	Ru-**7a**	*n*-Bu	CH_3	50	72	70
7		Ru-**7b**			25	98	74
8	**1j**	Ru-**7a**	$CH_3(CH_2)_5$	CH_3	50	80	89
9		Ru-**7b**			25	99	89
10	**1k**	Ru-**7a**	$CH_3(CH_2)_6$	CH_3	50	55	91
11		Ru-**7b**			25	99	93
12	**1l**	Ru-**7a**	$CH_3(CH_2)_7$	CH_3	50	56	91
13		Ru-**7b**			25	80	93
14	**1m**	Ru-**7b**	$CH_3(CH_2)_9$	CH_3	25	37	93
15	**8a**	Ru-**7b**	$CH_3(CH_2)_5$	CH_3CH_2	25	55	91
16	**8b**	Ru-**7b**	$CH_3(CH_2)_4$	CH_3CH_2	25	61	77
17	**1b**	Ru-**7b**	*c*-C_6H_{11}	CH_3	25	99	85
18	**1n**	Ru-**7b**	*c*-$C_6H_{11}CH_2$	CH_3	25	40	82

with $[RuCl_2(\eta^6\text{-}C_6H_6)]_2$ in H_2O–DMF (3 : 1). The representative results are shown in Table 3.3. Although the yields vary to a certain extent, the enantiomeric excesses are quite consistent in the series, around 90% ee in favor of the *S*-configuration. It should be noted that high-level differentiation of two sp^3-carbons with similar steric bulkiness was achieved in several cases (Entries 1, 11, and 15). Unfortunately, X-ray crystallographic analysis of the complexes was not carried out. Binding of a substrate into the cavity of β-cyclodextrin derivatives was supported by NOE experiments.

3.3 Asymmetric Hydrogenation of α,β-Unsaturated Ketones

3.3.1 Alkenyl Alkyl Ketones

The key issue of the enantioselective hydrogenation of α,β-unsaturated ketones (alkenyl alkyl ketones) is differentiation between sp^3- and sp^2-carbon which are located on each side of the carbonyl group. In this point of view, the situation is similar to that of hydrogenation of aromatic ketones (alkyl aryl ketones). However, it is not always easy to reduce the carbonyl group keeping the carbon–carbon double bond unchanged, since the C=C double-bond moieties of the conjugate alkenes (substrates) and allylic alcohols (products) are both readily reduced with conventional hydrogenation catalysts. The asymmetric (transfer) hydrogenation catalyzed by Ru–diphosphine/diamine systems provides the best way to perform this transformation.

Results of the asymmetric hydrogenation of benzylideneacetone (**10a**), which is a typical substrate as α,β-unsaturated ketone, catalyzed by Ru–diphosphine/diamine complexes are shown in Table 3.4 [4, 19–22]. The sense of enantioselectivity was the same as that observed in the reaction of alkyl aryl ketones, and this indicated that the reaction proceeds through the same mechanism as depicted in Scheme 3.4 and Figure 3.3. When the hydrogenation is conducted with $RuCl_2$[(*S*)-xylbinap][(*S*)-daipen] as a precatalyst (see Scheme 3.4), it is recommended to use K_2CO_3 instead of the strong base *t*-BuOK for activation of the precatalyst in order to avoid base-promoted oligomerization and/or decomposition of base-labile substrates (Entry 1). The use of (*S*)-XylP-Phos/(*S*,*S*)-DPEN–$RuCl_2$ complex gave a similar result (Entry 2). The strong base *t*-BuOK was preferably used for achieving higher reaction rate when the hydrogenation was catalyzed by the Ru complex with XylPHANEPHOS, XylSDP, or C_3*-XylTunePhos as a diphosphine ligand (Entries 3–5).

An aliphatic enone, 3-nonen-2-one (**10b**), was prone to polymerize in the presence of alkoxide base and required the use of K_2CO_3 to obtain the allylic alcohol product (*R*)-**11b** in high yield. While the hydrogenation of **10a**, a typical substrate, gave the allylic alcohol (*R*)-**11a** in equally high ee independent of the catalytic systems listed in Table 3.4, **10b** was hydrogenated with higher enantioselectivity catalyzed by the XylBINAP/DAIPEN system than the XylP-Phos/DPEN system (Scheme 3.8) [4, 19].

Hydrogenation of **10b** using the RuH(η^1-BH_4) complex (*S*,*S*)-**12** without an additional base proceeded smoothly to afford the desired product in close to perfect ee (Scheme 3.9) [11].

The XylBINAP/DAIPEN system is also effective in the asymmetric hydrogenation of sterically more-hindered α,β-unsaturated ketones **10c** and **10d** to give the corresponding secondary alcohols in moderate to high ee (Scheme 3.10) [4].

More recently, highly enantioselective hydrogenation of sterically hindered enones catalyzed by Ru complexes with a rigid chiral diamine and an achiral diphosphine has been reported [23]. A TON as high as 50 000 was achieved in the hydrogenation of a typical substrate **10e** (Scheme 3.11). The stairlike rigid

Table 3.4 Asymmetric hydrogenation of benzylideneacetone with Ru–diphosphine/diamine complexes.[a)]

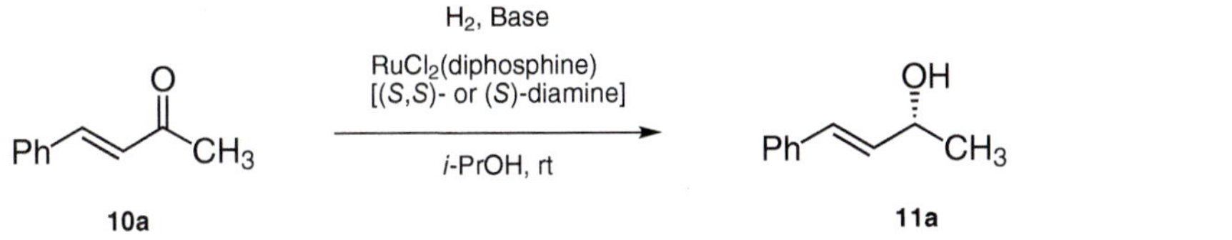

Entry	Diphosphine	Diamine	S/C	Base	H_2 (atm)	Time (h)	Yield (%)	ee (%)	References
1	(*S*)-XylBINAP	(*S*)-DAIPEN	100 000	K_2CO_3	80	43	100	97	[4]
2	(*S*)-XylP-Phos	(*S,S*)-DPEN	10 000	K_2CO_3	34	15	>99.9	97.1	[19]
3	(*R*)-XylPHANPHOS	(*S,S*)-DPEN	3000	*t*-BuOK	5.5	1	>99	97	[20]
4	(*R*)-XylSDP	(*S,S*)-DPEN	5000	*t*-BuOK	50	3	100	96	[21]
5	(*S*)-C_3^*-XylTunePhos	(*S*)-DAIPEN	10 000	*t*-BuOK	10	4	>99.9	97.4	[22]

a) For simplicity, the table lists results with Ru(II)–(*S,S*)- or (*S*)-diamine complexes, though some experiments were actually performed with the corresponding enantiomers, in which the configuration of the product was also opposite to that shown in the table.

Source: Ohkuma et al. [4], Wu and Chan [19], and Li et al. [22].

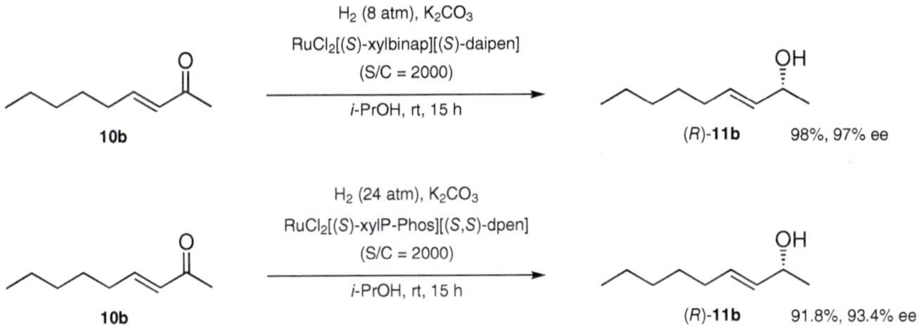

Scheme 3.8 Asymmetric hydrogenation of 3-nonen-2-one. Sources: Ohkuma et al. [4] and Wu and Chan [19].

chiral cyclic BIDN crafts a flat and open chiral environment at the *N,N′*-ligated portion of the corresponding Ru-phosphine/amine complex, whereas the two *trans*-located and free-rotated phenyl groups of the DPEN lead the equatorially oriented *N,N′*-ligated five-membered ring to be more crowded. This orientation is considered to lead to the sterically hindered enones entering the chiral pocket easily (Figure 3.7, see also Figure 3.3).

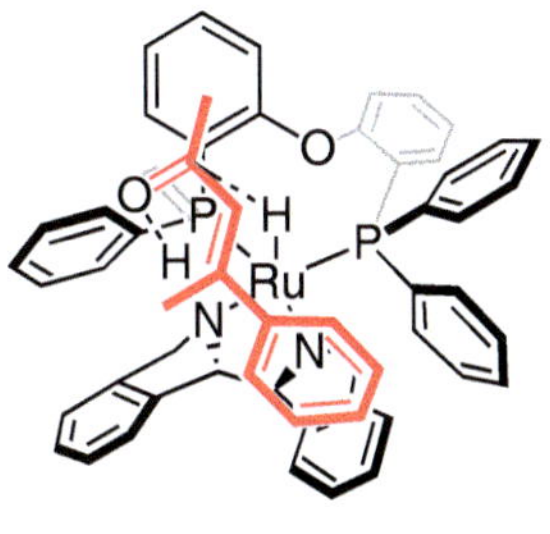

Figure 3.7 Favored transition state model for Ru–DPEphos/(*S,S*)-BIDN complex.

Asymmetric reduction of chalcones, a kind of alkenyl aryl ketones, to the allylic alcohols in high

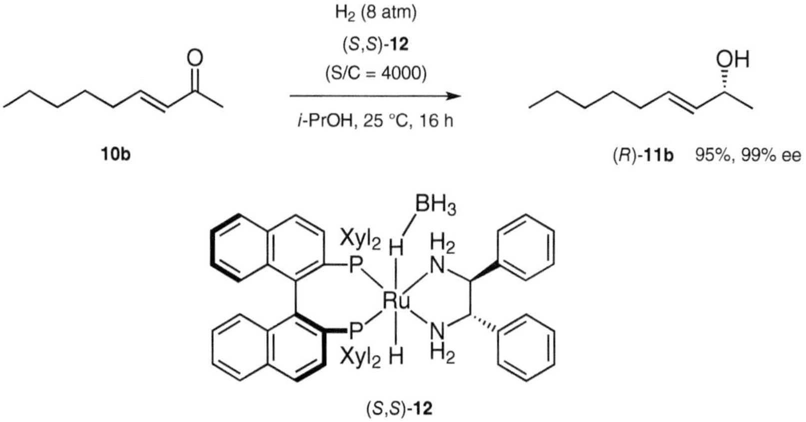

Scheme 3.9 Asymmetric hydrogenation of 3-nonen-2-one using Ru–BH_4 complex without external base. Source: Based on Ohkuma et al. [11].

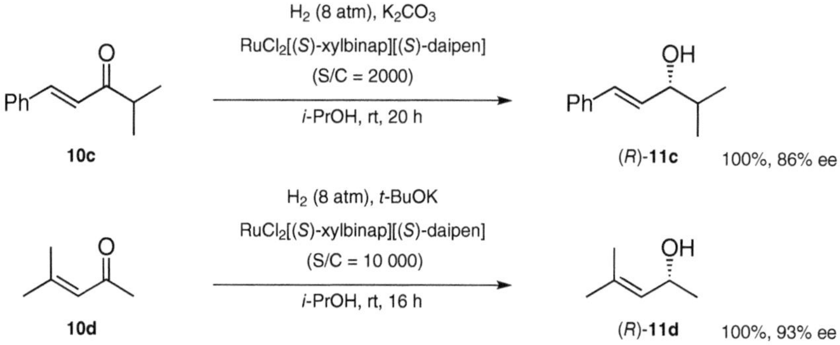

Scheme 3.10 Asymmetric hydrogenation of hindered α,β-unsaturated ketones. Source: Based on Ohkuma et al. [4].

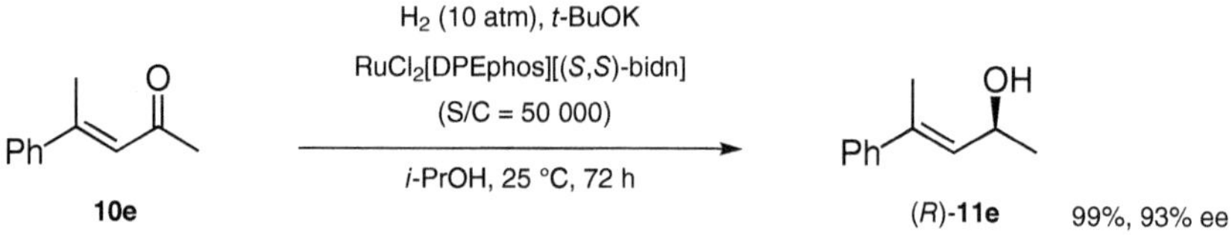

Scheme 3.11 Asymmetric hydrogenation of α,β-unsaturated ketone with high catalyst turnover.

enantioselectivity is a quite difficult subject, because precise discrimination between alkenyl (sp^2-carbon) and aryl (sp^2-carbon) groups is required. Even enzymatic methods have not achieved this asymmetric transformation. For instance, the hydrogenation of chalcone using the Ru–XylBINAP/DPEN catalytic system gave the corresponding alcohol in only 42% ee [5]. This difficult problem was circumvented using the Ru–TolBINAP/DMAPEN catalytic system described in the former section [24].

Table 3.5 Asymmetric hydrogenation of alkenyl aryl ketones with the Ru–TolBINAP/DMAPEN system.

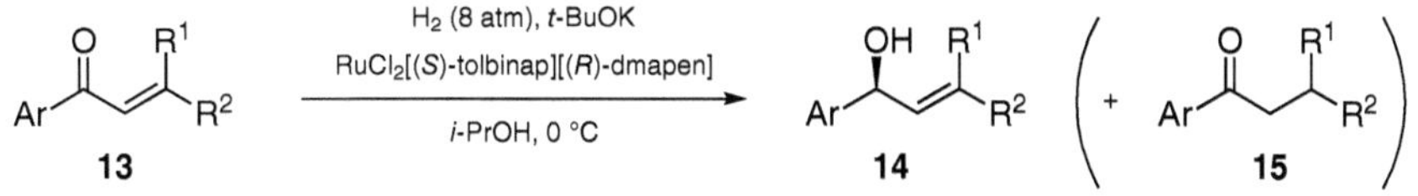

Entry	Substrate	Ar	R^1	R^2	S/C	Time (h)	Yield[a] (%)	ee (%)
1	**13a**	Ph	H	Ph	1000	5	99 (1)	97
2[b]	**13a**	Ph	H	Ph	10 000	3	99 (1)	97
3	**13b**	4-$CH_3C_6H_4$	H	Ph	1000	4	>99 (<1)	98
4	**13c**	4-FC_6H_4	H	Ph	1000	3	96 (1)	96
5[c]	**13d**	2-Naphthyl	H	Ph	1000	5	96 (1)	95
6[c][d]	**13e**	2-Furyl	H	Ph	1000	8	98 (<1)	92
7	**13f**	Ph	H	4-$CH_3C_6H_4$	1000	13	93 (1)	97
8	**13g**	Ph	H	4-ClC_6H_4	1000	5	99 (<1)	97
9[e]	**13h**	Ph	CH_3	Ph	1000	43	97 (<1)	89
10	**13i**	Ph	H	*t*-Bu	1000	8	86 (<1)	97

a) Yield of the saturated ketone **15** is shown in parentheses.
b) Under 40 atm H_2.
c) Reaction in a 2 : 1 mixture of 2-propanol and DMF.
d) 1 equiv of PPh_3 to the Ru complex was added.
e) Under 50 atm H_2.

The representative data are shown in Table 3.5. Hydrogenation of chalcones substituted by electron-donating or -withdrawing groups **13a–c** proceeded very well with the Ru–(*S*)-TolBINAP/(*R*)-DMAPEN catalytic system in 2-propanol (Entries 1–4). Formation of saturated ketones and alcohols was efficiently suppressed when this reaction was conducted at 0 °C. The 2′-naphthyl substrate **13d** was also a good substrate for this reaction (Entry 5). For hydrogenation of 2′-furyl enone **13e**, an addition of triphenylphosphine (1 equiv to Ru) was required to eliminate the undesired saturated by-products (Entry 6). Triphenylphosphine is believed to inhibit the coordination of the allylic alcohols **14** to the ruthenium center responsible for the isomerization to the saturated ketones **15**. Substituents on the β-phenyl group (R^2 in the Scheme) had little effect on the enantioselectivity of the hydrogenation (Entries 1, 7, and 8). The β,β-disubstituted enone, **13h**, was a difficult substrate to hydrogenate because of steric hindrance at the β-position. Complete conversion with a substrate/catalyst molar ratio (S/C) of 1000 is achieved after 43 hours under 50 atm of H_2, affording **14h** in 89% ee (Entry 9). The *tert*-butyl-substituted enone **13i**, though this is not categorized in chalcones, was also hydrogenated with an excellent enantioselectivity, while the reactivity was relatively low (Entry 10).

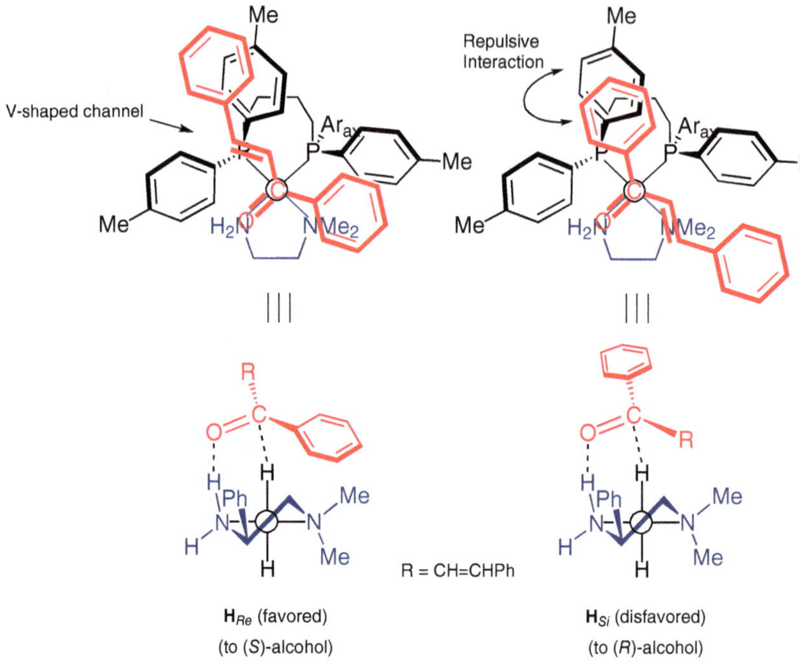

Figure 3.8 Enantioface selection in the asymmetric hydrogenation of **13a** by Ru(II)–(*S*)-TolBINAP/(*R*)-DMAPEN complex.

The stereoselectivity is reasonably explained using TS models depicted in Figure 3.8, which are closely related to those in Figure 3.3. Ketone **13a** approaches the catalyst reaction site with the *Re* face or *Si* face. The *Re*-face-selected TS, $\mathbf{H}_{Re}$, affording (*S*)-**14a** is more favored than the *Si*-face-selected TS, $\mathbf{H}_{Si}$, resulting in (*R*)-**14a**. The "sickle-shaped" vinylic group of **13a** fits well with the "V-shaped channel" of TolBINAP's Ar_{ax}–P–Ar_{eq} (Ar = 4-$CH_3C_6H_4$) structure in the TS $\mathbf{H}_{Re}$, while TS $\mathbf{H}_{Si}$ sustains serious nonbonded repulsion between the widespread phenyl group of **13a** and the narrow TolBINAP's V-shaped channel.

Recently, a related complex, $RuCl_2$[(*S*)-dm-segphos][(*S*)-dmapen], turned out to show excellent catalytic activity to double asymmetric hydrogenation of linear β,β-disubstituted α,β-unsaturated ketones to the γ-substituted secondary alcohols in high diastereo- and enantioselectivities (Scheme 3.12) [25]. A mechanism including cooperative action of two independent catalytic cycles is postulated. One of these cycles contributes to the enantioselective hydrogenation of the enone into the chiral

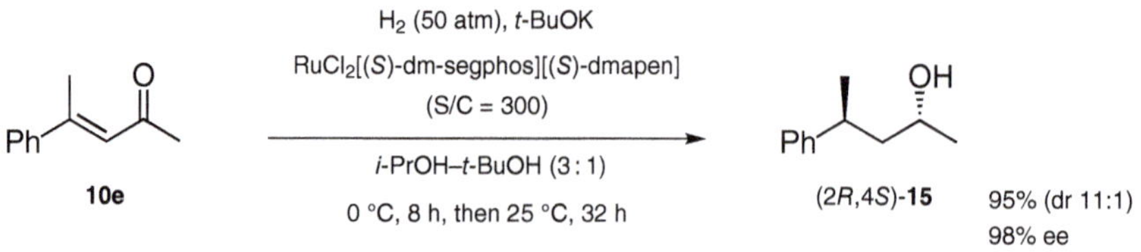

Scheme 3.12 Double asymmetric hydrogenation of α,β-unsaturated ketone. Source: Based on Arai et al. [25].

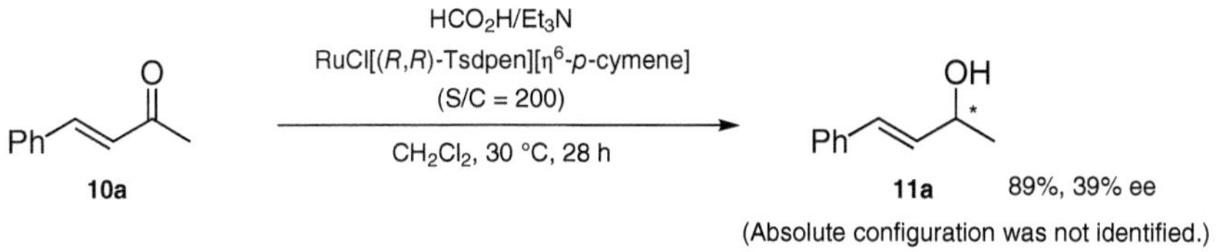

Scheme 3.13 Asymmetric transfer hydrogenation of α,β-unsaturated ketone. Source: Based on Xue et al. [26].

allylic alcohol, and the other takes part in the diastereoselective hydrogenation of the allylic alcohols into the γ-substituted secondary alcohols.

Compared to hydrogenation, there have been much less examples of ATH of alkenyl alkyl ketones that show high enantioselectivity. It was reported that transfer hydrogenation of benzylideneacetone (**10a**) with a typical Ru(II)–*p*-cymene/TsDPEN catalyst gave the allylic alcohol **11a** in unsatisfactory selectivity (Scheme 3.13) [26].

An ether-tethered Ru–arene complex **16**, which was found to be a highly active catalyst for ATH, was applied to the reduction of 1-Cyclohexenyl methyl ketone (**17**); however, there remained plenty room for improvement in the enantioselectivity (Scheme 3.14) [27].

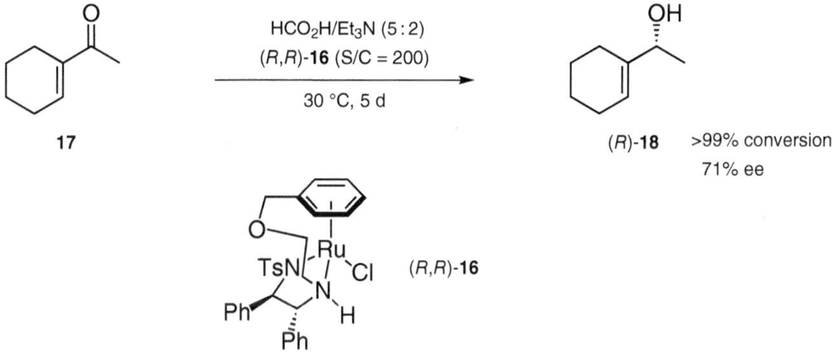

Scheme 3.14 Asymmetric transfer hydrogenation of 1-cyclohexenyl methyl ketone with ether-tethered Ru–arene complex. Source: Based on Parekh et al. [27].

3.3.2 Alkynyl Alkyl Ketones

Catalytic asymmetric hydrogenation of alkynyl ketones, which requires differentiation between sp^3- and sp-carbon groups, is difficult due to producing the side products mainly caused by undesirable 1,4-addition reactions because of intrinsic serious instability of alkynyl ketones under basic conditions. At the moment, few practical methods are reported on the enantioselective preparation of propargylic alcohols through the hydrogenation of alkynyl ketones. Hydrogenation of alkynyl ketones with Ru(OTf)[(*S*,*S*)-Tsdpen](η^6-*p*-cymeme) ((*S*,*S*)-**19**) or Cp*Ir(OTf)[(*S*,*S*)-Msdpen] ((*S*,*S*)-**20**) without base is likely the most suitable method to obtain chiral propargylic alcohols in high enantiomeric purity [28]. The representative results are summarized in Table 3.6. Asymmetric hydrogenation

Table 3.6 Asymmetric hydrogenation of alkynyl ketones with Ru(OTf)[(*S*,*S*)-Tsdpen](η⁶-*p*-cymeme) or Cp*Ir(OTf)[(*S*,*S*)-Msdpen] precatalyst.

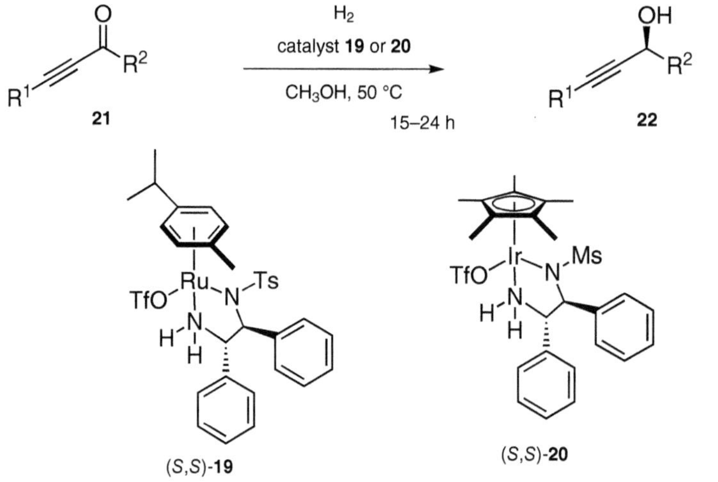

Entry	Substrate	Precatalyst	R^1	R^2	S/C	H_2 (atm)	Yield (%)	ee (%)
1	**21a**	**19**	TBS	CH_3	1000	10	94	97
2[a)]	**21a**	**19**	TBS	CH_3	5000	50	96	95
3	**21a**	**20**	TBS	CH_3	1000	10	73	93
4	**21b**	**19**	TBS	C_2H_5	1000	20	91	95
5	**21c**	**19**	TBS	n-C_9H_{19}	1000	20	94	96
6	**21d**	**19**	TBS	c-C_6H_{11}	1000	10	94	96
7	**21e**	**19**	TBS	Ph	100	10	16	9
8[b)]	**21f**	**19**	TBS	CH_2Cl	200	10	94	96
9	**21g**	**19**	TBDPS	CH_3	1000	10	98	95
10	**21h**	**19**	TIPS	CH_3	1000	10	98	96
11	**21i**	**19**	TMS	n-C_9H_{19}	200	10	63	93
12	**21j**	**19**	Ph	CH_3	200	10	78	92
13	**21k**	**19**	4-ClC_6H_4	CH_3	200	10	92	88
14	**21l**	**19**	4-$CH_3C_6H_4$	CH_3	200	10	88	94
15	**21m**	**19**	t-Bu	CH_3	200	20	61[c)]	89
16	**21n**	**19**	1-Cyclohexenyl	CH_3	200	10	78	95

a) This reaction using 8.46 mmol of **21a** was conducted for 40 hours.
b) Reaction at 30 °C.
c) A small amount (c. 4–5%) of impurity was included.

of **21a** with ((*S*,*S*)-**19**) in methanol under 10 atm of H_2 at 50 °C was completed in 15 hours to afford the *S* propargylic alcohol **22a** in 97% ee (Entry 1). The reaction with an S/C of 5000 under 50 atm of H_2 for 40 hours gave **22a** in 95% ee almost quantitatively (Entry 2). The isoelectronic complex ((*S*,*S*)-**20**) exhibited relatively lower reactivity and enantioselectivity for this reaction (Entry 3). In the hydrogenation of the ethyl and *n*-nonyl ketones (**21b** and **21c**), a condition of 20 atm of H_2 was required to achieve a high yield of >90% in 15 hours (Entries 4 and 5). The reaction rate and the enantioselectivity were significantly decreased in the hydrogenation of the phenyl ketone **21e** (Entry 7). The hydrogenation of other alkynyl ketones gave similar results under the regular conditions or with slightly increased catalyst loadings (Entries 6, 8–16). The sense of enantioselectivity was consistent throughout these substrates.

Catalytic ATH of alkynyl ketones is also a good method to obtain chiral propargylic alcohols with high enantiomeric excess. It was reported that the chiral 16-electron Ru complexes **23** are excellent catalysts for the carbonyl-selective asymmetric reduction with 2-propanol [29]. The asymmetric reduction, normally with a 0.1–1 M 2-propanol solution of a ketone, proceeds under neutral conditions at room temperature with an S/C of 100–200, giving the corresponding propargylic alcohols in up to 98% ee and in >99% yield. The Ru catalysts **23** can be conveniently generated *in situ* by mixing the 18-electron precursors **24** and KOH in a 1 : 1.2 molar ratio. The selected data are listed in Table 3.7. Various acetylenic ketones **21** can be reduced to chiral propargylic alcohols **22** in good yields. The ee values are consistently high regardless of the bulkiness of R^2 substituents from methyl to *tert*-butyl. 2-Propanol is the best hydrogen donor. Attempted reduction of **21a** using a 1 : 1 formic acid/triethylamine mixture in THF containing (*S*,*S*)-**23b** (S/C = 200, THF, 28 °C, 20 hours) gave (*S*)-**22a** in 91% ee but only 55% yield.

Recently, a chiral iridium complex was found to show excellent catalytic activity for the ATH of alkynyl ketones in the presence of both sodium formate and ethanol as hydrogen sources without the addition of any base (Scheme 3.15) [30].

The ATH of **21j** was performed under the conditions, (*S*)-**25**, HCO_2Na, in EtOH at 60 °C, to give (*R*)-**22j** of 96% ee in almost quantitative yield. Considering that ethanol is a renewable resource and a feedstock for the chemical industry, as well as an environmental- and human-friendly solvent, this protocol provides a practical and sustainable method for the preparation of optically active propargylic alcohols, although relatively large amount of the catalyst (S/C = 100) is required.

3.4 Asymmetric Hydrogenation of α-Aminoketones

Asymmetric hydrogenation of α-aminoketones is an important transformation to produce synthetically useful vicinal aminoalcohols in optically active form. As an early example, dimethylaminoacetone (**26a**) was successfully hydrogenated with axially chiral diphosphine–Ru complexes to afford the aminoalcohol **27a** in high ee (Scheme 3.16) [31]. An aromatic substrate **26b** was reduced as well as the aliphatic substrate.

Table 3.7 Asymmetric transfer hydrogenation of alkynyl ketones with Ru arene complexes.

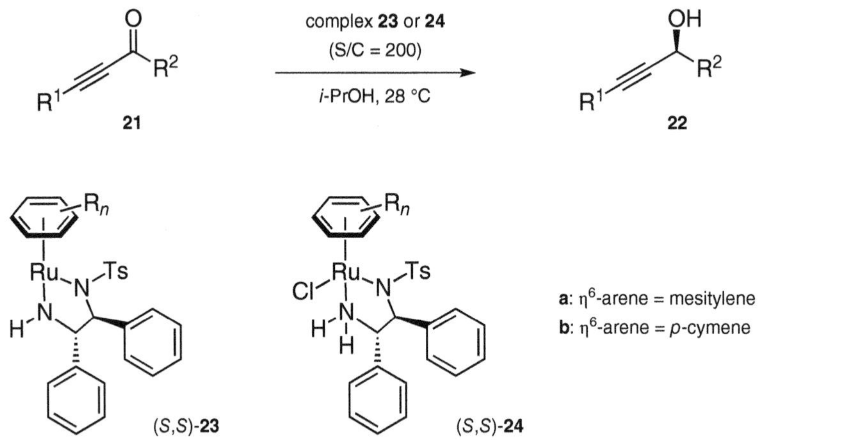

Entry	Substrate	Precatalyst	R^1	R^2	Method[a)]	Time (h)	Yield (%)	ee (%)
1	**21j**	(*S*,*S*)-**23a**	Ph	CH_3	A	20	87	98
2	**21j**	(*S*,*S*)-**23b**	Ph	CH_3	A	20	>99	97
3	**21j**	(*S*,*S*)-**24a**	Ph	CH_3	B	4	>99	97
4	**21j**	(*S*,*S*)-**24b**	Ph	CH_3	B	18	94	96
5	**21o**	(*S*,*S*)-**24a**	Ph	C_2H_5	B	12	97	97
6	**21p**	(*S*,*S*)-**24a**	Ph	*i*-Pr	B	5	98	99
7	**21q**	(*S*,*S*)-**24a**	Ph	*c*-C_6H_{11}	B	13	>99	98
8[b)]	**21r**	(*S*,*S*)-**24a**	Ph	*t*-Bu	B	13	84	98
9	**21s**	(*S*,*S*)-**24a**	*n*-Bu	CH_3	B	6	70	98
10	**21t**	(*S*,*S*)-**24a**	*n*-Bu	*i*-Pr	B	6	90	>99
11	**21u**	(*S*,*S*)-**23b**	TMS	CH_3	A	12	>99	98
12	**21v**	(*S*,*S*)-**23b**	TMS	*n*-C_4H_9	A	12	86	98
13[c)]	**21w**	(*S*,*S*)-**23b**	TMS	*n*-C_5H_{11}	A	15	98	99
14	**21x**	(*S*,*S*)-**23b**	TMS	*i*-Pr	A	12	>99	99

a) Method A, **21** : **23** = 200 : 1; method B, **21** : **24** : KOH = 200 : 1 : 1.2.
b) S/C = 100.
c) (*R*,*R*)-**23b** was used, giving (*R*)-**22x**.

Use of cationic iodide complex **28** instead of the neutral $Ru(OAc)_2$(binap) complex improved the ee up to 99.4%, though the reason was not clarified (Scheme 3.17) [32, 33].

Higher catalyst TON and lower H_2 pressure were achieved using $RuCl_2$ (diphosphine)(1,2-diamine) complexes with a slight sacrifice of the enantiofacial selectivity (Scheme 3.18) [34]. When 1.0 M solution of α-aminoketone **26a** in 2-propanol containing *trans*-$RuCl_2$[(*R*)-xylbinap][(*R*)-daipen] and *t*-C_4H_9OK

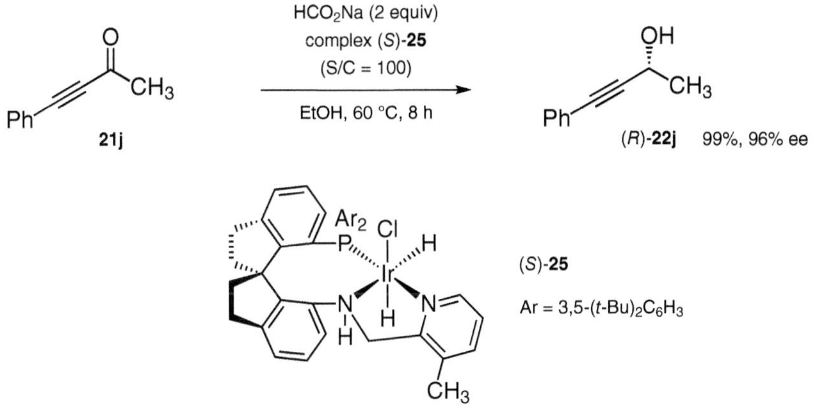

Scheme 3.15 Asymmetric transfer hydrogenation of alkynyl methyl ketone. Source: Based on Zhang et al. [30].

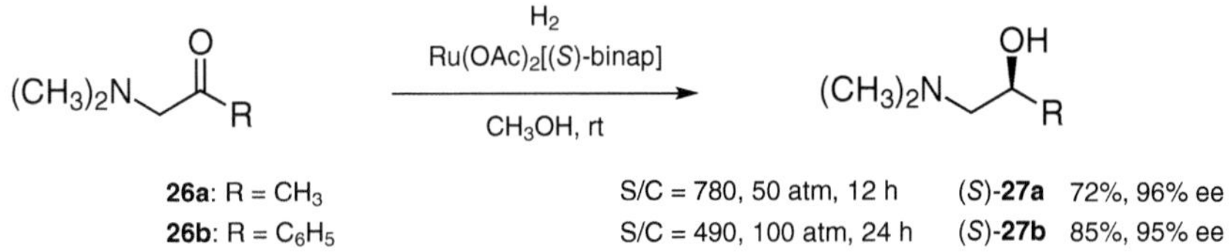

Scheme 3.16 Asymmetric hydrogenation of α-amino ketones: early examples. Source: Based on Kitamura et al. [31].

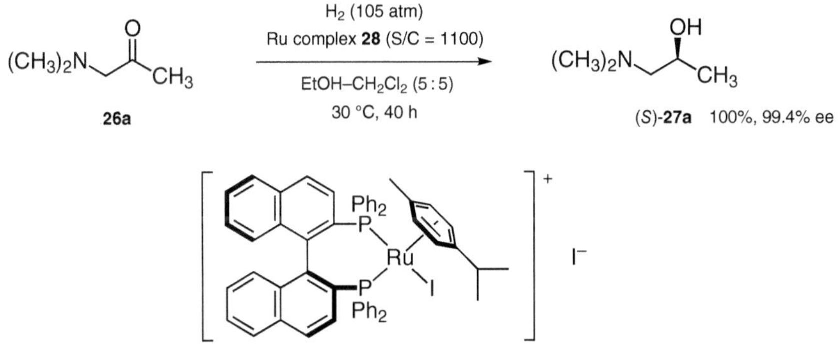

(S)-**28**

Scheme 3.17 Asymmetric hydrogenation of α-amino ketone with cationic ruthenium iodide complex. Source: Mashima et al. [32] and Mashima et al. [33].

(ketone:Ru:base molar ratio = 2000 : 1 : 16) was stirred under 8 atm of H_2 at 25 °C for four hours, the amino alcohol (*S*)-**27a** was produced in a 92% ee and 99% yield. Pure *S* amino alcohol was obtained via its crystalline hydrochloride. The Ru–diphosphine/diamine catalyst is much more reactive than the earlier devised diamine-free Ru–BINAP catalysts described above, which requires the assistance of heteroatom/Ru interaction and shows an opposite sense of asymmetric induction.

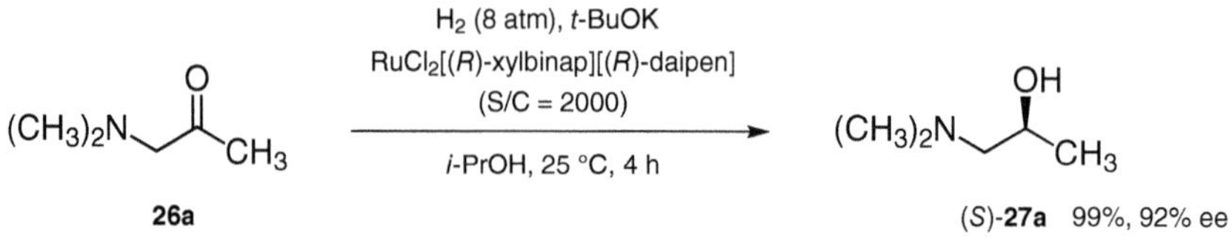

Scheme 3.18 Asymmetric hydrogenation of α-amino ketone with $RuCl_2$(diphosphine)(1,2-diamine) complex. Source: Based on Ohkuma et al. [34].

Compared to hydrogenation with gaseous hydrogen, transfer hydrogenation of α-aminoketones is much less explored. It is reported that transfer hydrogenation of a range of α-aminoketones with formic acid–triethylamine, catalyzed by RuCl[(*R*,*R*)-TsDPEN](η^6-*p*-cymene), produces the corresponding β-amino alcohols. While the excellent enantioselectivity was obtained in the reaction with aromatic amino ketones, the selectivity was unsatisfactory in the case of an aliphatic amino ketone (Scheme 3.19) [35].

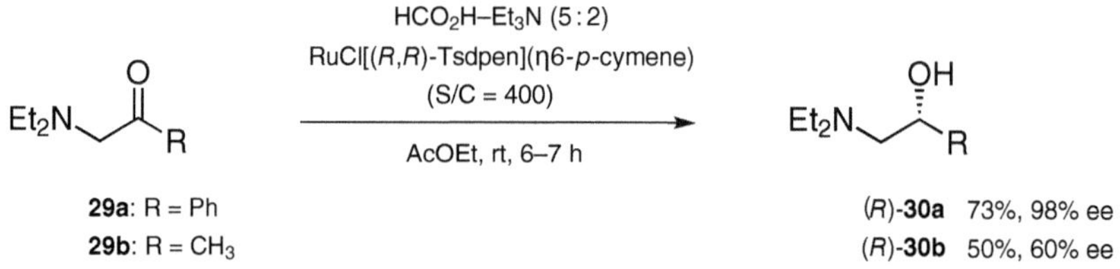

Scheme 3.19 Asymmetric transfer hydrogenation of α-amino ketones. Source: Based on Kosmalski et al. [35].

(*R*)-3-Quinuclidinol ((*R*)-**32**) is an important building block for the syntheses of muscarinic receptor ligands such as solifenacin (M_3 receptor antagonist). Asymmetric hydrogenation of **31** to (*R*)-**32** catalyzed by chiral metal complexes is a straightforward chemical procedure. Several practical methods have been developed to achieve this important transformation (Table 3.8) [14, 36–38]. The reaction with a Ru(II)–(*R*)-DM-SEGPHOS/(*S*)-DM-DAIPEN catalyst afforded (*R*)-**32** in 97% ee (S/C = 1000), but the catalytic activity was not sufficient for practical large-scale synthesis (Entry 1). A practical procedure for the synthesis of (*R*)-**32** at a 4.3-kg scale was achieved using Ru(II)–XylSkewPhos/PICA complex, though recrystallization was required to obtain enantiomerically pure **32** (Entry 2). Both high catalyst turnover and high ee were accomplished by a combination of BINAP and IPHANE, which was readily derived from a tartrate (Entry 3). Previously mentioned Ru complex **5** also gave a satisfactory result (Entry 4).

Recently, an Mn(I) complex bearing a lutidine-based chiral PNN ligand has been found to show high activities and good to excellent enantioselectivities in the asymmetric hydrogenation of α-aminoketones (Scheme 3.20) [39]. Hydrogenation of ketone **33** with Mn complex **35** gave amino alcohol **34**, which is the key intermediate for the synthesis of phenylephrine, in 95% ee. It should be noted that high catalytic performance was achieved using a chiral catalyst made of an earth-abundant metal.

Table 3.8 Asymmetric hydrogenation of 3-quinuclidinone with the Ru–diphosphine/diamine systems.[a)]

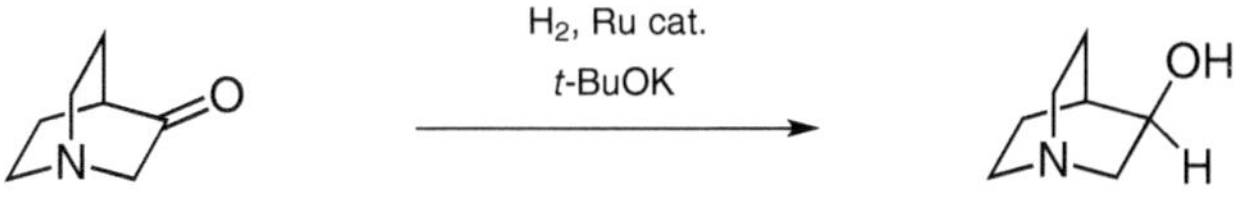

31 (R)-32

Entry	Precatalyst	S/C	H_2 (atm)	Conditions	Yield (%)	ee (%)	References
1	$RuCl_2$[(*R*)-dm-segphos][(*S*)-dm-daipen]	1000	30	*i*-PrOH, 30 °C, 16 h	97.5	97	[36]
2	$RuBr_2$[(*S*,*S*)-xylskew-phos](pica)	100 000	15	EtOH, 30–45 °C, 4 h	>99 (82)[b)]	88 (>99)[b)]	[37]
3	$RuCl_2$[(*S*)-binap][(*R*)-iphane]	50 000	50	*i*-PrOH, 25 °C, 24 h	99	97	[38]
4	RuCl[(*R*)-daipena][(*R*)-dm-segphos] ((R_N,R_P)-**5**)	100 000	10	*i*-PrOH, 5–30 °C, 6 h	92	95	[14]

a) For simplicity, the table lists results with the Ru(II) complex that gives (*R*)-**32**, though some experiments were actually performed with the enantiomeric complex, in which the configuration of the product was also opposite to that shown in the table.
b) The numerical value after recrystallization is shown in parentheses.
Source: Matsumura et al. [14], Takenaka [36], and Arai et al. [38].

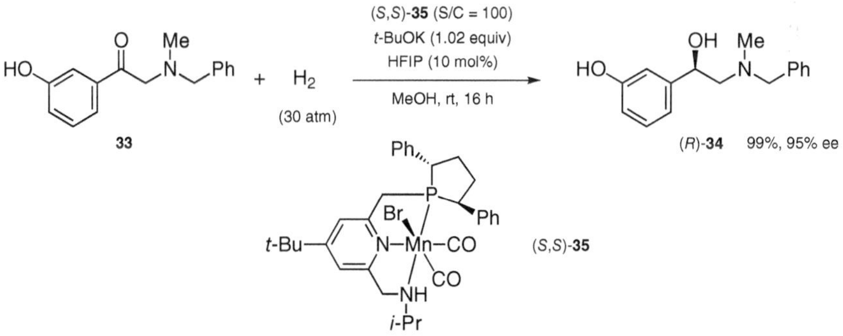

Scheme 3.20 Asymmetric hydrogenation of α-aminoketone with Mn(I) complex bearing a lutidine-based chiral PNN ligand. Source: Based on Zhang et al. [39].

3.5 Asymmetric Hydrogenation of α-hydroxyketones

Asymmetric hydrogenation of α-hydroxyketones is one of the reliable procedures to obtain optically active 1,2-diols that are useful chiral building blocks. However,

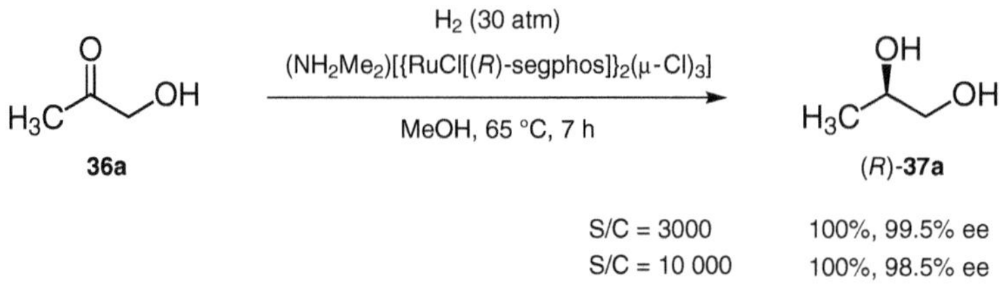

Scheme 3.21 Asymmetric hydrogenation of acetol with Ru(II)–SEGPHOS catalyst. Source: Based on Saito et al. [40].

this asymmetric transformation is still difficult even with modern, well-defined molecular catalysts, and not so many examples are found in the literature. As an early successful example, hydrogenation of hydroxyacetone (acetol, **36a**) with a Ru(II)–SEGPHOS catalyst afforded (*R*)-1,2-propanediol (**37a**) in excellent enantioselectivity (Scheme 3.21) [40].

A broad range of α-hydroxyketones were hydrogenated in highly enantioselective manner using Ru(II) complex in combination with thiophene-based modular phosphine ligands (Table 3.9) [41]. A series of sterically less-demanding 1,2-diols could be obtained in good yields and with excellent enantioselectivities. Not only linear substrates (**36a–d**, Entries 1–6) but also branched substrates (**36e**, **f**, Entries 7 and 8) were reduced to the corresponding diols with high optical purity. This method could be applicable to sterically more demanding *tert*-butyl ketone **36g** (Entry 9), cycloalkyl ketone **36h** (Entry 10), and α-hydroxyacetophenone (**36i**) (Entry 11).

In addition to the abovementioned Ru(II)–diphosphine catalytic system, Ru(II)–arene/TsDPEN or isoelectronic Cp*Ir/MsDPEN complexes were found to show excellent catalytic activity. When hydrogenation of α-hydroxyacetophenone (**36i**) using Cp*Ir(OTf)[(*S,S*)-Msdpen] as a precatalyst in methanol under 10 atm of H_2 at 60 °C was conducted for 15 hours, (*R*)-1-phenyl-1,2-ethanediol [(*R*)-**37i**] was produced in 96% ee and 97% yield (Scheme 3.22) [42]. It is worth noting that Ru(OTf)[(*S,S*)-TsDPEN](η^6-*p*-cymene), which is an efficient catalyst for asymmetric hydrogenation of α-chloro aromatic ketones [43], was virtually inert for the reaction of the electronically related α-hydroxyketones.

The catalyst system was applied to the asymmetric hydrogenation of a series of α-hydroxyketones (Table 3.10). The reactivity was markedly influenced by the electronic properties of the substrates. Thus, a hydroxyketone **36k** bearing an electron-donating MeO group at the 4′ position was not converted at all under the standard reaction conditions, while the hydrogenation of **36j** substituted by an electron-attracting CN group at the 4′ position proceeded smoothly (Entries 2 and 3). Complete conversion was achieved with a high level of enantioselectivity in the hydrogenation of 4′-F- and 4′-Cl-substituted ketones, **36l** and **36m**, at an S/C of 3000 under otherwise identical conditions. 2-Thienyl ketone **36n** was converted to the chiral alcohol **37n** with nearly perfect enantioselectivity, leaving the heteroaromatic ring intact. This catalytic system was also effective in hydrogenation of hydroxyacetone (**36a**), the simplest aliphatic α-hydroxy ketone, producing (*R*)-1,2-propanediol [(*R*)-**37a**] in 80% ee.

Table 3.9 Asymmetric hydrogenation of α-hydroxyketones with the Ru–thiophene-based diphosphine system.

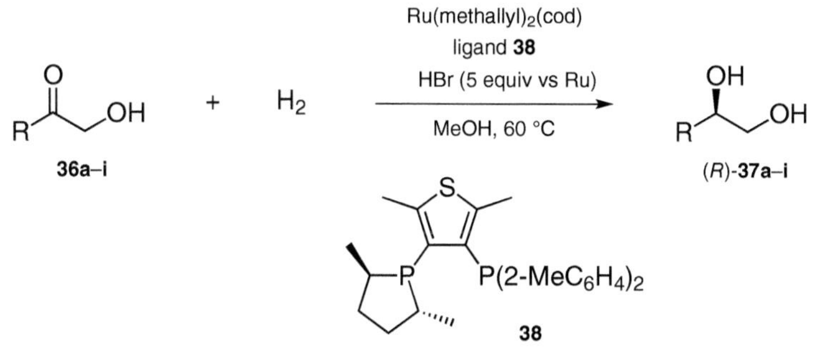

Entry	R	S/C	H_2 (atm)	Time (h)	Yield (%)	ee (%)
1	CH_3 (**36a**)	100	80	12	78[a]	96
2	CH_3 (**36a**)	1000	80	17	80	95
3	CH_3 (**36a**)	10 000	80	48	85	94
4	$CH_3(CH_2)_2$ (**36b**)	100	80	12	74[a]	94
5	$CH_3(CH_2)_4$ (**36c**)	100	50	12	82	97
6	$Ph(CH_2)_3$ (**36d**)	100	50	12	81	96
7	$(CH_3)_2CH$ (**36e**)	100	50	12	77[a]	96
8	$(CH_3)_2CHCH_2$ (**36f**)	100	50	12	81	91
9	*t*-Bu (**36g**)	100	50	36	93	93
10	*cyclo*-C_5H_9 (**36h**)	100	50	12	84	95
11	Ph (**36i**)	100	50	12	90	98

a) Yield of doubly acylated product.
Source: Based on Kadyrov et al. [41].

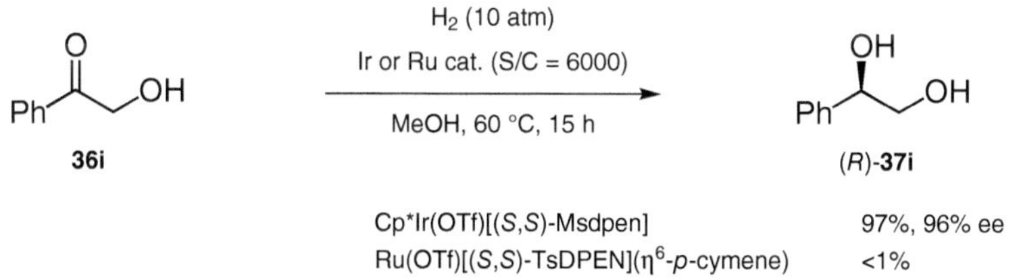

Scheme 3.22 Asymmetric hydrogenation of α-hydroxy ketone with Ru(II)–arene/TsDPEN or isoelectronic Cp*Ir/MsDPEN complexes. Source: Based on Ohkuma et al. [42].

In marked contrast to the inactivity of Ru(II)–arene/TsDPEN complex in the reaction mentioned above, a similar complex which contains a linking group (tether) between the TsDPEN unit and the arene ring on the ruthenium atom was found to catalyze the hydrogenation nicely to afford the corresponding diol in almost perfect enantioselectivity (Scheme 3.23) [44].

Table 3.10 Asymmetric hydrogenation of α-hydroxyketones with the Cp*Ir/MsDPEN complex.

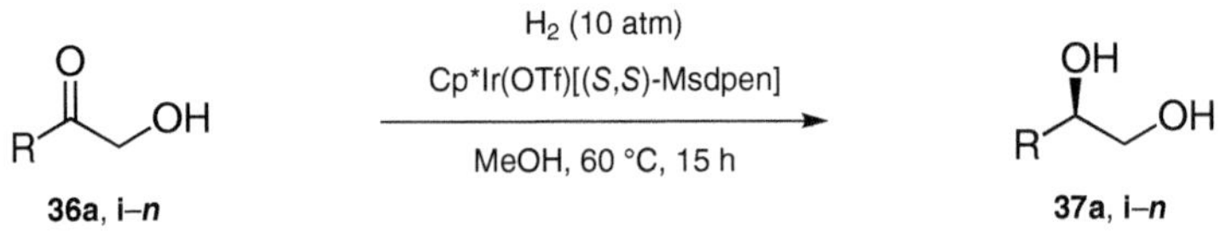

Entry	R	S/C	Yield[a] (%)	ee (%)
1	C_6H_5 (**36i**)	6000	97 (94)	96
2	4-NCC_6H_4 (**36j**)	1000	>99 (94)	94
3	4-$MeOC_6H_4$ (**36k**)	10 000	<1	nd[b]
4	4-FC_6H_4 (**36l**)	3000	>99 (93)	96
5	4-ClC_6H_4 (**36m**)	3000	>99 (91)	96
6	2-Thienyl (**36n**)	100	>99 (97)	99
7	CH_3 (**36a**)	100	97 (90)	80

a) Determined by ^{1}H NMR analysis. Isolated yield is stated in parentheses.
b) Not determined.

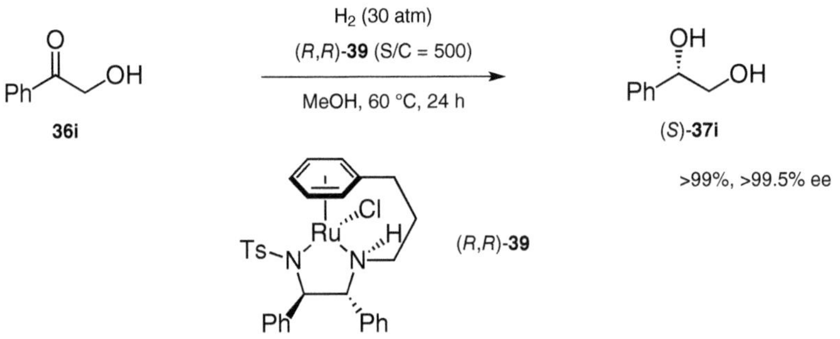

Scheme 3.23 Asymmetric hydrogenation of α-hydroxy ketone with tethered Ru(II)–arene/TsDPEN complex. Source: Based on Jolley et al. [44].

Transfer hydrogenation of α-hydroxyketones is scarcely explored. It is reported that transfer hydrogenation of hydroxyacetone (**36a**) to 1,2-propanediol (**37a**) using Ru(II)–arene/TsDPEN is bound to silica gel (Scheme 3.24) [45].

3.6 Asymmetric Hydrogenation of α-Oxophosphonates

Catalytic enantioselective hydrogenation of α-oxophosphonates is less common compared to the above examples. It was not until 2009 that the first example was reported by Beletskaya and coworkers. Asymmetric hydrogenation of α-oxophosphonates gave the corresponding α-hydroxyphosphonates in almost

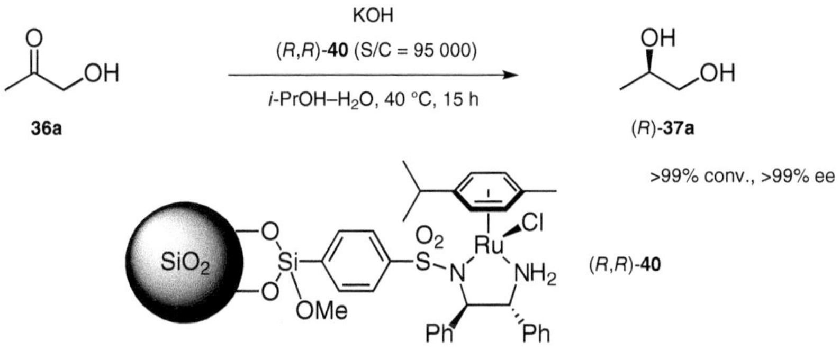

Scheme 3.24 Asymmetric transfer hydrogenation of acetol with silica-supported Ru(II)-arene/TsDPEN complex. Source: Based on Yue et al. [45].

quantitative yields by employing $Pd(OCOCF_3)_2$/(*R*)-MeO-BIPHEP catalytic system, though the enantiomeric purity of the products was not satisfactory (Scheme 3.25) [46].

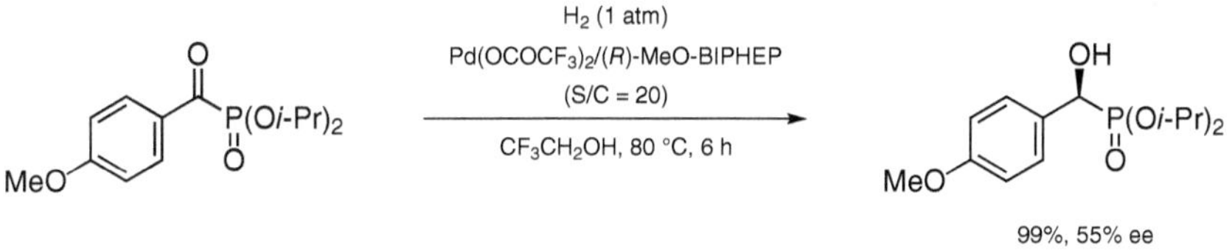

Scheme 3.25 Asymmetric hydrogenation of α-oxophosphonate with $Pd(OCOCF_3)_2$/(*R*)-MeO-BIPHEP catalyst. Source: Based on Goulioukina et al. [46].

More recently, the enantiomeric purity of the products was largely improved using a catalytic system that comprised $[Rh(cod)_2]BF_4$/chiral phosphine–phosphoramidite ligands (Scheme 3.26) [47]. Thus, hydrogenation of aromatic α-oxophosphonate **41a** afforded α-hydroxyphosphonate **42a** in 87% ee, whereas hydrogenation of alkyl α-oxophosphonate **41b** resulted in the formation of **42b** in 50% ee.

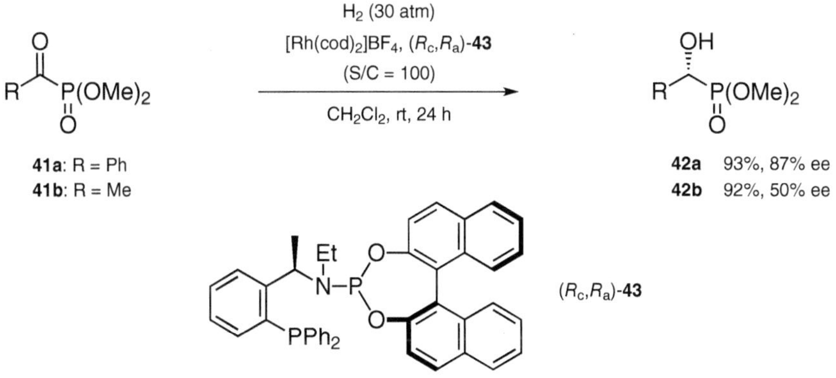

Scheme 3.26 Asymmetric hydrogenation of α-oxophosphonate with $[Rh(cod)_2]BF_4$/chiral phosphine–phosphoramidite catalyst. Source: Based on Li et al. [47].

Unlike the asymmetric reduction of functionalized ketones described above, reduction of α-oxophosphonates is achieved in better enantioselectivity via transfer hydrogenation than those by hydrogenation with H_2. Ru(II)-catalyzed transfer hydrogenation of aliphatic α-oxophosphonates **41c–e** proceeded to provide enantiopure products **42c–e** in high yield, while the reduction of aromatic α-oxophosphonate **41a** under the same reaction conditions provided **42a** in 84% ee (Scheme 3.27) [48].

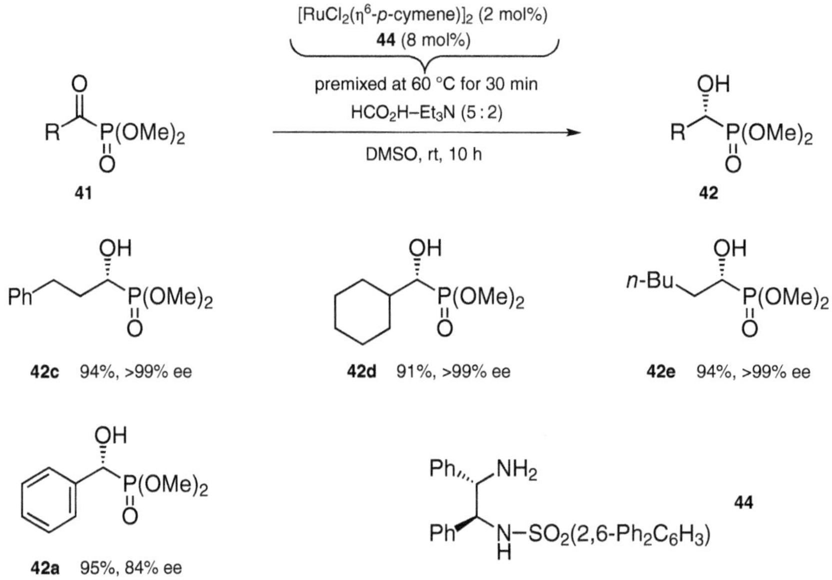

Scheme 3.27 Asymmetric transfer hydrogenation of α-oxophosphonates. Source: Based on Corbett and Johnson [48].

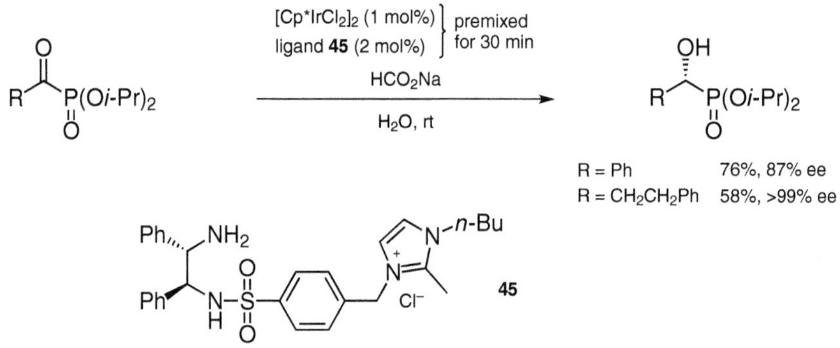

Scheme 3.28 Asymmetric transfer hydrogenation of α-oxophosphonates in aqueous solution. Source: Based on Sun et al. [49].

Ir(III)-catalyzed ATH of α-oxophosphonates in aqueous solution was achieved using imidazolium ion-tethered TsDPEN ligands (Scheme 3.28) [49]. Also in this case, aliphatic α-oxophosphonates gave superior enantioselectivity to aromatic α-oxophosphonates.

3.7 Summary and Conclusions

Among various catalytic systems for asymmetric hydrogenation, the Ru(II)–diphosphine/diamine catalytic system is applied to a wide range of asymmetric hydrogenation of ketones, enabled by the structural diversity of these types of Ru catalysts. The development of chiral Ru(II)–diphosphine/PICA catalysts expanded the substrate scope to *tert*-alkyl ketones. Hardly accessible asymmetric hydrogenation of alkenyl ketones was achieved using the Ru(II)–TolBINAP/DMAPEN catalyst with a unique shape-selected chiral environment. A range of alkynyl ketones were successfully hydrogenated in a highly enantioselective manner by employing a catalytic amount of Ru(OTf)(TsDPEN)(η^6-arene) or the isoelectronic Cp*Ir(OTf)(MsDPEN) under base-free conditions. A series of α-functionalized ketones were also enantioselectively hydrogenated by the use of an appropriate combination of the ligands. Though there has been less exploration of transfer hydrogenation to these substrates, some interesting achievements were reported in specific cases.

References

1 Ohkuma, T. and Noyori, R. (1999). Hydrogenation of carbonyl groups. In: *Comprehensive Asymmetric Catalysis*, Chapter 6.1, vol. 1 (eds. E.N. Jacobsen, A. Pfaltz and H. Yamamoto), 199–246. Berlin, Germany: Springer.
2 Ohkuma, T., Kitamura, M., and Noyori, R. (2000). Hydrogenation of ketones. In: *Catalytic Asymmetric Synthesis*, Chapter 1.4 (ed. I. Ojima), 34–83. Weinheim, Germany: Wiley-VCH.
3 Hedberg, C. (2008). Carbonyl hydrogenation. In: *Modern Reduction Methods*, Chapter 5 (eds. P.G. Andersson and I.J. Minslow), 109–134. Weinheim, Germany: Wiley-VCH.
4 Ohkuma, T., Koizumi, M., Doucet, H. et al. (1998). *J. Am. Chem. Soc.* 120: 13529–13530.
5 Noyori, R. and Ohkuma, T. (2001). *Angew. Chem., Int. Ed.* 40: 40–73.
6 Osawa, T. (1985). *Chem. Lett.*: 1609–1612.
7 Wei, W.-L., Hao, S.-J., Zhou, J. et al. (2004). *Polym. Adv. Technol.* 15: 287–290.
8 Jiang, Q., Jiang, Y., Xiao, D. et al. (1998). *Angew. Chem. Int. Ed.* 37: 1100–1103.
9 Doucet, H., Ohkuma, T., Murata, K. et al. (1998). *Angew. Chem. Int. Ed.* 37: 1703–1707.
10 Sandoval, C.A., Ohkuma, T., Muñiz, K., and Noyori, R. (2003). *J. Am. Chem. Soc.* 125: 13490–13503.
11 Ohkuma, T., Koizumi, M., Muñiz, K. et al. (2002). *J. Am. Chem. Soc.* 124: 6508–6509.

12 Ohkuma, T., Sandoval, C.A., Strinivasan, R. et al. (2005). *J. Am. Chem. Soc.* 127: 8288–8289.
13 Li, W., Hou, G., Wang, C. et al. (2010). *Chem. Commun.* 46: 3979–3981.
14 Matsumura, K., Arai, N., Hori, K. et al. (2011). *J. Am. Chem. Soc.* 133: 10696–10699.
15 Garbe, M., Junge, K., Walker, S. et al. (2017). *Angew. Chem. Int. Ed.* 56: 11237–11241.
16 Schlatter, A. and Woggon, W.-D. (2008). *Adv. Synth. Catal.* 354: 995–1000.
17 Takehara, J., Hashiguchi, S., Fujii, A. et al. (1996). *Chem. Commun.*: 233–234.
18 Noyori, R. and Hashiguchi, S. (1997). *Acc. Chem. Res.* 30: 97–102.
19 Wu, J. and Chan, A.S.C. (2006). *Acc. Chem. Res.* 39: 711–720.
20 Burk, M.J., Hems, W., Herzberg, D. et al. (2000). *Org. Lett.* 2: 4173–4176.
21 Xie, J.-H. and Zhou, Q.-L. (2008). *Acc. Chem. Res.* 41: 581–593.
22 Li, W., Sun, X., Zhou, L. et al. (2009). *J. Org. Chem.* 74: 1397–1399.
23 Chen, X., Zhou, H., Zhang, K. et al. (2014). *Org. Lett.* 16: 3912–3915.
24 Arai, N., Azuma, K., Nii, N., and Ohkuma, T. (2008). *Angew. Chem. Int. Ed.* 47: 7457–7460.
25 Arai, N., Satoh, H., Komatsu, R., and Ohkuma, T. (2017). *Chem. Eur. J.* 23: 8806–8809.
26 Xue, D., Chen, Y.-C., Cui, X. et al. (2005). *J. Org. Chem.* 70: 3584–3591.
27 Parekh, V., Ramsden, J.A., and Wills, M. (2012). *Catal. Sci. Technol.* 2: 406–414.
28 Arai, N., Satoh, H., Utsumi, N. et al. (2013). *Org. Lett.* 15: 3030–3033.
29 Matsumura, K., Hashiguchi, S., Ikariya, T., and Noyori, R. (1997). *J. Am. Chem. Soc.* 119: 8738–8739.
30 Zhang, Y.-M., Yuan, M.-L., Liu, W.-P. et al. (2018). *Org. Lett.* 20: 4486–4489.
31 Kitamura, M., Ohkuma, T., Inoue, S. et al. (1988). *J. Am. Chem. Soc.* 110: 629–631.
32 Mashima, K., Kusano, K.-h., Ohta, T. et al. (1989). *J. Chem. Soc., Chem. Commun.*: 1208–1210.
33 Mashima, K., Kusano, K.-h., Sato, N. et al. (1994). *J. Org. Chem.* 59: 3064–3076.
34 Ohkuma, T., Ishii, D., Takeno, H., and Noyori, R. (2000). *J. Am. Chem. Soc.* 122: 6510–6511.
35 Kosmalski, T., Wojtczak, A., and Zaidlewicz, M. (2009). *Tetrahedron: Asymmetry* 20: 1138–1143.
36 Takenaka, M. (2005). *Chem. Abstr.* 143: 422509; Jpn. Kokai Tokkyo Koho 2005306804 A 20051104, 2005.
37 Tsutsumi, K., Katayama, T., Utsumi, N. et al. (2009). *Org. Process Res. Dev.* 13: 625–628.
38 Arai, N., Akashi, M., Sugizaki, S. et al. (2010). *Org. Lett.* 12: 3380–3383.
39 Zhang, L., Tang, Y., Han, Z., and Ding, K. (2019). *Angew. Chem. Int. Ed.* 58: 4973–4977.
40 Saito, T., Yokozawa, T., Ishizaki, T. et al. (2001). *Adv. Synth. Catal.* 343: 264–267.
41 Kadyrov, R., Koenigs, R.M., Brinkmann, C. et al. (2009). *Angew. Chem. Int. Ed.* 48: 7556–7559.
42 Ohkuma, T., Utsumi, N., Watanabe, M. et al. (2007). *Org. Lett.* 9: 2565–2567.

43 Ohkuma, T., Tsutsumi, K., Utsumi, N. et al. (2007). *Org. Lett.* 9: 255–257.
44 Jolley, K.E., Zanotti-Gerosa, A., Hancock, F. et al. (2012). *Adv. Synth. Catal.* 354: 2545–2555.
45 Yue, C.-J., Gu, L.-P., Zhuang, Y.-F., and Liu, B.-L. (2015). *Russ. J. Appl. Chem.* 88: 1207–1218.
46 Goulioukina, N.S., Bondarenko, G.N., Bogdanov, A.V. et al. (2009). *Eur. J. Org. Chem.*: 510–515.
47 Li, Q., Hou, C.-J., Liu, Y.-J. et al. (2015). *Tetrahedron: Asymmetry* 26: 617–622.
48 Corbett, M.T. and Johnson, J.S. (2013). *J. Am. Chem. Soc.* 135: 594–597.
49 Sun, M., Campbell, J., Kang, G. et al. (2016). *J. Organomet. Chem.* 810: 12–14.

4

Asymmetric (Transfer) Hydrogenation of Aryl and Heteroaryl Ketones

Jian-Hua Xie and Qi-Lin Zhou

Nankai University, Institute of Elemento-organic Chemistry, College of Chemistry, Weijin Road 94, Tianjin 300071, China

4.1 Introduction

Chiral alcohols are versatile building blocks for the synthesis of pharmaceuticals, agrochemicals, flavors and fragrances, and natural products. Considerable efforts have been devoted to the development of highly efficient asymmetric catalysis for the synthesis of various chiral alcohols over the past decades [1]. Ketones are readily available bulk chemicals and represent one of the most common families of unsaturated compounds. Hydrogen (H_2) is a clean, cheap, and abundant resource. Transition-metal-catalyzed asymmetric hydrogenation of ketones with hydrogen is recognized as one of the most clean, effective, atom-economical, and environmentally benign methods for the preparation of enantiomerically enriched alcohols (Figure 4.1a). In addition, catalytic asymmetric transfer hydrogenation of ketones with hydrogen donors such as 2-propanol or formate, originated from Meerwein–Ponndorf–Verley reduction, is also an operationally simple and safe method for preparation of chiral optically active alcohols (Figure 4.1b). In the past few decades, transition-metal-catalyzed asymmetric hydrogenation and transfer hydrogenation of ketones have been intensively studied. A variety of chiral metal catalysts such as rhodium, ruthenium, iridium, iron, and manganese complexes with the chiral ligands such as monophosphines, diphosphines, phosphine–oxazolines, diamines, and tridentate phosphine-containing ligands have been developed [2–6]. Many of these chiral metal catalysts showed high activity and enantioselectivity in the hydrogenation of ketones. Some of them have been successfully applied in industrial preparation of chiral alcohols. In this chapter, we review the recent advances of the transition-metal-catalyzed asymmetric hydrogenation and transfer hydrogenation of aryl and heteroaryl ketones. Because chiral metal catalysts play a critical role in the asymmetric hydrogenation and transfer hydrogenation of ketones, and almost every breakthrough in this field is related to the development of chiral catalysts, this chapter is organized from the perspective of the development of chiral transition metal catalysts.

Asymmetric Hydrogenation and Transfer Hydrogenation, First Edition.
Edited by Virginie Ratovelomanana-Vidal and Phannarath Phansavath.

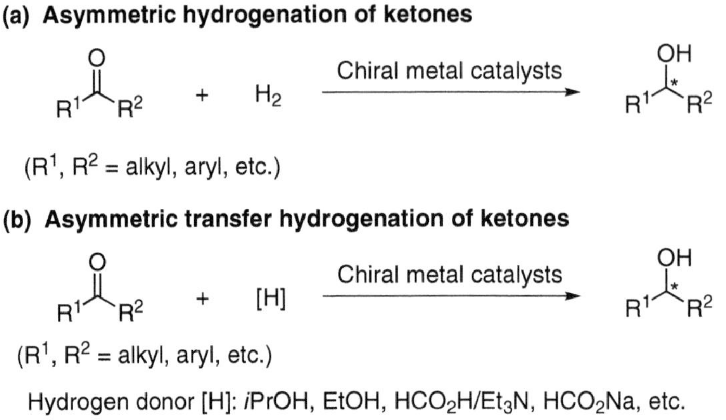

Figure 4.1 Asymmetric hydrogenation (a) and transfer hydrogenation (b) of ketones.

4.2 Asymmetric Hydrogenation of Aryl and Heteroaryl Ketones

In 1970, Schrock and Osborn reported the first example of catalytic hydrogenation of ketones using cationic rhodium complexes of phosphine ligands [7]. Subsequently, Scorrano and Takegami et al. independently used chiral phosphine ligands in this reaction, achieving the asymmetric hydrogenation of ketones [8, 9]. Since then, rhodium- and ruthenium-catalyzed asymmetric hydrogenation of ketones has received intensive studies by employing chiral phosphine ligands including BINAP ligands [10, 11]. However, although a variety of functionalized ketones can be hydrogenated to chiral secondary alcohols in high enantioselectivity by chiral rhodium and ruthenium catalysts, the asymmetric hydrogenation of simple ketones such as aryl alkyl ketones is still a challenge. In addition, as reported by Takaya and coworkers in 1993, the chiral iridium complexes of BINAP and achiral aminophosphine ligands (e.g. (*R*)-**1**) showed high enantioselectivity (up to 95% ee) for the hydrogenation of aryl ketones, but limited to the cycloalkanone substrates (Figure 4.2) [12]. A breakthrough for the catalytic asymmetric hydrogenation of aryl ketones was achieved by Noyori and coworkers in 1995, when they developed highly efficient chiral $RuCl_2$–BINAP/diamine catalysts [13]. After this work, they soon found that $RuCl_2$–(*S*)-Tol-BINAP/(*S*,*S*)-DPEN ((*S*,*S*,*S*)-**2**) had a super activity (turnover numbers, TONs, 2 400 000) in the hydrogenation of acetophenone (Figure 4.2) [14]. The activity of the catalyst (*S*,*S*,*S*)-**2** was extremely high, representing the highest level at that time; however, the enantioselectivity (80% ee) was not very satisfactory at high TON. The enantioselectivity was improved by the combination of (*S*)-Xyl-BINAP with (*S*)-DaiPEN to form the catalyst $RuCl_2$–(*S*)-Xyl-BINAP/(*S*)-DaiPEN ((*S*,*S*)-**3**) [15, 16]. The hydrogenation of acetophenone catalyzed by (*S*,*S*)-**3** at TONs of 100 000 afforded (*R*)-1-phenylethanol in 99% ee. With catalyst (*S*,*S*)-**3**, various *ortho*-, *meta*-, and *para*-substituted acetophenones have been hydrogenated in high enantioselectivity. The high performance of Noyori's $RuCl_2$–BINAPs/diamine

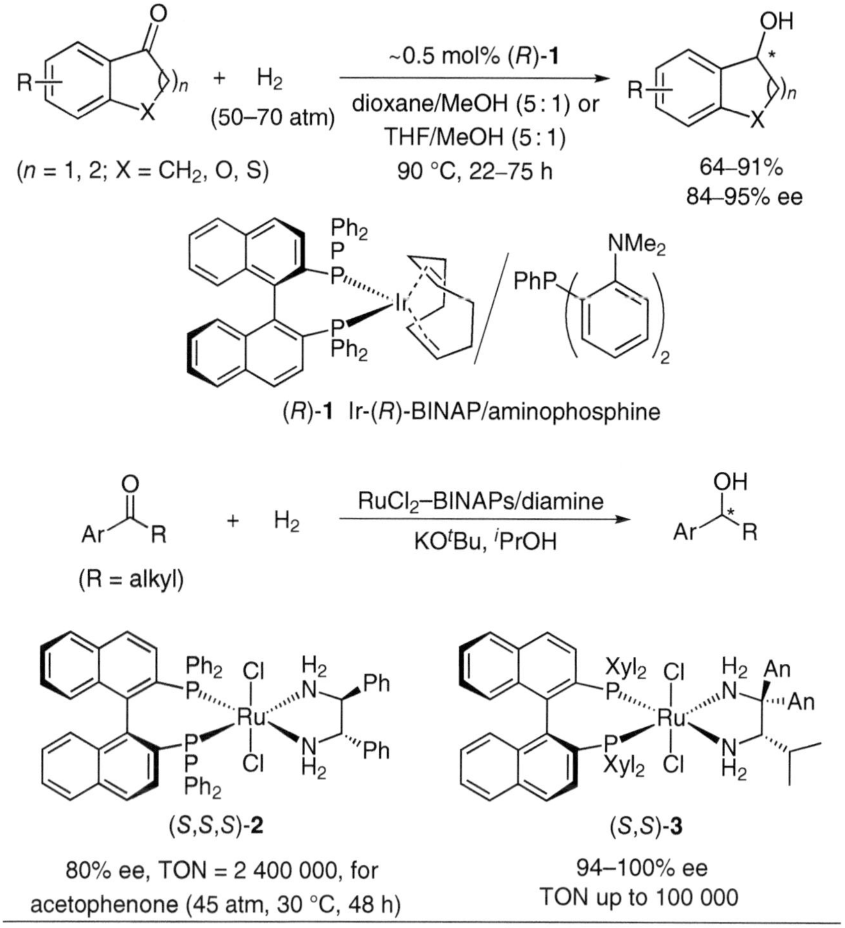

Note: Xyl = 3,5-dimethylphenyl; An = 4-methoxyphenyl

Figure 4.2 The early typical examples of asymmetric hydrogenation of aryl ketones.

catalysts attracted considerable attention in the following years, resulting in the development of a wide range of chiral ruthenium-diphosphine/diamine catalysts.

To date, many efficient chiral transition metal catalysts such as Ru, Ir, Fe, and Mn complexes of chiral ligands including monophosphines, diphosphines, phosphine–oxazolines, and tridentate phosphine-containing ligands have been developed for the asymmetric hydrogenation of aryl ketones. Among these chiral catalysts, the iridium catalyst of chiral tridentate spiro pyridine-aminophosphine SpiroPAP ligands developed by Xie and Zhou [17] and the lutidine-based chiral pincer manganese catalysts recently developed by Ding and coworkers [18] showed extremely high activity and enantioselectivity. Transition-metal-catalyzed asymmetric hydrogenation of aryl ketones has become one of the most appealing and useful methods for the preparation of optically active alcohols and is also utilized in the production of chiral pharmaceuticals.

4.2.1 Chiral Ruthenium Catalysts

Since Noyori and coworkers reported Ru–BINAP/diamine catalysts and achieved high enantioselectivity in the asymmetric hydrogenation of aryl ketones [13], a variety of chiral ruthenium catalysts have been developed in the following years. In addition to the ruthenium-diphosphine/diamine catalysts, the ruthenium complexes of phosphine–oxazoline ligands, Ts-diamine ligands, tridentate phosphine-containing ligands, etc. were also developed as the catalysts for the asymmetric hydrogenation of aryl ketones.

4.2.1.1 Chiral Ruthenium-Diphosphine/Diamine Catalysts

Noyori-type ruthenium-diphosphine/diamine catalysts contain two types of ligands. In theory, changing any of them with chiral or achiral diphosphine ligand or diamine ligand can form new chiral ruthenium catalysts. Mechanism studies showed that the ruthenium-diphosphine/diamine catalyzes the hydrogenation of ketones through a metal–ligand bifunctional pathway (Figure 4.3) [19, 20]. Hydrogen is activated by a catalyst and transferred to the ketone group via a precyclic six-membered ring between the catalyst and the substrate ($H^{\delta-}$—Ru—N—$H^{\delta+}$ with $C^{\delta+}$—$O^{\delta-}$). In consequence, at least one N—H moiety, preferably an NH_2 group, is required in this type of catalysts.

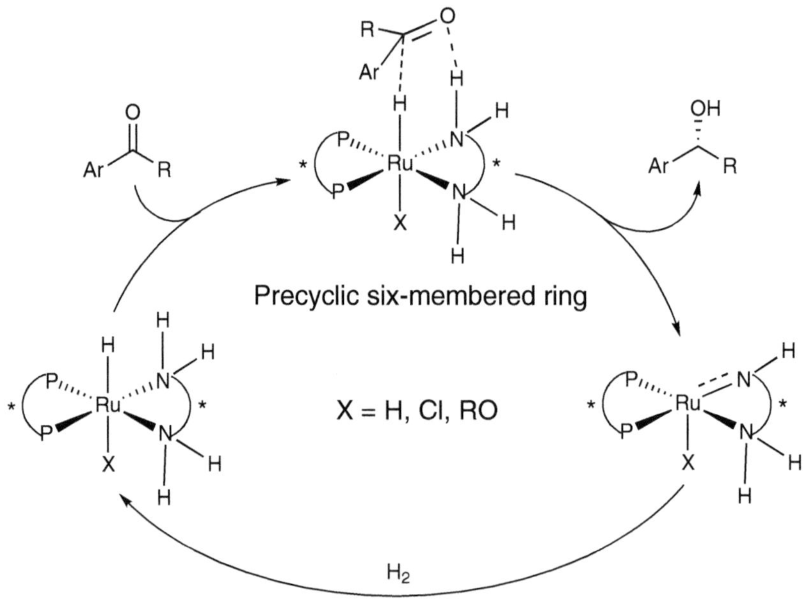

Figure 4.3 Metal–ligand bifunctional mechanism for asymmetric hydrogenation of aryl ketones. Source: Yamakawa et al. [19] and Noyori et al. [20].

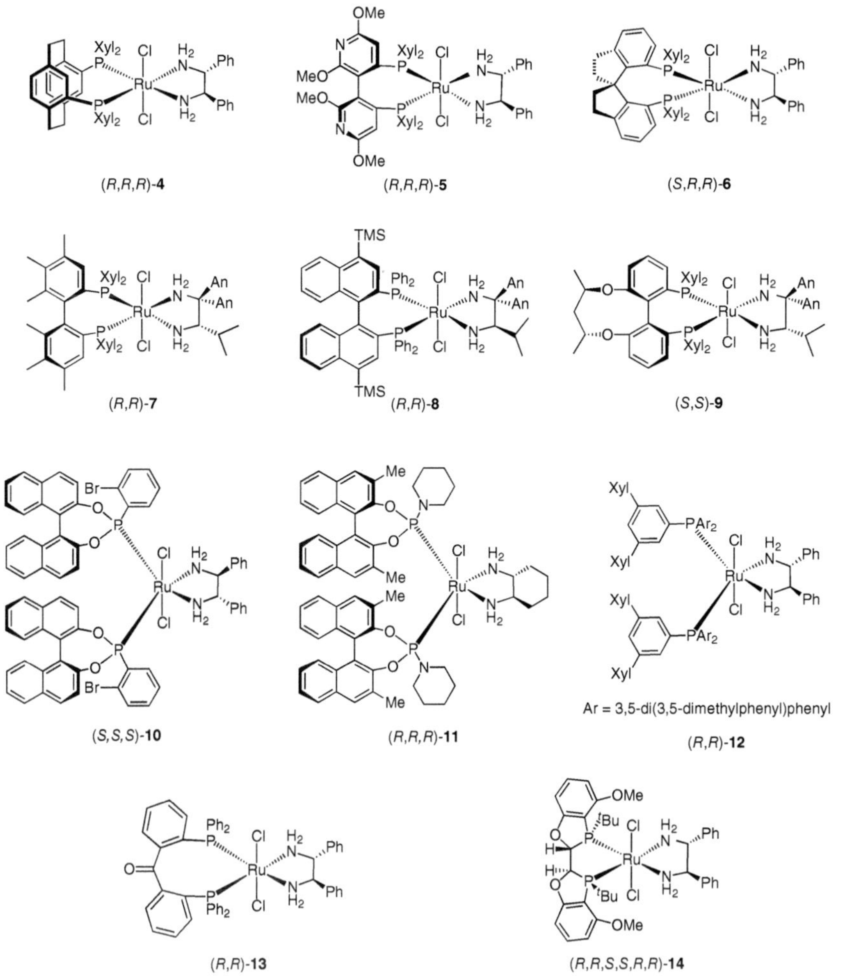

Figure 4.4 Ruthenium-diphosphine/diamine catalysts having different phosphine ligands.

In 2000, Burk et al. introduced a planar-chiral diphosphine ligand Xyl-PhanePhos into ruthenium-diphosphine/diamine complexes and developed the catalyst $RuCl_2$–Xyl-Phanephos/DPEN (e.g. (*R*,*R*,*R*)-**4**, Figure 4.4) [21]. The catalyst (*R*,*R*,*R*)-**4** provides 98–99.5% ee with TONs up to 40 000 for the asymmetric hydrogenation of a range of aryl and heteroaryl ketones. In 2002, Chan and coworkers [22] developed the catalyst $RuCl_2$–(*R*)-Xyl-P-Phos/(*R*,*R*)-DPEN ((*R*,*R*,*R*)-**5**) containing an axial-chiral diphosphine ligand Xyl-P-Phos and diamine DPEN (Figure 4.4) and achieved 93.3 to >99.9% ee in the asymmetric hydrogenation of substituted acetophenones with up to 100 000 of TONs. Later on, the catalyst (*R*,*R*,*R*)-**5** was also demonstrated to be highly efficient for the hydrogenation of heteroaryl ketones (TONs up to 10 000, 97–99.0% ee) [23]. At the same time, Zhou and coworkers reported chiral ruthenium catalysts with spiro diphosphine ligands SDPs and found that the $RuCl_2$–(*S*)-Xyl-SDP/(*R*,*R*)-DPEN ((*S*,*R*,*R*)-**6**) (Figure 4.4) can catalyze the asymmetric hydrogenation of aryl and heteroaryl ketones in 98–99.5% ee with up to 100 000 of TONs [24]. It is worth mentioning that the nature of the aryl groups on the phosphorus atoms of chiral diphosphine ligands is important for the aforementioned ruthenium catalysts, and the 3,5-dimethylphenyl groups (Xyl) on the phosphorus atoms are generally required for obtaining high TON and high enantioselectivity (Figure 4.4).

Many other ruthenium-diphosphine/diamine catalysts require chiral diamine DaiPEN, which is more expensive than the diamine DPEN, for achieving high TON and high enantioselectivity. For example, the catalyst $RuCl_2$–(*S*)-Xyl-HexaPHEMP/(*S*)-DaiPEN ((*S*,*S*)-**7**) is more active than the corresponding ruthenium catalyst with (*S*,*S*)-DPEN ligand (Figure 4.4) [25]. The ruthenium catalyst $RuCl_2$–(*R*)-4,4′-TMS-BINAP/(*R*)-DaiPEN ((*R*,*R*)-**8**) (Figure 4.4) bearing a diphosphine 4,4′-substituted-BINAP and a diamine DaiPEN affords 96.1–99.8% ee with up to 1 000 000 of TONs for the hydrogenation of aryl ketones [26, 27]. The ruthenium catalyst (*S*,*S*)-**9** (Figure 4.4) containing a C_3*-TunePhos ligand developed by Zhang and coworkers in 2009 also shows high enantioselectivity (97.0–99.8% ee) with TONs up to 945 000 for the hydrogenation of aryl and heteroaryl ketones [28].

Introducing two chiral/achiral monophosphines and other types of diphosphines into the ruthenium-diphosphine/diamine complexes leads to several efficient chiral ruthenium catalysts for the asymmetric hydrogenation of aryl ketones. In 2004, Wills and coworkers found that the ruthenium catalyst (*S*,*S*,*S*)-**10** (Figure 4.4) containing two BINOL-based monophosphonite ligands (*S*)-BrXuPhos gives 75–99% ee with TONs up to 8000 for the hydrogenation of aryl ketones [29, 30]. The ruthenium catalyst (*R*,*R*,*R*)-**11** (Figure 4.4) with two monophosphoramidite ligands (*R*)-PipPhos developed by de Vries and coworkers shows modest to high enantioselectivity (65–97% ee) [31]. In 2005, Ding and coworkers introduced a series of achiral bulky triaryl phosphine ligand PAr_3 into the ruthenium complexes with diamine (*R*,*R*)-DPEN and found that the catalyst (*R*,*R*)-**12** (Figure 4.4) gives 87–97% ee with TONs up to 10 000 for the hydrogenation of aryl and heteroaryl ketones [32]. This result indicates a new approach for the design of chiral ruthenium catalysts. Following this work, the ruthenium catalyst (*R*,*R*)-**13** (Figure 4.4) bearing an achiral benzophenone-based bisphosphine DPBP ligand and a diamine (*R*,*R*)-DPEN independently developed by Ding and coworkers [33] and Mikami et al. [34] also shows good to high enantioselectivity (up to 10 000 of TONs, 90–99% ee) for the hydrogenation of aryl ketones. In addition, although it gives relatively lower enantioselectivity in the hydrogenation of simple aryl and heteroaryl ketones (80–92% ee), the ruthenium catalyst (*R*,*R*,*S*,*S*,*R*,*R*)-**14** (Figure 4.4) bearing a chiral bisdihydrobenzooxaphosphole (*R*,*R*,*S*,*S*)-MeO-BIBOP ligand shows high enantioselectivity (94–98% ee) with up to 100 000 of TONs for the hydrogenation of aryl and heteroaryl cyclic ketones [35].

New chiral ruthenium-diphosphine/diamine catalysts also can be achieved by changing diamine ligands (Figure 4.5). To solve the problem of slow reaction rate and poor enantioselectivity of the hydrogenation of benzo-fused cyclic ketones such as tetralone, Noyori and coworkers [36] replaced the 1,2-chiral diamine ligands

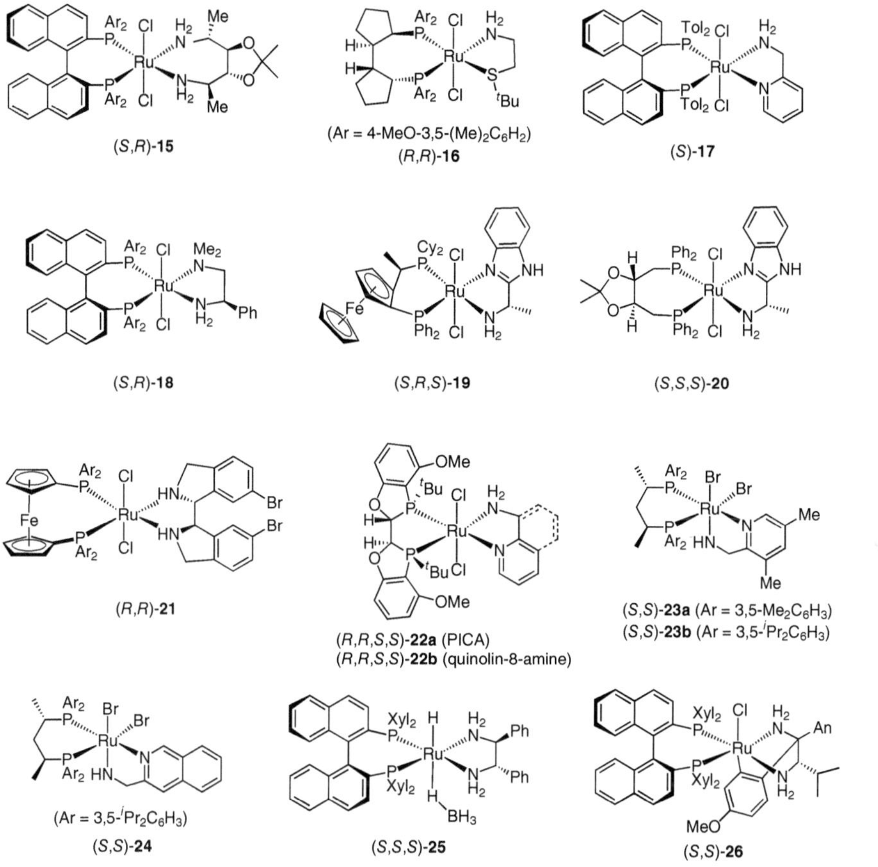

Figure 4.5 Ruthenium-diphosphine/diamine catalysts having different diamine ligands.

with chiral 1,4-diamine in the chiral ruthenium-diphosphine/diamine complexes to develop the catalyst $RuCl_2$–BINAP/IPHAN (3,4-*O*-isopropylidene-hexane-2,5-diamine) (*S*,*R*)-**15** (Figure 4.5). The catalyst (*S*,*R*)-**15** is highly efficient for the asymmetric hydrogenation of α-tetralone-type cyclic ketones, providing the corresponding chiral cyclic alcohols in 80–99% ee with TONs up to 55 000. In contrast, the chiral ruthenium catalyst (*S*,*S*)-**3** bearing a DaiPEN ligand gives a low enantioselectivity (30% ee) for the hydrogenation of 1-tetralone. In the same year, Genov and Ager [37] developed ruthenium catalysts containing 2-(alkylthio)amines which have only one NH_2 group and are efficient for the hydrogenations of aryl and heteroaryl ketones. For example, the catalyst (*R*,*R*)-**16** (Figure 4.5) bearing an achiral 2-*tert*-butylthioethylamine (ThioEA) gives 93% ee with TONs up to 1000 in the asymmetric hydrogenation of 3′,5′-bis(trifluoromethyl)acetophenone. The catalyst (*R*,*R*)-**16** represents the first example of chiral ruthenium catalysts bearing an achiral amine ligand for the asymmetric hydrogenation of ketones. Then, Noyori and coworkers [38] reported Tol-BINAP-based chiral ruthenium catalyst (*S*)-**17** (Figure 4.5) with an achiral and unsymmetric NH_2/pyridine hybrid ligand, α-picolyamine (PICA). The catalyst (*S*)-**17** shows 90–98% ee with TONs up to 2400 for the hydrogenation of aryl *tert*-butyl ketones. Subsequently, they found that the catalyst (*S*)-**17** is also efficient for the hydrogenation of aryl *tert*-butyldimethylsilyl ketones, providing the corresponding α-hydroxysilanes in 95–96% ee with TONs up to 10 000 [39]. Encouraged by these results, Ohkuma and coworkers used 2-dimethylamino-1-phenylethyl amine (DMAPEN), an unsymmetrical amine with one NH_2 group as ligand, and developed BINAP-based chiral ruthenium catalyst (*S*,*R*)-**18** (Figure 4.5) [40]. The catalyst (*S*,*R*)-**18** gives moderate to good enantioselectivity for the hydrogenations of aryl alkyl ketones (81–95% ee) [41] and aryl vinyl ketones (89–98% ee) with TONs up to 10 000 [42].

The NH_2–benzimidazole BIMA ligand was applied in Josiphos-based ruthenium catalysts by Sandoval and coworkers in 2009 [43]. In the presence of suitable additives such as triphenylphosphine, the catalyst (*S*,*R*,*S*)-**19** (Figure 4.5) shows high enantioselectivity (82–99% ee) and high TONs (up to 50 000) for the hydrogenation of aryl and heteroaryl ketones. Subsequently, they combined NH_2–benzimidazole BIMA ligand with DIOP ligand to develop a more efficient ruthenium catalyst (*S*,*S*,*S*)-**20** (Figure 4.5) [44]. Without triphenylphosphine additive, catalyst (*S*,*S*,*S*)-**20** affords 96–99% ee with TONs up to 100 000 for the hydrogenation of aryl ketones including α-tetralone-type cyclic ketones and 2′-chlorobenzophenone.

Huang and coworkers combined achiral diphosphine DPPF ligand with chiral NH-type 1,2-diamine 1,1′-biisoindoline ligand ((*R*,*R*)-BIDN) to develop a new chiral ruthenium catalyst (*R*,*R*)-**21** (Figure 4.5) [45]. This catalyst has good to excellent enantioselectivity (88–99% ee) with TONs up to 100 000 for hydrogenation of aryl and heteroaryl ketones. This result indicates that the NH-type diamines are also suitable ligands for chiral ruthenium-diphosphine/diamine catalysts. In the synthesis of potential cholesteryl transfer protein (CETP) inhibitors, Rodríguez et al. [35] found that the combination of a chiral bisdihydrobenzooxaphosphole (*R*,*R*,*S*,*S*)-MeO-BIBOP ligand with achiral PICA ligand is superior to that with chiral diamine (*R*,*R*)-DPEN ligand, resulting in the catalyst (*R*,*R*,*S*,*S*)-**22a** (Figure 4.5). The catalyst (*R*,*R*,*S*,*S*)-**22a** gives up to 98% ee with TONs up to 19 600 for the hydrogenation of heteroaryl cyclic ketone. In addition, the catalyst (*R*,*R*,*S*,*S*)-**22b** with an achiral quinolin-8-amine ligand shows 96% ee and up to 100 000 of TONs for the hydrogenation of α-tetralone-type cyclic ketones, but the enantioselectivity is low for the hydrogenation of simple aryl and heteroaryl ketones. In the investigation of asymmetric hydrogenation of poly-substituted aryl ketones, Ohkuma and coworkers [46] found that the ruthenium–Skewphos/PICA complexes (*S*,*S*)-**23** and (*S*,*S*)-**24** (Figure 4.5) are efficient catalysts. The catalysts (*S*,*S*)-**23b** and (*S*,*S*)-**24** (Figure 4.5) afford up to 99% ee with TONs up to10 000 in the asymmetric hydrogenation of a series of poly-substituted aryl ketones. In an industrial-scale (50 kg) hydrogenation of 2′,6′-dichloro-3′-fluoroacetophenone, the catalyst (*S*,*S*)-**23a** gives the corresponding chiral alcohol, a key intermediate for the synthesis of the medicine crizotinib, in 96% isolated yield and 98% ee with TONs of 10 000.

Typical chiral $RuCl_2$–diphosphine/diamine catalysts require strong base to generate active catalysts, limiting the scope of the substrates. In order to avoid such a drawback Noyori and coworkers developed BH_4^- anion-bounded ruthenium–BINAP/diamine complexes such as (*S*,*S*,*S*)-**25** (Figure 4.5) by the reaction of $RuCl_2$–BINAP/diamine complexes with excess sodium borohydride (25 equiv) [47]. Since such ruthenium complexes have been activated by sodium borohydride, they can hydrogenate ketones under base-free conditions. With catalyst (*S*,*S*,*S*)-**25**, several base-sensitive aryl ketones such as 3-(dimethylamino)-1-phenylpropan-1-one have been hydrogenated to the corresponding chiral alcohols in high yields (99–100%) and 97–99% ee with TONs up to 100 000. The chiral ruthenium catalyst (*S*,*S*)-**26** (Figure 4.5) with a unique ruthenabicyclo[2.2.1] skeleton was reported by Ohkuma and coworkers in 2011 [48]. This new type of catalyst exhibits a remarkably high activity in the asymmetric hydrogenation of ketones. Its turnover frequency (TOF) reaches 35 000 min^{-1} with enantioselectivity up to >99% ee, and the scope of substrate is superior to that of the chiral ruthenium-diphosphine/diamine catalyst (*S*,*S*)-**3**. Moreover, this chiral ruthenium catalyst is very stable and its ruthenabicyclic structure maintained during the hydrogenation.

4.2.1.2 Chiral Arene–Ruthenium-Diamine Catalysts

Chiral arene–ruthenium-diamine complexes such as (η^6-arene)Ru–Ts-DPEN complexes are another type of metal-NH bifunctional catalysts, but their catalytic performances differ from that of $RuCl_2$–diphosphine/diamine complexes. In 2001, Ikariya and coworkers reported that the Cp*Ru-1,2-diamine complex (*S*)-**27** (Figure 4.6) is an efficient catalyst for the asymmetric transfer hydrogenation of aryl ketones [49]. With catalyst (*S*)-**27**, several aromatic ketones were hydrogenated to chiral alcohols in moderate enantioselectivities (64–85% ee), but only one example (3-methyl-1-phenylbutan-1-one) gives 95% ee. Subsequently, Brandt and Andersson et al. introduced quincorine-amine containing one primary and one tertiary

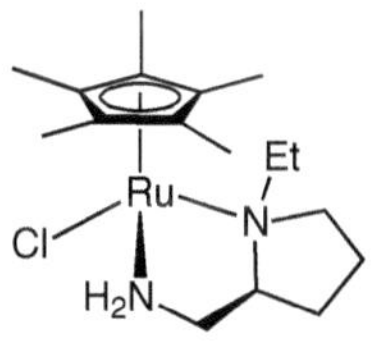

(*S*)-**27**

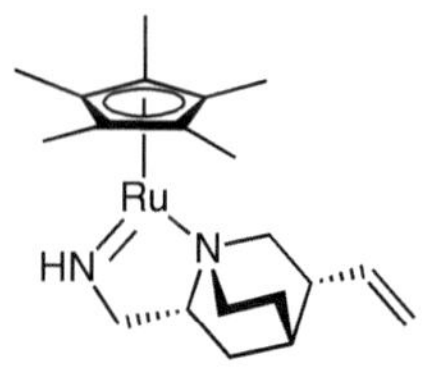

(1*S*,2*R*,4*S*,5*R*)-**28**

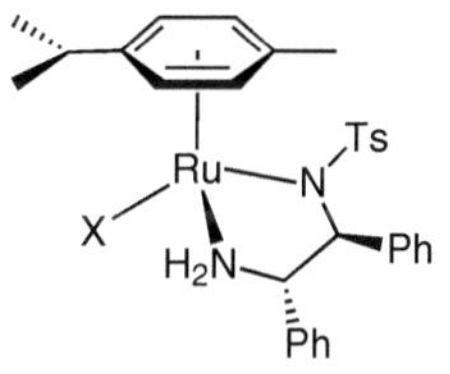

(*S*,*S*)-**29a** X = Cl
(*S*,*S*)-**29b** X = OTf

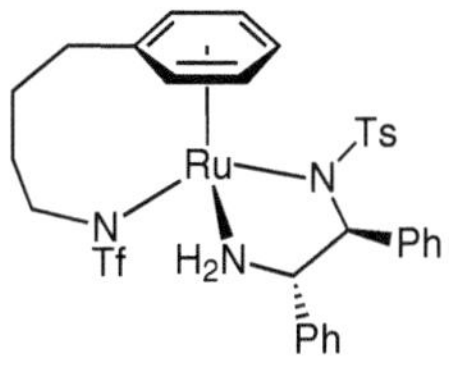

(*S*,*S*)-**30**

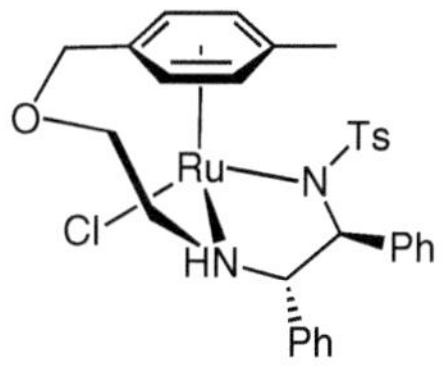

(*S*,*S*)-**31**

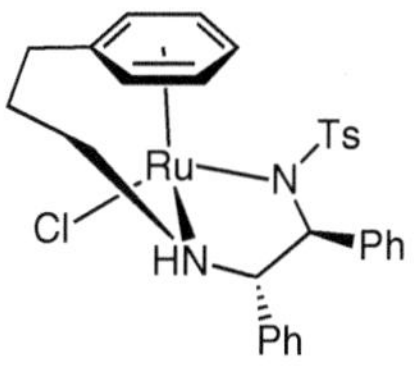

(*S*,*S*)-**32**

Figure 4.6 Chiral arene–ruthenium–diamine complexes.

amino groups, and found that its RuCp*-1,2-diamine complex (1*S*,2*R*,4*S*,5*R*)-**28** (Figure 4.6) was an efficient catalyst for the asymmetric hydrogenation of aromatic ketones (up to 90% ee) [50]. In 2006, Noyori and coworkers reported that the ruthenium complexes (*S*,*S*)-**29a** and **29b** (Figure 4.6) containing a chiral Ts-DPEN ligand showed high enantioselectivity for the hydrogenation of 4-chromanones, a type of challenging substrates which have acidic α-H of the carbonyl group [51]. The catalyst (*S*,*S*)-**29b** with a triflate anion has higher activity than the catalyst (*S*,*S*)-**29a** with a chloride anion. For example, the catalyst (*S*,*S*)-**29b** affords 100% yield with 97% ee, while the catalyst (*S*,*S*)-**29a** affords 64% yield with 95% ee in the hydrogenation of chroman-4-one with 0.03% mol catalyst. In 2008, Ikariya et al. reported that the well-defined triflylamide-tethered η^6-arene/*N*-Ts-DPEN ruthenium catalyst (*S*,*S*)-**30** (Figure 4.6) afforded high enantioselectivities (91–98% ee) for the hydrogenation of aromatic ketones with TONs up to 1000 [52]. Subsequently, they found that the (η^6-arene)Ru–Ts-DPEN catalyst (*S*,*S*)-**31** (Figure 4.6) chelating with an oxo-chain tether affords as high as >99% ee and up to 5000 of TONs for the hydrogenation of aryl cycloketones including 4-chromanone [53]. The all-carbon chain-tethered ruthenium catalyst (*S*,*S*)-**32** (Figure 4.6) was developed by Wills and coworkers in 2012 for asymmetric hydrogenation of aromatic ketones, giving up to 99.5% ee and 1000 of TONs for the hydrogenation of acyclic/cyclic ketones including 4-chromanone [54].

4.2.1.3 Chiral Ruthenium–Phosphine–Oxazoline Catalysts

Asymmetric hydrogenation of aromatic ketones catalyzed by ruthenium complexes of chiral phosphine–oxazoline ligands was initially described by Naud et al. in 2006 [55]. They reported that chiral ruthenium catalyst (S_a,*S*)-**33** (Figure 4.7) *in situ* generated from $RuCl_2(PPh_3)_3$ and chiral phosphine–oxazoline ligand bearing a ferrocene backbone is highly efficient for the hydrogenation of aromatic ketones. This ruthenium catalyst can tolerate high substrate concentration and provides a pilot process (140 kg, S/C = 20 000, 20 atm H_2, and 95–96% ee) for the synthesis of optically active 3,5-bistrifluoromethyl phenyl ethanol, a chiral building block for an NK-1 receptor antagonist [56]. In 2011, Weissensteiner and coworkers reported that the ruthenium catalyst (R_a,*R*,*R*)-**34** (Figure 4.7) of a phosphine–oxazoline ligand containing ferrocenylethyl backbone also gave excellent enantioselectivities (97–99% ee) for asymmetric hydrogenation of aromatic ketones, albeit with relative higher catalyst loading (4 mol%) [57]. In 2012, Zhang and coworkers reported that chiral bimetallic ruthenium catalyst (S_a,*S*)-**35** (Figure 4.7) bearing a chiral

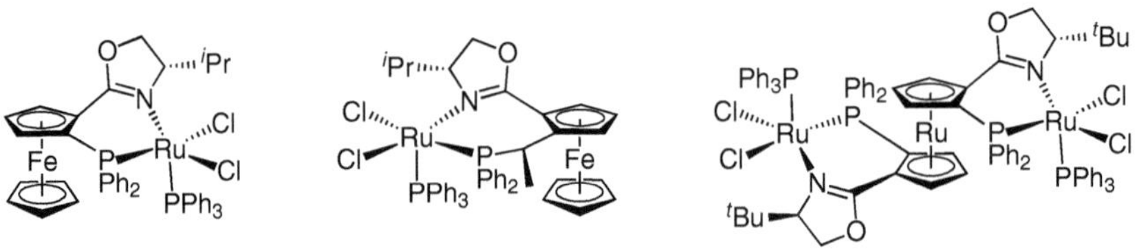

(S_a,*S*)-**33** (R_a,*R*,*R*)-**34** (S_a,*S*)-**35**

Figure 4.7 Chiral ruthenium catalysts with phosphine–oxazoline ligands.

phosphine–oxazoline ligand (RuPHOX) with a ruthenocenyl backbone shows excellent enantioselectivity (92–99.7% ee) for the asymmetric hydrogenation of acyclic aromatic and heteroaromatic ketones and cyclic aromatic ketones [58]. Subsequently, they found that the catalyst (S_a,*S*)-**35** is very stable to moisture and air and is highly efficient for the asymmetric hydrogenation of β-amino aromatic ketones (quantitative yield, up to 99.9% ee, and up to 2000 of TONs), providing an effective method for the syntheses of the key chiral intermediates of chiral drugs such fluoxetine, tomoxetine, and nisoxetine [59].

4.2.1.4 Chiral Ruthenium Catalysts Containing Tridentate Pincer Ligands

The chiral ruthenium catalysts bearing tridendate pincer ligands also showed high enantioselectivity for the asymmetric hydrogenation of aryl ketones. In 2007, Clark et al. reported that the chiral ruthenium catalyst (*R*,*R*)-**36a** (Figure 4.8) with a chiral tridentate PNN ligand derived from *trans*-1,2-diaminocyclohexane gives moderate to high enantioselectivity (up to 94% ee) for the asymmetric hydrogenation of *tert*-butyl aryl ketones with TONs up to 200 [60, 61]. Subsequently, they changed the terminal amino group of the PNN ligand to hydroxy group and found that the corresponding chiral catalyst (*R*,*R*)-**36b** has higher enantioselectivity (up to 97% ee) for hydrogenation of *tert*-alkyl aryl ketones [62]. The chiral ruthenium catalyst (S_p,*S*,*R*)-**37** (Figure 4.8) containing a CNN-type ligand 1-(pyridine-2-yl)methylamine-type (MePyme) and a Josiphos diphosphine ligand, developed by Baratta et al. in 2007 [63], showed 87–99% ee with TONs of 5000 for the hydrogenation of aryl ketones [64]. The similar catalyst (S_a,*S*)-**38** (Figure 4.8) containing an achiral 2-aminomethylbenzo[*h*]quinolone gave higher enantioselectivity (90–99% ee) and higher activity (TONs up to 10 000) [65]. These results indicated that the metal–carbon bond affords the chiral ruthenium catalysts high thermal stability and prevents deactivation of catalyst. In 2010, Zhang and coworkers found that chiral ruthenium catalyst (*S*,*R*)-**39** (Figure 4.8) with a chiral tridentate NNN-type Inda-ambox ligand gives good to high enantioselectivity (80–97% ee) with TONs up to 100 for the asymmetric hydrogenation of aryl ketones [66]. The NH moiety of the Inda-ambox ligand in the catalyst (*S*,*R*)-**39** is crucial for obtaining high activity and enantioselectivity.

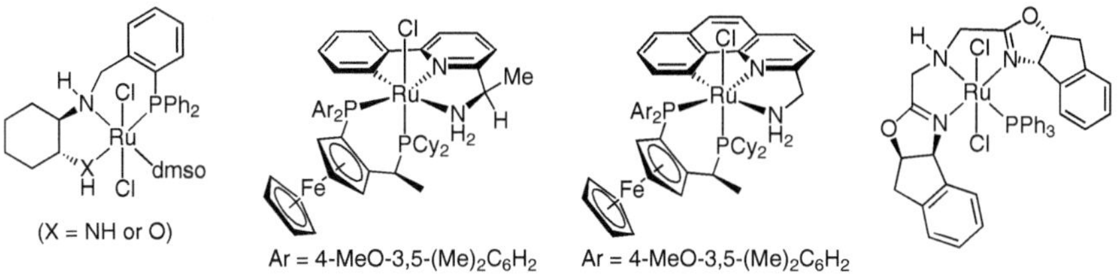

(*R*,*R*)-**36a** X = NH
(*R*,*R*)-**36b** X = O

(S_p,*S*,*R*)-**37**

(S_p,*S*)-**38**

(*S*,*R*)-**39**

Figure 4.8 Chiral ruthenium complexes of tridentate pincer ligands.

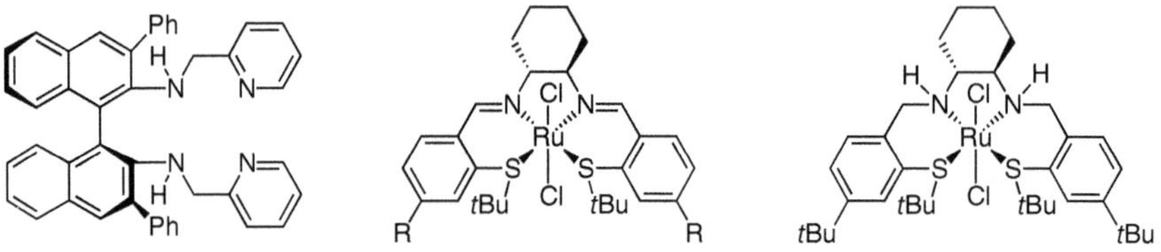

(*S*)-**40** (*S*)-BINAN-Py (*R*,*R*)-**41a** R = H (*R*,*R*)-**41b** R = *t*Bu (*R*,*R*)-**42**

Figure 4.9 Chiral ruthenium complexes of tetradentate ligands.

4.2.1.5 Chiral Ruthenium Catalysts Containing Tetradentate Ligands

The chiral ruthenium complexes of tetradentate ligands with or without NH group have been studied as catalysts for the asymmetric hydrogenation of aryl ketones. In 2006, Kitamura and coworkers reported that the chiral ruthenium catalyst having a chiral tetradentate NNNN-type ligand (*S*)-**40** (Figure 4.9) with a binaphthyl skeleton gives up to 99% ee for the hydrogenation of aromatic ketones (TONs of 100) [67]. The catalyst Ru-(*S*)-**40** can be used under base-free conditions. The ruthenium catalyst (*R*,*R*)-**41b** (Figure 4.9) bearing a chiral diimine-type SNNS ligand was developed by Mezzetti and Santoro et al. in 2013. It showed good to high enantioselectivity (83–95% ee) for the hydrogenation of aromatic ketones (TONs up to 2000) [68]. The bulky *tert*-butyl groups on the sulfur atoms of the ligand in the catalyst (*R*,*R*)-**41** are necessary for achieving high enantioselectivity. The catalyst (*R*,*R*)-**41a** which has no *tert*-butyl groups on the phenyl rings of the ligand gives extremely high activity with TONs up to 900 000 for the hydrogenation of acetophenone, albeit the enantioselectivity is low (61–68% ee). Furthermore, the ruthenium catalyst (*R*,*R*)-**42** (Figure 4.9) derived by saturating the imine moieties of the catalyst (*R*,*R*)-**41b** to two NH groups also provides high activity for the asymmetric hydrogenation of acetophenone (TONs up to 100 000) with a moderate enantioselectivity (61% ee).

4.2.2 Chiral Iridium Catalysts

Chiral iridium catalysts have been widely used for the asymmetric hydrogenations of olefins and imines. Recent progress in asymmetric hydrogenations indicated that chiral iridium catalysts are also highly efficient for the asymmetric hydrogenation of ketones. Therefore, a great attention has been paid to develop efficient chiral iridium catalysts for asymmetric hydrogenation of ketones in the past decade.

In 2007, Manoury and Peruzzini et al. described that the chiral iridium catalysts (*S*)-**43** (Figure 4.10) with planar-chiral phosphine-thioether ligands are efficient for the hydrogenation of aryl ketones with moderate to high enantioselectivity [69]. However, the substrate scope is very narrow; only two acetophenone-type ketones (4-Me, 4-F) have enantioselectivity up to 90% ee. In 2011, Xie and Zhou et al. reported that the chiral spiro iridium catalyst Ir-(*S*)-SpiroAP ((*S*)-**44**, Figure 4.10) with a chiral spiro aminophosphine SpiroAP ligand gives high to excellent enantioselectivity (89–97% ee) with TONs up to 10 000 for the hydrogenation of aryl and heteroaryl ketones including α-tetralone [70]. The catalyst was found to lose

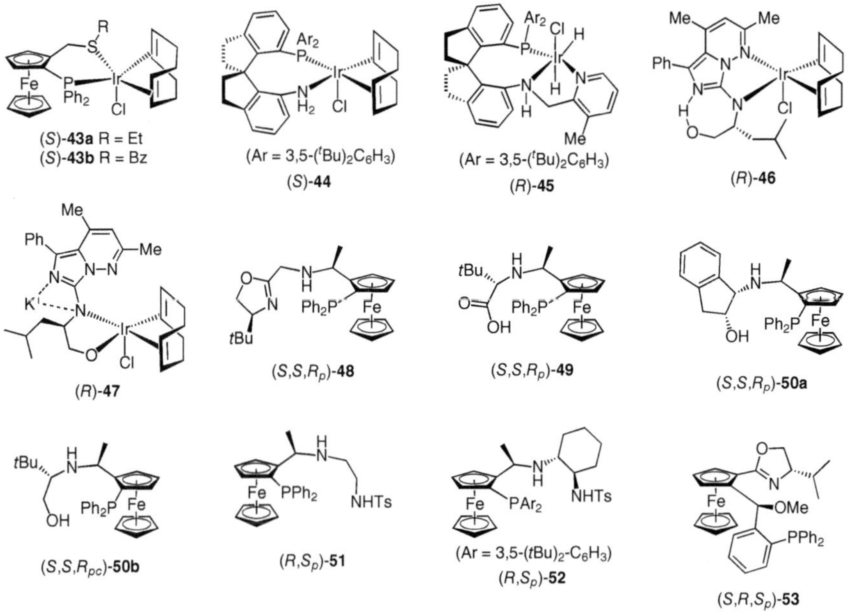

Figure 4.10 Chiral iridium complexes and related chiral ligands.

the activity by forming an iridium complex having two SpiroAP ligands in the hydrogenation reaction. To overcome the deactivation of the iridium catalyst, they introduced an additional coordinating pyridyl moiety to the SpiroAP ligand, making it a tridentate pyridine-aminophosphine ligand SpiroPAP. This tridentate ligand stabilizes the catalyst (*R*)-**45** (Figure 4.10) by preventing the coordination of the second ligand [17]. The catalyst (*R*)-**45** shows an extremely high efficiency for the asymmetric hydrogenation of aromatic ketones, providing chiral alcohols in excellent enantioselectivity (96–99.9% ee) with TONs up to 4 550 000. This iridium catalyst is successfully applied in the enantioselective syntheses of chiral drugs such as (*S*)-rivastigmine [71], (*S*)-crizotinib [72], and montelukast sodium [73]. Kempe and coworkers found in 2011 that chiral iridium complex (*R*)-**46** (Figure 4.10) with an imidazo[1,5-*b*]pyridazine-type ligand also gives excellent enantioselectivity (up to 99% ee) for the asymmetric hydrogenation of aromatic ketones [74]. This catalyst also shows high efficiency, and TONs reaches 174 000. The catalyst (*R*)-**46** converted to complex (*R*)-**47** (Figure 4.10) under the reaction condition in which iridium is coordinated with a nitrogen atom and an oxygen atom. The complex (*R*)-**47** is believed to be the active catalyst in the hydrogenation reaction.

Due to the excellent performances of the aforementioned chiral iridium catalysts, especially chiral spiro iridium catalyst Ir-SpiroPAP ((*R*)-**45**), the search for new efficient chiral iridium catalysts for asymmetric hydrogenation of ketones has attracted intensive attention over the past several years. In 2016, Zhang and coworkers reported that the chiral iridium catalyst containing a tridentate ferrocene-based aminophosphoxazoline ligand (S,S,R_p)-**48** ((S,S,R_p)-f-amphox, Figure 4.10) affords excellent enantioselectivity (94 to >99.9% ee) for the asymmetric hydrogenation of aryl and heteroaryl ketones [75]. The TONs of the catalyst Ir-(*S*,*S*,R_p)-**48** are up to 1 000 000. Subsequently, they developed chiral tridentate aminophosphine ligands based on Ugi's amine and found that the iridium complexes of these ligands give comparable activity and enantioselectivity for asymmetric hydrogenation of ketones. For example, the chiral iridium catalyst (*S*,*S*,R_p)-**49** containing a tridentate aminophosphine acid ligand ((*S*,*S*,R_p)-f-ampha, Figure 4.10) shows 96–99.7% ee with TONs up to 500 000 for the asymmetric hydrogenation of aryl ketones [76]. The iridium catalysts (*S*,*S*,R_p)-**50a** and **50b** (Figure 4.10) having tridentate aminophosphine alcohol ligands give 90–99.7% ee and up to 200 000 of TONs for the asymmetric hydrogenation of aryl ketones [77]. Control experiments and density functional theory (DFT) calculations show that the carboxylic group in (*S*,*S*,R_p)-**49** and the hydroxyl group in (*S*,*S*,R_p)-**50a** play a key role for achieving high activity and enantioselectivity. The iridium catalyst (*R*,S_p)-**51** (Figure 4.10) with a ferrocene-based aminophosphine–sulfonamide ligand (f-Amphamide, Figure 4.10) developed by Zhang and coworkers [78] in 2018 shows 92–99% ee for the asymmetric hydrogenation of aryl and heteroaryl ketones with TONs up to 200 000. A series of tridentate ferrocene-based aminophosphine–sulfonamide ligands (*R*,S_p)-**52** (f-diaphos, Figure 4.10) were synthesized by Zhong [79]. The iridium complex of (*R*,S_p)-**52** gives 99% ee for the hydrogenation of aryl and heteroaryl ketones and 82–98% ee for the hydrogenation of diaryl ketones. Xie and Zhang et al. found that the iridium catalyst with ferrocene-based phosphine–oxazoline ligand (*S*,*R*,S_p)-**53** (Figure 4.10) shows good to excellent enantioselectivity (80–99% ee) for the hydrogenation of aryl ketones and diaryl ketones with TON up to 20 000 [80]. These results indicate that the iridium complex of phosphine–oxazoline ligand can catalyze the asymmetric hydrogenation of ketones even without NH group.

4.2.3 Other Chiral Metal Catalysts

Except for chiral ruthenium and iridium catalysts, other chiral metal catalysts such as Os, Fe, and Mn catalysts also have been developed for the asymmetric hydrogenation of aromatic ketones. In 2008, Barratta et al. developed chiral osmium catalyst (S_p,*S*,*R*)-**54** (Figure 4.11) containing a Josiphos diphosphine ligand by simply changing the central metal of chiral ruthenium complex (S_p,*S*,*R*)-**37** to osmium [81]. The catalyst (S_p,*S*,*R*)-**54** gives up to 98% ee with TON up to 50 000 for the asymmetric hydrogenation of aryl ketones. Subsequently, they found that the chiral osmium catalyst (S_p,*S*,*R*)-**55** (Figure 4.11), which is also obtained by the replacement of the ruthenium atom of the corresponding complex analogue to (S_p,*S*)-**38** with osmium atom, shows comparable activity (TON up to 10 000) and enantioselectivity (up to 99% ee) with the catalyst (S_p,*S*,*R*)-**54** [65]. Chiral osmium catalyst (R_a,*R*,*R*)-**56** (Figure 4.11) bearing a diphosphine ligand (R_a)-xyl-BINAP and a diamine ligand (*R*,*R*)-DPEN also shows good enantioselectivity and slightly lower activity for the hydrogenation of aryl ketones [82]. In addition, the chiral osmium catalyst (R_a,*R*,*R*)-**57** (Figure 4.11) containing a diphosphine ligand (R_a)-Xyl-MeOBiphep and a diamine ligand (*R*,*R*)-DPEN gives higher activity (TONs

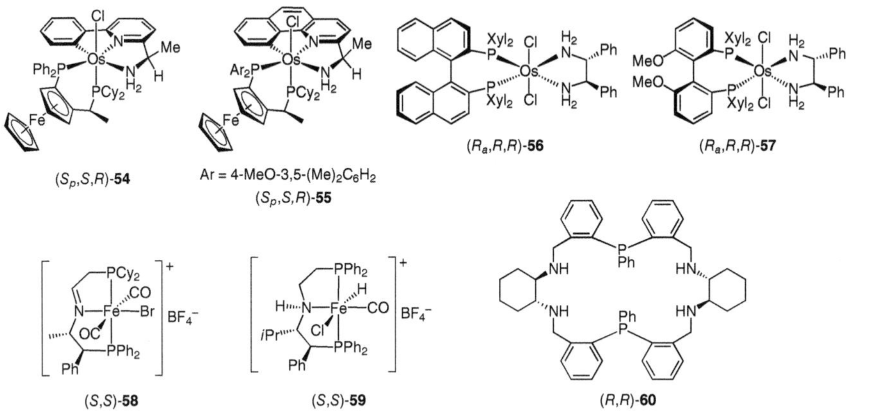

Figure 4.11 Chiral osmium and iron catalysts and related chiral ligands.

up to 200 000) but lower enantioselectivity (90% ee) in the hydrogenation of aryl ketones [82].

The first example of iron-catalyzed asymmetric hydrogenation of aryl ketones was reported by Morris and coworkers in 2008 [83]. Since then, many chiral iron catalysts have been developed, but only a few of them gave enantioselectivity higher than 90% ee in the asymmetric hydrogenation of aryl ketones [84]. In 2014, Morris and coworkers found that the chiral iron dicarbonyl complex (*S*,*S*)-**58** (Figure 4.11) with an imine-type PNP ligand shows up to 85% ee with TONs up to 5000 for the asymmetric hydrogenation of aryl and heteroaryl ketones after *in situ* activation with $LiAlH_4$ and 2-methyl-2-butanol [85]. Later on, they developed a series of chiral iron catalysts containing chiral pincer ligands. The iron catalyst (*S*,*S*)-**59** (Figure 4.11) having a chiral amino-type PNP ligand showed 62–96% ee with TONs up to 1000 [86]. The chiral iron complex of $Fe_3(CO)_{12}$ and a 22-membered macrocyclic phosphine ligand (*R*,*R*)-**60** (Figure 4.11), which was developed by Xiao and Gao et al. in 2014, gave moderate to excellent enantioselectivities (60–99% ee) in the asymmetric hydrogenation of aryl ketones. However, this iron-catalyzed hydrogenation of ketones appears to be a heterogeneous reaction [87].

In recent years, chiral manganese complexes of pincer ligands have been applied in the asymmetric hydrogenation of aryl ketones. In 2017, Clarke and coworkers reported that the chiral manganese catalyst (*S*,R_p)-**61** (Figure 4.12) bearing a tridentate ferrocene-based PNN-type pincer ligand gives 61–91% ee for the asymmetric hydrogenation of aryl ketones (TON = 100) [88]. Also, in 2017, Beller and coworkers reported that the chiral manganese catalyst (*R*,*R*)-**62** (Figure 4.12) with a PNP-type pincer ligand affords moderate to good enantioselectivity in the hydrogenation of cyclic aryl ketones, such as α-tetralone (80% ee) and α-indanone (84% ee), and heteroaryl ketones (70–74% ee) [89]. However, very low enantioselectivities are observed for the hydrogenation of simple aryl ketones (18–20% ee). The chiral manganese complex (*R*,*R*)-**63** (Figure 4.12) of pyridine-based chiral PNN-type pincer ligand developed by Han and Ding et al. in 2019 shows high activity and enantioselectivity for the asymmetric hydrogenation of aryl ketones

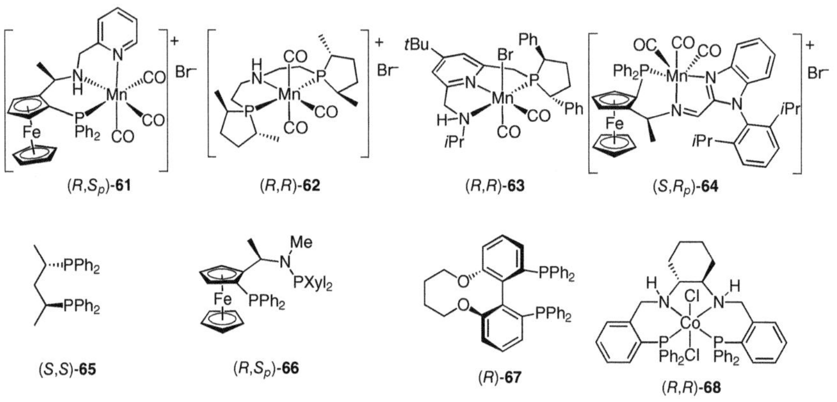

(*R*,S_p)-**61** (*R*,*R*)-**62** (*R*,*R*)-**63** (*S*,R_p)-**64**

(*S*,*S*)-**65** (*R*,S_p)-**66** (*R*)-**67** (*R*,*R*)-**68**

Figure 4.12 Other chiral metal catalysts and related chiral ligands.

[18]. With catalyst (*R*,*R*)-**63**, a wide range of aryl, heteroaryl ketones, benzo-fused cyclic ketones, and *ortho*-substituted benzophenones are hydrogenated to the corresponding chiral alcohols in 85–98% ee with TONs up to 9800. Zhong and coworkers found that the chiral manganese catalyst (*S*,R_p)-**64** (Figure 4.12) bearing a tridentate ferrocene-based imidazole–iminophosphine PNN-type ligand gives up to 99% ee for the asymmetric hydrogenation of *ortho*-substituted benzophenones [90]. The activity of catalyst (*S*,R_p)-**64** is very high, and the TON reaches 13 000 for the hydrogenation of (2-bromo-3-methylphenyl)(4-methoxyphenyl)methanone. This protocol has been applied in the gram-scale synthesis of the key intermediate of chiral medicine (*S*)-neobenodine. The chiral manganese catalyst (*S*,R_p)-**64** was also evaluated in the asymmetric hydrogenation of diaryl ketones without *ortho*-substitutions, and very low enantioselectivity was observed (<20% ee).

Copper-catalyzed asymmetric hydrogenation of aryl ketones was firstly reported by Shimizu et al. in 2007 [91]. They found that the chiral copper catalyst generated *in situ* from [Cu(NO_3)(P(3,5-xylyl)$_3$] and (*S*,*S*)-BDPP ((*S*,*S*)-**65**, Figure 4.12) gives good enantioselectivity (81–91% ee) for the hydrogenation of aryl ketones with *ortho*-substitutions (Me, MeO, CF_3) and moderate enantioselectivity (56–64% ee) for the hydrogenation of aryl ketones without *ortho*-substitutions (TON = 500). Subsequently, Hatcher and Johnson et al. reported in 2013 that the chiral copper catalyst containing ferrocene-based diphosphine ligand *N*-Me-3,5-Xylyl-BoPhoz ((*R*,S_p)-**66**) (Figure 4.12) is highly enantioselective. Employing the copper catalyst *in situ* generated from Cu(OAc)$_2$, P(3,5-xylyl)$_3$, and (*R*,S_p)-**66**, a range of aryl and heteroaryl ketones were hydrogenated to chiral alcohols with 66–96% ee (TON = 66) [92]. In addition, chiral palladium and cobalt complexes are also applied in the asymmetric hydrogenation of aryl ketones. In 2011, Bao and Zhou et al. reported that in the presence of Brønsted acids such as salicylic acid in 2,2,2-trifluoroethanol (TFE), the chiral palladium complex of chiral diphosphine ligand (*R*)-**67** ((*R*)-C_4-TunePhos) (Figure 4.12) affords up to 88% ee for the asymmetric hydrogenation of aryl ketones (TON = 50) [93]. In 2016, Li and coworkers described that the chiral cobalt catalyst (*R*,*R*)-**68** (Figure 4.12) containing a semioxidated tetradentate PNNP-type ligand provides moderate to high enantioselectivity in the hydrogenation of aryl ketones (TON = 50) [94].

4.3 Asymmetric Transfer Hydrogenation of Aryl and Heteroaryl Ketones

Transition-metal-catalyzed asymmetric transfer hydrogenation of ketones was first reported in the early 1980s [95–97]. Since then, it has received increasing attention due to its operational simplicity, safety, and no need of special instruments for high pressure. However, although many chiral metal catalysts have been developed in the first decade, the enantioselectivities of the asymmetric transfer hydrogenation of ketones were poor until 1990s. In 1993, Evans et al. found that the chiral samarium complex (*R*,*R*)-**69** (Figure 4.13) bearing an amino-alcohol ligand affords 92–97% ee for asymmetric hydrogenation of aryl ketones with 2-propanol as hydrogen donor [98]. However, the high catalyst loading (5 mol%) makes this catalyst unpractical. In 1995, Noyori and coworkers reported that the chiral ruthenium complex (*S*,*S*)-**29a** with a diamine ligand (*S*,*S*)-Ts-DPEN ((1*S*,2*S*)-*N*-(*p*-toluenesulfonyl)-1,2-diphenylethylenediamine) gives 72–98% ee for the asymmetric transfer hydrogenation of aryl ketones with 2-propanol at a low catalyst loading (0.5–0.2 mol%) [99]. Later, they found that this type of chiral ruthenium catalyst shows higher efficiency if a mixture of formic acid and triethylamine is used as hydrogen donor. The loading of catalyst (*S*,*S*)-**29a** can be reduced to 0.1 mol% without significant loss of activity and enantioselectivity [100]. These landmark works have further stimulated investigations on the development of chiral ligands and catalysts for the asymmetric transfer hydrogenation of aryl ketones. As a result, a wide range of chiral ligands and catalysts have been developed. Typical examples include chiral ruthenium complexes (*S*,*S*)-**70** (Figure 4.13) bearing a tetradentate diphosphine–diamine ligand developed by Noyori and coworkers in 1996 [101]

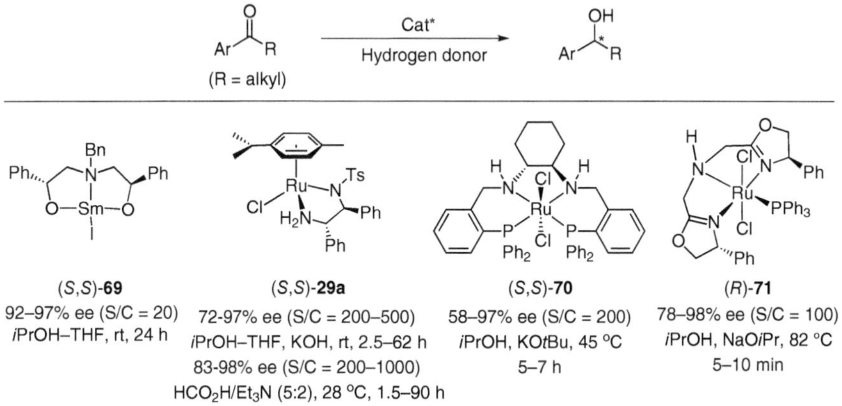

Figure 4.13 The early examples of chiral catalysts for asymmetric transfer hydrogenation of aryl ketones.

and (*R*)-**71** (Figure 4.13) containing a tridentate bis(oxazolinylmethyl)amine ligand developed by Zhang and coworkers in 1998 [102]. These chiral ruthenium catalysts also showed good to high enantioselectivities for the asymmetric transfer hydrogenation of aryl ketones with 2-propanol as hydrogen donor.

After more than 30 years of development, the asymmetric transfer hydrogenation of aryl ketones has become a useful method for the synthesis of chiral alcohols. A large number of chiral transition-metal catalysts have been synthesized and used in the asymmetric transfer hydrogenation of aryl ketones. In the following discussions, we focus on the development of chiral catalysts in the asymmetric transfer hydrogenation of aryl and heteroaryl ketones, especially the important progress in the recent decade.

4.3.1 Chiral Ruthenium Catalysts

Although many chiral transition-metal catalysts have been developed for the asymmetric transfer hydrogenation of aryl ketones, the efficient catalysts are dominated by chiral ruthenium catalysts, in particular, chiral arene–ruthenium complexes bearing chiral N-sulfonylated 1,2-diamines [3, 103]. In addition, chiral-ruthenium complexes bearing chiral amino-alcohol ligands, diphosphine ligands, tridentate NNN- and CNN-type of pincer ligands, etc. are also widely used in the asymmetric transfer hydrogenation of aryl ketones.

4.3.1.1 Chiral Arene Ruthenium–N-Sulfonylated 1,2-Diamine Complexes

Since initiated by Noyori and coworkers in 1995 [99], chiral arene ruthenium–N-sulfonylated 1,2-diamine complexes have received considerable attention due to their excellent catalytic activity and enantioselectivity for the asymmetric transfer hydrogenation of ketones. A wide range of arene ruthenium complexes with modified N-sulfonylated 1,2-diamine ligands have been developed over the past decades.

In 2004, Wills and coworkers reported the tethered chiral Ru–*N*-Ts-DPEN complex (*S*,*S*)-**72** (Figure 4.14), which is obtained by connecting the *N*-Ts-DPEN ligand and the aromatic ring of Noyori's Ru–Ts-DPEN catalyst through a sulfonyl linker [104]. The tethered chiral ruthenium complex (*S*,*S*)-**72** gives 90–96% ee and 84–98% ee for the asymmetric transfer hydrogenations of aryl ketones and benzo-fused cyclic ketones, respectively (TONs up to 1000). Later, they designed a "reverse-tethered" Ru–Ts-DPEN complex (*S*,*S*)-**73a** (Figure 4.14), which has a three-carbon-atom linker and found it to be more effective than the catalyst (*S*,*S*)-**72** [105]. The reaction rate of (*S*,*S*)-**73a** exceeds those reported for the untethered chiral Ru–Ts-DPEN complexes, and the catalyst loading of (*S*,*S*)-**73a** can be lowered to 0.01 mol%. By systematically tailoring the length of the tether, Wills and coworkers prepared the catalyst (*S*,*S*)-**73b** with a "four-carbon" tether, which has a fast rate in the asymmetric transfer hydrogenations of aryl ketones and ketopyridines (TONs up to 10 000) [106]. In 2013, the tethered chiral Ru–Ts-DPEN complexes with electron-rich arene rings were developed by an "arene-exchange" approach. The catalyst (*S*,*S*)-**73c** with a *para*-methoxy group at the arene ring was found to be efficient for the transfer hydrogenations

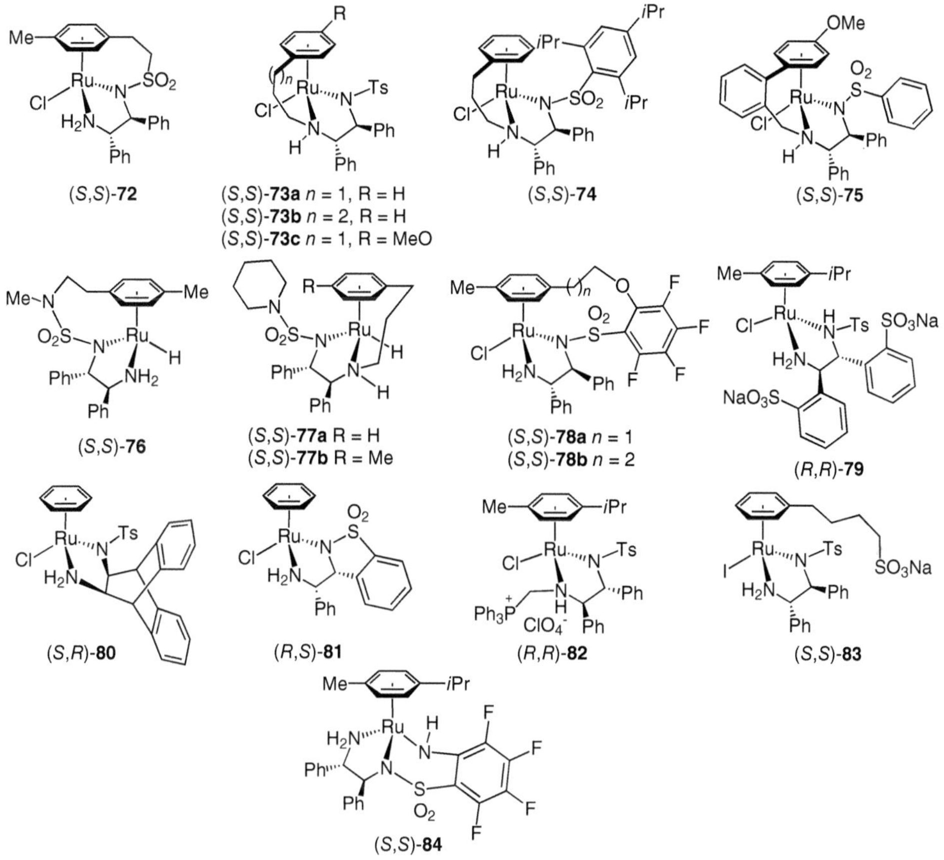

Figure 4.14 Chiral arene ruthenium complexes of *N*-Ts-diamine ligands.

of aryl, heteroaryl, and benzo-fused cyclic ketones. When the hydrogenation was performed at a higher temperature, a faster reaction rate was observed [107]. The tethered ruthenium catalysts (*S*,*S*)-**73a** and **73c** are also effective for the asymmetric transfer hydrogenation of electron-rich aryl ketones containing amine and methoxy groups on the phenyl ring (up to 98% ee) [108]. Changing the steric hindrance and electronic property of the arene sulfonyl group of the tethered ruthenium Ts-DPEN complexes has a significant effect on the reactivity and enantioselectivity of the catalysts [109]. For example, the catalyst (*S*,*S*)-**74** (Figure 4.14) having a 2,4,6-(iPr)$_3C_6H_2SO_2$ requires a longer reaction time and/or an increase of temperature to complete the reaction. The benzyl-tethered arene-Ru–Ts-DPEN complexes such as (*S*,*S*)-**75** (Figure 4.14) were less effective than the parent ruthenium complex (*S*,*S*)-**73c** although comparable enantioselectivities were observed [110]. In addition, Wakeham et al. found that the tethered ruthenium catalyst (*S*,*S*)-**73a** is also effective for the asymmetric transfer hydrogenation of aryl ketones with *cis*-1,4-butenediol as hydrogen donor, providing chiral alcohols with 75 to >99% ee [111].

In 2011, Ikariya and coworkers reported that the *oxo*-tethered Ru–Ts-DPEN complex (*R*,*R*)-**31** exhibits excellent performance in terms of activity and selectivity for both asymmetric hydrogenation (see Section 4.2.1.2) and transfer hydrogenation of simple ketones under neutral conditions without any cocatalyst [52]. With (*R*,*R*)-**31** as catalyst and HCO_2H/Et_3N (5 : 2) as hydrogen donor, a wide range of aryl, heteroaryl, and benzo-fused cyclic ketones are hydrogenated to the corresponding chiral alcohols with 94 to >99% yield and 84 to >99% ee. The catalytic activity of (*R*,*R*)-**31** can be maintained after 4 days of reaction, and the TON is up to 30 000, being the highest activity among the series of arene-Ru–Ts-DPEN catalysts. The catalyst (*R*,*R*)-**31** is also effective for the asymmetric transfer hydrogenation of unsymmetrical benzophenones, producing chiral benzhydrols with good to excellent enantioselectivity (77 to >99% ee) [112]. Following this work, Zhou and coworkers found that chiral ruthenium (*R*,*R*)-**31** can catalyze the asymmetric transfer hydrogenation of *ortho*-substituted aryl *N*-heteroaryl ketones and non-*ortho*-substituted *N*-oxide of aryl *N*-heteroaryl ketones with HCO_2Na/H_2O as hydrogen donor in 70–99.9% ee [113].

The sulfamoyl-tethered chiral ruthenium–N-sulfamoylated 1,2-diamine complex (*S*,*S*)-**76** (Figure 4.14) is introduced by Mohar and coworkers in 2013, which gives up to >99.9% ee for the transfer hydrogenation of 1-naphthyl ketones (TON = 1000) [114]. Later, they reversed the tether and developed chiral ruthenium–N-sulfamoylated 1,2-diamine complexes (*S*,*S*)-**77a** and **77b** (Figure 4.14) and found them to be effective catalysts for the asymmetric transfer hydrogenation of aryl, heteroaryl, and benzo-fused cyclic ketones (TONs up to 1000, 86 to >99.9% ee) [115]. In 2018, Kayaki and coworkers developed *oxo*-tethered ruthenium complexes (*S*,*S*)-**78a** and **78b** (Figure 4.14) from *N*-pentafluorobenzenesulfonyl-1,2-diamine (Fs-DPEN) and obtained 88–96% ee in asymmetric transfer hydrogenation of aryl ketones (TON up to 1000) [116].

The modification of the chiral diamine ligands of Noyori's Ru–Ts-DPEN catalysts also led to several efficient chiral ruthenium catalysts for the asymmetric transfer hydrogenation of aryl ketones. In 2003, Deng and coworkers synthesized a water-soluble chiral N-tosylated *ortho*-sulfonated 1,2-diphenylethlenediamine ligand and its ruthenium complex (*R*,*R*)-**79** (Figure 4.14), which gave 81–98% ee for the asymmetric transfer hydrogenation of aryl, heteroaryl, and benzo-fused cyclic ketones with HCO_2Na as hydrogen donor in aqueous media (TON = 100) [117]. The chiral ruthenium catalyst (*R*,*R*)-**79** can be reused without loss of enantioselectivity, albeit its activity slightly declines. The "roofed" ruthenium catalyst (*S*,*R*)-**80** (Figure 4.14) with an *N*-Ts-*cis*-1,2-diamine ligand based on a rigid and steric bulky 1,10-dihydro-9,10-ethanoanthracene backbone, developed by Ishizuka and coworkers in 2005, also showed good to high enantioselectivities (66–99% ee) for the asymmetric transfer hydrogenation of aryl and benzo-fused cyclic ketones (TON = 200) [118]. The ruthenium catalyst (*R*,*S*)-**81** (Figure 4.14) containing a *syn*-3-(α-aminobenzyl)-benzo-γ-sultam ligand, developed by Mohar and coworkers in 2016, affords excellent enantioselectivity (99% ee) for the asymmetric transfer hydrogenation of benzo-fused cyclic ketones (TONs up to 1000) [119]. In 2015, Charette and coworkers modified Noyori's Ru–Ts-DPEN catalyst with a tetraarylphosphonium and made it to be a recyclable ruthenium catalyst (*R*,*R*)-**82** (Figure 4.14) [120]. The catalyst (*R*,*R*)-**82** gives 91–98% ee for the hydrogenation of aryl, heteroaryl, and benzo-fused cyclic ketones and can be reused in four cycles without any significant loss of activity. The water-soluble ruthenium catalyst (*S*,*S*)-**83** (Figure 4.14), developed by Gebbink and coworkers in 2012 by introducing an alkyl sulfonate tail to the arene group of Noyori's Ru–Ts-DPEN catalyst, gives 94% ee for the asymmetric transfer hydrogenation of acetophenone [121]. However, although the enantioselectivity remained, the activity of the catalyst decreased significantly in second recycling run of reaction. The tridentate azametallacycles ruthenium catalyst (*S*,*S*)-**84** (Figure 4.14), developed by Dub and Kayaki et al., gives 93–98% ee for the transfer hydrogenation of aryl ketones (TON = 200), which are comparable with those obtained by the catalyst Ru–Fs-DPEN [122].

To expand the scope of asymmetric transfer hydrogenations of aryl ketones, Dong and coworkers used Noyori's (*R*,*R*)-**85** (Figure 4.15) to reduce 1,4- or 1,5-keto alcohols to the corresponding chiral diols [123]. In the reaction, (*R*,*R*)-**85** hydrogenated the ketone group of the substrate, then the primary alcohol is dehydrogenated to aldehyde, and a hemiacetal is subsequently formed and further dehydrogenated by (*R*,*R*)-**85** to afford the desired lactone. Thus, the final result is that the use of catalyst (*R*,*R*)-**85** enables the asymmetric transfer hydrogenation of aryl ketone with itself as hydrogen donors. In 2016, Vyas and Bhanage reported that the ruthenium catalyst (*R*,*R*)-**29a** is effective for the asymmetric transfer hydrogenation of dibenzo-fused-azepine-diones with formic acid/trimethylamine as hydrogen donor, providing the corresponding chiral alcohols with up to 99% ee (TON = 50) [124]. The chiral ruthenium catalyst (*R*,*R*)-**86** (Figure 4.15) has been applied in the enantioselective synthesis of (+)-erythro mefloquine [125]. The asymmetric transfer hydrogenation of pyridyl ketone with (*R*,*R*)-**86** at a catalyst loading of 0.1 mol% affords the corresponding chiral diaryl methanol in 91% yield and 98% ee.

4.3.1.2 Chiral Ruthenium Catalysts with Other Bidentate Ligands

In addition to the chiral arene–ruthenium–*N*-sulfonyl-1,2-diamine catalysts, many other chiral ruthenium catalysts containing bidentate ligands, such as 1,2-amino alcohols, aminothiols, and phosphine–oxazolines, have also been studied in the asymmetric transfer hydrogenation of ketones. In 1996, Noyori and coworkers reported that arene–ruthenium complexes of chiral β-amino alcohols give high enantioselectivity for the reduction of aryl ketones with 2-propanol as hydrogen donor [126]. For example, the ruthenium catalyst generated *in situ* from $[RuCl_2(\eta^6\text{-}C_6Me_6)]_2$ and chiral amino alcohol (*S*,*S*)-**87** (Figure 4.16) provides 94% yield and 92% ee for the asymmetric transfer hydrogenation of acetophenone (TON = 200). Subsequently, Andersson and coworkers found that the chiral ruthenium complex of 2-*aza*-norboryl amino alcohol **88** (Figure 4.16) shows high activity and enantioselectivity for aryl and pyridyl ketone substrates [127]. The hydrogenation of acetophenone with 2-propanol gives 96% ee at a low catalyst loading (0.02 mol%). Good to high enantioselectivities (89–99% ee) are observed for the hydrogenation of pyridyl ketones (TONs up to 1000). In 2002, Adolfsson and coworkers described that the ruthenium catalyst coordinated with *N*-Boc-protected

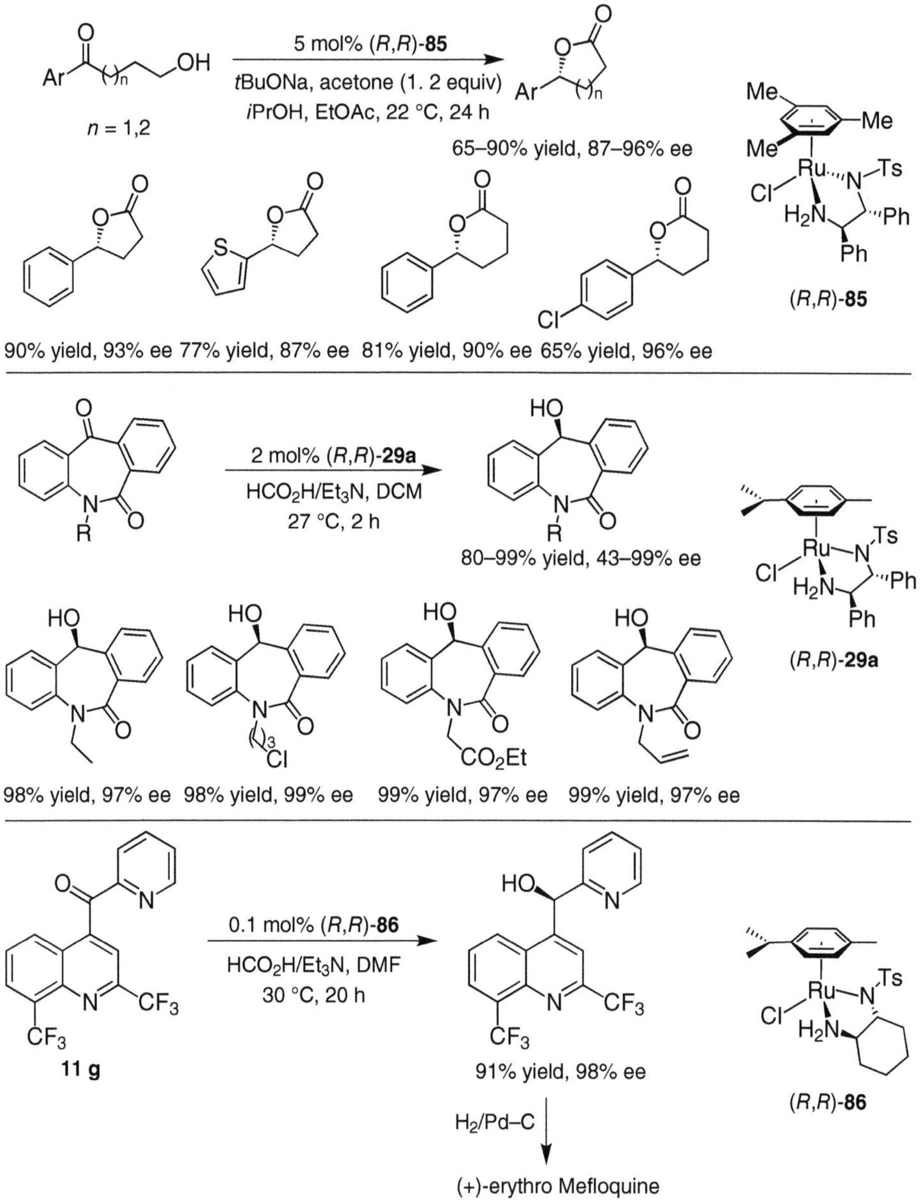

Figure 4.15 Chiral arene ruthenium *N*-Ts-diamine complexes-catalyzed asymmetric transfer hydrogenations.

L-valine-modified 1,2-amino alcohol (*S*,*R*)-**89** (Figure 4.16) gives 84–96% ee for the transfer hydrogenation of a series of aryl ketones (TON = 200) [128, 129]. Several years later, they found that the *N*-Boc-protected L-alanine-modified 1,2-amino alcohol (*S*,*S*)-**90** (Figure 4.16) shows high enantioselectivity (95–99% ee) for the reduction of electron-rich aryl ketones [130]. In 2011, Diéguez, Pàmies, and

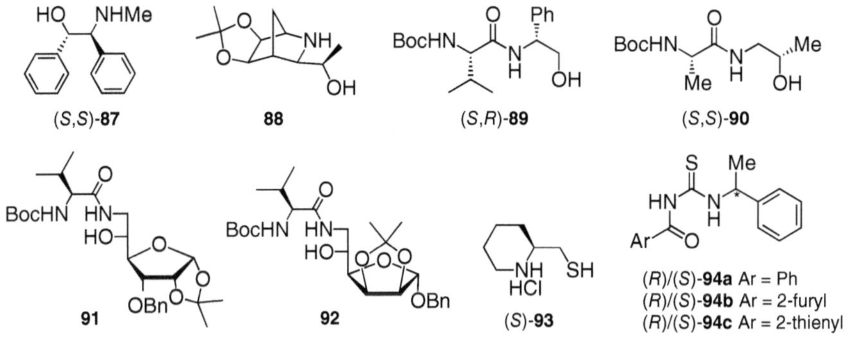

Figure 4.16 Chiral amino-alcohol and related chiral ligands.

Adolfsson and coworkers prepared a series of *N*-Boc-protected L-alanine-modified sugar amino-alcohol ligands and found that the furanoside-based amino alcohol **91** (Figure 4.16) derived from D-glucose shows excellent enantioselectivity (98 to >99% ee) for the asymmetric transfer hydrogenation of aryl ketones (TON = 200) [131]. Later, they found that the *N*-Boc-protected L-alanine-modified sugar amino-alcohol ligand **92** (Figure 4.16) derived from D-mannose gave up to 99% ee for a wide range of aryl ketones, heteroaryl, and benzo-fused cyclic ketones (TON = 400) [132]. In addition, the pipecoline-derived chiral protic aminothiol ligand (*S*)-**93** (Figure 4.16), introduced by Ikariya and coworkers in 2009, shows good to high enantioselectivities (75–98% ee) for a wide range of aryl, heteroaryl, and benzo-fused cyclic ketones [133]. Chiral acylthiourea ligands (*R*)-/(*S*)-**94** (Figure 4.16), developed by Karvembu and coworkers in 2014 from (*S*)-/(*R*)-1-phenylethylamine, also show moderate to high enantioselectivity (43–99% ee) for the asymmetric transfer hydrogenation of a range of aryl ketones (TON = 200) [134, 135].

The chiral ruthenium complexes of diphosphine ligand and diamine ligand, initially developed by Noyori et al. for the asymmetric hydrogenation of ketones, are also proven to be highly efficient catalysts for the asymmetric transfer hydrogenation of aryl ketones. In 2004, Baratta et al. found that 2-(aminomethyl)pyridine (ampy) has a strong ligand acceleration effect in the transfer hydrogenation catalyzed by cyclometalated ruthenium complex of (2,6-dimethylphenyl)diphenylphosphane [136]. Subsequently, they evaluated several chiral diphosphine ligands in combination with 2-(aminomethyl)pyridine and found that the chiral ruthenium catalyst (*S*,*S*)-**95** (Figure 4.17) with Skewphos ligand and the catalyst (*R*,S_p)-**96a** (Figure 4.17) with Josiphos ligand afforded high activity and good to high enantioselectivities (83–94% ee) for the transfer hydrogenation of aryl and heteroaryl ketone at low catalyst loading (TONs up to 10 000) [137]. In order to improve the enantioselectivity of the catalysts, they introduced Pyme ligand and found that the matched chiral ruthenium catalyst (*R*,*R*,S_p)-**96b** has very high enantioselectivity (95–99% ee) for the transfer hydrogenation of aryl ketones (TON = 2000) [63]. Recently, Xu and Xing et al. reported that the chiral ruthenium catalyst (*R*)-**97** (Figure 4.17) bearing an achiral diphosphine ligand and a chiral 1-(pyridine-2-yl)methylamine

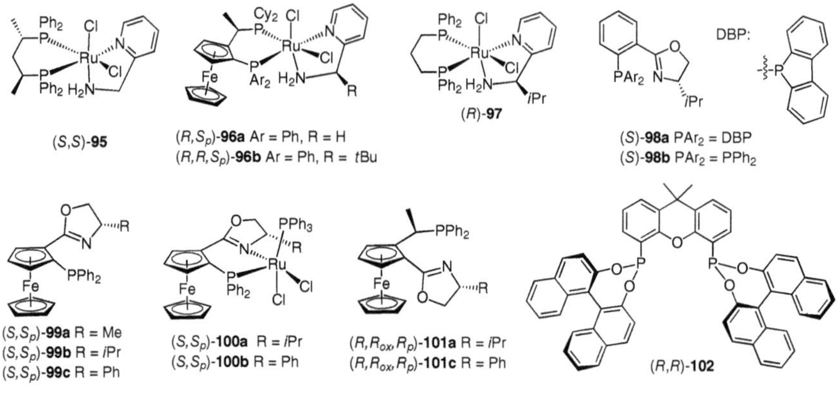

Figure 4.17 Chiral ruthenium complexes of diphosphine/amine and related ligands.

ligand provides 82–97% ee for the asymmetric transfer hydrogenation of aryl and heteroaryl ketones (TON = 1000) [138].

The aforementioned chiral ruthenium catalysts with chiral bidentate ligands contain at least an NH or NH_2 moiety. However, the ruthenium complexes with bidentate chiral ligands such as phosphine–oxazoline ligands and diphosphine ligands, which have no NH or NH_2 moiety, also can be used as catalysts for the asymmetric transfer hydrogenation of aryl ketones. The chiral phosphine–oxazoline ligands are introduced into ruthenium-catalyzed asymmetric transfer hydrogenation of ketones by Langer and Helmchen in 1996 [139]. With the ruthenium catalyst generated *in situ* from $RuCl_2(PPh_3)_3$ and chiral phosphine–oxazoline ligand (*S*)-**98** (Figure 4.17), three aryl ketones were hydrogenated to chiral alcohols with up to 94% ee in 2-propanol with TONs of 1000. Subsequently, Sammakia and Stangeland described that the chiral ruthenium catalyst (S,S_p)-**99c** (Figure 4.17) with a chiral ferrocene-based phosphine–oxazoline ligand affords 84–95% ee in the transfer hydrogenation of aryl ketones (TON = 500) [140]. Almost at the same time, Dai and coworkers reported that the ruthenium complex of ligand (S,S_p)-**99b** and its diastereoisomer show high activity and moderate enantioselectivity (59–85% ee) for the transfer hydrogenation of aryl ketones and α-tetralone (TONs up to 50 000) [141]. In 1999, Uemura and Hidai et al. obtained the molecular structure of ruthenium complex (S,S_p)-**100a** (Figure 4.17) of chiral ferrocene-based phosphine–oxazoline ligand and found that the isolated chiral ruthenium complex affords excellent enantioselectivities (96.6 to >99.7% ee) for the asymmetric transfer hydrogenation of aryl ketones and 2-acetofurone (TON = 200) [142]. In 2012, Weissensteiner and coworkers developed a series of chiral ruthenium complexes (R,R_{ox},R_p)-**101** (Figure 4.17) of ferrocene-based phosphine–oxazoline ligands and found they have comparable activity and lower enantioselectivity (50–98% ee) than those obtained with chiral ruthenium complex of (S,S_p)-**100a** in the asymmetric transfer hydrogenation of aryl ketones [143].

Although the chiral ruthenium complexes derived from chiral diphosphine ligands combined with 2-(aminomethyl)pyridine ligands are demonstrated to be highly effective catalysts for asymmetric transfer hydrogenation of aryl ketones, the chiral ruthenium complexes of chiral diphosphonite ligands are rarely used in this reaction. In 2006, Reetz and Li reported that the BINOL-derived diphosphonite (*R*,*R*)-**102** (Figure 4.17) is a suitable ligand for the ruthenium-catalyzed asymmetric transfer hydrogenation of aryl ketones [144]. With the catalyst Ru-(*R*,*R*)-**102**, a series of aryl ketones are hydrogenated to chiral alcohols with 93–99% ee (TON = 200).

4.3.1.3 Chiral Ruthenium Catalysts Containing Tridentate and Tetradentate Ligands

Chiral ruthenium complexes of tridentate and tetradentate ligands are also demonstrated to be effective catalysts for the asymmetric transfer hydrogenation of aryl ketones. After the development of chiral ruthenium catalyst (*R*)-**71** bearing tridentate bis(oxazolinylmethyl)amine ligand for the asymmetric transfer hydrogenation of aryl ketones [3, 103], Zhang et al. reported chiral ruthenium catalysts having tridentate PNP- and NPN-type pincer ligands. But these catalysts only show low to moderate level of enantioselectivity in the asymmetric transfer hydrogenation of aryl ketones [145–147]. Following these works, Gimeno and coworkers described in 2004 that chiral ruthenium complexes (*S*,*S*)-**103** (Figure 4.18) of Ph-pybox ligand are active catalysts for the asymmetric transfer hydrogenation of acetophenone with high enantioselectivity (up to 95% ee) [148]. With these ruthenium catalysts a variety of aryl ketones are reduced to chiral alcohols with good to high enantioselectivities (70–94% ee). In 2007, Beller and coworkers reported that the ruthenium complex (*R*,*R*)-**104** (Figure 4.18) of pyridine-bisimidazoline ligand shows moderate to excellent enantioselectivities for the asymmetric transfer hydrogenation of aryl ketones [149]. The ruthenium catalyst (*R*,*R*)-**104**, generated *in situ* from $[RuCl_2(C_6H_6)]_2$, gives higher enantioselectivity comparing to the catalyst generated from $[RuCl_2(PPh_3)_3]$. In 2011, Yu and coworkers developed chiral ruthenium complexes (*R*)-**105** (Figure 4.18) of benzimidazolyl-oxazolyl ligands [150]. These chiral ruthenium complexes bearing unsymmetrical tridentate NNN-type pincer ligands exhibit high catalytic activity in asymmetric transfer hydrogenation of aryl ketones. Using chiral ruthenium complex (*R*)-**105b** as catalyst, a wide range of aryl ketones

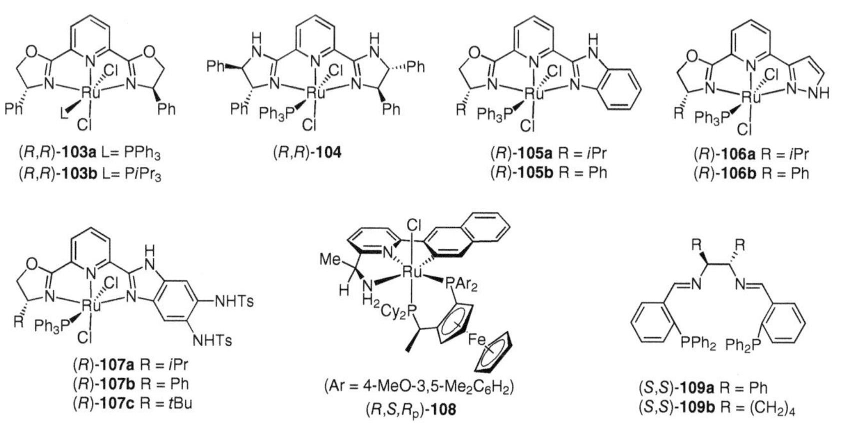

Figure 4.18 Chiral ruthenium complexes of tridentate ligands and related chiral ligands.

are reduced to chiral alcohols in 79–97% ee with TONs up to 1000. Subsequently, they used a pyrazolyl to replace the benzimidazolyl moiety in ruthenium complexes (*R*)-**105** and developed a new type of chiral ruthenium catalysts (*R*)-**106** (Figure 4.18) [151]. The catalyst (*R*)-**106a** gives comparable enantioselectivity (73–99% ee) to the catalyst (*R*)-**105b** under similar reaction conditions but has relatively lower catalytic activity. The catalysts (*R*)-**105** were then modified by introducing NHTs functional group on the benzimidazolyl ring, resulting in the chiral ruthenium catalysts (*R*)-**107** (Figure 4.18). The catalyst (*R*)-**107a** exhibits excellent activity and enantioselectivity for the asymmetric transfer hydrogenation of aryl ketones (TON = 1000, 85 to >99.9% ee) [152].

The Josiphos-based ruthenium complex (*S*,*R*,R_p)-**37** bearing a chiral CNN-type 1-(pyridine-2-yl)methylamine (Pyme) ligand, developed by Baratta et al. in 2009, gives high activity and enantioselectivity (93–99% ee) for the asymmetric transfer hydrogenation of aryl and heteroaryl ketones (TONs up to 50 000) [65]. In 2010, they further found that the chiral ruthenium catalyst (*R*,*S*,S_p)-**108** (Figure 4.18) containing an (*R*, S_p)-Josiphos ligand and an (*S*)-2-(1-aminoethyl)-6-(2-naphthyl)pyridine ligand also shows high activity and enantioselectivity for the asymmetric transfer hydrogenation of aryl and heteroaryl ketones (TONs up to 20 000, 86–99% ee) [153]. In addition, the ruthenium catalysts (*S*,*S*)-**109** (Figure 4.18), generated *in situ* from $Ru_3(CO)_{12}$ and chiral tetradentate diiminodiphosphine ligands (Figure 4.18), are effective for the asymmetric transfer hydrogenation of aryl ketones (TON = 200, up to 99% ee) [154].

4.3.2 Chiral Rhodium and Iridium Catalysts

The earlier examples of rhodium- and iridium-catalyzed asymmetric transfer hydrogenation of aryl ketones were achieved using chiral diphosphine ligands with 2-propanol as hydrogen donor, but with low enantioselectivity (<50% ee) [155–158]. Later, Randaccio and coworkers improved the enantioselectivity up to 80% ee in the reduction of *tert*-butyl phenyl ketone by employing chiral iridium complex (*S*)-**110** (Figure 4.19) of Schiff base ligand [159]. In 1991, Pfaltz and coworkers found that the chiral iridium catalyst (*S*)-**111** (Figure 4.19) bearing a chiral bisoxazoline ligand gives up to 91% ee for the transfer hydrogenation of aryl ketones [160]. In 1998, Mashima and Tani et al. significantly improved the enantioselectivity of rhodium- and iridium-catalyzed asymmetric transfer

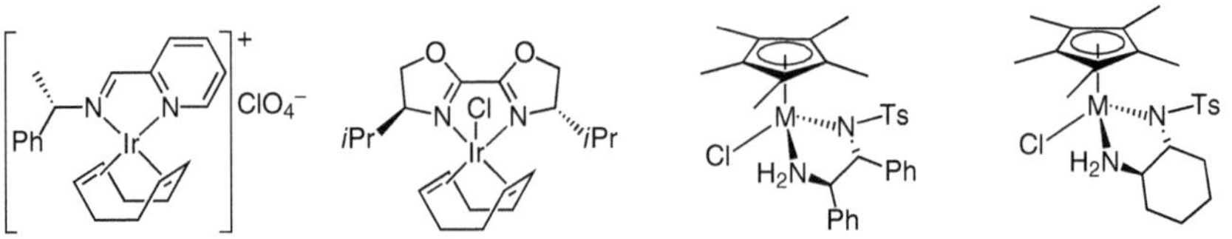

(S)-**110** (S,S)-**111** (R,R)-**112a** M = Rh (R,R)-**112b** M = Ir (R,R)-**113a** M = Rh (R,R)-**113b** M = Ir

Figure 4.19 The early examples of chiral iridium and rhodium catalysts for asymmetric transfer hydrogenation.

hydrogenation of aryl ketones by introducing chiral rhodium catalyst (*R,R*)-**112a** and iridium catalyst (*R,R*)-**112b** (Figure 4.19) bearing a chiral *N*-Ts-diamine ligand (*R,R*)-*N*-Ts-DPEN (up to 99% ee and 96% ee, respectively) [161]. Simultaneously, Ikariya and Noyori et al. described that the chiral rhodium catalyst (*R,R*)-**113a** and iridium catalyst (*R,R*)-**113b** (Figure 4.19) containing a chiral *N*-Ts-diamine ligand (1*R*,2*R*)-*N*-(*p*-toluenesulfonyl)-1,2-cyclohexanediamine ((*R,R*)-*N*-Ts-DACH) give slightly better enantioselectivity than the corresponding *N*-Ts-DPEN-based catalysts (*R,R*)-**112a** and **112b** in the asymmetric transfer hydrogenation of aryl ketones and benzo-fused cyclic ketones (up to >99% ee and 97% ee, respectively) [162].

4.3.2.1 Chiral Rhodium and Iridium Complexes Containing Diamine and Related Ligands

Since Ikariya and Noyori et al. reported that the chiral rhodium and iridium complexes of *N*-Ts-diamine ligands are effective catalysts for the asymmetric transfer hydrogenation of aryl ketones [162], a variety of chiral rhodium and iridium catalysts containing *N*-Ts-diamines and related ligands have been developed in the past decade. In 2005, Wills and coworkers reported that a tethered chiral rhodium complex (*R,R*)-**114a** (Figure 4.20) of *N*-Ts-1,2-diphenylethylenediamine ligand is an effective catalyst for the asymmetric transfer hydrogenation of aryl ketones with activity over the corresponding untethered rhodium complex [163]. Using formic acid/triethylamine as both solvent and hydrogen donor, the catalyst (*R,R*)-**114a** affords excellent enantioselectivity (95–99.9% ee) in the asymmetric transfer hydrogenation of aryl ketones and tetralone (TON = 200). Later, they found that the tethered chiral rhodium complex (*R,R*)-**115a** (Figure 4.20) of *N*-Ts-1,2-cyclohexanediamine ligand gives moderate to excellent enantioselectivity

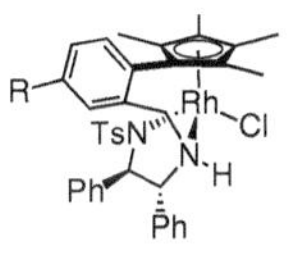

(*R,R*)-**114a** R = H
(*R,R*)-**114b** R = MeO
(*R,R*)-**114c** R = Me
(*R,R*)-**114d** R = F
(*R,R*)-**114e** R = CF_3

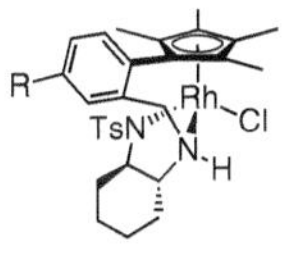

(*R,R*)-**115a** R = H
(*R,R*)-**115b** R = MeO

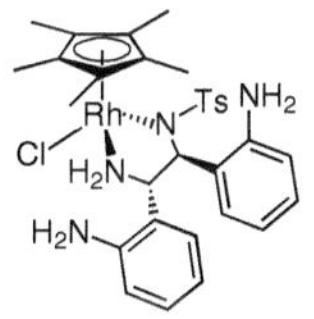

(*S,S*)-**116**

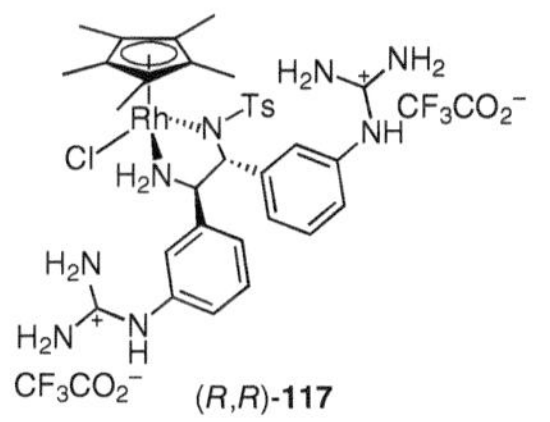

(*R,R*)-**117**

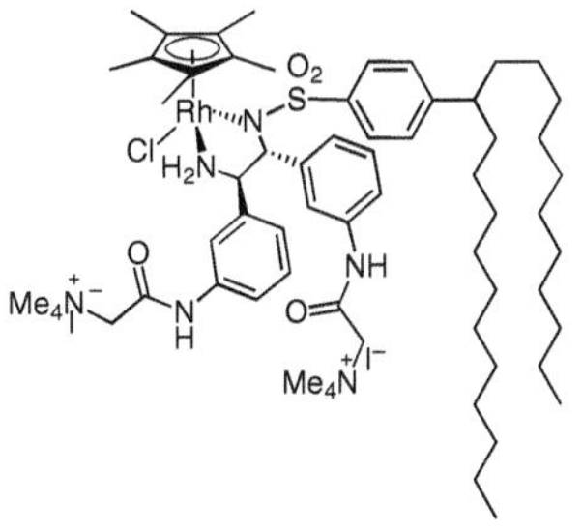

(*R,R*)-**118**

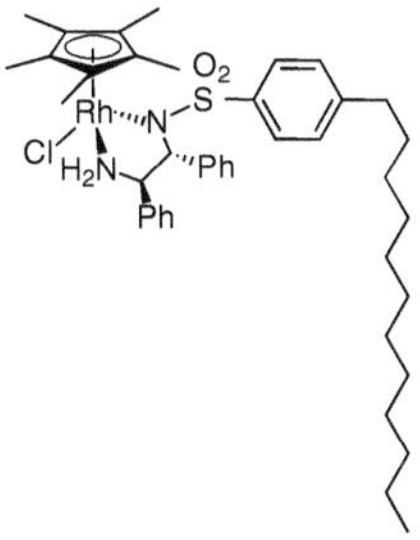

(*R,R*)-**119**

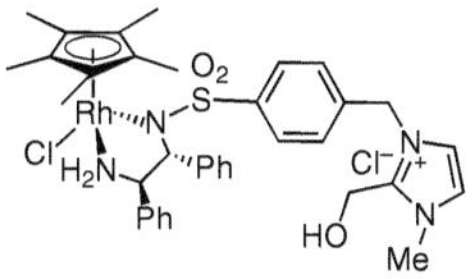

(*R,R*)-**120**

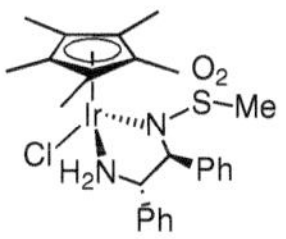

(*S,S*)-**121**

Figure 4.20 Chiral rhodium and iridium complexes of chiral *N*-Ts-diamine ligands.

(62–99.4% ee, TON = 200) for a range of aryl and heteroaryl ketones and tetralone [164]. With sodium formate in water as the reducing agent, the catalyst (*R*,*R*)-**115a** generally gives rapid reaction and high enantioselectivity in the asymmetric transfer hydrogenation. The most remarkable feature of the catalyst (*R*,*R*)-**115a** is its stability. When the catalyst loading was reduced to 0.001 mol%, it still provided 100% conversion and 98% ee for the hydrogenations of 2-acetylthiophene and 2-acetylfuran. Further modifications to the tethering phenyl ring of the catalysts (*R*,*R*)-**114a** [165, 166] and (*R*,*R*)-**115a** [167] lead to new chiral rhodium catalysts (*R*,*R*)-**114b–e** and (*R*,*R*)-**115b**, which also show good yields and moderate to excellent enantioselectivities (64 to >99% ee) for the hydrogenations of aryl, heteroaryl, and benzo-fused ketones (TON = 200). Although the properties of the substituents on the phenyl tethering ring of the catalyst (*R*,*R*)-**114** did not alter the stereochemical outcome of the reaction, the catalysts bearing an electron-donating group exhibited a higher activity than those having an electron-withdrawing group [166].

In 2005, Xiao and coworkers found that the chiral rhodium catalyst (*R*,*R*)-**113a** with a chiral diamine ligand *N*-Ts-DACH has superior activity and enantioselectivity (0.25 hours, 95% ee) than the corresponding ruthenium catalyst (*R*,*R*)-**86** (2 hours, 85% ee) for the asymmetric transfer hydrogenation of acetophenone, while the iridium catalyst (*R*,*R*)-**113b** has slightly lower enantioselectivity (1 hour, 93% ee) [168]. With the chiral rhodium catalyst (*R*,*R*)-**113a**, a wide range of aryl, heteroaryl, and benzo-fused ketones were hydrogenated to optically active alcohols in high yields with good to excellent enantioselectivities (77–99% ee). In addition, the reduction of 2-acetylfuran with (*R*,*R*)-**113a** affords 1-(furan-2-yl)ethan-1-ol with 99% ee and 1000 of TONs within 1.5 hour in the open air. They also evaluated the chiral metal catalysts with chiral diamine ligand *N*-Ts-DPEN for asymmetric transfer hydrogenation of acetophenone using formate in water as reductant. The results showed that the rhodium catalyst (*R*,*R*)-**112a** (0.5 hour, 97% ee) is superior to the iridium catalyst (*R*,*R*)-**112b** (3 hour, 93% ee) and ruthenium catalyst (*R*,*R*)-**29a** (1 hour, 95% ee) [169]. With the rhodium catalyst (*R*,*R*)-**112a** and iridium catalyst (*R*,*R*)-**112b**, a series of aryl, heteroaryl, and benzo-fused cyclic ketones were reduced with up to 99% ee.

In 2007, Deng and coworkers reported that the chiral rhodium catalyst (*S*,*S*)-**116** (Figure 4.20) containing a water-soluble diamine ligand *N*-Ts-DPEN with two *ortho*-amino groups on the phenyl rings gives 88–98% ee for the asymmetric transfer hydrogenation of aryl, heteroaryl, and benzo-fused cyclic ketones. Comparable results (82 to >99% ee) are obtained by the chiral rhodium catalyst (*R*,*R*)-**117** (Figure 4.20) with two *meta*-guanidinium groups on the phenyl rings of the diamine ligand [170]. Recently, they developed a chiral double-chain surfactant-*N*-Ts-DPEN-type ligand and found that its rhodium complex (*S*,*S*)-**118** (Figure 4.20) can self-assemble into vesicular aggregates in water and gives up to 99% yield and 99% ee for asymmetric transfer hydrogenation of acetophenone. When the catalyst loading was lowered to 0.1 mol%, the reaction still gave 97% ee with 94% conversion [171]. The lipophilic rhodium catalyst (*R*,*R*)-**119** (Figure 4.20) bearing a chiral diamine ligand *N*-Ts-DPEN with a $C_{12}H_{25}$-tail, reported by Adolfsson et al. in 2008, shows moderate to high enantioselectivity (76–97% ee) for the transfer hydrogenation of a range of aryl ketones in water with formate as hydrogen donor in the presence of sodium dodecylsulfonate (TON = 100) [172]. In 2014, Ni and coworkers tethered an imidazolium to the *N*-Ts-DPEN ligand and developed a water-soluble chiral rhodium catalyst (*R,R*)-**120** (Figure 4.20) [173]. Using catalyst (*R,R*)-**120**, a range of aryl, heteroaryl, and benzo-fused ketones were hydrogenated to chiral alcohols in 67–98% ee with 54–100% conversions (TON = 500). In addition, Zhou and coworkers demonstrated that the *N*-Ms-DPEN-derived chiral iridium catalyst (*S,S*)-**121** (Figure 4.20) is highly enantioselective (up to 98% ee) for the transfer hydrogenation of non-*ortho*-substituted aryl *ortho*-*N*-oxide heteroaryl ketones in the mixed solvent of water and isopropanol [174]. This asymmetric transfer hydrogenation was successfully utilized to the enantioselective synthesis of bepotastine besilate.

In 2006, Adolfsson and coworkers found that the Boc-amino acid thioamide ligand (*S,S*)-**122** (Figure 4.21), derived from (*S*)-valine and (*S*)-1-phenylethanamine, is effective for the rhodium-catalyzed asymmetric transfer hydrogenation of aryl ketones in 2-propanol [175]. With the rhodium catalyst generated *in situ* from $[RhCl_2Cp^*]_2$ and (*S,S*)-**122**, several aryl ketones and 3-pyridyl ketones are hydrogenated to the corresponding chiral alcohols with 91–97% ee (TON = 200). However, the corresponding Boc-amino amide ligand provides lower enantioselectivity. By screening a series of *N*-Boc-amino acid-modified 1,2-amino-alcohol ligands they found that the ligand (*S,S*)-**123** (Figure 4.21) having an *N*-Boc-protected L-phenylalanine moiety provides up to 98% ee in the rhodium-catalyzed hydrogenation of aryl ketones (TON = 100) [176]. Subsequently, they systematically evaluated amino acid-based thioamides, hydroxamic acids, and hydrazides as chiral ligands in the rhodium-catalyzed asymmetric transfer hydrogenation of acetophenone [177]. The catalysts containing thioamide ligands derived from L-valine selectively generate *R*-product ((*S,S*)-**122**, 95% ee, Figure 4.21), and the L-valine/hydroxamic acid and L-valine/hydrazide-based ligands selectively form (*S*)-product ((*S*)-**124**, 97% ee; (*S*)-**125**, 91% ee, Figure 4.21). Adolfsson and Diéguez et al. developed a series of *N*-Boc-protected amino amide and thioamide ligands from easily accessible

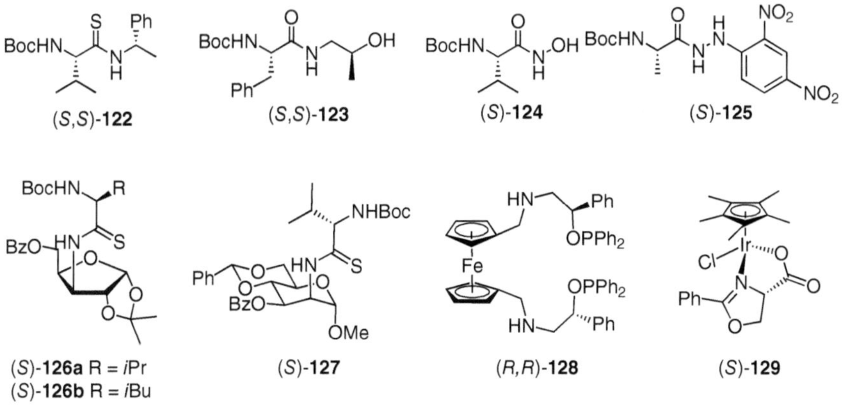

Figure 4.21 Chiral ligands and iridium complexes derived from amino acids and amino alcohols.

D-xylose and D-glucose in 2012 and found that the furanoside-based thioamide ligands (*S*)-**126a** and **126b** (Figure 4.21) show high enantioselectivity in the rhodium-catalyzed transfer hydrogenation of aryl ketones (TON = 200, up to 99% ee) [178]. They also evaluated pyranoside-based *N*-Boc-protected amino acid amide and thioamide ligands in 2014. Among these ligands, the thioamide ligand (*S*)-**127** (Figure 4.21) containing an *N*-Boc-protected L-valine provides moderate to excellent enantioselectivities (55–99% ee) for the rhodium-catalyzed transfer hydrogenation of aryl ketone [179]. The C_2-symmetric ferrocenyl phosphinite ligand (*R*,*R*)-**128** (Figure 4.21), developed by Baysal and coworkers in 2015, is found to be effective for iridium- and rhodium-catalyzed asymmetric transfer hydrogenation of aryl ketones (up to 98% ee and 99% ee, respectively) [180, 181]. Recently, Chen and Xiao reported that the chiral iridium catalyst (*S*)-**129** (Figure 4.21) bearing an oxazoline-carboxylate ligand shows good to excellent enantioselectivities for the asymmetric transfer hydrogenation of aryl ketones [182]. They found that both the nature of the amine and its ratio with the reductant HCO_2H have effect on the enantioselectivity of the hydrogenation. Among the tested amines, the HCO_2H—*i*$PrNH_2$ (2 : 1) mixture gives the highest enantioselectivity. With this reductant, a range of aryl ketones are hydrogenated to chiral alcohols with 90–99% ee (TON = 100).

4.3.2.2 Chiral Rhodium and Iridium Catalysts Containing Other Ligands

Chiral tetradentate PNNP-type ligands derived from chiral 1,2-diamines are well studied for asymmetric transfer hydrogenation of aryl ketones [183]. In addition to the aforementioned chiral ruthenium complexes of PNNP-type ligands, chiral rhodium and iridium complexes of tetradentate diaminodiphosphine ligands such as (*S*,*S*)-**130** (Figure 4.22) also show high catalytic activity and enantioselectivity in the asymmetric transfer hydrogenation of aryl ketones. In 2004, Gao and coworkers reported that the chiral iridium catalysts, generated *in situ* from $[IrHCl_2(COD)]_2$ and (*S*,*S*)-**130**, give enantioselectivity up to 99% ee for the asymmetric transfer hydrogenation of aryl ketones [184]. Lowering the catalyst loading to 0.025 or 0.013 mol% gave no significant loss in enantioselectivity. They also found that the chiral rhodium and iridium catalysts, generated *in situ* from the metal–hydride precursors and PNNP-type ligands, exhibit excellent activity and good enantioselectivity in the asymmetric transfer hydrogenation of aryl ketones in 2-propanol without base [185]. The best results are obtained by the chiral iridium catalyst generated from $[IrH(CO)(PPh_3)_3]$ and ligand (*S*,*S*)-**130** (up to 97% ee). The catalyst loading can be reduced to 0.05 mol%, albeit the enantioselectivity was slightly lowered [186]. The dimerized diaminodiphosphine ligand (*R*,*R*,*R*,*R*)-**131** (Figure 4.22) [187] and macrocycle ligand (*R*,*R*,*R*′,*R*′)-**132** (Figure 4.22) [188] also give high enantioselectivity (up to 99% ee) in the iridium- and rhodium-catalyzed asymmetric transfer hydrogenation of aryl alkyl ketones.

In 2015, Xie and Zhou et al. reported that the chiral spiro iridium catalyst (*R*)-**45** bearing a tridentate spiro pyridine-aminophosphine SpiroPAP ligand is highly efficient for the asymmetric transfer hydrogenation of aryl ketones with ethanol as hydrogen donor [189]. With the catalyst (*R*)-**45**, a series of aryl ketones were reduced to chiral alcohols in high yields and enantioselectivity (TON = 1000, 90–98% ee).

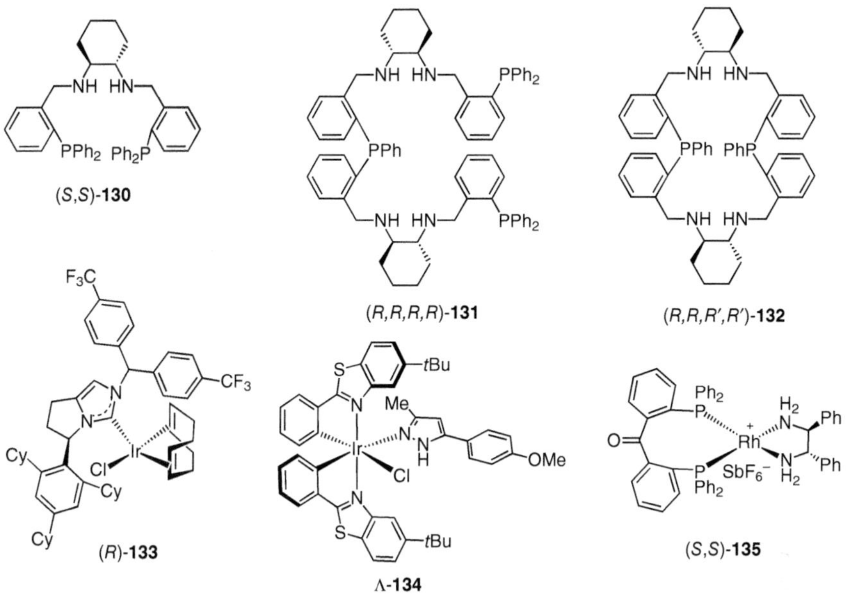

Figure 4.22 Chiral multidentate aminophosphine ligands and chiral iridium and rhodium complexes.

Yoshida and Yanagisawa and coworkers reported in 2015 that the chiral iridium catalyst (*R*)-**133** (Figure 4.22) bearing a chiral bicyclic *N*-heterocyclic carbene ligand affords high enantioselectivity (up to 98% ee) with TONs up to 4500 for the asymmetric transfer hydrogenation of aryl ketones [190]. The chiral bis-cyclometalated iridium complex Λ-**134** (Figure 4.22) with metal-centered chirality, developed by Gong and Meggers et al. gave 51–99% ee with TONs up to 25 500 for a wide range of aryl, heteroaryl, and benzo-fused ketones [191]. The pyrazole coligand significantly improved the catalytic activity and enantioselectivity of the iridium catalyst Λ-**134**. The hydrogenation is proposed to undergo via metal–ligand cooperative mechanism (bifunctional catalysis). In addition, Mikami et al. reported in 2006 that the chiral rhodium catalyst (*S*,*S*)-**135** (Figure 4.22) bearing an achiral benzophenone-based diphosphine ligand and a chiral 1,2-diphenylethylenediamine also affords high enantioselectivity (up to 99% ee) in the asymmetric transfer hydrogenation of aryl ketones [192].

4.3.3 Other Chiral Metal Catalysts

In addition to aforementioned chiral ruthenium, rhodium, and iridium catalysts, other chiral transition-metal catalysts are also used in the asymmetric transfer hydrogenation of aryl ketones. For example, the iron catalysts containing chiral multidentate aminophosphine ligands and the osmium and manganese complexes containing chiral tridentate pincer ligands have shown high activity and enantioselectivity in this transformation.

4.3.3.1 Chiral Iron Catalysts

In 2004, Gao and coworkers reported that chiral iron complex generated *in situ* from $[Et_3NH][HFe_3(CO)_{11}]$ and chiral tetradentate diaminophosphine ligand (*S*,*S*)-**130** (Figure 4.22) is an efficient catalyst for the asymmetric transfer hydrogenation of aryl ketones [193]. Higher enantioselectivities are obtained for aryl alky ketones with a bulkier alkyl group (Et, 72% ee; *i*Pr, 78% ee; and *t*Bu, 93% ee at TONs of 100). This represents the first example of the iron-catalyzed asymmetric transfer hydrogenation of ketones. Later, they found that the iron complex of macrocycle aminophosphine ligand (*R*,*R*,*R*′,*R*′)-**132** (Figure 4.22) gives excellent enantioselectivity (90–99% ee) for a range of aryl and heteroaryl ketones [194].

In 2008, Morris and coworkers reported that the chiral iron catalyst (*R*,*R*)-**136a** (Figure 4.23) with diiminodiphosphine ligand is efficient for the asymmetric transfer hydrogenation of aryl ketones under mild conditions, but with low enantioselectivity (for acetophenone: TON = 400, 29% ee) [83]. Later, they found that the iron catalyst (*R*,*R*)-**137b** (Figure 4.23) with diiminodiphosphine ligand significantly improves the activity and selectivity of the reaction (for acetophenone: TON = 2000, 82% ee) [195]. The catalysts (*R*,*R*)-**136b** and (*R*,*R*)-**137a** are also efficient for the asymmetric transfer hydrogenation of aryl ketones [196]. A systematical evaluation on the diamine unit of the ligand in the catalysts (*R*,*R*)-**138** (Figure 4.23) showed that the activity of the catalyst increases with the size of the R group on the backbone of ligand [197]. The steric and electronic properties of the aryl groups on the phosphine atoms of ligands also show significant effect on both activity and enantioselectivity. The catalyst (*R*,*R*)-**138d** with 4-methylphenyl groups has higher activity, and the catalyst

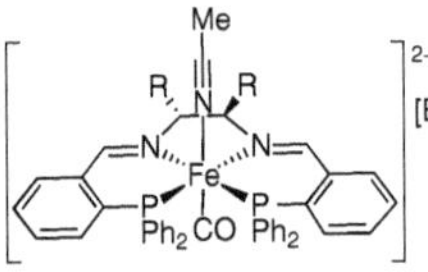

(*R*,*R*)-**136a** R = $(CH_2)_4$
(*R*,*R*)-**136b** R = Ph

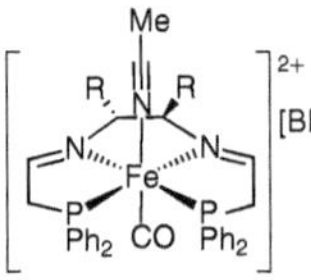

(*R*,*R*)-**137a** R = $(CH_2)_4$
(*R*,*R*)-**137b** R = Ph

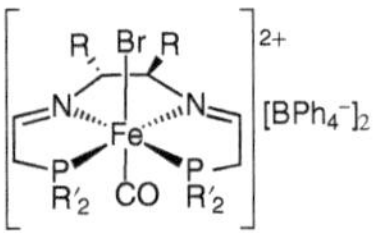

(*R*,*R*)-**138a** R = $(CH_2)_4$, R′ = Ph
(*R*,*R*)-**138b** R = Ph, R′ = Ph
(*R*,*R*)-**138c** R = 4-$MeOC_6H_4$, R′ = Ph
(*R*,*R*)-**138d** R = Ph, R′ = 4-MeC_6H_4
(*R*,*R*)-**138e** R = Ph, R′ = 3,5-$(Me)_2C_6H_3$
(*R*,*R*)-**138f** R = Ph, R′ = Et

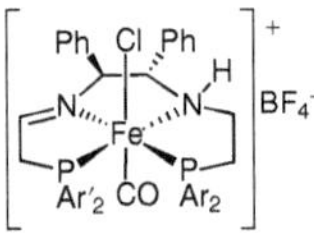

(*S*,*S*)-**139a** Ar = Ar′ = Ph
(*S*,*S*)-**139b** Ar = Ph, Ar′ = 4-MeC_6H_4
(*S*,*S*)-**139c** Ar = Ar′ = 3,5-$(Me)_2C_6H_3$

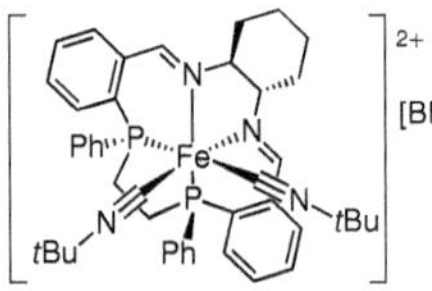

(*S*,*S*,*S*,*S*)-**140**

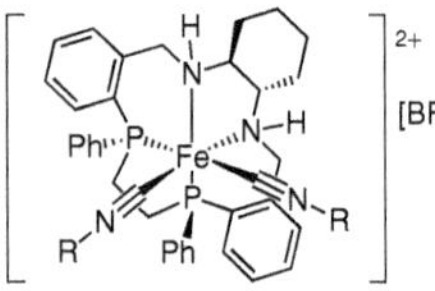

(*S*,*S*,*S*,*S*)-**141a** R = CEt_3
(*S*,*S*,*S*,*S*)-**141b** R = N*i*Pr_2

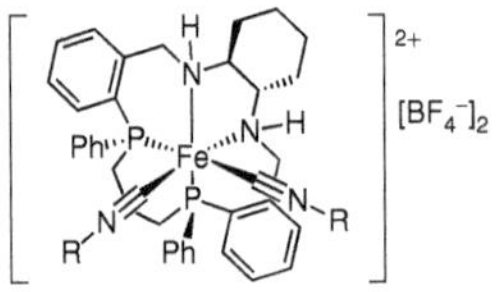

(*S*,*S*,*S*,*S*)-**142a** R = CEt_3
(*S*,*S*,*S*,*S*)-**142b** R = N*i*Pr_2

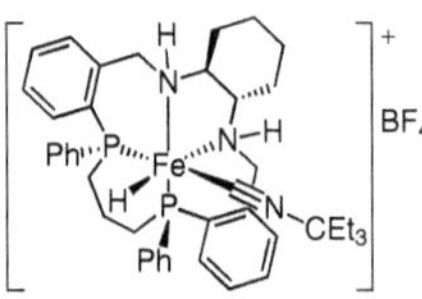

(*S*,*S*,*S*,*S*)-**143**

Figure 4.23 Chiral iron complexes of tetradentate PNNP-type ligands.

(*R*,*R*)-**138e** with 3,5-dimethylphenyl groups on the phosphorus atoms has higher enantioselectivity [198]. However, the replacement of the phenyl groups on the phosphorus atoms of ligand by alkyl groups such as ethyl, isopropyl, and cyclohexyl leads to a decrease or complete loss of activity of the catalyst [199]. The chiral iron catalysts (*S*,*S*)-**139** (Figure 4.23) bearing partially saturated amine(imine)diphosphine ligands also have been studied. They show lower activity and enantioselectivity than the corresponding catalysts **138** [200].

In 2014, Mezzetti and coworkers reported that the diamagnetic bis(acetonitrile) iron catalyst (*S*,*S*,*S*,*S*)-**140** (Figure 4.23) bearing a C_2-sysmmetric diiminephosphine-type 14-membered N_2P_2 macrocycle ligand affords up to 91% ee in the asymmetric transfer hydrogenations of aryl, heteroaryl, and benzo-fused cyclic ketones (TON = 100) [201]. The chiral iron catalysts bearing a diaminophosphine-type 14-membered N_2P_2 macrocycle ligand were also investigated [202]. It was found that the iron catalysts (*S*,*S*,*S*,*S*)-**141a** and **141b** (Figure 4.23) have excellent enantioselectivities (86–99% ee and 87–99% ee, respectively) for the hydrogenation of aryl ketones (TON = 1000). In particular, these catalysts are highly efficient for the hydrogenation of industrially useful substrates such as *tert*-butyl phenyl ketone (99.7% yield, 99.4% ee) [203]. Later, they changed the linker between the phosphines and developed a new family of chiral iron catalysts (*S*,*S*,*S*,*S*)-**142a** and **142b** (Figure 4.23) bearing a chiral diaminophosphine-type 15-membered N_2P_2 macrocycle ligand. These catalysts showed excellent enantioselectivity (up to 99.6% ee) for a broad scope of aryl and heteroaryl ketones (TON = 1000) [204]. The chiral iron hydride catalyst (*S*,*S*,*S*,*S*)-**143** (Figure 4.23) is able to catalyze the asymmetric transfer hydrogenation of aryl ketones under base-free condition [205].

4.3.3.2 Chiral Osmium Catalysts

Chiral osmium complex (S_p,*S*,*R*)-**54** (Figure 4.11) containing a CNN-type ligand, developed by Barratta et al. in 2008, is a highly efficient catalyst for asymmetric transfer hydrogenation of aryl ketones (TONs up to 20 000, 90–97% ee) [81]. Higher activity and enantioselectivity were obtained by osmium catalyst (*R*,*S*,R_p)-**144** (Figure 4.24) containing a 2-naphthyl moiety on the chiral CNN pincer ligand [153]. For example, the asymmetric transfer hydrogenation of 1-(3,5-dimethoxyphenyl)ethanone catalyzed by 0.002 mol% of (*R*,*S*,R_p)-**144** yields the chiral alcohol with 97% ee. In 2013, Gamasa and coworkers showed that the chiral osmium catalyst (*S*,*S*)-**145** (Figure 4.24) bearing a *i*Pr-pybox ligand gives 90–94% ee for the asymmetric transfer hydrogenation of aryl ketones [206]. The chiral osmium catalyst (*S*,*S*)-**146** (Figure 4.24) bearing a diamine (*S*,*S*)-Ts-DPEN ligand, reported by Sadler and coworkers in 2015, is also highly effective for the asymmetric transfer hydrogenation of aryl ketones (TONs up to 200, 96–98% ee) [207].

4.3.3.3 Other Chiral Metal Catalysts

Recently, the chiral manganese, nickel, and cobalt complexes of tridentate and multidentate aminophosphine ligands have also been used as catalysts in the asymmetric transfer hydrogenation of aryl ketones. For example, the chiral manganese catalyst (*R*,*R*,S_p)-**147** (Figure 4.24) bearing a ferrocene-based tridentate PNP-type

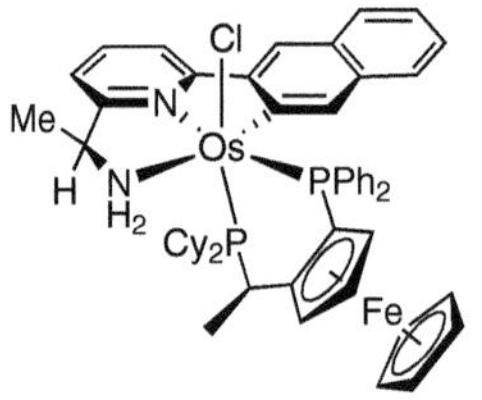

(R,S,S_p)-**144**

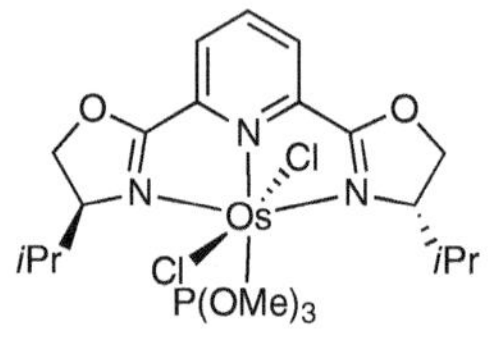

(S,S)-**145**

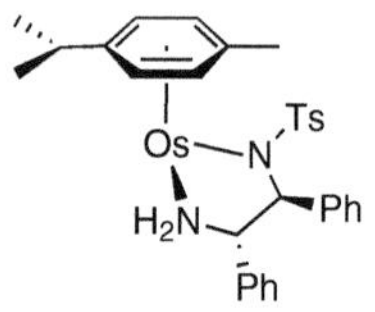

(S,S)-**146**

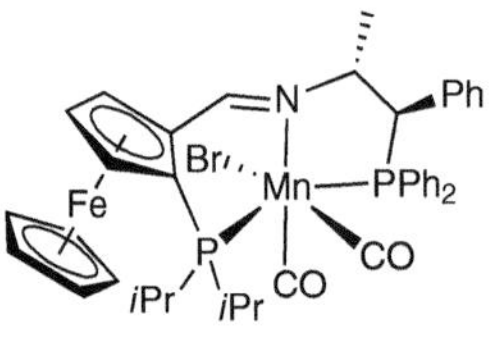

(R,R,S_p)-**147**

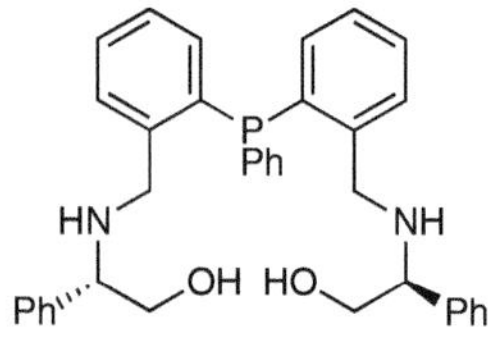

(S,S)-**148**

Figure 4.24 Other chiral metal catalysts and related chiral ligands.

iminophosphine ligand, developed by Zirakzadeh and Kirchner et al. in 2017, gives up to 86% ee in the asymmetric transfer hydrogenation of aryl ketones [208]. The nickel catalyst (*R*,*R*)-**148** (Figure 4.24) having a chiral multidentate aminophosphine ligand, developed by Dong and Gao et al. in 2012, can reduce aryl ketones to chiral alcohols with up to 84% ee [209]. Gao and coworkers also found that the cobalt catalyst formed with (*S*,*S*)-**130** (Figure 4.22) bearing a chiral diaminodiphosphine ligand affords up to 63% ee in the asymmetric transfer hydrogenation of aryl ketones [210].

4.4 Summary

Transition-metal-catalyzed asymmetric hydrogenation and asymmetric transfer hydrogenation of aryl ketones with hydrogen or other hydrogen sources such as alcohols and formates have been intensively studied during the past decades. Many chiral transition-metal catalysts containing various chiral ligands have been developed for the asymmetric hydrogenation and transfer hydrogenation of aryl ketones, providing a wide range of chiral secondary alcohols in high enantioselectivity. Among the efficient chiral catalysts in this area, the chiral ruthenium complexes are predominant, especially those ligated with chiral diphosphine/diamine ligands (for asymmetric hydrogenation) and chiral *N*-tosylethylenediamine ligands (for asymmetric transfer hydrogenation). Several other types of chiral catalysts including iridium and osmium complexes of tridentate ligands also show high activity and enantioselectivity for the asymmetric hydrogenation and/or asymmetric transfer hydrogenation of aryl ketones. The most prominent example is the chiral iridium catalyst bearing a chiral spiro pyridine-aminophosphine SpiroPAP ligand, which shows extremely high activity and enantioselectivity for the asymmetric hydrogenation of aryl ketones. The chiral iron and manganese catalysts have also been demonstrated to be highly efficient and enantioselective for the asymmetric hydrogenation and transfer hydrogenation of aryl ketones. For example, the chiral manganese catalysts bearing a tridentate PNN type of pincer ligand afford very high activity and enantioselectivity in the hydrogenation of aryl ketones, which are comparable with chiral precious metal catalysts.

Although a remarkable progress has been made and several chiral metal catalysts exhibited comparable or even superior efficiency to enzyme, there are still many challenges existing in the transition-metal-catalyzed asymmetric hydrogenation and transfer hydrogenation of aryl and heteroaryl ketones to be solved. The efficient catalysts for the enantioselective hydrogenation of the aryl ketones having sterically bulky alkyl groups and the diaryl ketones having no-*ortho* substituents are scarce. The chiral abundant metal catalysts are often associated with a narrow substrate scope and are far from meeting the needs of practical applications. Although asymmetric transfer hydrogenation is operationally simple and safe, its efficiency and substrate range are not as good as asymmetric hydrogenation. To address these problems, the development of highly efficient chiral metal catalysts, especially inexpensive abundant metal catalysts, is still the focus in the area of asymmetric hydrogenation and asymmetric transfer hydrogenation of aryl and heteroaryl ketones.

References

1 Caprio, V. and Williams, V.J.M.J. (eds.) (2009). *Catalysis in Asymmetric Synthesis*. Chichester: Wiley-VCH.
2 Štefane, B. and Požgan, F. (2016). *Top. Curr. Chem.* 374: 18. https://doi.org/10.1007/s41061-016-0015-5.
3 Wang, D. and Astruc, D. (2015). *Chem. Rev.* 115: 6621–6686.
4 Foubelo, F., Nájera, C., and Yus, M. (2015). *Tetrahedron: Asymmetry* 26: 769–790.
5 Xie, J.-H., Bao, D.-H., and Zhou, Q.-L. (2015). *Synthesis* 47: 460–471.
6 Požgan, F. and Štefane, B. (2012). *Hydrogenation: Asymmetric Hydrogenation and Transfer Hydrogenation of Ketones* (ed. I. Karamé), 31–68. Rijeka, Croatia: InTech.
7 Schrock, R.R. and Osborn, J.A. (1970). *Chem. Commun.*: 567–568.
8 Bonvicini, P., Levi, A., Modena, G., and Scorrano, G. (1972). *J. Chem. Soc., Chem. Commun.*: 1188–1189.
9 Tanaka, M., Watanabe, Y., Mitsudo, T.-a. et al. (1973). *Chem. Lett.* 2: 239–240.
10 Noyori, R., Ohkuma, T., Kitamura, M. et al. (1987). *J. Am. Chem. Soc.* 109: 5856–5858.
11 Noyori, R. and Takaya, H. (1990). *Acc. Chem. Res.* 23: 345–350.
12 Zhang, X., Taketomi, T., Yoshizumi, T. et al. (1993). *J. Am. Chem. Soc.* 115: 3318–3319.
13 Ohkuma, T., Ooka, H., Hashiguchi, S. et al. (1995). *J. Am. Chem. Soc.* 117: 2675–2676.
14 Doucet, H., Ohkuma, T., Murata, K. et al. (1998). *Angew. Chem. Int. Ed.* 37: 1703–1707.

15 Ohkuma, T., Koizumi, M., Doucet, H. et al. (1998). *J. Am. Chem. Soc.* 120: 13529–13530.

16 Noyori, R. and Okuma, T. (2001). *Angew. Chem. Int. Ed.* 40: 40–73.

17 Xie, J.-H., Liu, X.-Y., Xie, J.-B. et al. (2011). *Angew. Chem. Int. Ed.* 50: 7329–7332.

18 Zhang, L., Tang, Y., Han, Z., and Ding, K. (2019). *Angew. Chem. Int. Ed.* 58: 4973–4977.

19 Yamakawa, M., Ito, H., and Noyori, R. (2000). *J. Am. Chem. Soc.* 122: 1466–1478.

20 Noyori, R., Kitamura, M., and Ohkuma, T. (2004). *Proc. Natl. Acad. Sci. U. S. A.* 101: 5356–5362.

21 Burk, M.J., Hems, W., Herzberg, D. et al. (2000). *Org. Lett.* 2: 4173–4176.

22 Wu, J., Chen, H., Kwok, W. et al. (2002). *J. Org. Chem.* 67: 7908–7910.

23 Wu, J., Ji, J.-X., Guo, R. et al. (2003). *Chem. Eur. J.* 9: 2963–2968.

24 Xie, J.-H., Wang, L.-X., Fu, Y. et al. (2003). *J. Am. Chem. Soc.* 125: 4404–4405.

25 Henschke, J.P., Burk, M.J., Malan, C.G. et al. (2003). *Adv. Synth. Catal.* 345: 300–307.

26 Hu, A., Ngo, H.L., and Lin, W. (2004). *Org. Lett.* 6: 2937–2940.

27 Ngo, H.L. and Lin, W. (2005). *J. Org. Chem.* 70: 1177–1187.

28 Li, W., Sun, X., Zhou, L. et al. (2009). *J. Org. Chem.* 74: 1397–1399.

29 Xu, Y., Alcock, N.W., Clarkson, G.J. et al. (2004). *Org. Lett.* 6: 4105–4107.

30 Xu, Y., Clarkson, G.C., Docherty, G. et al. (2005). *J. Org. Chem.* 70: 8079–8087.

31 Stegink, B., van Boxtel, L., Lefort, L. et al. (2010). *Adv. Synth. Catal.* 352: 2621–2628.

32 Jing, Q., Zhang, X., Sun, J., and Ding, K. (2005). *Adv. Synth. Catal.* 347: 1193–1197.

33 Jing, Q., Sandoval, C.A., Wang, Z., and Ding, K. (2006). *Eur. J. Org. Chem.*: 3606–3616.

34 Mikami, K., Wakabayashi, K., and Aikawa, K. (2006). *Org. Lett.* 8: 1517–1519.

35 Rodríguez, S., Qu, B., Fandrick, K.R. et al. (2014). *Adv. Synth. Catal.* 356: 301–307.

36 Ohkuma, T., Hattori, T., Ooka, H. et al. (2004). *Org. Lett.* 6: 2681–2683.

37 Genov, D.G. and Ager, D.J. (2004). *Angew. Chem. Int. Ed.* 43: 2816–2819.

38 Ohkuma, T., Sandoval, C.A., Srinivasan, R. et al. (2005). *J. Am. Chem. Soc.* 127: 8288–8289.

39 Arai, N., Suzuki, K., Sugizaki, S. et al. (2008). *Angew. Chem. Int. Ed.* 47: 1770–1773.

40 Arai, N., Ooka, H., Azuma, K. et al. (2007). *Org. Lett.* 9: 939–941.

41 Ooka, H., Arai, N., Azuma, K. et al. (2008). *J. Org. Chem.* 73: 9084–9093.

42 Arai, N., Azuma, K., Nii, N., and Ohkuma, T. (2008). *Angew. Chem. Int. Ed.* 47: 7457–7460.

43 Li, Y., Ding, K., and Sandoval, C.A. (2009). *Org. Lett.* 11: 907–910.

44 Li, Y., Zhou, Y., Shi, Q. et al. (2011). *Adv. Synth. Catal.* 353: 495–500.

45 Zhu, Q., Shi, D., Xia, C., and Huang, H. (2011). *Chem. Eur. J.* 17: 7760–7763.

46 Utsumi, N., Arai, N., Kawaguchi, K. et al. (2018). *ChemCatChem* 10: 3955–3959.

47 Ohkuma, T., Koizumi, M., Muñiz, K. et al. (2002). *J. Am. Chem. Soc.* 124: 6508–6509.
48 Matsumura, K., Arai, N., Hori, K. et al. (2011). *J. Am. Chem. Soc.* 133: 10696–10699.
49 Ito, M., Hirakawa, M., Murata, K., and Ikariya, T. (2001). *Organometallics* 20: 379–381.
50 Hedberg, C., Källström, K., Arvidsson, P.I. et al. (2005). *J. Am. Chem. Soc.* 127: 15083–15090.
51 Ohkuma, T., Utsumi, N., Tsutsumi, K. et al. (2006). *J. Am. Chem. Soc.* 128: 8724–8725.
52 Ito, M., Endo, Y., and Ikariya, T. (2008). *Organometallics* 27: 6053–6055.
53 Touge, T., Hakamata, T., Nara, H. et al. (2011). *J. Am. Chem. Soc.* 133: 14960–14963.
54 Jolley, K.E., Zanotti-Gerosa, A., Hancock, F. et al. (2012). *Adv. Synth. Catal.* 354: 2545–2555.
55 Naud, F., Malan, C., Spindler, F. et al. (2006). *Adv. Synth. Catal.* 348: 47–50.
56 Hansen, K.B., Chilenski, J.R., Desmond, R. et al. (2003). *Tetrahedron: Asymmetry* 14: 3581–3587.
57 Schuecher, R., Zirakzadeh, A., Mereiter, K. et al. (2011). *Organometallics* 30: 4711–4719.
58 Guo, H., Liu, D., Butt, N.A. et al. (2012). *Tetrahedron* 68: 3295–3299.
59 Wang, J., Liu, D., Liu, Y., and Zhang, W. (2013). *Org. Biomol. Chem.* 11: 3855–3861.
60 Clarke, M.L., Diaz-Valenzuela, M.B., and Slawin, A.M.Z. (2007). *Organometallics* 26: 16–19.
61 Díaz-Valenzuela, M.B., Philips, S.D., France, M.B. et al. (2009). *Chem. Eur J.* 15: 1227–1232.
62 Philips, S.D., Fuentes, J.A., and Clarke, M.L. (2010). *Chem. Eur J.* 16: 8002–8005.
63 Baratta, W., Chelucci, G., Herdtweck, E. et al. (2007). *Angew. Chem. Int. Ed.* 46: 7651–7654.
64 Baratta, W., Chelucci, G., Magnolia, S. et al. (2009). *Chem. Eur. J.* 15: 726–732.
65 Baratta, W., Fanfoni, L., Magnolia, S. et al. (2010). *Eur. J. Inorg. Chem.*: 1419–1423.
66 Li, W., Hou, G., Wang, C. et al. (2010). *Chem. Commun.* 46: 3979–3981.
67 Huang, H., Okuno, T., Tsuda, K. et al. (2006). *J. Am. Chem. Soc.* 128: 8716–8717.
68 Patchett, R., Magpantay, I., Saudan, L. et al. (2013). *Angew. Chem. Int. Ed.* 52: 10352–10355.
69 Roux, E.L., Malacea, R., Manoury, E. et al. (2007). *Adv. Synth. Catal.* 349: 309–313.
70 Xie, J.-B., Xie, J.-H., Liu, X.-Y. et al. (2011). *Chem. Asian J.* 6: 899–908.
71 Yan, P.-C., Zhu, G.-L., Xie, J.-H. et al. (2013). *Org. Process Res. Dev.* 17: 307–312.
72 Qian, J.-Q., Yan, P.-C., Che, D.-Q. et al. (2014). *Tetrahedron Lett.* 55: 1528–1531.

73 Zhu, G.-L., Zhang, X.-D., Yang, L.-J. et al. (2016). *Org. Process Res. Dev.* 20: 81–85.

74 Irrgang, T., Friedrich, D., and Kempe, R. (2011). *Angew. Chem. Int. Ed.* 50: 2183–2186.

75 Wu, W., Liu, S., Duan, M. et al. (2016). *Org. Lett.* 18: 2938–2941.

76 Yu, J., Long, J., Yang, Y. et al. (2017). *Org. Lett.* 19: 690–693.

77 Yu, J., Duan, M., Wu, W. et al. (2017). *Chem. Eur. J.* 23: 970–975.

78 Liang, Z., Yang, T., Gu, G. et al. (2018). *Chin. J. Chem.* 36: 851–856.

79 Ling, F., Nian, S., Chen, J. et al. (2018). *J. Org. Chem.* 83: 10749–10761.

80 Wang, Y., Yang, G., Xie, F., and Zhang, W. (2018). *Org. Lett.* 20: 6135–6139.

81 Baratta, W., Ballico, M., Chelucci, G. et al. (2008). *Angew. Chem. Int. Ed.* 47: 4362–4365.

82 Baratta, W., Barbato, C., Magnolia, S. et al. (2010). *Chem. Eur. J.* 16: 3201–3206.

83 Sui-Seng, C., Freutel, F., Lough, A.J., and Morris, R.H. (2008). *Angew. Chem. Int. Ed.* 47: 940–943.

84 Kallmeier, F. and Kempe, R. (2018). *Angew. Chem. Int. Ed.* 57: 46–60.

85 Lagaditis, P.O., Sues, P.E., Sonnenberg, J.F. et al. (2014). *J. Am. Chem. Soc.* 136: 1367–1380.

86 Smith, S.A.M., Lagaditis, P.O., Lüpke, A. et al. (2017). *Chem. Eur. J.* 23: 7212–7216.

87 Li, Y., Yu, S., Wu, X. et al. (2014). *J. Am. Chem. Soc.* 136: 4031–4039.

88 Widegren, M.B., Harkness, G.J., Slawin, A.M.Z. et al. (2017). *Angew. Chem. Int. Ed.* 56: 5825–5828.

89 Garbe, M., Junge, K., Walker, S. et al. (2017). *Angew. Chem. Int. Ed.* 56: 11237–11241.

90 Ling, F., Hou, H., Chen, J. et al. (2019). *Org. Lett.* 21: 3937–3941.

91 Shimizu, H., Igarashi, D., Kuriyama, W. et al. (2007). *Org. Lett.* 9: 1655–1657.

92 Krabbe, S.W., Hatcher, M.A., Bowman, R.K. et al. (2013). *Org. Lett.* 15: 4560–4563.

93 Zhou, X.-Y., Wang, D.-S., Bao, M., and Zhou, Y.-G. (2011). *Tetrahedron Lett.* 52: 2826–2829.

94 Zhang, D., Zhu, E.-Z., Lin, Z.-W. et al. (2016). *Asian J. Org. Chem.* 5: 1323–1326.

95 Zassinovich, G., Mestroni, G., and Gladiali, S. (1992). *Chem. Rev.* 92: 1051–1069.

96 Bianchi, M., Matteoli, U., Menchi, G. et al. (1980). *J. Organomet. Chem.* 198: 73–80.

97 Matteoli, U., Frediani, P., Bianchi, M. et al. (1981). *J. Mol. Catal.* 12: 265–319.

98 Evans, D.A., Nelson, S.G., Gagne, M.R., and Muci, A.R. (1993). *J. Am. Chem. Soc.* 115: 9800–9801.

99 Hashiguchi, S., Fujii, A., Takehara, J. et al. (1995). *J. Am. Chem. Soc.* 117: 7562–7563.

100 Fujii, A., Hashiguchi, S., Uematsu, N. et al. (1996). *J. Am. Chem. Soc.* 118: 2521–2522.

101 Gao, J.-X., Ikariya, T., and Noyori, R. (1996). *Organometallics* 15: 1087–1089.

102 Jiang, Y., Jiang, Q., and Zhang, X. (1998). *J. Am. Chem. Soc.* 120: 3817–3818.

103 Ito, J.-I. and Nishiyama, H. (2014). *Tetrahedron Lett.* 55: 3133–3146.

104 Hannedouche, J., Clarkson, G.J., and Wills, M. (2004). *J. Am. Chem. Soc.* 126: 986–987.

105 Hayes, A.M., Morris, D.J., Clarkson, G.J., and Wills, M. (2005). *J. Am. Chem. Soc.* 127: 7318–7319.

106 Cheung, F.K., Lin, C., Minissi, F. et al. (2007). *Org. Lett.* 9: 4659–4662.

107 Soni, R., Jolley, K.E., Clarkson, G.J., and Wills, M. (2013). *Org. Lett.* 15: 5110–5113.

108 Soni, R., Hall, T.H., Mitchell, B.P. et al. (2015). *J. Org. Chem.* 80: 6784–6793.

109 Hodgkinson, R., Jurčík, V., Zanotti-Gerosa, A. et al. (2014). *Organometallics* 33: 5517–5524.

110 Soni, R., Jolley, K.E., Gosiewska, S. et al. (2018). *Organometallics* 37: 48–64.

111 Wakeham, R.J., Morris, J.A., and Williams, J.M.J. (2015). *ChemCatChem* 7: 4039–4041.

112 Touge, T., Nara, H., Fujiwhara, M. et al. (2016). *J. Am. Chem. Soc.* 138: 10084–10087.

113 Wang, B., Zhou, H., Lu, G. et al. (2017). *Org. Lett.* 19: 2094–2097.

114 Kišić, A., Stephan, M., and Mohar, B. (2013). *Org. Lett.* 15: 1614–1617.

115 Kišić, A., Stephan, M., and Mohar, B. (2015). *Adv. Synth. Catal.* 357: 2540–2546.

116 Mitsunami, A., Ikeda, M., Nakamura, H. et al. (2018). *Org. Lett.* 20: 5213–5218.

117 Ma, Y., Liu, H., Chen, L. et al. (2003). *Org. Lett.* 5: 2103–2106.

118 Matsunaga, H., Ishizuka, T., and Kunieda, T. (2005). *Tetrahedron Lett.* 46: 3645–3648.

119 Rast, S., Modec, B., Stephan, M., and Mohar, B. (2016). *Org. Biomol. Chem.* 14: 2112–2120.

120 Zimbron, J.M., Dauphinais, M., and Charette, A.B. (2015). *Green Chem.* 17: 3255–3259.

121 Virboul, M.A.N. and Gebbink, R.J.M.K. (2012). *Organometallics* 31: 85–91.

122 Dub, P.A., Matsunami, A., Kuwata, S., and Kayaki, Y. (2019). *J. Am. Chem. Soc.* 141: 2661–2667.

123 Murphy, S.K. and Dong, V.M. (2013). *J. Am. Chem. Soc.* 135: 5553–5556.

124 Vyas, V.K. and Bhanage, B.M. (2016). *Org. Chem. Front.* 3: 614–619.

125 Hems, W.P., Jackson, W.P., Nightingale, P., and Bryant, R. (2012). *Org. Process Res. Dev.* 16: 461–463.

126 Takehara, J., Hashiguchi, S., Fujii, A. et al. (1996). *Chem. Commun.*: 233–234.

127 Nordin, S.J.M., Roth, P., Tarnai, T. et al. (2001). *Chem. Eur. J.* 7: 1431–1436.

128 Pastor, I.M., Västilä, P., and Adolfsson, H. (2002). *Chem. Commun.*: 2046–2047.

129 Pastor, I.M., Västilä, P., and Adolfsson, H. (2003). *Chem. Eur. J.* 9: 4031–4045.

130 Wettergren, J., Bøgevig, A., Portier, M., and Adolfsson, H. (2006). *Adv. Synth. Catal.* 348: 1277–1282.

131 Coll, M., Pàmies, O., Adolfsson, H., and Diéguez, M. (2011). *Chem. Commun.* 47: 12188–12190.

132 Margalef, J., Slagbrand, T., Tinnis, F. et al. (2016). *Adv. Synth. Catal.* 358: 4006–4018.

133 Ito, M., Shibata, Y., Watanable, A., and Ikariya, T. (2009). *Synlett*: 1621–1626.

134 Sheeba, M.M., Tamizh, M.M., Farrugia, L.J. et al. (2014). *Organometallics* 33: 540–550.
135 Sheeba, M.M., Preethi, S., Nijamudheen, A. et al. (2015). *Catal. Sci. Tecnol.* 5: 4790–4799.
136 Baratta, W., Da Ros, P., Del Zotto, A. et al. (2004). *Angew. Chem. Int. Ed.* 43: 3584–3588.
137 Baratta, W., Herdtweck, E., Siega, K. et al. (2005). *Organometallics* 24: 1660–1669.
138 Chen, F., He, D., Chen, L. et al. (2019). *ACS Catal.* 9: 5562–5566.
139 Langer, T. and Helmchen, G. (1996). *Tetrahedron Lett.* 37: 1381–1384.
140 Sammakia, T. and Stangeland, E.L. (1997). *J. Org. Chem.* 62: 6104–6105.
141 Du, X.-D., Dai, L.-X., Hou, X.-L. et al. (1998). *Chin. J. Chem.* 16: 90–93.
142 Nishibayashi, Y., Takei, I., Uemura, S., and Hidai, M. (1999). *Organometallics* 18: 2291–2293.
143 Zirakzadeh, A., Schuecker, R., Gorgas, N. et al. (2012). *Organometallics* 31: 4241–4250.
144 Reetz, M.R. and Li, X. (2006). *J. Am. Chem. Soc.* 128: 1044–1045.
145 Jiang, Q., Van Plew, D., Murtuza, S., and Zhang, X. (1996). *Tetrahedron Lett.* 37: 797–800.
146 Jiang, Y., Jiang, Q., Zhu, G., and Zhang, X. (1997). *Tetrahedron Lett.* 38: 215–218.
147 Jiang, Y., Jiang, Q., Zhu, G., and Zhang, X. (1997). *Tetrahedron Lett.* 38: 6565–6568.
148 Cuervo, D., Gamasa, M.P., and Gimeno, J. (2004). *Chem. Eur. J.* 10: 425–432.
149 Enthaler, S., Hagemann, B., Bhor, S. et al. (2007). *Adv. Synth. Catal.* 349: 853–860.
150 Ye, W., Zhao, M., Du, W. et al. (2011). *Chem. Eur. J.* 17: 4737–4741.
151 Ye, W., Zhao, M., and Yu, Z. (2012). *Chem. Eur. J.* 18: 10843–10846.
152 Chai, H., Liu, T., and Yu, Z. (2017). *Organometallics* 36: 4136–4144.
153 Barratta, W., Benedetti, F., Zotto, A.D. et al. (2010). *Organometallics* 29: 3563–3570.
154 Zhang, H., Yang, C.-B., Li, Y.-Y. et al. (2003). *Chem. Commun.*: 142–143.
155 Spogliarich, R., Zassinovich, G., Kaspar, J., and Graziani, M. (1982). *J. Mol. Catal.* 16: 359–361.
156 Kvintovics, P., Bakos, J., and Heil, B. (1985). *J. Mol. Catal.* 32: 111–114.
157 Krause, H.W. and Bhatnagar, A.K. (1986). *J. Organomet. Chem.* 302: 265–267.
158 Spogliarich, R., Kašpar, J., Graziani, M., and Morandini, F. (1986). *J. Organomet. Chem.* 306: 407–412.
159 Zassinovich, G., Bettella, R., Mestroni, G. et al. (1989). *J. Organomet. Chem.* 370: 187–202.
160 Müller, D., Umbricht, G., Weber, B., and Pfaltz, A. (1991). *Helv. Chim. Acta.* 74: 232–240.
161 Mashima, K., Abe, T., and Tani, K. (1998). *Chem. Lett.*: 1199–1200.
162 Murata, K., Ikariya, T., and Noyori, R. (1999). *J. Org. Chem.* 64: 2186–2187.

163 Matharu, D.S., Morris, D.J., Kawamoto, A.M. et al. (2005). *Org. Lett.* 7: 5489–5491.
164 Matharu, D.S., Morris, D.J., Clarkson, G.J., and Wills, M. (2006). *Chem. Commun.*: 3232–3234.
165 Echeverria, P.-G., Férard, C., Phansavath, P., and Ratovelomanana-Vidal, V. (2015). *Catal. Commun.* 62: 95–99.
166 Zheng, L.-S., Llopis, Q., Echeverria, P.-G. et al. (2017). *J. Org. Chem.* 82: 5607–5615.
167 Matharu, D.S., Martins, J.E.D., and Wills, M. (2008). *Chem. Asian J.* 3: 1374–1383.
168 Wu, X., Vinci, D., Ikariya, T., and Xiao, J. (2005). *Chem. Commun.*: 4447–4449.
169 Wu, X., Li, X., Zanotti-Gerosa, A. et al. (2008). *Chem. Eur. J.* 14: 2209–2222.
170 Tang, Y., Li, X., Lian, C. et al. (2011). *Tetrahedron: Asymmetry* 22: 1530–1535.
171 Li, J., Lin, Z., Huang, Q. et al. (2017). *Green Chem.* 19: 5367–5370.
172 Ahfford, K., Lind, J., Mäler, L., and Adolfsson, H. (2008). *Green Chem.* 10: 832–835.
173 Kang, G., Lin, S., Shiwakoti, A., and Ni, B. (2014). *Catal. Commun.* 57: 111–114.
174 Liu, Q., Wang, C., Zhou, H. et al. (2018). *Org. Lett.* 20: 971–974.
175 Zaitsev, A.B. and Adolfsson, H. (2006). *Org. Lett.* 8: 5129–5132.
176 Wettergren, J., Zaitsev, A.B., and Adolfsson, H. (2007). *Adv. Synth. Catal.* 349: 2556–2562.
177 Ahfford, K., Ekström, J., Zaitsev, A.B. et al. (2009). *Chem. Eur. J.* 15: 11197–11209.
178 Coll, M., Ahlford, K., Pàmies, O. et al. (2012). *Adv. Synth. Catal.* 354: 415–427.
179 Coll, M., Pàmies, O., and Diéguez, M. (2014). *Adv. Synth. Catal.* 356: 2293–2302.
180 Ak, B., Aydemir, M., Durap, F. et al. (2015). *Tetrahedron: Asymmetry* 26: 1307–1313.
181 Ak, B., Aydemir, M., Durap, F. et al. (2015). *Inorg. Chim. Acta* 438: 42–51.
182 Zhou, G., Aboo, A.H., Robertson, C.M. et al. (2018). *ACS Catal.* 8: 8020–8026.
183 Li, Y.-Y., Dong, Z.-R., Zhang, J.-N., and Gao, J.-X. (2011). *Sci. Sin. Chim.* 41: 654–662.
184 Chen, J.-S., Li, Y.-Y., Dong, Z.-R. et al. (2004). *Tetrahedron Lett.* 45: 8415–8418.
185 Dong, Z.-R., Li, Y.-Y., Chen, J.-S. et al. (2005). *Org. Lett.* 7: 1043–1045.
186 Xing, Y., Chen, J.-S., Dong, Z.-R. et al. (2006). *Tetrahedron Lett.* 47: 4501–4503.
187 Tao, M., Wu, F., Li, T. et al. (2017). *Chin. Chem. Lett.* 28: 97–100.
188 Zhang, W.-J., Ruan, S.-H., Shen, W.-Y. et al. (2019). *Catal. Commun.* 119: 153–158.
189 Liu, W.-P., Yuan, M.-L., Yang, X.-H. et al. (2015). *Chem. Commun.* 51: 6123–6125.
190 Yoshida, K., Kamimura, T., Kuwabara, H., and Yanagisawa, A. (2015). *Chem. Commun.* 51: 15442–15445.
191 Tian, C., Gong, L., and Meggers, E. (2016). *Chem. Commun.* 52: 4207–4210.
192 Mikami, K., Wakabayashi, K., Yusa, Y., and Aikawa, K. (2006). *Chem. Commun.*: 2365–2367.
193 Chen, J.S., Chen, L.L., Xing, Y. et al. (2004). *Acta Chim. Sinica* 62: 1745–1750.

194 Yu, S., Shen, W., Li, Y. et al. (2012). *Adv. Synth. Catal.* 354: 818–822.

195 Mikhailine, A., Lough, A.J., and Morris, R.H. (2009). *J. Am. Chem. Soc.* 131: 1394–1395.

196 Meyer, N., Lough, A.J., and Morris, R.H. (2009). *Chem. Eur. J.* 15: 5605–5610.

197 Mikhailine, A.A. and Morris, R.H. (2010). *Inorg. Chem.* 49: 11039–11044.

198 Sues, P.E., Lough, A.J., and Morris, R.H. (2011). *Organometallics* 30: 4418–4431.

199 Lagaditis, P.O., Lough, A.J., and Morris, R.H. (2010). *Inorg. Chem.* 49: 10057–10066.

200 Zou, W., Lough, A.J., Li, Y.F., and Morris, R.H. (2013). *Science* 342: 1080–1083.

201 Bigler, R. and Mezzetti, A. (2014). *Org. Lett.* 16: 6460–6463.

202 Bigler, R., Huber, R., and Mezzetti, A. (2015). *Angew. Chem. Int. Ed.* 54: 5171–5174.

203 Bigler, R. and Mezzetti, A. (2016). *Org. Process Res. Dev.* 20: 253–261.

204 Bigler, R., Huber, R., Stöckli, M., and Mezzetti, A. (2016). *ACS Catal.* 6: 6455–6464.

205 De Luca, L. and Mezzetti, A. (2017). *Angew. Chem. Int. Ed.* 56: 11949–11953.

206 Vega, E., Lastra, V.E., and Gamasa, M.P. (2013). *Inorg. Chem.* 52: 6193–6198.

207 Coverdale, J.P.C., Sanzhez-Cano, C., Clarkson, G.J. et al. (2015). *Chem. Eur. J.* 21: 8043–8046.

208 Zirakzadeh, A., de Aguiar, S.R.M.M., Stöger, B. et al. (2017). *ChemCatChem* 9: 1744–1748.

209 Dong, Z.-R., Li, Y.-Y., Yu, S.-L. et al. (2012). *Chin. Chem. Lett.* 23: 533–536.

210 Li, Y.-Y., Yu, S.-L., Shen, W.-Y., and Gao, J.-X. (2015). *Acc. Chem. Res.* 48: 2587–2598.

5

Asymmetric (Transfer) Hydrogenation of Substituted Ketones Through Dynamic Kinetic Resolution

Pierre-Georges Echeverria, Tahar Ayad, Phannarath Phansavath and Virginie Ratovelomanana-Vidal

PSL University, Chimie ParisTech-CNRS, Institute of Chemistry for Life & Health Sciences, CSB2D Team, 11 rue Pierre et Marie Curie, Paris, France

5.1 Introduction

Because enantiomerically enriched alcohols are highly valuable and versatile intermediates for the preparation of pharmaceutical agents, agrochemicals, fragrances, or advanced materials, the development of efficient and straightforward approaches to these compounds has been the object of numerous studies. In this regard, transition-metal-catalyzed asymmetric homogeneous hydrogenation (AH) [1] and asymmetric transfer hydrogenation (ATH) [2] of ketones appear as very efficient methods to synthesize chiral alcohols with high levels of stereoselectivity. Initial work carried out in this area used ruthenium(II) catalysts associated to the BINAP diphosphine ligand developed by Noyori and coworkers for the AH of functionalized ketones [3]. On the other hand, Noyori et al. developed Ru complexes bearing an *N*-(*p*-toluenesulfonyl)-1,2-diphenylethylenediamine (TsDPEN) ligand for the ATH of nonfunctionalized ketones, using either isopropanol or formate salts as the hydrogen donors [4]. Since these initial reports, extensive efforts have been displayed toward the synthesis of modified Ru(II) complexes or the development of other transition-metal catalysts such as rhodium and iridium derivatives bearing various chiral ligands for both the AH and ATH of ketones. In this context, combination of both methods with a dynamic kinetic resolution (DKR) process allows a highly efficient route to chiral alcohols bearing two or more stereogenic centers from substituted ketones. Particularly, in 1989, the pioneering works of Noyori et al. [5] and Genêt and coworkers [6] led to the first examples of homogeneous enantioselective ruthenium-catalyzed hydrogenations of racemic α-acetamido β-keto esters via DKR using, respectively, BinapRu(II) and ChiraphosRu(II) catalysts. The hydrogenation reaction of racemic 2-acylamino-3-oxobutyrate delivered the corresponding *syn* L- and D-threonine derivatives with high levels of diastereo- and enantioselectivities (up to 98% ee, *syn*/*anti* up to 99 : 1). This chapter presents progress achieved from 2007 to May 2019 in the field of asymmetric (transfer) hydrogenation of substituted ketones through DKR [7] with a particular focus on the development of new

Asymmetric Hydrogenation and Transfer Hydrogenation, First Edition.
Edited by Virginie Ratovelomanana-Vidal and Phannarath Phansavath.

catalytic systems and highlights the utility of such homogeneous catalytic processes with examples of industrial applications and total syntheses of biorelevant targets.

5.2 α-Substituted Ketones

In 2007, researchers from Merck published the enantioselective synthesis of Taranabant, a drug used for the treatment of obesity employing an Ru-catalyzed enantioselective reduction of ketone via DKR as a key step. After thorough evaluation of the reaction conditions, the *anti*-compound was obtained with excellent stereoselectivity (dr = 9 : 1, ee = 94%) on 6.6 kg scale (Scheme 5.1) [8].

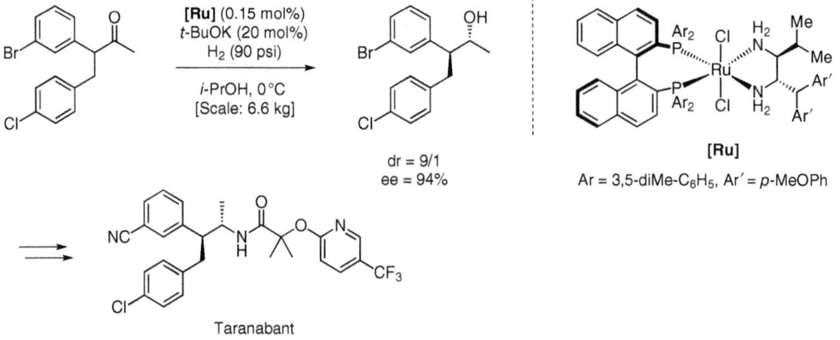

Scheme 5.1 Enantioselective synthesis of Taranabant. Source: Based on Chen et al. [8].

Along with the synthesis of enantiopure α-hydroxy acetals, Ohkuma and coworkers reported an efficient protocol for the ruthenium-mediated AH of α-amido ketones via DKR. With this method, perfect *syn*-selectivity was observed with excellent enantioselectivity (up to 99% ee). The same system was applied to α-alkoxy ketones leading to *anti* compounds (*anti*/*syn* up to 97 : 3) with excellent enantioselectivity (up to 98% ee). To explain the diastereocontrol, the authors suggested a Felkin–Anh-type transition state simply based on group size difference (Scheme 5.2) [9].

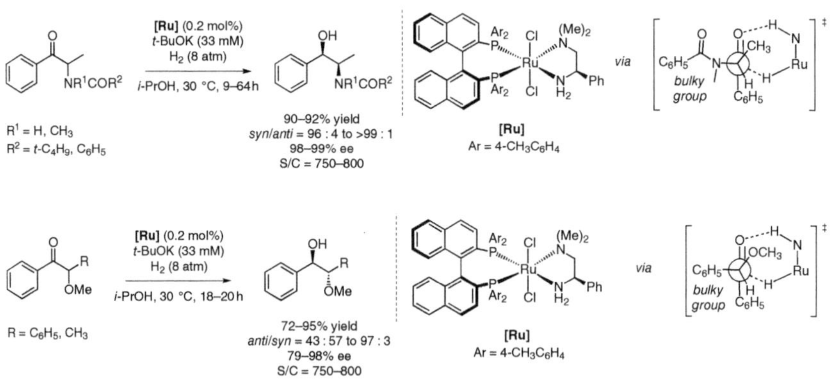

Scheme 5.2 AH of α-branched aromatic ketones. Source: Arai and coworkers [9].

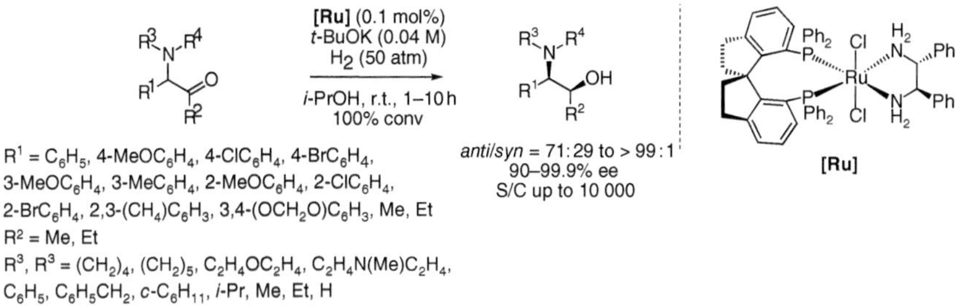

Scheme 5.3 Enantio- and diastereoselective synthesis of chiral amino alcohols via ruthenium-catalyzed AH of α-amino aliphatic ketones.

In 2009, Zhou et al. reported the AH of acyclic racemic α-amino aliphatic ketones via DKR using [$RuCl_2$\{(*S*)-SDPs\}\{(*R*,*R*)-DPEN\}], ((*S*)-SDPs: (*S*)-(−)-7,7′-Bis(diphenylphosphino)-2,2′,3,3′-tetrahydro-1,1′-spirobiindene; DPEN: (1*R*,2*R*)-(+)-1,2-Diamino-1,2-diphenylethane), catalyst, and *t*-BuOK under 50 atm of H_2 in isopropanol at room temperature (Scheme 5.3). With this system, a large variety of α-*N*,*N*-dialkylamino aliphatic ketones were hydrogenated with excellent diastereoselectivity (up to >99 : 1) and enantioselectivity (up to 99.9% ee) [10].

The same year, Hamada et al. published an Ni-catalyzed AH of nonactivated substituted aromatic α-aminoketone hydrochlorides catalyzed by homogeneous chiral nickel catalyst in the presence of NaBArF in toluene. *Anti*-products were obtained in perfect diastereoselectivity (up to >99 : 1) and good enantioselectivity (up to 96% ee). Interestingly, they observed a strong solvent effect: when the reaction was carried out in acetic acid on enantiopure α-aminoketone, racemization was suppressed and *anti* or *syn* products were obtained, depending on the ligand configuration (Scheme 5.4) [11].

The group of Itsuno demonstrated the AH of racemic α-(*N*-benzoyl-*N*-methylamino)propiophenone through DKR using polymer-immobilized catalyst. They showed that a combination of (*S*)-BINAP and (*R*,*R*)-supported diamine was a matched pair to obtain high stereoselectivity. They also observed that the nature of the cross-linking agent (degree and structure) affected the catalytic activity. Finally, the catalytic system was easily recycled at least five times without any loss of activity and stereoselectivity (Scheme 5.5) [13].

An Ru-catalyzed enantioselective synthesis of chiral cyclic and acyclic β-aryloxy alcohols through DKR was described by the group of Zhou with excellent levels

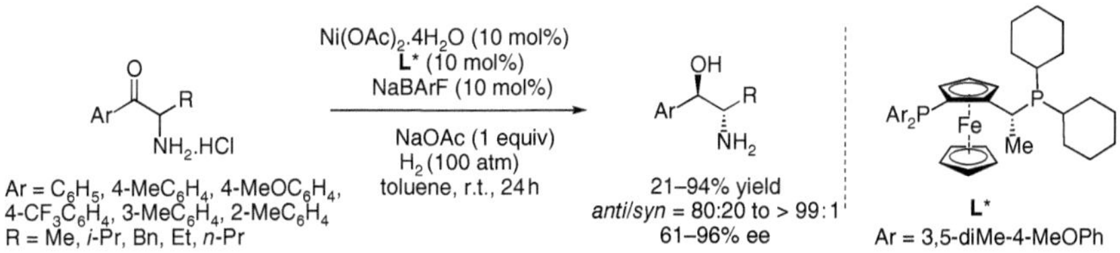

Scheme 5.4 Homogeneous chiral nickel-catalyzed AH of substituted aromatic α-aminoketone hydrochlorides through DKR. Source: Based on Hibino et al. [11].

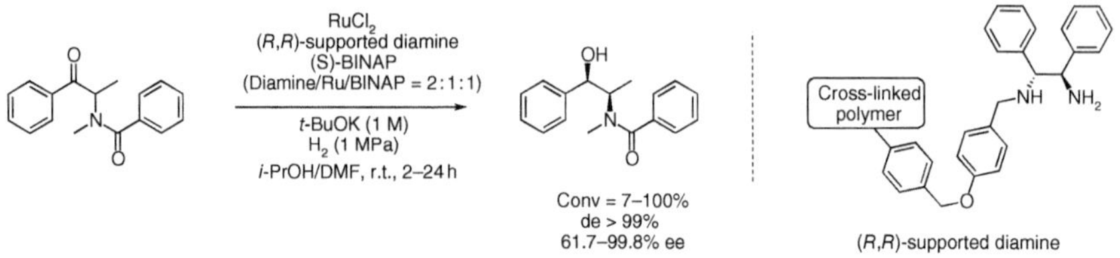

Scheme 5.5 Polymer-immobilized catalyst for AH of racemic α-(*N*-benzoyl-*N*-methylamino)propiophenone. Source: Based on Chiwara et al. [12].

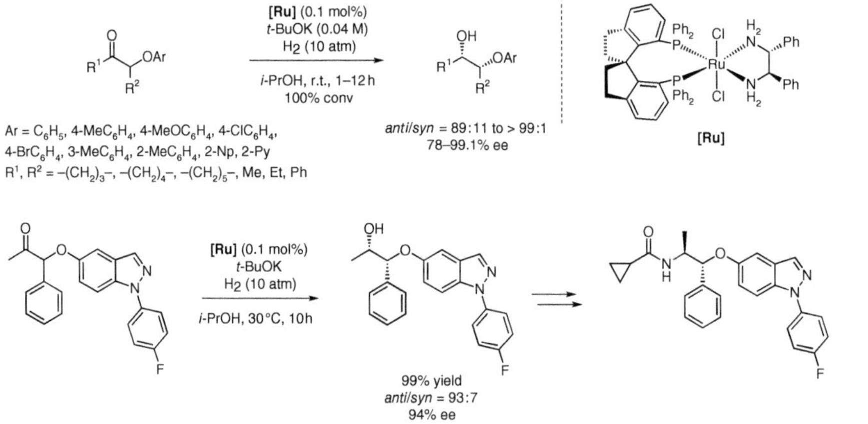

Scheme 5.6 Enantioselective synthesis of chiral β-aryloxy alcohols by AH via DKR. Source: Based on Bai et al. [12].

of stereoselectivity (*anti*/*syn* up to >99 : 1 and ee up to 99.1%). This method was successfully applied to the enantioselective synthesis of potential nonsteroidal glucocorticoid modulators (Scheme 5.6) [12].

Asymmetric hydrogenation of diverse aryl heterocycloalkyl ketones catalyzed by [$RuCl_2$((*S*)-Binap)((*R*)-DMAPEN)] through DKR was developed in 2011 by the group of Ohkuma (Scheme 5.7) providing the corresponding alcohols in 95–100%

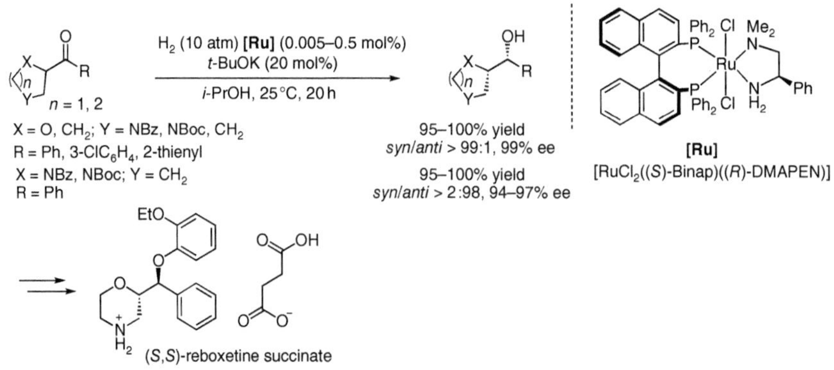

Scheme 5.7 Ru-catalyzed asymmetric hydrogenation of aryl heterocycloalkyl ketones.

yields with diastereoselectivities, depending on the X group present on the heterocyclic scaffolds (up to >99 : 1) and enantioselectivities up to 99%. Hydrogenation of ketones led to *syn* selectivities (X = O or CH_2), whereas high *anti* selectivities were obtained for heterocycles bearing a bulkier NBz or NBoc substituent. The hydrogenation could be carried out with catalyst loading up to S/C = 20 000 without sacrificing the catalytic efficiency. This method was used to synthesize (*S*,*S*)-reboxetine succinate, a selective norepinephrine uptake inhibitor [14].

In 2012, Chung et al. from Merck Research Laboratories reported a practical Ru-catalyzed DKR asymmetric hydrogenation to access a new glucagon receptor antagonist drug candidate for the treatment of type 2 diabetes (Scheme 5.8). After screening the hydrogenation conditions, [$RuCl_2$((*S*)-xyl-Segphos)((*S*)-DAIPEN)], **[Ru]**, was selected among several catalysts for further development producing the targeted alcohol in 94% yield with excellent *anti* diastereoselectivity (>99% dr) and enantioselectivity (>98.5%). The hydrogenation reaction was successfully performed on 110 kg scale at S/C = 5000 [15].

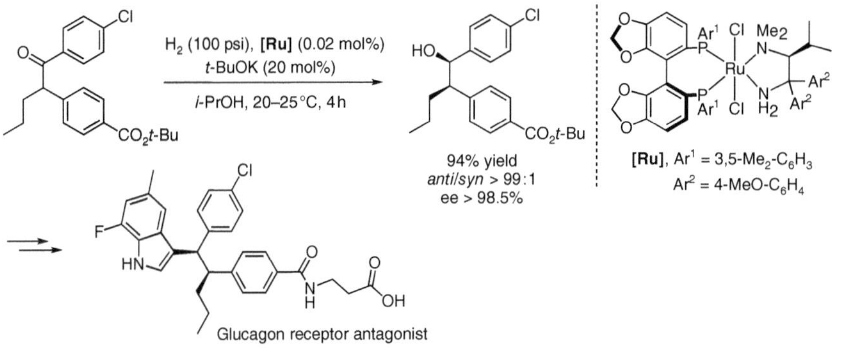

Scheme 5.8 Ru-catalyzed DKR/AH to access a new glucagon receptor antagonist.

In 2013, Chen and coworkers developed a concise synthesis of (+)-conocarpan and (+)-obtusafuran, natural dihydrobenzofuran neolignan derivatives, based on asymmetric hydrogenations via DKR (Scheme 5.9). Optimizations of the reaction conditions to access (+)-conocarpan demonstrated that 0.1 mol% of [$RuCl_2$((*S*)-xyl-Segphos)((*S*)-DAIPEN)] catalyst in the presence of *t*-BuOK

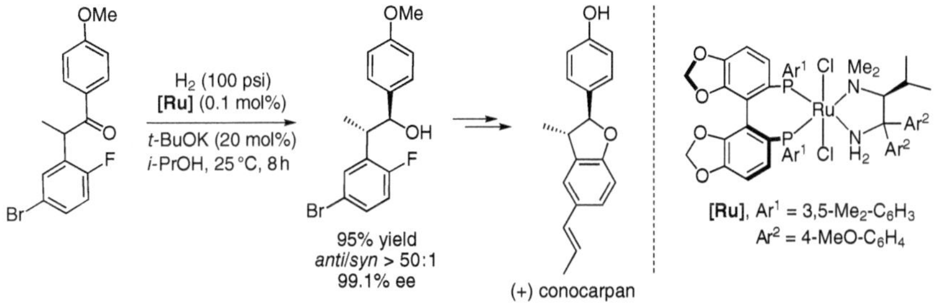

Scheme 5.9 Concise synthesis of (+)-conocarpan and (+)-obtusafuran.

facilitated the epimerization, delivering the corresponding alcohol in 95% yield with high *anti* selectivities (99.1% ee, >50 : 1 dr) [16].

Zhang and coworkers reported in 2017 a spectacular highly diastereo- and enantioselective synthesis of chiral amino alcohols, which are important structural motifs found in many natural products and pharmaceuticals, through Ir-f-amphox-catalyzed AH via DKR [17]. After optimization of the reaction parameters, a wide variety of racemic α-amino β-unfunctionalized *N*,*N*-disubstituted ketones were smoothly converted into the corresponding α-amino β-unfunctionalized alcohols in quantitative yields with excellent diastereo- and enantioselectivities (all products >99 : 1 dr, and >99% ee with turnover number (TON) up to 100 000). Importantly, this process provides a powerful low cost and extremely green synthetic strategy for the synthesis of key chiral intermediates of the preclinical antitumor agent (*S*,*S*)-R116010 (Scheme 5.10).

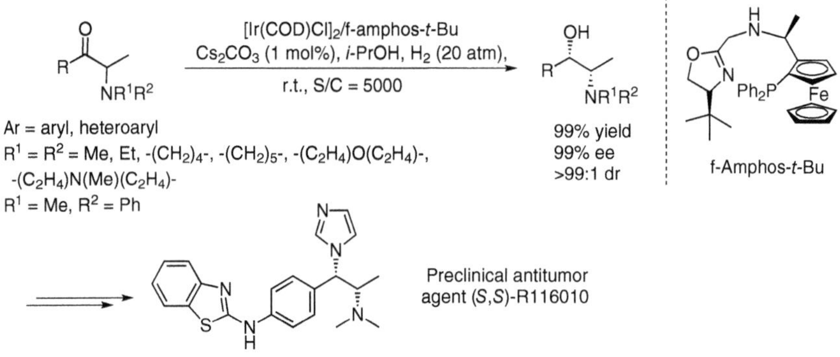

Scheme 5.10 Asymmetric synthesis of α-amino β-unfunctionalized alcohols.

In 2015, Chung, Scott, and coworkers from Merck & Co Process Chemistry Department described a convergent synthetic route to Omarigliptin, a long-acting DPP-4 inhibitor for the treatment of type 2 diabetes. The process development for manufacturing scale involved an Ru-catalyzed ATH via DKR of a racemic *N*-Boc-α-substituted ketone to produce the desired *anti* 1,2-amino alcohol (Scheme 5.11). The best catalytic performance was achieved using 0.1 mol% of the oxo-tethered Ru(II) catalyst, [RuCl((*R*,*R*)-TsDENEB)], (*R*,*R*)-**[Ru]**, in the presence of 5 equiv of HCO_2H/Et_3N (5 : 2) as the hydrogen source and 3 equiv of DABCO allowing to improve the diastereoselectivity (96 : 4 dr) and enantioselectivity

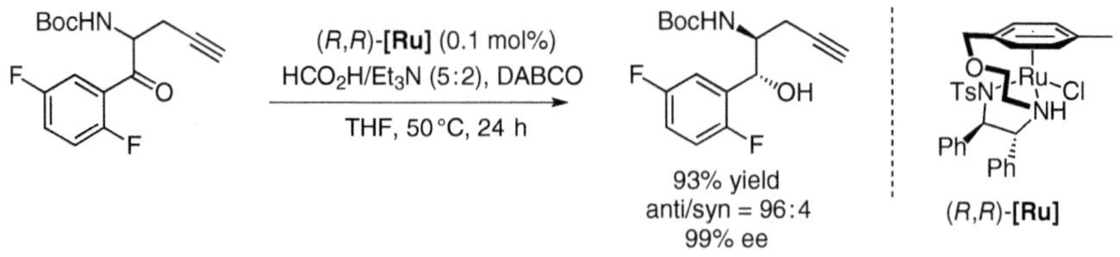

Scheme 5.11 Convergent synthetic route to Omarigliptin.

(>99% ee) with 93% yield. This synthetic route is amenable to multikilogram-scale production [18].

5.3 α-Substituted Cyclic Ketones

Zhou and coworkers devised a highly enantioselective synthesis of *cis*-cyclic amino alcohols using AH by DKR (Scheme 5.12). The reaction was performed using [$RuCl_2${(*S*)-SDPs}{(*R*,*R*)-DPEN}], **[Ru]**, to deliver the corresponding *N*,*N*-disubstituted *cis*-α-aminocycloalkanols with perfect diastereoselectivity (*cis*/*trans* >99 : 1) and excellent enantioselectivity (up to 99.9% ee) [19].

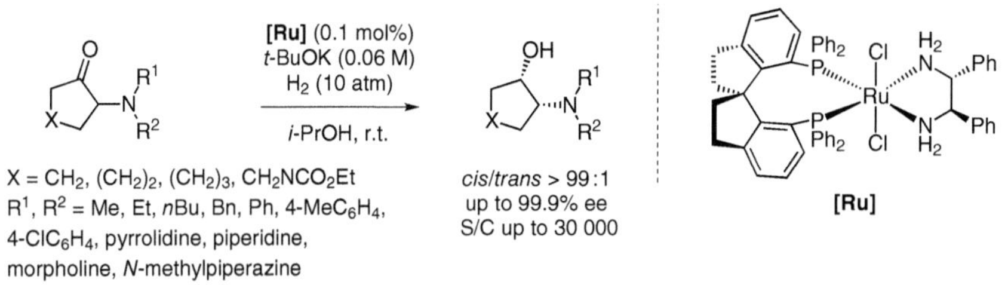

Scheme 5.12 Enantioselective synthesis of *cis*-α-aminocycloalkanols by AH via DKR.

In 2012, the same group published a ruthenium-catalyzed hydrogenation of racemic α-arylcyclohexanones through DKR using an $RuCl_2$-SDPs/diamine catalyst (SDPs=7,7′-bis(diarylphosphino)-1,1′spirobiindane). A range of enantiomerically enriched *cis* α-arylcyclohexanols were obtained in 68–98% yields (*cis*/*trans* >99 : 1) and enantioselectivities varying from 57% to 99% (Scheme 5.13). Enantioselectivities and reactivities of the hydrogenation decreased with α-arylcyclohexanones bearing *ortho*-substituents. Based on this reaction, (−)-α-lycorane, (−)-CP 55940, a potent cannabinoid receptor agonist, and tetrahydrocannabinol derivatives were prepared [20].

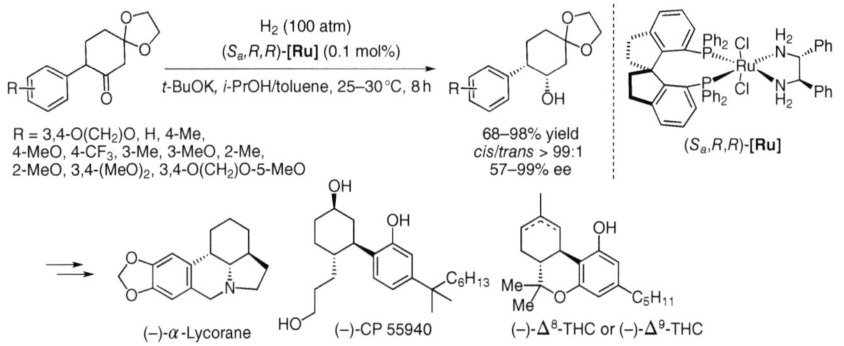

Scheme 5.13 Preparation of (−)-α-lycorane and (−)-CP 55940.

Xie and Zhou employed a highly efficient ruthenium-catalyzed asymmetric hydrogenation of racemic α-aryloxy cyclohexanones to access (−)-galanthamine, and (−)-lycoramine alkaloids, active as acetylcholinesterase inhibitors, using $RuCl_2$-(*S*)-SDP-(*R*,*R*)-DPEN, (S_a,*R*,*R*)-**[Ru]** (Scheme 5.14). The corresponding *cis* alcohols (*cis*/*trans* >99 : 1) were prepared in 99% yield and 97% ee value [21].

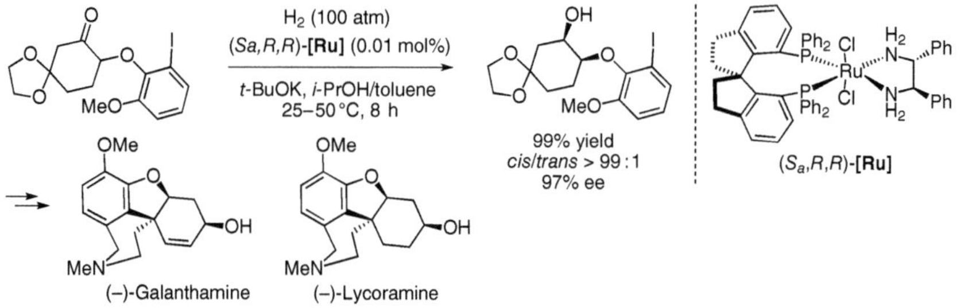

Scheme 5.14 Synthesis of (−)-galanthamine and (−)-lycoramine.

Zhang and coworkers demonstrated that Ir/f-amphol complex is an effective catalyst for the AH of racemic methyl-substituted benzo-fused five- to seven-membered cyclic ketones involving a DKR process [22]. As shown in Scheme 5.15, the corresponding enantioenriched cyclic alcohols were obtained with good to excellent results, depending on the size of the benzo-fused ring.

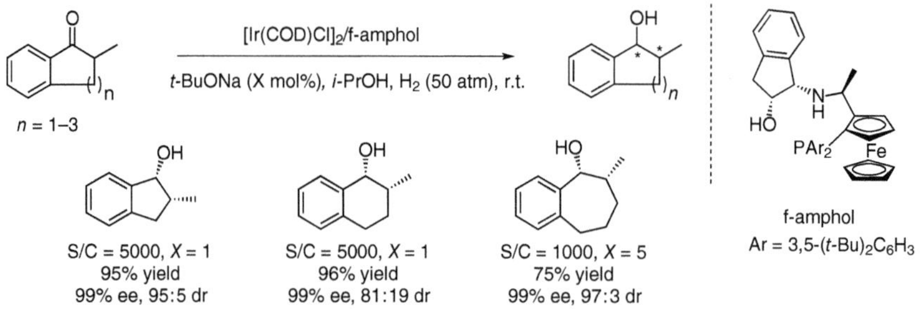

Scheme 5.15 Asymmetric synthesis of benzo-fused five to seven-membered cyclic alcohols.

In 2018, Kozlowski and coworkers described the first examples of Cu-mediated AH of 2-substituted-1-tetralones and related heteroaryl ketones via DKR involving a new class of tunable heterophosphole dimeric ligands [23]. Mechanistic studies revealed that the reaction proceeded via *in situ* formation of a novel heteroligated dimeric copper hydride transition state incorporating two Cu atoms, where different ligands can be coordinated to each Cu. Under optimized conditions using 5 mol% of Cu complex containing the electron-poor azaphosphole *N*-BABIPhos in combination with the electron-rich P(3,5-xylyl) phosphine in *t*-AmylOH, various enantioenriched tetralols that possess two contiguous stereogenic centers were obtained in moderate to excellent yields and selectivities at room temperature (Scheme 5.16).

Efforts by Lassaletta, Fernández, and coworkers established ATH via DKR as an efficient process to deliver *cis*-α-substituted cycloalkanols (Scheme 5.17). Various cyclic α-substituted ketones were reacted in the presence of 0.5 mol% of chiral catalyst [RuCl(η^6-*p*-cymene)(*S*,*S*)-TsDPEN] in the azeotropic mixture HCO_2H/Et_3N to

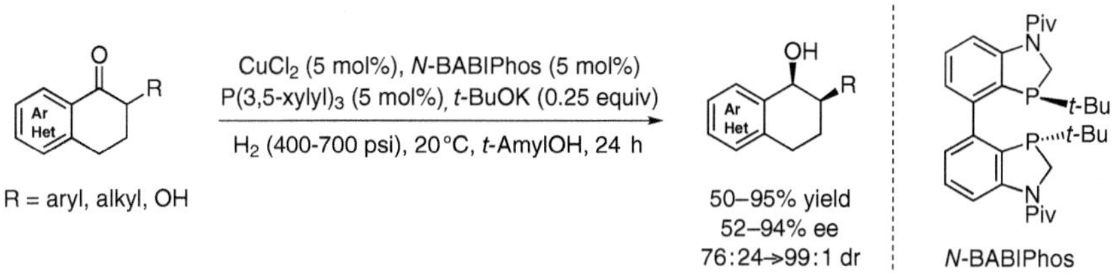

Scheme 5.16 Asymmetric synthesis of two-substituted-1-tetralols.

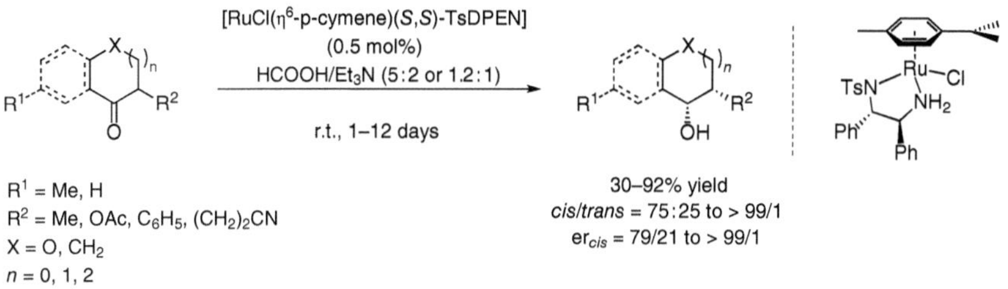

Scheme 5.17 Enantioselective synthesis of *cis*-α-substituted cycloalkanols and *trans*-cycloalkyl amines.

furnish the targeted products with a high level of stereoselectivity (*cis*/*trans* up to >99 : 1 and er up to >99 : 1) [24].

In 2016, Bhanage et al. presented the Ru-catalyzed DKR/ATH of *rac*-α-heteroaryl amino bicyclic alkanones in the presence of HCO_2H/Et_3N (5 : 2) as the hydrogen source in a highly enantio- and diastereoselective manner to access biologically important *cis*-β-heteroaryl amino cycloalkanols. However, monocyclic substrates were obtained with low to good enantiocontrol (Scheme 5.18) [25]. In 2018, the same group further reported Ru/prolinamide-catalyzed DKR/ATH of *rac*-α-heteroaryl- and sulfonyl-substituted monocyclic alkanones using sodium formate as the

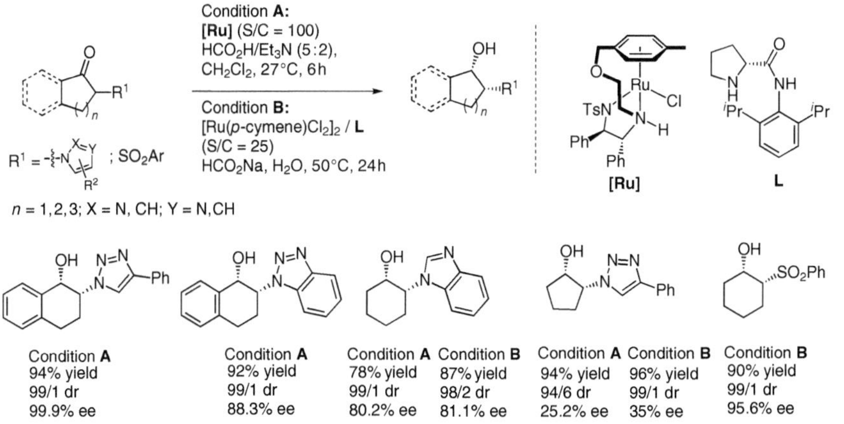

Scheme 5.18 Ru-catalyzed DKR/ATH of *rac*-α-heteroaryl amino mono- or bicyclic alkanones. Source: Vyas and Bhanage [25].

hydrogen source and water as a solvent. The method showed high stereoselectivities for six- or seven-membered substrates, but five-membered substrates gave low enantioselectivities (Scheme 5.18).

In 2017, Metz and co-worker [26] reported a simple method for preparing enantioenriched isoflavanones such as (*R*)-dihydrodaidzein, through ruthenium-catalyzed DKR/ATH of easily available racemic substrates followed by oxidation of the resulting isoflavanols (Scheme 5.19).

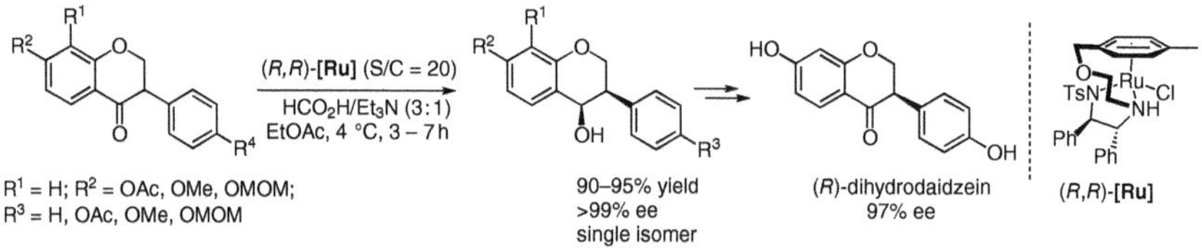

Scheme 5.19 Ruthenium-catalyzed DKR/ATH approach to enantioenriched isoflavanones. Source: Based on Metz and co-worker [26].

In 2016, Fang et al. reported a new strategy for the total synthesis of daldinin A and concentricolide using Ru-catalyzed DKR/ATH of the corresponding ketones in the presence of sodium formate as the hydrogen source under mild conditions. Daldinin A could be transformed into daldinins B and C. The protocol showed a mild, facile, and practical access to a variety of *syn* dihydrobenzofuran and phthalide motifs in enantiomerically pure form (Scheme 5.20) [27].

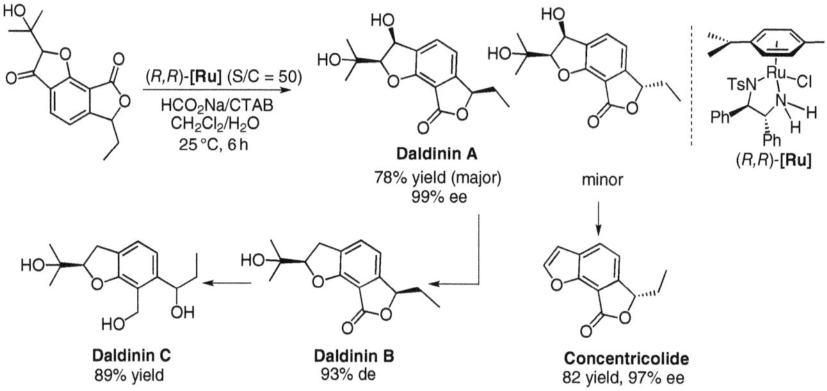

Scheme 5.20 Total synthesis of daldinin A and concentricolide using Ru-catalyzed DKR/ATH. Source: Based on Fang et al. [27].

In 2017, Fang et al. reported Ru-catalyzed DKR/ATH of *rac*-2,3-dihydrobenzofuran α-substituted ketones to obtain *cis*-2,3-dihydrobenzofuran-3-ols in CH_2Cl_2/water under mild conditions (Scheme 5.21). This mild, efficient route proceeded with excellent stereoselectivities and offers an interesting alternative to the asymmetric hydrogenation of benzofurans [29]. Before Fang's work, McErlean and coworkers reported in 2016 the preparation of compound **A** (78% yield, 73% ee, after

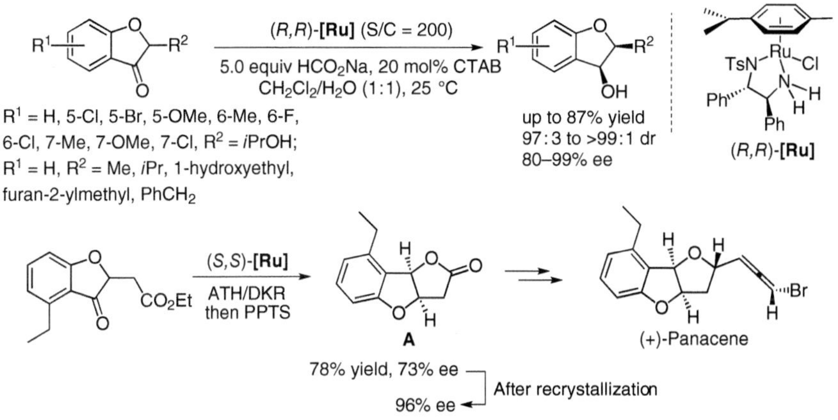

Scheme 5.21 Ru-catalyzed DKR/ATH of *rac*-2,3-dihydrobenzofuran α-substituted ketones. Source: Based on Alnafta et al. [28].

recrystallization in 96% ee) through ATH/DKR, and the latter was used in the total synthesis of (+)-Panacene (Scheme 5.21) [28].

In 2018, Fang and coworkers depicted an effective method for the synthesis of *cis*-2,3-dihydrobenzofuran-3-ol derivatives through ATH of benzofuran-3-(2H)-ones using Ru(II)complex-mediated DKR [30]. To achieve high catalytic performance, the reaction required to be carried out in methanol with an optimum temperature of 65 °C in the presence of 0.5 mol% of [RuCl(*p*-cymene){(*R*,*R*)-TsDPEN}] as a catalyst in combination with cetyltrimethylammonium bromide as phase-transfer catalyst and HCO_2Na as a hydrogen donor. Under these conditions, several dihydrobenzofuranols containing three stereocenters were obtained in low to high yields (31–90%), with high to excellent asymmetric inductions (98 : 2 to >99 : 1 dr, 85–99% ee), albeit the substrate scope amenable to this transformation was rather limited (Scheme 5.22).

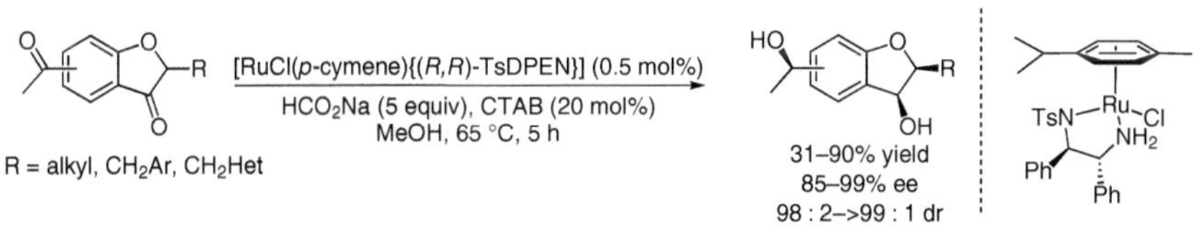

Scheme 5.22 Asymmetric synthesis of *cis*-2,3-dihydrobenzofuran-3-ol derivatives.

In 2016, Bhanage and coworkers achieved a highly stereoselective synthesis of biologically relevant *cis*-β-heteroaryl amino cycloalkanols through Ru(II)-catalyzed ATH relying upon the kinetic conversion of stereolabile 2-substituted 1-cycloalkanones [25a]. Optimized conditions were established using an Ru-(*R*,*R*)-TsDPEN-tethered complex as a catalyst in combination with HCO_2H/Et_3N (5 : 2) as a hydrogen source at 27 °C in dichloromethane. Through this protocol, different β-heteroaryl amino cycloalkanols with two contiguous chiral centers were obtained in yields up to 96% and generally with high to excellent

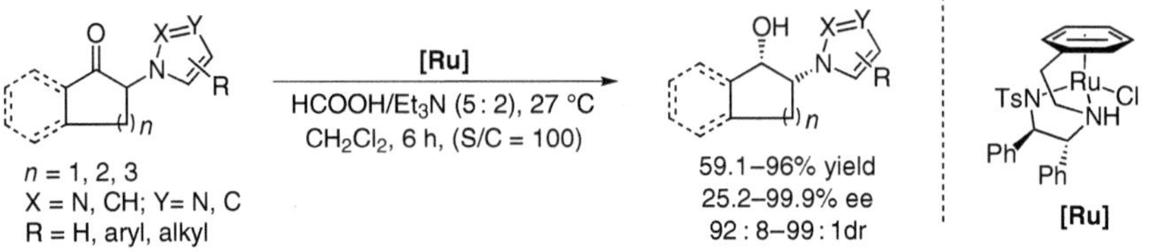

Scheme 5.23 Asymmetric synthesis of *cis*-β-heteroaryl amino cycloalkanols.

selectivity, depending on the substitution pattern on the heteroaryl groups as well as the cycloalkanone ring size (Scheme 5.23). For example, it turned out that the cyclopentanone substituted by an imidazole nucleus at the two-position was not a suitable substrate for this reaction (25.2% ee). Moreover, the usefulness of this methodology has been demonstrated by the synthesis of an antileishmanial agent and a chiral ionic liquid.

In 2019, Zhang, Guo, and coworkers described an efficient synthetic route for the construction of valuable chiral carbocyclic nucleosides through ATH via DKR of rac-α-(purin-9-yl)cyclopentones [31]. Optimal results were obtained by performing the reaction in dioxane at room temperature in the presence of 1 mol% of *N*-pentafluorophenyl sulfonyl-DPEN-based Ru(II) complex and an azeotropic (1 : 1) mixture of HCO_2H and Et_3N as the hydrogen source. Under these conditions, a broad range of optically active *cis*-β-(purin-9-yl)cyclopentanols were obtained in high yields (78–97%), with moderate to excellent enantioselectivities (70–99% ee) and uniformly high diastereoselectivities (*cis*/*trans* >20 : 1) (Scheme 5.24).

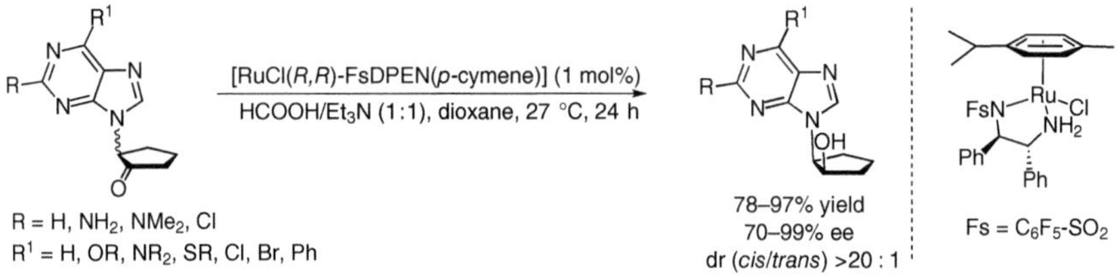

Scheme 5.24 Asymmetric synthesis of *cis*-β-(purin-9-yl)cyclopentanols.

In 2013, McErlean and coworkers reported a scalable and flexible enantioselective synthesis of synthetic strigolactone (+)-GR24 (Scheme 5.25) based on ruthenium-catalyzed ATH. Both enantiomers of GR24 could be accessed using this strategy. Particularly, DKR of a racemic indanone using 4 mol% of Noyori's complex (*S*,*S*)-**[Ru]** and HCO_2H/*i*-Pr_2NEt as the hydrogen source provided after acid-mediated epimerization the desired chiral lactone (*syn*/*anti* >99 : 1) in 80% yield and 92% ee which was upgraded to >99% after a single recrystallization. The same sequence performed using (*R*,*R*)-**[Ru]** as a catalyst delivered the corresponding (−)-GR24 and epi-(−)-GR24 strigolactone with equal facility [32].

In 2017, Mohar et al. demonstrated the effectiveness of the *ansa*-Ru(II) complex developed in their group, containing the enantiopure piperidino-SO_2DPEN$(CH_2)_4$ (η^6-Ph) ligand in the asymmetric DKR-transfer hydrogenation of α-(*N*-sulfonylimino)

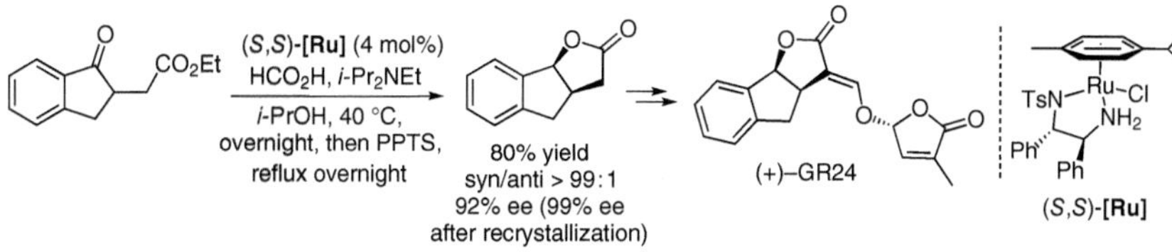

Scheme 5.25 Scalable and flexible enantioselective synthesis of synthetic strigolactone (+)-GR24.

and α-(*N*-sulfonylamino) aryl ketones [33]. Remarkably, this protocol could be performed with a very low catalyst loading (S/C = 10 000) in an HCO_2H/Et_3N mixture at 60 °C, leading to enantioenriched functionalized benzo-γ- and benzo-δ-sultams in excellent yields (97–100%), with high diastereomeric ratios (97 : 3 to >99 : 9 dr) and perfect enantioselectivities (99–99.9% ee) (Scheme 5.26).

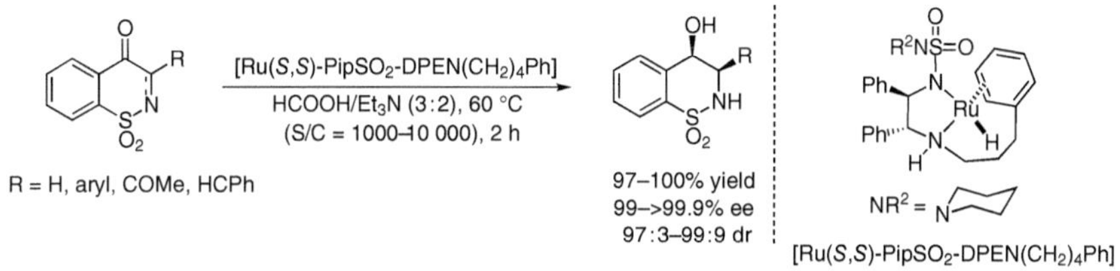

Scheme 5.26 Asymmetric synthesis of functionalized benzo-γ- and benzo-δ-sultams.

In 2019, our group reported the first synthesis of optically active *cis*-3-(hydroxymethyl)-chroman-4-ols from 3-formylchromone derivatives through Rh(III)-catalyzed ATH combined with a DKR process under mild conditions (Scheme 5.27) [34]. Intensive experiments revealed that the best results were obtained in dichloromethane at 30 °C using 0.5 mol% of the in-house developed TsDPEN-based tethered Rh(III) complex in the presence of the formic acid/triethylamine combination (5 : 2). Under these conditions, a wide range of 3-formylchromones that possess either electron-donating or -withdrawing groups on the phenyl ring were smoothly reduced, in a single operation, to the corresponding 3-(hydroxymethyl)chromanols in good isolated yields (72–92%) with uniformly high enantioselectivities (97–99% ee) and excellent levels of diastereoselectivities (92 : 8–98 : 2 dr).

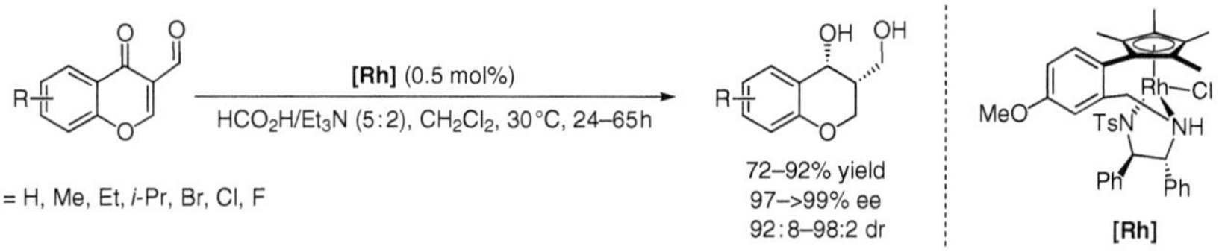

Scheme 5.27 Asymmetric synthesis of *cis*-3-(hydroxymethyl)-chroman-4-ols. Source: Modified from He et al. [34].

Ruthenium-catalyzed ATH of 2-aroyl-1-tetralones was described in 2011 by Zhang and coworkers using $[RuCl_2(p\text{-cymene})]_2$ combined with (1*S*,2*S*)-TsDPEN

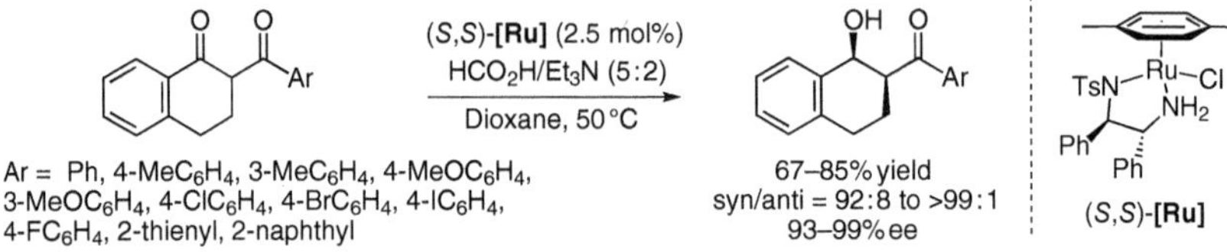

Scheme 5.28 Ruthenium-catalyzed asymmetric transfer hydrogenation of 2-aroyl-1-tetralones.

(TsDPEN = *N*-(tosyl)-1,2-diphenylethylenediamine), i.e. (*S,S*)-**[Ru]** complex in HCO_2H/Et_3N (5 : 2) as the hydrogen source (Scheme 5.28). The ATH/DKR reaction afforded the corresponding 2-aroyl-1-tetralols in yields ranging from 67% to 85% with diastereomeric ratios up to >99 : 1 (*syn*/*anti*) and ee values up to 99% ee. Conversions <5% were observed with *ortho*-substituted phenyl groups (*o*-ClC_6H_4, 2,4-$Cl_2C_6H_3$). Better results were obtained with phenyl groups bearing an electron-donating substituent at the *para* or *meta* position of the aryl ring. Finally, significantly lower stereoselectivities (50% ee, 72/28 dr) were achieved with a cyclohexyl diketone derivative demonstrating the crucial role of the fused ring system [35].

A highly stereoselective synthesis of α-CF_3-substituted 1,3-diols from α-CF_3C(O)-substituted benzofused five- to seven-membered cyclic ketones via Ru(II)-catalyzed ATH under DKR process was disclosed by Mohar and coworkers in 2016 using an *ansa*-Ru(II) complex containing the enantiopure piperidino-SO_2DPEN-$(CH_2)_3$(η^6-Ph) ligand (Scheme 5.29) [36]. In particular, by applying milder conditions (S/C = 2000, 40 °C), a single DKR occurred to furnish the monoreduced alcohol intermediate with high levels of selectivity, while running the reaction at 60 °C with a higher catalyst loading (S/C = 500–1000) delivered various α-CF_3-substituted 1,3-diols bearing three contiguous stereocenters in high yield (86–97%) and excellent asymmetric induction (>99.9 dr, 98 to >99.9% ee), through an unprecedented double DKR.

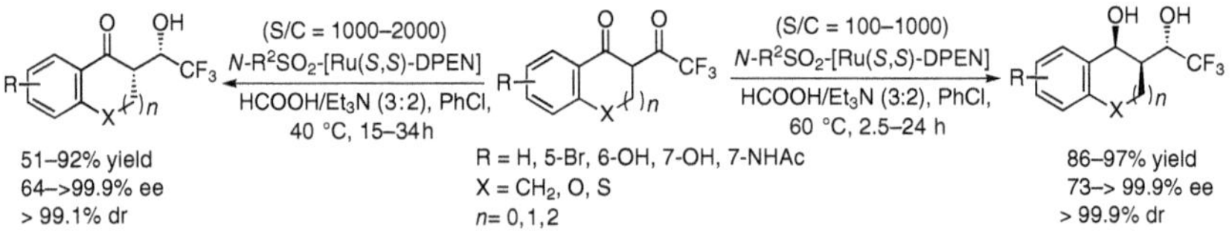

Scheme 5.29 Asymmetric synthesis of benzofused cyclic α-CF_3-substituted 1,3-diols.

5.4 α,α′-Disubstituted Cyclic Ketones

In 2013, Zhou, Xie, and coworkers disclosed the ruthenium-catalyzed hydrogenation of racemic α,α′-disubstituted cyclic ketones through DKR to produce chiral alcohols with three contiguous stereocenters in a one-step procedure (Scheme 5.30). $RuCl_2$(S_a)-Xyl-SDP-(*R,R*)-DPEN, (S_a,*R,R*)-**[Ru]**, catalyzed at S/C = 1000 the

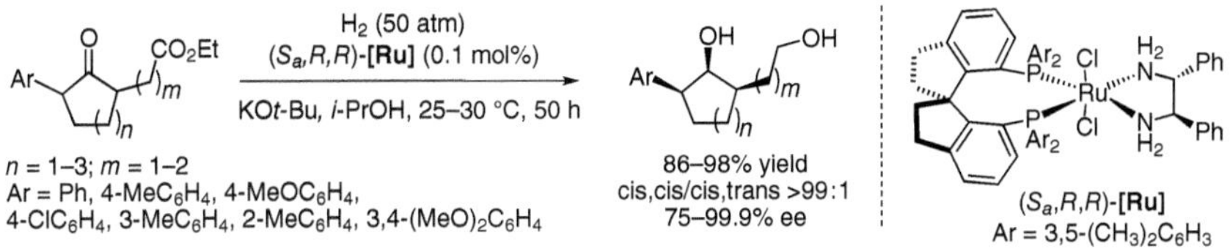

Scheme 5.30 Ruthenium-catalyzed hydrogenation of racemic α,α′-disubstituted cyclic ketones.

hydrogenation of α-ethoxycarbonyl-alkyl-α′-arylcyclic ketones to the corresponding diols in up to 98% yield with excellent *cis,cis* selectivities (*cis,cis*/*cis,trans*: >99 : 1) and ee values up to 99.9% ee. Surprisingly, both the keto and ester groups were reduced under 50 atm of hydrogen pressure at room temperature for 50 hours. Lower enantioselectivity (75% ee) was attained with a five-membered ring, whereas six- and seven-membered rings led to higher enantioinductions (respectively, up to 99.9% ee and 96% ee). The authors demonstrated that both the aryl and ester groups were crucial to get high enantioselectivity. This method was used for the asymmetric total synthesis of (+)-γ-lycorane [37].

5.5 α,β-Disubstituted Cyclic Ketones

In 2016, Zhou, Xie, and coworkers devoted their efforts to the development of iridium-catalyzed transfer hydrogenation of racemic 2-(3-methoxy-2,5-dimethylphenyl)-3-(ethoxycarbonyl)cyclopentanone via DKR to access the chiral aryl-substituted cyclopentane core framework of the naturally occurring hamigeran family bearing three contiguous stereocenters. Chiral iridium complex (*R*)-**[Ir]**, having a spiro pyridine-aminophosphine ligand, was the best catalyst delivering the desired optically active cyclopentanol in 97% with 99% ee and >99 : 1 *trans*/*cis* diastereoselectivity (Scheme 5.31) [38].

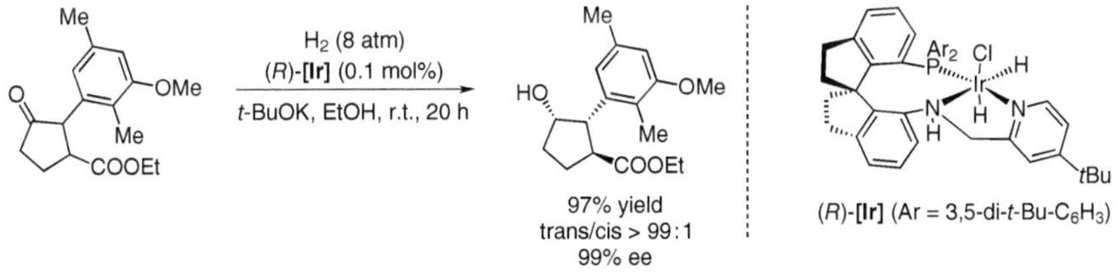

Scheme 5.31 ATH of 2-(3-methoxy-2,5-dimethylphenyl)-3-(ethoxycarbonyl) cyclopentanone. Source: Modified from Lin et al. [38].

A practical and effective method for the construction of chiral chromanols with 1,3-distereocenters from β-substituted chromanones has been achieved by Ashley and coworkers [39]. This procedure tolerated a wide variety of functional groups and involves a highly selective ketone ATH step combined with an efficient DBU-mediated racemization of the β-stereocenter through a conjugate elimination/

conjugate addition process. Under optimal conditions, the reaction performed well using either Ru-TsDENEB or TsDPEN-mesitylene complexes at 40 °C with DBU/HCO_2H as a hydrogen source, delivering synthetically useful chroman derivatives in high yields (71–96%) with moderate to excellent diastereoselectivities (2.3 : 1 to >50 : 1 dr) and high levels of enantioselectivities (93 to >99% ee) (Scheme 5.32).

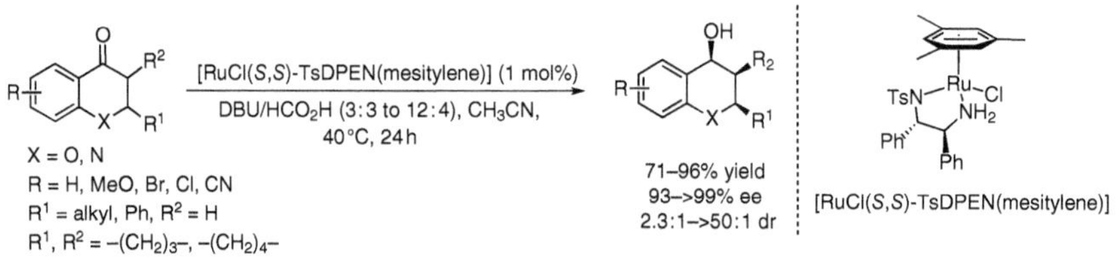

Scheme 5.32 Asymmetric synthesis of chromanols with 1,3-distereocenters.

Moreover, it turned out that the *ansa*-Ru[PipSO$_2$DPEN(CH$_2$)$_4$Ph] complex was also an effective catalyst for the construction of indanol derivatives bearing multiple contiguous stereocenters via a DKR–ATH process [40]. As shown in Scheme 5.33, the reaction was conducted at 60 °C in HCO_2H/Et_3N (3 : 2) mixture as a hydrogen donor at low catalyst loading (S/C = 1000) allowed the conversion of various functionalized racemic 2,3-disubstituted 1-indanones, which possess two stereolabile centers, into the corresponding 2,3-disubstituted-1-indanols in stereopure forms (up to >99% dr, up to 99.9% ee) and moderate to high isolated yields (42–97% yields).

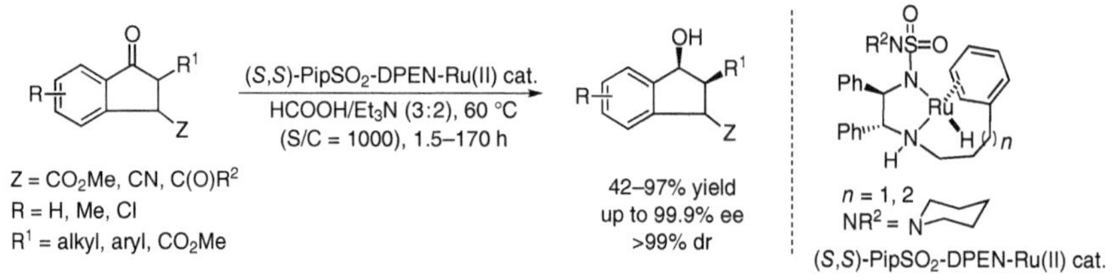

Scheme 5.33 Asymmetric synthesis of 2,3-disubstituted-1-indanols.

5.6 α-Substituted β-Keto Esters

5.6.1 α-Amino β-Keto Esters

In 2004, Hamada [41] and Genêt and Vidal [42] reported simultaneously the synthesis of *anti* α-amino-β-hydroxyesters via Ru-catalyzed AH through DKR of α-amino-β-ketoesters. Following these pioneering results, Hamada and coworkers described thorough investigations toward substituent effects and mechanistic investigations. They showed that aromatic ketones inhibited the stereochemical outcome of the reaction, and only racemic products were obtained. Interestingly,

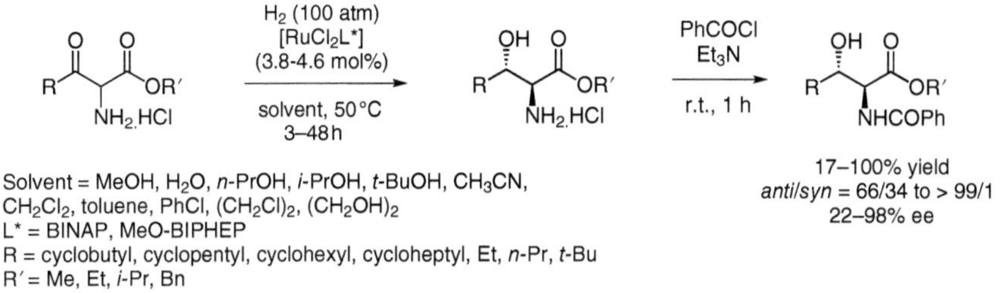

Scheme 5.34 *Anti*-selective AH of α-amino-β-keto esters through DKR. Source: Modified from Makino et al. [43].

according to deuterated experiments, the authors highlighted that the enol form was hydrogenated during this process (Scheme 5.34) [43].

To overcome the limitations associated with Ru catalysts, the group of Hamada devised two generations of Ir catalysts for the synthesis of *anti* α-amino-β-hydroxyesters. While the use of the first-generation catalyst [44] involved harsh reaction conditions (high hydrogen pressure (100 bar) and high catalyst loading), the authors designed a new Ir catalyst which allowed the use of milder conditions by replacing the additive NaI by NaBArF [45]. These optimized DKR hydrogenation reaction conditions were applied to a series of diversely substituted aryl α-amino-β-keto esters bearing both electron-donating and -withdrawing groups on the aromatic ring with perfect diastereoselectivity (*anti*/*syn* >99/1) and excellent enantioselectivity (up to 96% ee). The authors also pointed out that the enantioselectivity was higher when the hydrogen pressure was lower. Finally, these conditions also allowed the transformation of alkyl α-amino-β-keto esters with high stereoselectivity (Scheme 5.35).

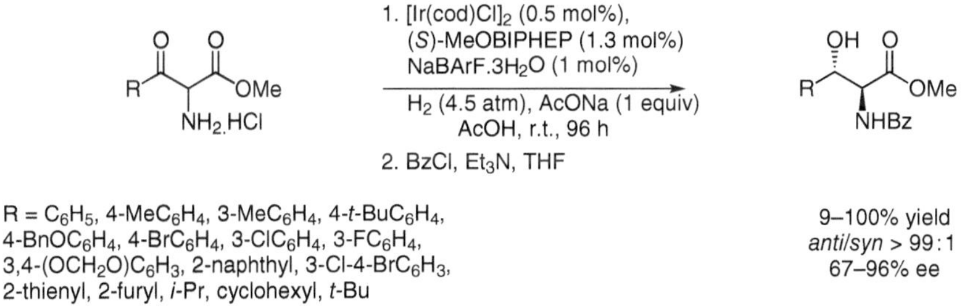

Scheme 5.35 Ir-catalyzed AH of α-amino-β-keto ester hydrochlorides.

In 2008, Hamada and coworkers displayed an Ni-catalyzed AH of various β-keto-α-amino acid esters through DKR using $Ni(OAc)_2.4H_2O$ and a chiral ferrocenylphosphine in a mixture of TFE/AcOH (1 : 4) under 100 atm of H_2 (Scheme 5.36). Under optimized reaction conditions, a large variety of *anti*-β-hydroxy-α-amino acid esters were obtained with perfect diastereoselectivity (>99/1) and good to excellent enantioselectivity (up to 95% ee) [46].

In 2009, our group reported the first asymmetric synthesis of gymnangiamide, a cytotoxic pentapeptide. The key feature of this approach relied on AH through DKR

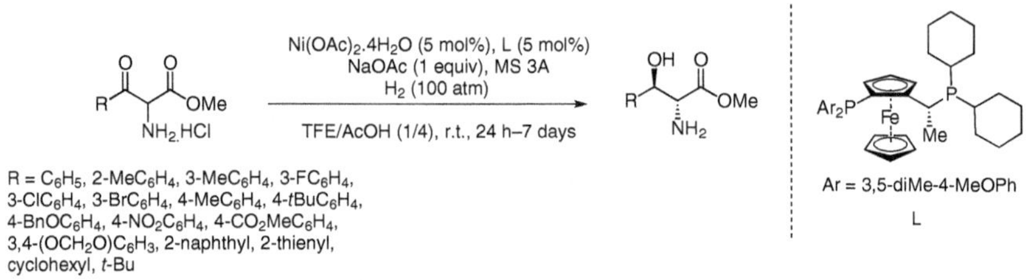

Scheme 5.36 Ni-catalyzed AH of α-amino-β-keto ester hydrochlorides through DKR.

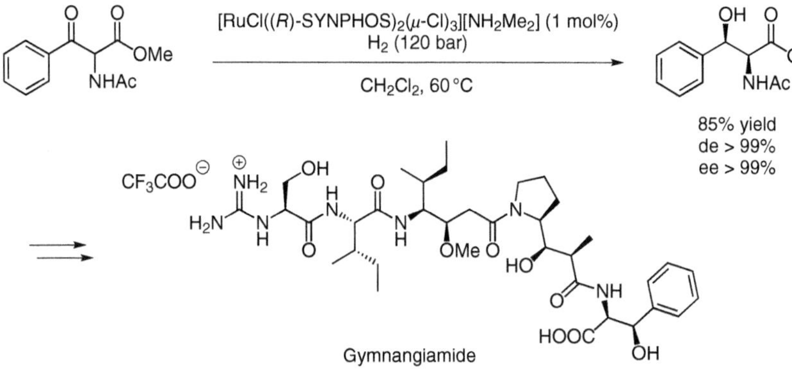

Scheme 5.37 Asymmetric total synthesis and stereochemical revision of gymnangiamide.

using 1 mol% of [RuCl((*R*)-SYNPHOS)$_2$(μ-Cl)$_3$][NH$_2$Me$_2$] under 120 bar of hydrogen in dichloromethane at 60 °C (Scheme 5.37). The two contiguous stereocenters were established with perfect stereoselectivity (de > 99% and ee > 99%) [47].

Our group also elaborated a flexible, concise, and enantioselective synthesis of the naturally occurring bioactive ceramide symbioramide. This convergent route involved the Ru(II)-SYNPHOS-catalyzed hydrogenation of readily accessible racemic α-amino β-keto-ester derivatives through a DKR process (Scheme 5.38) affording the corresponding *anti* and *syn* aminoalcohols in enantioselectivities up to 98% ee and diastereoselectivities up to 98%. This flexible method enabled an efficient access to seven structural isomers of symbioramide as well with high asymmetric induction [48].

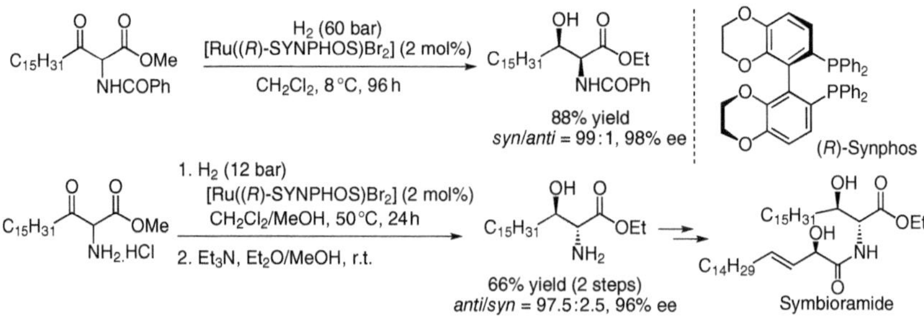

Scheme 5.38 A flexible approach to symbioramide and isomers.

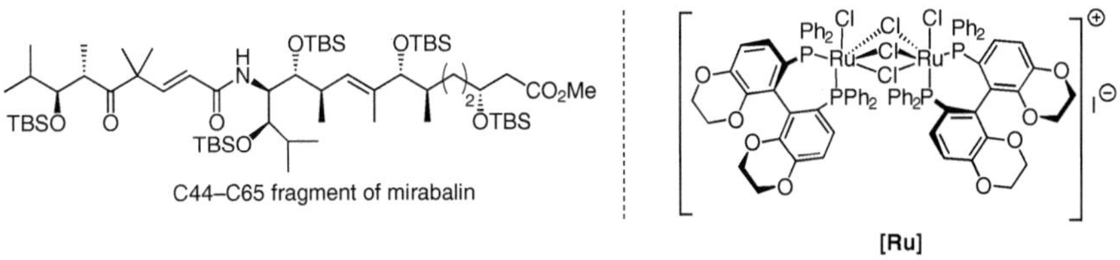

Scheme 5.39 Synthesis of the C44–C65 fragments of mirabalin.

Continuing a long-established interest in metal-catalyzed reductions/DKR process, we reported the synthesis of the C44–C65 fragments of mirabalin, a cytotoxic macrolide isolated in 2008 from the marine sponge *Siliquariaspongia mirabilis* (Scheme 5.39). The key feature of this approach relied on the Ru(II)-SYNPHOS-catalyzed DKR of readily accessible racemic α-amino β-keto ester hydrochlorides under mild reaction conditions, using 1 mol% of the in-house developed Ru-SYNPHOS catalyst **[Ru]**. This process, scaled to 25 g, provided an efficient access to the corresponding *N*-protected *anti* amino alcohol in 92% yield and high levels of enantio- and diastereoselectivities (97% de, 98% ee) [49].

In 2004, Hamada [41] and our group [42] described the first examples of *anti*-selective Ru-catalyzed asymmetric hydrogenation of α-amino β-keto ester hydrochlorides through a DKR process. In 2014, we reported the Ir-SYNPHOS-catalyzed hydrogenation of α-amino β-keto ester hydrochlorides using 1.5 mol% of the cationic dinuclear iridium(III) catalyst $[(Ir(H)[(S)\text{-SYNPHOS}])_2(\mu\text{-}I)_3]I$, **[Ir]**, and AcONa in acetic acid (Scheme 5.40). The reaction provided access to diversely functionalized enantioenriched anti α-amino β-hydroxy ester hydrochloride derivatives containing alkyl, electron-donating, and -withdrawing substituents on the aryl groups with excellent anti diastereoselectivities (*anti*/*syn* up to 99 : 1) and enantioinductions (up to 90% ee) [50].

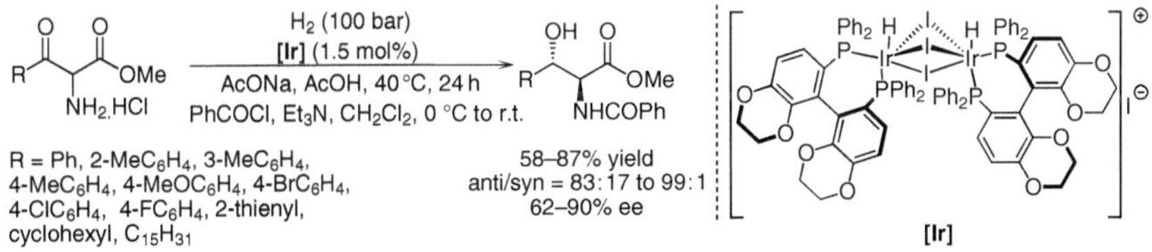

Scheme 5.40 Ir-SYNPHOS-catalyzed AH of α-amino β-keto ester hydrochlorides.

Because chiral *anti* β-hydroxy-α-amino acid derivatives are important structural motifs found in a wide variety of bioactive active and natural products, several stereoselective approaches to access such compounds have been reported, from which AH and ATH of α-substituted-β-keto esters, amides, and phosphonates are attractive methods. In 2010, Somfai and coworkers detailed the synthesis of chiral *anti* β-hydroxy-α-amino acids using 10 mol% of (*S*,*S*)-**[Ru]** complex prepared from $[RuCl_2(benzene)]_2$ in combination with (*S*,*S*)-BnDPAE, in an azeotropic mixture of HCO_2H/Et_3N (5 : 2) as the hydrogen donor. The Ru-catalyzed ATH

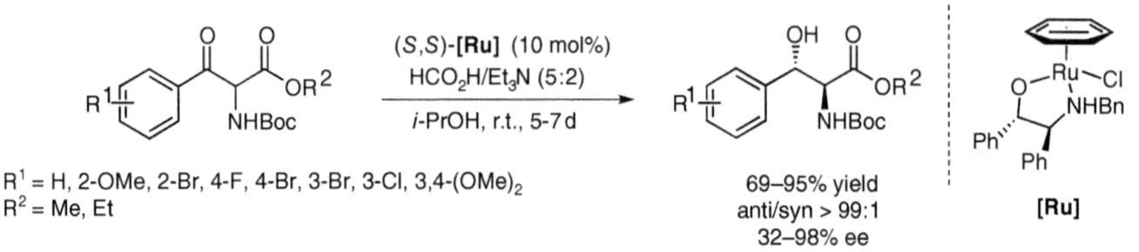

Scheme 5.41 Synthesis of chiral *anti* β-hydroxy-α-amino acids. Source: Based on Seashore-Ludlow et al. [51].

proceeded in 69–95% yield with excellent diastereoselectivities (anti/syn > 99 : 1) and enantioselectivities (up to 98% ee) (Scheme 5.41) [51].

In 2011, Liu, Shultz, and coworkers from Merck Research Laboratories published a highly enantioselective Ru-catalyzed transfer hydrogenation providing enriched *anti* aryl β-hydroxy α-amino ester derivatives via DKR using [RuCl(*p*-cymene)((*R*,*R*)-$C_6F_5SO_2$DPEN)] complex, (*R*,*R*)-**[Ru]**, having an electron-deficient perfluorinated ligand (Scheme 5.42). The ATH of aryl β-keto α-amino ester was conducted in HCO_2H/Et_3N (5 : 2) as the hydrogen donor, with a slow addition of HCO_2H over five hours, to achieve the best stereoselectivities (42 : 1 dr, 91% ee). A series of diversely functionalized *anti* aryl β-hydroxy α-amino esters bearing both electron-donating and -withdrawing substituents on the aromatic ring were successfully synthesized in yields up to 99%, enantiofacial discrimination up to >97% ee, and diastereoselectivities up to >99 : 1 dr. The absolute stereochemical assignment of the *anti* products was established by chemical derivatization and vibrational circular dichroism spectroscopy [52].

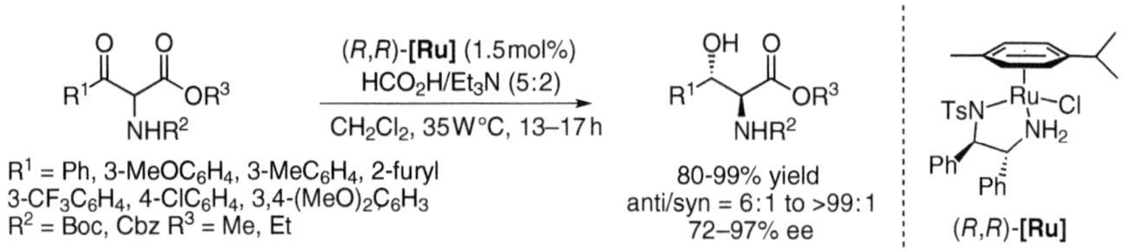

Scheme 5.42 Ru-catalyzed ATH of aryl β-keto α-amino esters.

In 2012, Somfai and coworkers succeeded in developing an operationally simple and enantioselective synthesis of *anti* β-hydroxy α-amido esters through ATH/DKR in aqueous media using HCO_2Na in emulsions and tetrabutylammonium iodide (Scheme 5.43). The Ru-catalyzed transfer hydrogenation of α-amido β-keto esters had a broad substrate scope including aryl-, heteroaryl-, alkenyl-, and alkyl-substituted compounds. This new emulsion-based approach proceeded in good yields utilizing the preformed catalyst (*S*,*S*)-**[Ru]** at S/C = 33 for three to five days with high diastereo- and enantioinductions (*anti*/*syn* up to 95 : 5, up to 98% ee) [53].

The same group reported the ATH in nondegassed water using Tween 20 [polyoxyethylene (20) sorbitan monolaurate] as a neutral surfactant to overcome solubility issues. The Tween 20 was the best for [$RuCl_2$(benzene)$_2$] and (*S*,S)-BnDPAE

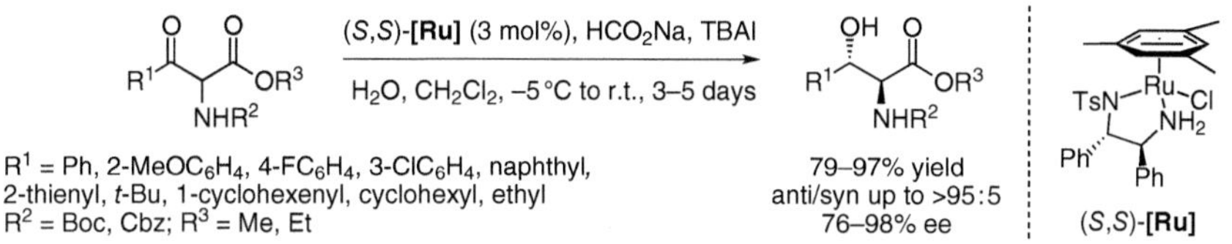

Scheme 5.43 Synthesis of anti β-hydroxy α-amido esters by ATH in emulsions.

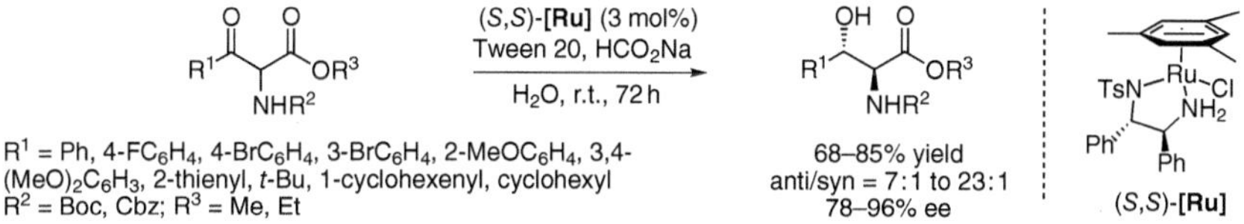

Scheme 5.44 Asymmetric hydrogen transfer in water. Source: Based on Seashore-Ludlow et al. [54].

catalyst combination. The procedure led to good yields ranging from 68% to 85%, *anti*/*syn* diastereoselectivities up to 23 : 1, and ee values up to 96% ee for a wide range of substrates (Scheme 5.44) [54].

In 2015, our group disclosed a flexible and atom-economical synthesis of the four stereoisomers of (+)-(1*R*,2*R*)-thiamphenicol, used for its antibacterial activities against several Gram-positive and -negative microorganisms, through both AH/DKR and ATH/DKR processes of a racemic α-amido β-ketoester (Scheme 5.45). ATH catalyzed by [RuCl(η^6-mesitylene)((*S*,*S*)-TsDPEN)] or [RuCl(η^6-mesitylene)((*R*,*R*)-TsDPEN)] complex with HCO_2H/Et_3N (5 : 2) as the hydrogen donor delivered at 50 °C, respectively, the *anti* (2*R*,3*R*)- and (2*S*,3*S*)-isomers, in 77% and 95% isolated yields and in high *anti* diastereoselectivities (*anti*/*syn* = 97 : 3) and enantioselectivities up to 94% ee. The ruthenium-catalyzed asymmetric hydrogenation, conducted under 120 bar of hydrogen pressure at 50 °C using the *in situ* generated Ru(II)-SYNPHOS developed in our group, provided the corresponding (2*S*,3*R*) and (2*R*,3*S*) *syn* alcohols in high yields and stereoselectivities (*syn*/*anti* > 99 : 1, 90% ee). Notably, the complementarity of AH and ATH was demonstrated through the efficient access to all *syn* and *anti* stereoisomers of the antibacterial thiamphenicol [55].

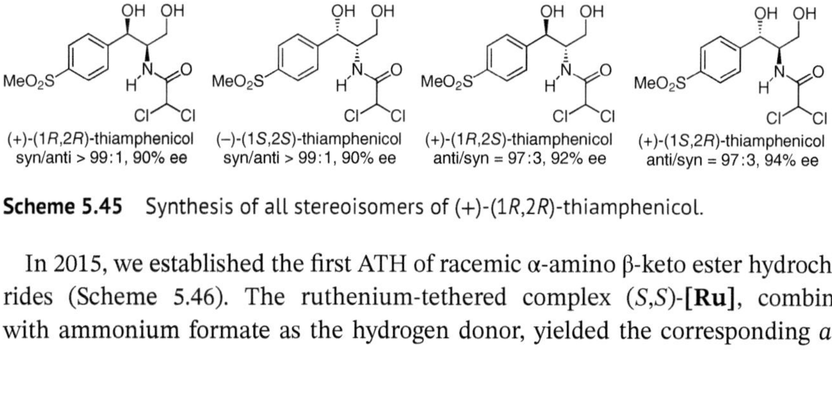

Scheme 5.45 Synthesis of all stereoisomers of (+)-(1*R*,2*R*)-thiamphenicol.

In 2015, we established the first ATH of racemic α-amino β-keto ester hydrochlorides (Scheme 5.46). The ruthenium-tethered complex (*S*,*S*)-**[Ru]**, combined with ammonium formate as the hydrogen donor, yielded the corresponding *anti*

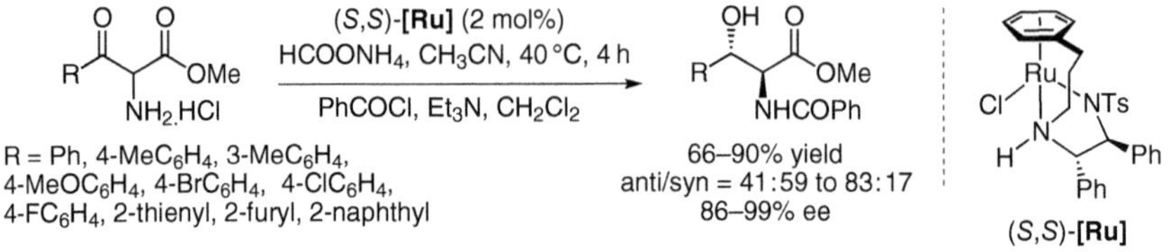

Scheme 5.46 ATH of racemic α-amino β-keto ester hydrochlorides.

alcohols through DKR in good yields up to 90%, diastereo- and enantioselectivities (*anti*/*syn* up to 83 : 17, up to 99% ee). This ATH proceeded with high functional group tolerance with α-amino β-keto ester hydrochlorides having both electron-donating and -withdrawing substituents on the aryl moieties. Interestingly, heteroaromatic-substituted α-amino β-keto ester hydrochlorides underwent the ATH in yields up to 79% delivering predominantly the *syn* isomers with an excellent level of enantioselectivity (>98% ee) albeit moderate diastereoselectivities (*syn*/*anti* up to 68 : 32) [56].

In 2016, Chen and coworkers disclosed the asymmetric synthesis of florfenicol, a fluorinated analog of thiamphenicol having a broad spectrum of antibacterial activity, using Ru-chloramphenicol base ligand-mediated ATH of *N*-boc α-amino-β-ketoester combined with a DKR process [57]. The reaction carried out in water at 25 °C using 10 mol% of catalyst in the presence of 20 mol% of Tween 20 and HCO_2Na proceeded smoothly to furnish the *anti*-(2*S*,3*S*)-α-Boc-amino-β-hydroxyl ester bearing two adjacent stereocenters in 85% yield, 92% diastereoselectivity, and acceptable enantioselectivity of 78% ee that could be improved to 85% after recrystallization (Scheme 5.47). It should be noted that by salting out the reaction with 20% citric acid, followed by neutralization with saturated $NaHCO_3$ and back-extraction with dichloromethane, the chloramphenicol base ligand was recovered in 94% yield.

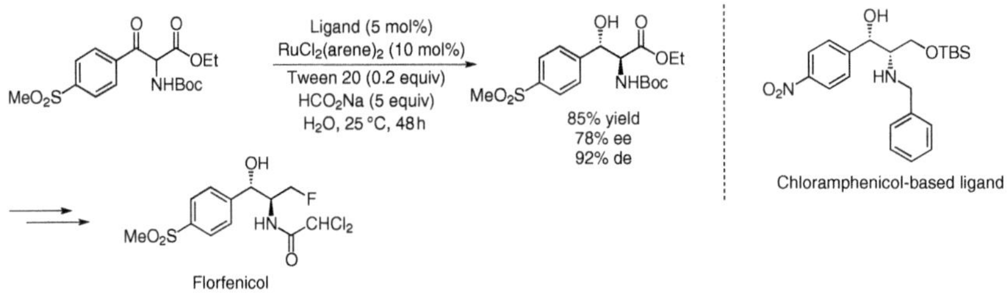

Scheme 5.47 Asymmetric synthesis of *anti*-(2*S*,3*S*)-α-Boc-amino-β-hydroxyl ester.

In 2017, Wang et al. developed a highly efficient method to access enantiomerically pure *syn* aryl β-hydroxy α-dibenzylamino esters through the ATH/DKR of aryl α-dibenzylamino β-keto esters, in high yields (78–98%) with excellent diastereoselectivities (>20 : 1 dr) and enantioselectivities (94 to 99% ee). Furthermore, this method was applied for the gram-scale preparation of droxidopa (Scheme 5.48) [58].

Touge and coworkers reported in 2019 the first example of ATH in a pipes-in-series flow reactor using oxo-tethered Ru(II)complex-mediated DKR on a production scale

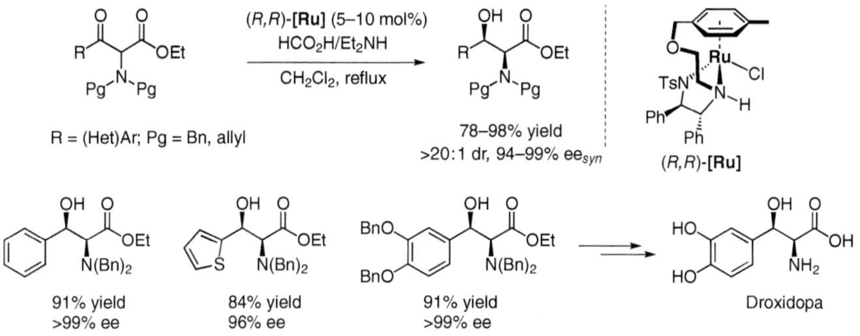

Scheme 5.48 Synthesis of *syn* aryl β-hydroxy α-dibenzylamino esters. Source: Based on Sun et al. [58].

[59]. Under optimized reaction conditions using 0.1 mol% RuCl(*R*,*R*)-TsDENEB catalyst under 0.95 MPa N_2 at 85 °C in the presence of HCO_2H/Et_3N (2 : 0.3), the methyl 2-acetamido-3-oxooctadecanoate was reduced to the corresponding *anti* β-hydroxy α-amido ester, with excellent yield (96.2%) and acceptable diastereoselectivity (69.4% de). Further $NaBH_4$ reduction, followed by NaOH-mediated deacetylation and subsequent condensation with methyl stearate, gave 58 kg of the target ceramide *D*-erythro-CER[NDS] with near-perfect enantiocontrol (>99% de, >99% ee) after crystallization. It is noteworthy that this robust process was efficiently performed at a >50-kg scale with no isolation of intermediates and no expensive reagents (Scheme 5.49).

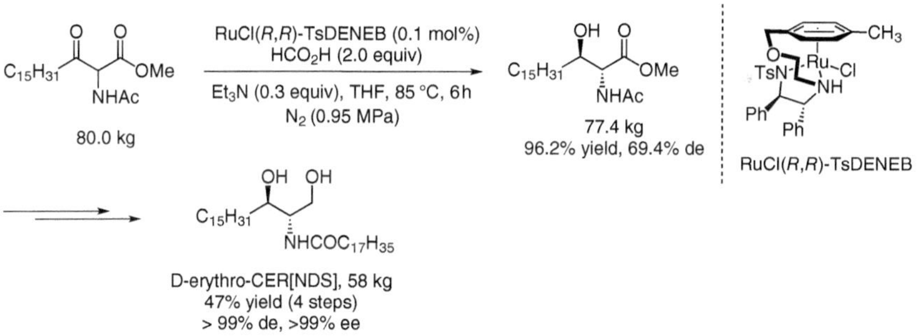

Scheme 5.49 Asymmetric synthesis of *anti* β-hydroxy α-amido ester.

5.6.2 Other α-Substituted β-Keto Esters

In 2010, our group succeeded in developing the hemisynthesis of taxotere based on Ru-catalyzed asymmetric hydrogenation. A DKR was efficiently applied on an α-chloro-β-ketoester to furnish *anti*-α-chloro-β-hydroxyester in excellent diastereoselectivity (96% de) and enantioselectivity (92% ee). This optimized process provided the key *anti* configuration of the two contiguous stereocenters (Scheme 5.50) [60].

In 2011, Zhang and coworkers achieved the asymmetric hydrogen-transfer hydrogenation reaction of 2-chloro-3-oxo esters using [$RuCl_2$(p-cymene){(*S*,*S*)-TsDPEN}],

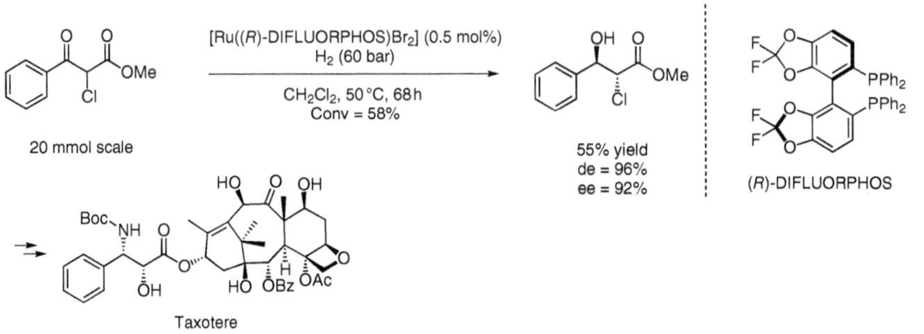

Scheme 5.50 Hemisynthesis of Taxotere through Ru-DIFLUORPHOS-catalyzed AH/DKR. Source: Based on Prévost et al. [60].

(*S*,*S*)-**[Ru]**, and HCO_2H/Et_3N as the hydrogen source through a DKR process under mild conditions to access diverse *syn* 2-chloro 3-hydroxy esters in yields ranging from 50% to 85% and good diastereo- and enantioselectivities (*syn*/*anti* up to 88 : 12, 59–98% ee). This stereoselective approach allowed the preparation of *syn* alcohols as key intermediates for the synthesis bioactive molecules (Scheme 5.51) [61].

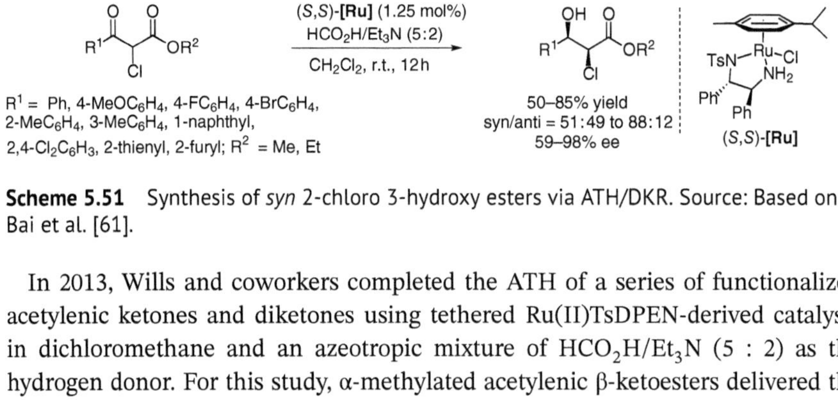

Scheme 5.51 Synthesis of *syn* 2-chloro 3-hydroxy esters via ATH/DKR. Source: Based on Bai et al. [61].

In 2013, Wills and coworkers completed the ATH of a series of functionalized acetylenic ketones and diketones using tethered Ru(II)TsDPEN-derived catalysts in dichloromethane and an azeotropic mixture of HCO_2H/Et_3N (5 : 2) as the hydrogen donor. For this study, α-methylated acetylenic β-ketoesters delivered the *syn* products, and the DKR proceeded in high yields (76–99%) and diastereo- and enantioselectivities (*syn*/*anti* up to 31 : 1, 98–99% ee). Thioester derivatives were not reduced under these reaction conditions even at a catalyst loading of 10 mol% (Scheme 5.52) [62].

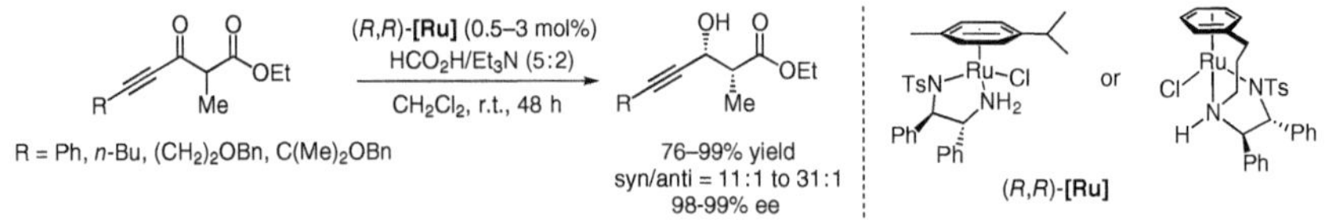

Scheme 5.52 Asymmetric transfer hydrogenation of functionalized acetylenic ketones. Source: Modified from Fang and Wills [62].

A highly enantioselective synthesis of *syn*-1,2-diols was reported in 2010 by our group. The pivotal operation relied on Ru-catalyzed transfer hydrogenation through

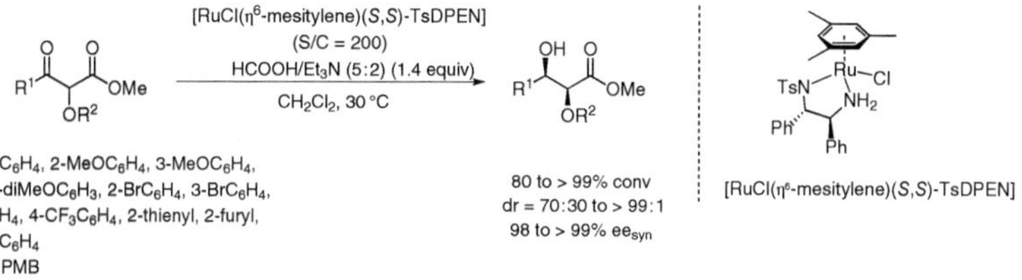

Scheme 5.53 Diastereo- and enantioselective synthesis of *syn*-1,2-diol derivatives through ATH via DKR.

DKR of α-alkoxy-substituted β-ketoesters (Scheme 5.53). High stereocontrol of two contiguous stereocenters was efficiently accomplished on a large range of substrates bearing either electron-donating or -withdrawing groups as well as heteroaryl rings with high levels of diastereoselectivities (dr = 70/30 to >99/1) and enantioselectivities (98 to >99% ee) [63].

Differentiated *syn* 1,2-diol derivatives are useful building blocks in organic synthesis and natural product synthesis. Our group outlined the first enantio- and diastereoselective Ru(II)-TsDPEN and Rh(III)-catalyzed ATH of racemic α-alkoxy β-ketoesters under mild reaction conditions at 30 °C using the azeotropic mixture of HCO_2H/Et_3N (5 : 2) as hydrogen source. This method had a broad scope and tolerated a range of diverse α-alkoxy β-ketoesters containing aryl and heteroaryl-, alkenyl-, alkynyl-, and alkyl-substituted ketones furnishing the corresponding *syn* α-hydroxy β-ketoesters with excellent diastereo- and enantiocontrol (*syn*/*anti* up to 99 : 1, up to 99% ee). No reaction occurred with alkenyl *ortho*-substituted aryl α-alkoxy β-ketoesters (Scheme 5.54) [64, 65].

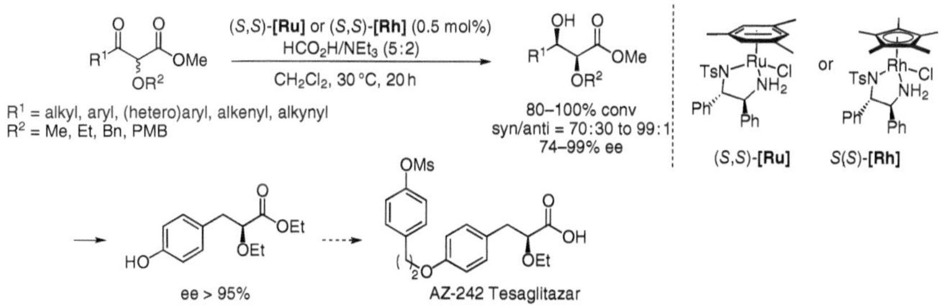

Scheme 5.54 Asymmetric transfer hydrogenation of racemic α-alkoxy β-ketoesters. Source: [64] Monnereau et al. [65].

In 2019, we reported a straightforward and atom-economical access to valuable α-methoxy β-hydroxyesters involving the ATH of racemic α-alkoxy β-ketoesters through DKR in ecofriendly solvents [66]. The reaction efficiently proceeded in an open flask with 0.5 mol% of a new *N*-pentafluorophenylsulfonyl-DPEN-based tethered Rh(III) complex in 2-MeTHF with formic acid/triethylamine (5 : 2) at 30 °C or in water with sodium formate (5.0 equiv) and cetyltrimethylammonium bromide (0.2 equiv) as surfactants at 40 °C (Scheme 5.55). Under these optimized conditions, a wide variety of monodifferentiated β-hydroxyester derivatives featuring

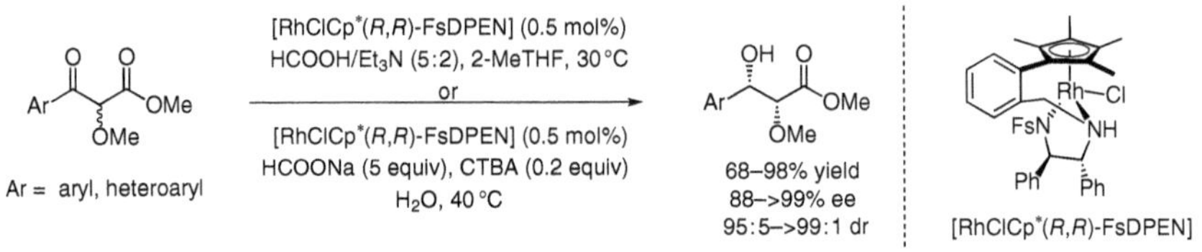

Scheme 5.55 Asymmetric synthesis of monodifferentiated β-hydroxyester derivatives.

aryl substituents that bear either electron-donating or -deficient groups as well as heteroaryl rings were obtained in high yields (up to 98%) with high levels of diastereoselectivities (up to >99 : 1), and excellent ee values (up to >99%).

In 2016, Mohar and coworkers developed the synthesis of γ-sultam-cored *N,N* ligands. The authors showed that enantiopure *syn* and *anti* 3-(α-aminobenzyl)-benzo-γ-sultam ligands were effective in the Ru-catalyzed ATH of aryl ketones using the azeotropic mixture of HCO_2H/Et_3N (5 : 2) as hydride donor (Scheme 5.56). Particularly, (*R,R*)-**[Ru]** prepared from $[RuCl_2(p\text{-cymene})]_2$ reduced the racemic 2- or 3-methoxycarbonyl-1-indanones into the corresponding chiral indanol derivatives with enantioselectivities up to >99% ee and high *cis* diastereoselectivities (*cis*/*trans* ranging from 95 : 5 to 97 : 3). Comparable results were observed with 2-methoxycarbonyl-substituted α-tetralones (99% ee, *cis*/*trans* = 98 : 2) although no diastereoselectivity was obtained (*cis*/*trans* = 50 : 50) for the ATH of 3- and 4-methoxycarbonyl-α-tetralones, while displaying excellent enantioselectivities (>99% ee) [67].

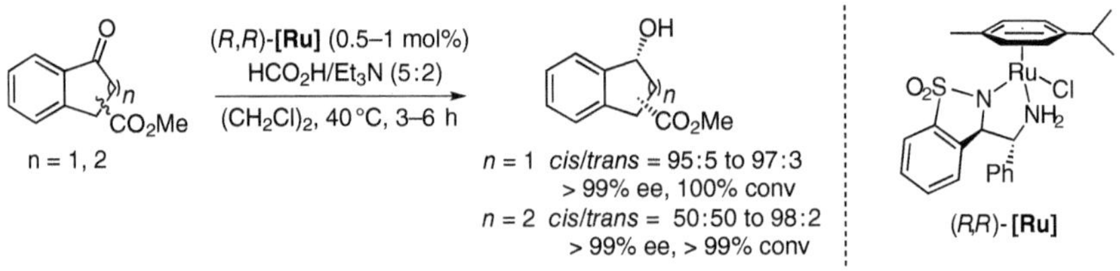

Scheme 5.56 Application of γ-sultam-cored *N,N* ligands in ATH of aryl ketones.

Synthesis of a key intermediate of fluorinated β-lactam antibiotics was achieved using ATH upon DKR. In 2009, the group of Mohar described the asymmetric reduction of ethyl 2-benzamidomethyl-4-fluoro-3-oxo-butanoate using $[RuCl(\eta^6\text{-}1,3,5\text{-}Et_3C_6H_3)(S,S)\text{-}Me_2NSO_2DPEN]$ as a catalyst and the mixture HCO_2H/Et_3N (5 : 2) as the hydrogen source (Scheme 5.57). With these optimized

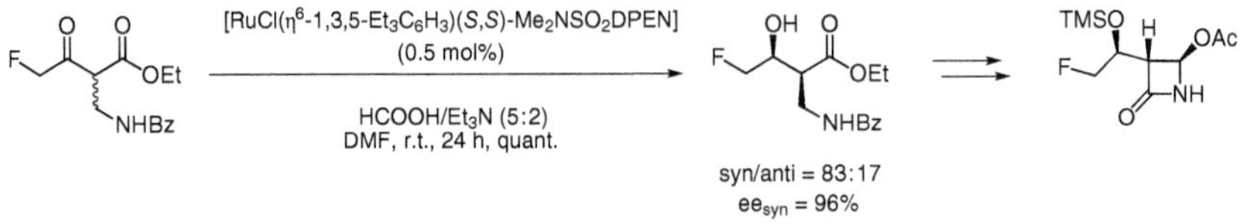

Scheme 5.57 Stereoselective synthesis of a key intermediate of fluorinated β-lactam antibiotic.

conditions, the *syn* isomer was obtained in 60% yield with good diastereoselectivity (*syn*/*anti* = 83 : 17) and excellent enantioselectivity (96% ee) [68].

Leveraging similar principles, the same authors reported the stereoselective synthesis of fluorinated analogs of sanfetrinem and LK-157. The key step of the synthesis relied on ATH via DKR of ethyl 2-benzamidomethyl-3-oxo-4,4,4-trifluorobutanoate using *in situ* generated Ru catalyst in the presence of HCO_2H/Et_3N (5 : 2). The reaction proceeded smoothly in DMF at room temperature to furnish *syn*-ethyl (2*S*,3*S*)-2-benzamidomethyl-3-hydroxy-4,4,4-trifluorobutanoate in 84% yield with excellent diastereoselectivity (*syn*/*anti* = 94 : 6) and perfect enantioselectivity (ee > 99%). The authors pointed out that the diastereoselectivity was highly dependent on the substrate concentration and the nature of the η^6-arene (Scheme 5.58) [69].

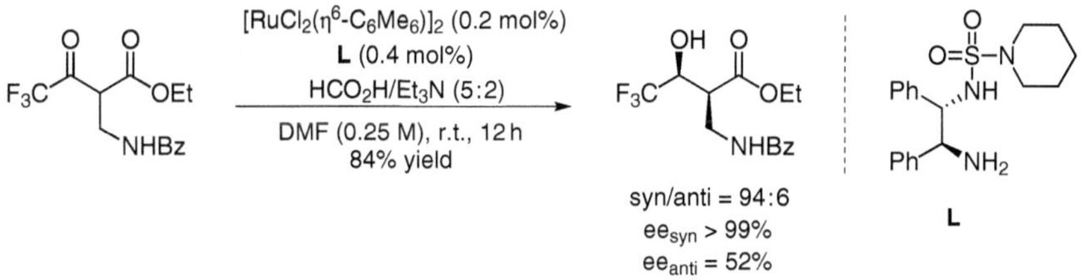

Scheme 5.58 ATH via DKR of ethyl 2-benzamidomethyl-3-oxo-4,4,4-trifluorobutanoate.

In 2013, Zhang and coworkers elaborated a convenient method to access β′-hydroxy-β-amino acids via asymmetric hydrogenation using *in situ* generated $[RuCl_2(p\text{-cymene})]_2$–(*S*)-DTBM-SunPhos as a catalyst. The DKR of β′-keto-β-amino esters proceeded with the highest levels of diastereo- (up to 98.8% de) and enantioinductions (up to 99.9% ee) when conducted in dichloromethane/2,2,2-trifluoroethanol (TFE) or 1,2-dichloroethane/TFE combinations as solvents. Interestingly, $[RuCl_2(p\text{-cymene})]_2$-(*S*)-DTBM-SunPhos complex exhibited better activity and selectivity in this mixed-solvent systems than the corresponding $[RuI_2(p\text{-cymene})]_2$-(*S*)-DTBM-SunPhos (Scheme 5.59) [70].

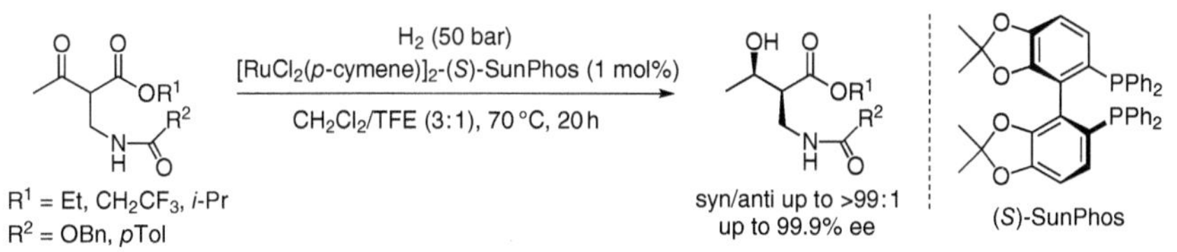

Scheme 5.59 Synthesis of β′-hydroxy-β-amino acids via Ru-catalyzed AH. Source: Based on Li et al. [70].

In 2016, Hu and coworkers described a new class of sterically demanding chiral ferrocenyl *P*,*N*,*N*-ligands which proved to be very effective in the Ir-catalyzed AH of numerous α-alkyl-substituted acyclic β-aryl-β-ketoesters containing alkyl, electron-donating, and -withdrawing aryl groups via DKR [71]. Under optimized reaction conditions, a low catalyst loading of 0.01 mol% allowed for the first time the synthesis of various enantioenriched *anti*-β-hydroxyesters, which are useful building blocks, in high yields with high diastereoselectivities and excellent enantioselectivities up to 99%. Importantly, this study revealed that the introduction of an additional stereocenter at the pyridinylmethyl position, which significantly increased the rigidity of these ligands, was crucial to achieve high levels of selectivity (Scheme 5.60).

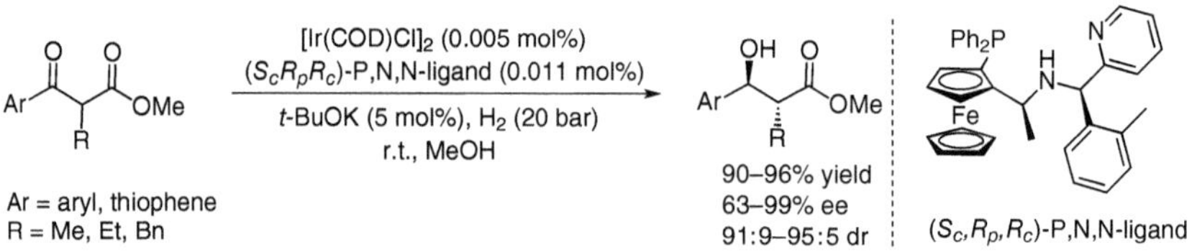

Scheme 5.60 Synthesis of α-substituted *anti*-β-hydroxyesters.

Chiral cyclopentanols are common and privileged motifs in natural products, pharmaceuticals, and many bioactive molecules. In 2017, Zhou and coworkers disclosed a highly efficient preparation of chiral cycloalkanols that bear three contiguous stereocenters through Ir-catalyzed AH of tetrasubstituted cyclic enones coupled with a DKR process [72]. This method involved a one-pot sequential reduction of the C=C bond of the substrates to first install the β-alkyl-substituted tertiary chiral center, followed by AH of the C=O bond of the resulting β-ketoesters enones via DKR. Using (*R*)-Ir-SpiroPAP catalyst, numerous cyclopentanols as well as cyclohexanols were obtained with excellent enantioselectivities (92 to >99.9% ee) and diastereoselectivities (dr 95 : 5 to >99 : 1). This practical method has been successfully applied to the synthesis of all stereoisomers of the antiulcer drug rosaprostol (Scheme 5.61).

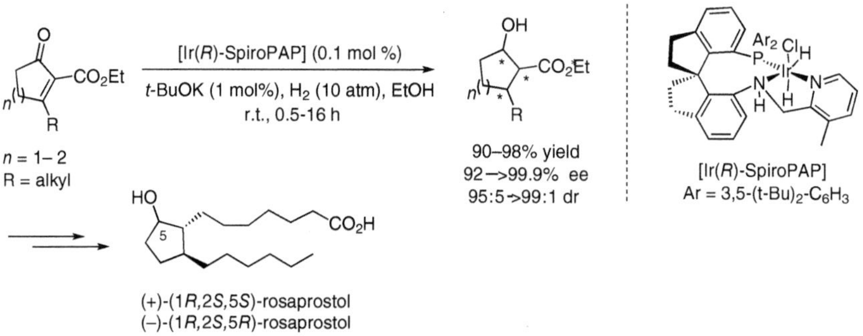

Scheme 5.61 Asymmetric synthesis of cyclopentanols and cyclohexanols.

5.7 α-Substituted β-Keto Amides

In 2013, scientists from Eli Lilly Company reported an efficient route to access serotonin norepinephrine reuptake inhibitors (SNRIs), active in the treatment of

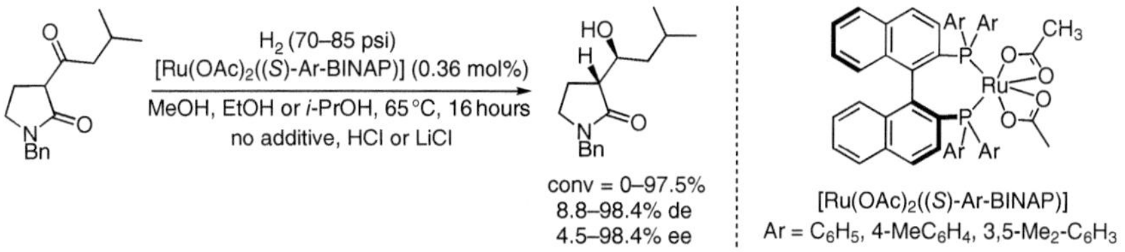

Scheme 5.62 Asymmetric syntheses of serotonin norepinephrine reuptake inhibitors.

chronic painful conditions. The strategy was based on enantio- and diastereoselective Ru-catalyzed hydrogenation of a β-keto-γ-lactam derivative via a DKR process (Scheme 5.62). Hydrogenation screening experiments demonstrated that higher stereoselectivities (96% ee, 94% de) were attained using [Ru(OAc)$_2$((*S*)-tol-BINAP)] complex compared to [Ru(OAc)$_2$((*S*)-BINAP)] or [Ru(OAc)$_2$((*S*)-xyl-BINAP)] in *i*-PrOH compared to EtOH or MeOH. The authors established that additives promoted the hydrogenation of β-keto-γ-lactam with catalytic amounts of HCl (6 mol%) and LiCl (1 mol%) that enabled higher yields up to 93%. Interestingly, studies employing online NMR and HPLC to observe the rate and equilibrium of the DKR hydrogenation showed that one enantiomer of the racemic β-keto-γ-lactam was reduced faster than the interconversion between the enantiomers [73].

With the idea to develop an efficient route to access optically active β-hydroxy-γ-lactams, a structural motif common to several bioactive compounds, the group of Maguire showed that Ru-tol-BINAP catalyst enabled the reduction of β-keto-γ-lactams through asymmetric DKR hydrogenation [74]. It was found that this method required the use of HCl and LiCl as additives and was generally applicable across a range of substrates, other than those with sterically demanding *tert*-butyl substituent close to the site of hydrogenation when isopropanol was used as solvent, leading to a wide range of substituted β-hydroxy-γ-lactams with high to excellent diastereoselectivities (up to >98% de) and selectivities (up to 97% ee) (Scheme 5.63).

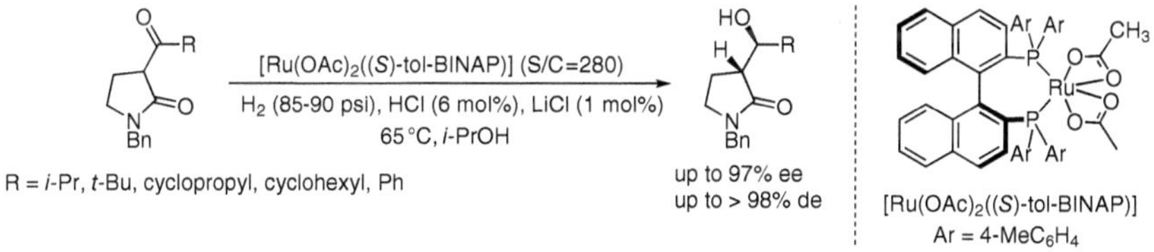

Scheme 5.63 Asymmetric synthesis of substituted β-hydroxy-γ-lactams.

In 2017, Zhou and coworkers obtained similar results in terms of both reactivity and selectivity for the Ir-catalyzed AH of racemic β-keto lactams via DKR under mild reaction conditions [75]. After intensive experiments, it was found that using 0.1 mol% of Ir-(*R*)-SpiroSAP as a catalyst in combination with 0.1 equiv of KOH as base in methanol, a series of racemic β-keto lactams including β-keto γ-, δ-, and ε-lactams were reduced to the corresponding chiral β-hydroxy lactams in high yields with excellent diastereo- and enantioselectivities

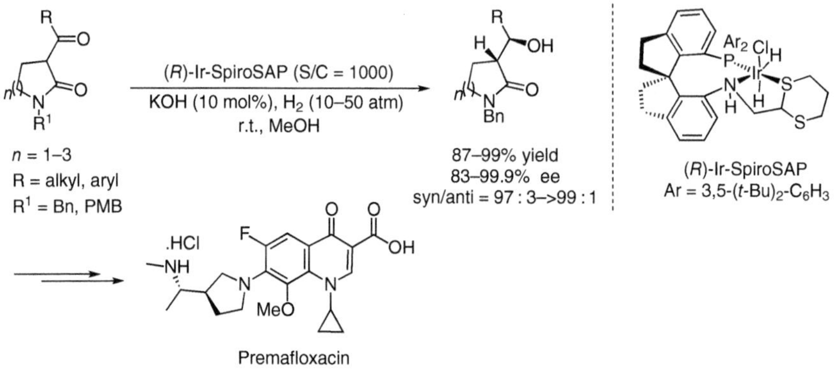

Scheme 5.64 Asymmetric synthesis of substituted β-hydroxy-γ-lactams.

with turnover numbers up to 5000. An example of the potential utility of this efficient method was demonstrated by the preparation of stereochemically pure (*S*)-*N*-methyl-1-((*R*)-pyrrolidin-3-yl)etha-1-amine, a chiral intermediate of fluoroquinolone antibiotic premafloxacin (Scheme 5.64).

In 2013, Lee and coworkers reported the Ru-catalyzed ATH of 2-benzoylmorpholin-3-ones. The reduced (2*R*,3*S*)- or (2*S*,3*R*)-2-(hydroxyphenylmethyl)morpholin-3-ones were produced using 0.5 mol% of the ruthenium complex (*R*,*R*)-[Ru] and HCO_2H/Et_3N (5 : 2) azeotropic mixture as the hydrogen donor in up to 98% yield with high diastereo- and enantioselectivities (anti/syn up to 99 : 1, 95–99% ee) (Scheme 5.65). This method was used to access all four stereoisomers of the antidepressant reboxetine [76].

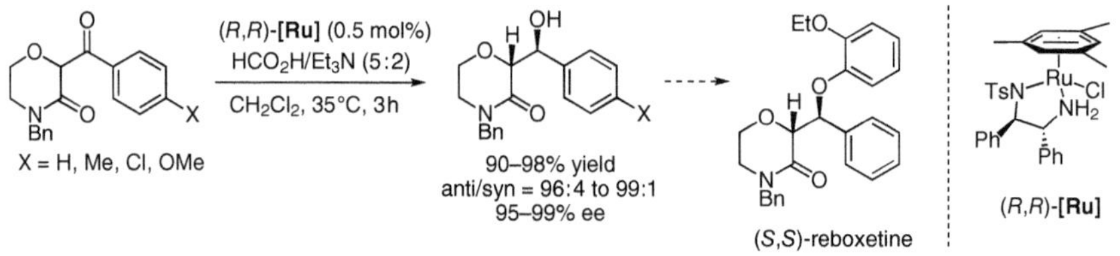

Scheme 5.65 Synthesis of (hydroxyphenylmethyl)morpholin-3-one derivatives.

In 2013, Kumaraswamy and coworkers reported the ATH/DKR reaction for the enantioenriched synthesis of a series of δ/γ-alkoxy-β-hydroxy-α-alkyl-substituted Weinreb amides (Scheme 5.66). The reduction was conducted using 1 mol% of [Ru(*p*-cymene)((*R*,*R*)-TsDPEN)], (*R*,*R*)-**[Ru]**, in isopropanol at 55 °C for 24 hours

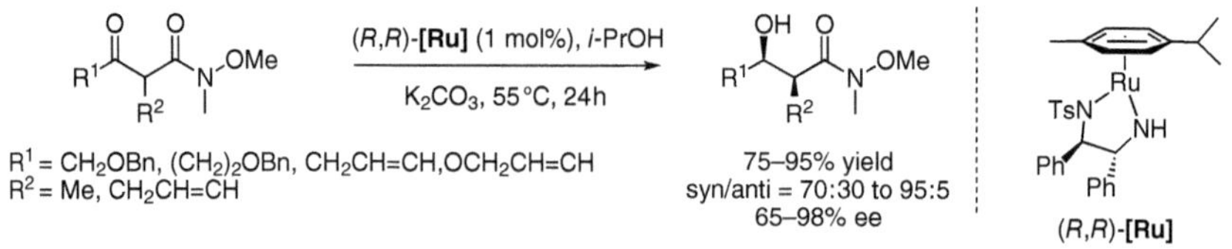

Scheme 5.66 Synthesis of δ/γ-alkoxy-β-hydroxy-α-alkyl-substituted Weinreb amides.

with 10 mol% of K_2CO_3 providing the *syn* alcohols with high stereoselectivities (*syn*/*anti* 70 : 30 to 95 : 5, up to 98% ee). Interestingly, the benzyloxy-tethered substrates without Weinreb amides led to the corresponding α-hydroxy-β-methylketo derivatives with high diastereoselectivities (*syn*/*anti* up to 92 : 8) and good ee values (up to 97% ee). To expand the utility of this method, this transformation was scaled to 10 g and applied to the synthesis of a key intermediate of natural product (−)-brevisamide [77].

In 2015, the same group also accomplished an Ru(II)-catalyzed ATH of a racemic acetylenic-β-keto-α-methyl Weinreb amide to access an advanced intermediate of the crocacin family bearing consecutive anti–anti–syn stereogenic centers. This approach to the natural product (+)-crocacin C allowed to install the key *syn* configuration of the stereotetrad motif (Scheme 5.67). The ATH/DKR was carried out at 50 °C using [RuCl(*p*-cymene)((*R*,*R*)-TsDPEN)] at S/C = 33 in HCO_2H/Et_3N (5 : 2). The enantioenriched acetylenic α-methylated β-hydroxy Weinreb amide was obtained in 60% yield, 92% ee, and syn diastereoselectivity (*syn*/*anti* 92 : 8). Of note, a better catalytic activity (94% yield, 98% ee, and *syn*/*anti* 94 : 6) was achieved at a lower catalyst loading (S/C = 100) for the same transformation using the isolated 16-electron Ru-TsDPEN amido complex (*R*,*R*)-**[Ru]** [78].

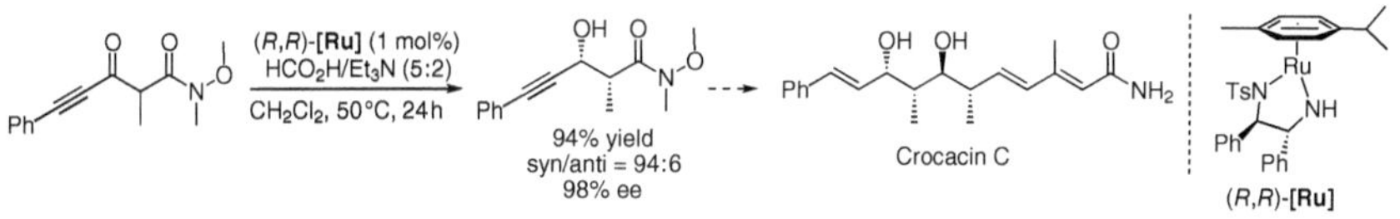

Scheme 5.67 Synthesis of acetylenic α-methylated β-hydroxy Weinreb amide.

In 2014, Wills and coworkers investigated the Ru-catalyzed ATH of *N*-benzyl 5-acetyluracil using (*R*,*R*)- or (*S*,*S*)-**[Ru]** complexes in HCO_2H/Et_3N (5 : 2) (Scheme 5.68). Interestingly, (*R*,*R*)-**[Ru]** gave the reduced product in 92% yield with a 4 : 1 diastereomeric ratio and 33% ee, while (*S*,*S*)-**[Ru]** led to comparable stereoselectivity delivering the same major diastereoisomer. The relative and absolute configurations of the diastereoisomers were not determined. This study suggested that conjugate addition occurred first, resulting in the formation of an enol intermediate, which would tautomerize to give racemic ketone for which the reduction may then proceed via a (dynamic)kinetic resolution process [79].

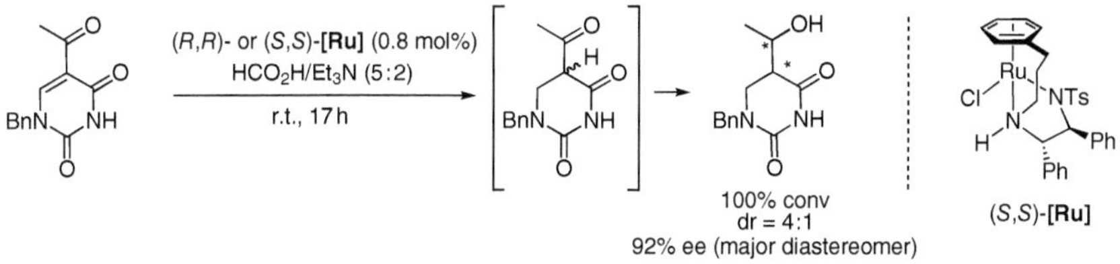

Scheme 5.68 Asymmetric transfer hydrogenation of *N*-benzyl 5-acetyluracil.

In 2019, Mohar and coworkers reported a highly efficient *trans*-selective ATH via DKR of α,β-dehydro-α-acetamido and α-acetamido benzocyclic ketones catalyzed by a new *ansa*-Ru(II) complex bearing an enantiomerically pure *ent-syn-N,N*-ULTAM$(CH_2)_3$Ph ligand [80]. Under optimized reaction conditions employing 0.5 mol% of *ansa*-[RuCl-*syn*-ULTAM-$(CH_2)_3$Ph] complex in chlorobenzene with an HCO_2H/Et_3N (3 : 2) mixture at 60 °C, a wide variety of *ent-trans*-β-amido alcohols were prepared in moderate to good yields with high diastereomeric ratios and excellent enantioselectivities (Scheme 5.69). Interestingly, stereochemically pure *trans*-products could be easily obtained after a single recrystallization from ethyl acetate or trituration with acetonitrile (all products >99% ee, dr >99 : 1).

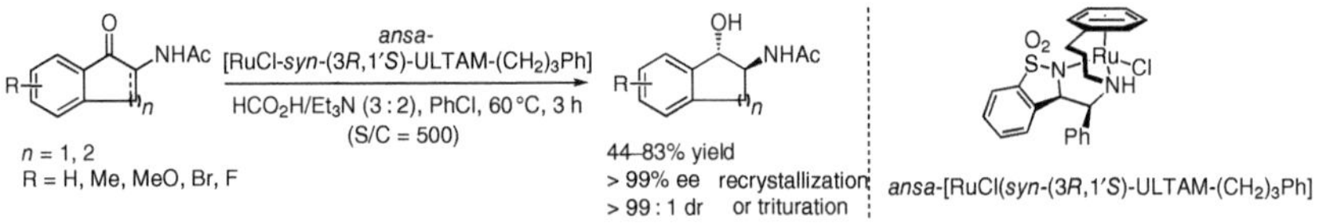

Scheme 5.69 Asymmetric synthesis of *trans*-β-amido alcohols.

5.8 α-Substituted β-Keto Sulfones, Sulfonamides, and Phosphonates

Because of their unique biological activities as well as their potential uses as peptide mimics, chiral β-hydroxy α-amino phosphonates have attracted considerable attention. Zhang and coworkers disclosed in 2012 the Ru-catalyzed asymmetric hydrogenation of α-substituted β-keto phosphonates employing [RuCl(benzene)((*S*)-SunPhos)]Cl as a catalyst delivering the corresponding syn β-hydroxy phosphonates with high diastereo- and enantioselectivities (up to 96 : 4 syn/anti, up to >99.8% ee (Scheme 5.70) [81].

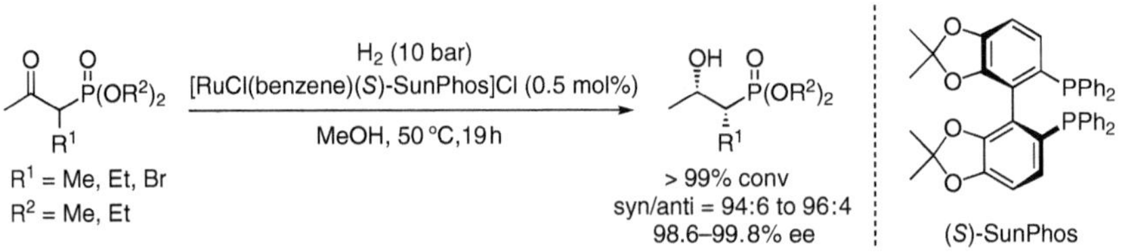

Scheme 5.70 Synthesis of chiral β-hydroxy α-amino phosphonates. Source: Based on Tao et al. [81].

In 2013, the same group published a convenient approach to synthesize β-hydroxy α-amino phosphonates through Ru-catalyzed hydrogenation/DKR process of α-amido β-keto phosphonates (Scheme 5.71). Remarkable stereoselectivities (up to 99 : 1 *syn/anti,* up to 99.8% ee) were obtained for the corresponding *syn* α-amido β-hydroxy phosphonates using [RuCl(benzene)((*S*)-SunPhos)]Cl as a catalyst. A dramatic enhancement of both diastereo- and enantioselectivities was

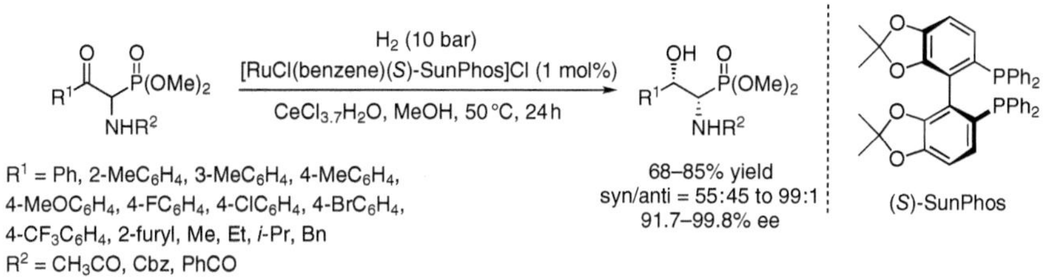

Scheme 5.71 Synthesis of β-hydroxy α-amino phosphonates using additives.

observed after addition of $CeCl_3{\cdot}7H_2O$, demonstrating the crucial role of additives in the stereochemical outcome of the reduction [82].

In 2014, Lee and coworkers focused on the ATH of readily available racemic 2-substituted α-alkoxy β-ketophosphonates using HCO_2H/Et_3N (1 : 5) as the hydrogen source and solvent, along with the well-defined chiral [RuCl((*R,R*)TsDPEN) mesitylene] catalyst, (*R,R*)-**[Ru]** (Scheme 5.72). The corresponding *syn* 2-substituted α-alkoxy-β-hydroxy phosphonates bearing 2-aryl-, 2-heteroaryl-, 2-alkyl-, and 2-alkenyl groups were prepared in yields ranging from 95% to 99% and excellent levels of diastereo- and enantioselectivities (*syn*/*anti* up to 99 : 1, up to 99% ee) [83].

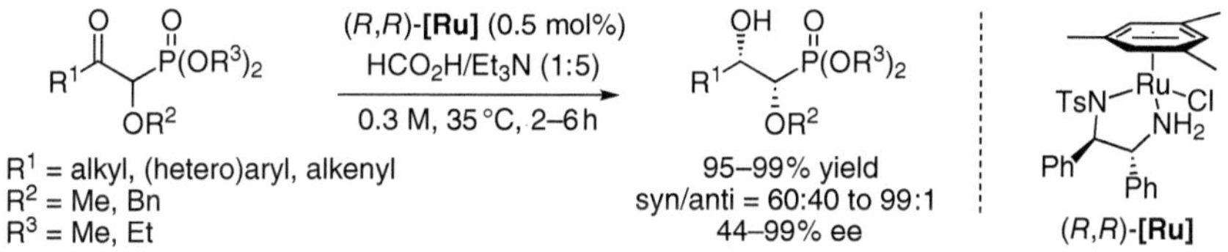

Scheme 5.72 Synthesis of *syn* 2-substituted α-alkoxy-β-hydroxy phosphonates.

In 2009, Zhang and coworkers developed an enantioselective synthesis of β-hydroxy sulfones using ATH with DKR (Scheme 5.73). Under the best operating conditions using *in situ* generated Noyori catalyst (1.3 mol%) and HCO_2H/Et_3N (5 : 2), a wide variety of optically active β-hydroxy sulfones were obtained in low to

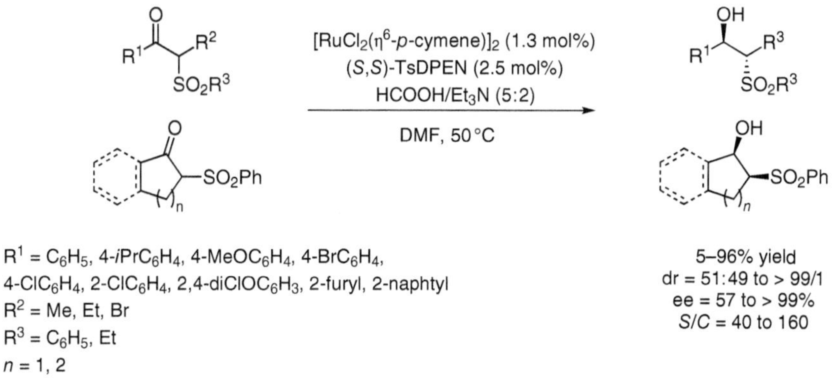

Scheme 5.73 Enantioselective synthesis of β-hydroxy sulfones through ATH/DKR.

excellent yields (5–96%), with moderate to excellent diastereoselectivities (dr up to >99/1) and moderate to excellent enantioselectivities (ee up to >99%) [84].

Because enantiomerically enriched β-hydroxy sulfonamides and β-hydroxy sulfones are key building blocks to access bioactive compounds and pharmaceuticals, Wang and coworkers realized in 2013 a cascade asymmetric hydrogenation/DKR of racemic cyclic β-keto sulfonamides and β-keto sulfones derived from α-indanone or α-tetralone (Scheme 5.74). The corresponding *cis* β-hydroxy sulfonamides and β-hydroxy sulfones were produced in methanol at room temperature under 40 atm of hydrogen pressure in yields of 92–97%, with excellent stereoselectivities (*cis*/*trans*: >99 : 1, 98% ee) utilizing 0.4 mol% of the cationic complex [Ru(OTf)(*p*-cymene)((*R*,*R*)-TsDPEN)], (*R*,*R*)-**[Ru]** [85].

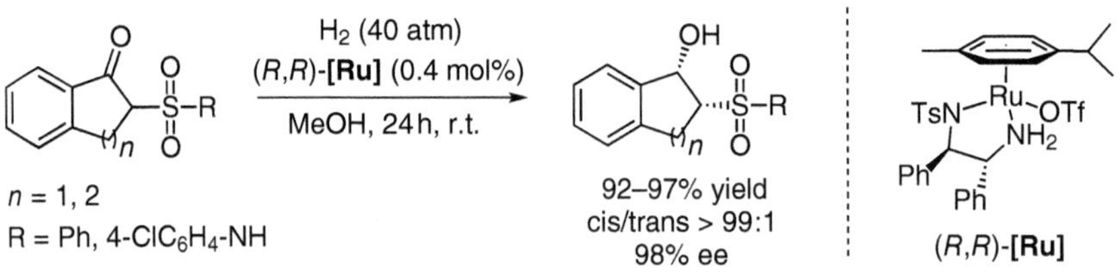

Scheme 5.74 Synthesis of β-hydroxy sulfonamides and β-hydroxy sulfones.

In 2011, Zhang and coworkers described the ATH through DKR of cyclic α-tetralone and α-indanone derivatives (Scheme 5.75). β-Ketosulfonamides underwent Ru-catalyzed ATH/DKR with HCO_2H/Et_3N (5 : 2) as the hydrogen source in dioxane at room temperature with high ee values (up to 99% ee) and diastereoselectivities (up to >99 : 1 dr) using (*S*,*S*)-**[Ru]** as a catalyst [86].

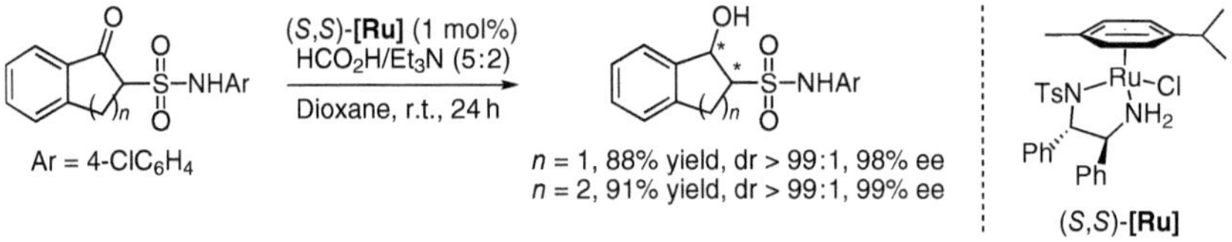

Scheme 5.75 Enantioselective synthesis of β-hydroxysulfonamides.

In 2018, Lv and Zhang et al. described an efficient ATH–DKR process for α-sulfonamide-β-ketones using a chiral Ru(II) catalyst with HCO_2H/Et_3N azeotropic mixture as the hydrogen donor under mild conditions, to provide the corresponding α-substituted β-hydroxy sulfonamides in good yields with excellent diastereo- and enantioselectivities (Scheme 5.76) [87].

Recently, Cheng and coworkers developed an operationally simple, one-pot two-step asymmetric organic transformation of α-bromoindenones into the corresponding optically active β-hydroxy sulfones, involving a bromide substitution with sodium arylsulfinate followed by *in situ* ATH of the resulting β-keto sulfones coupled with a DKR process [88]. Under the best operating conditions, using [$RuCl_2$(S,S)-TsDPEN(mesitylene)] as a catalyst, and HCO_2Na as a hydrogen source

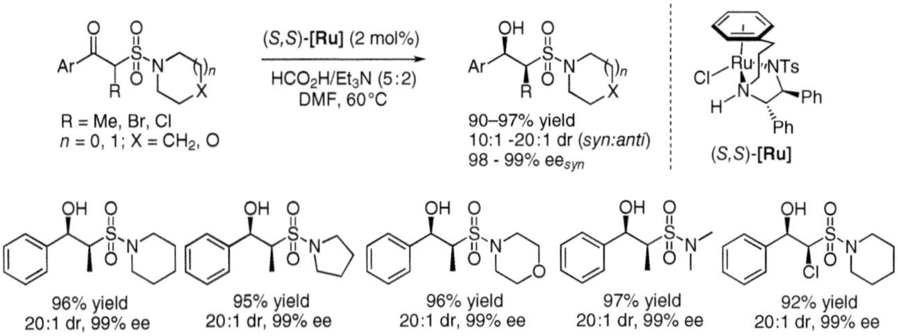

Scheme 5.76 ATH-DKR of α-sulfonamide-β-ketones. Source: Based on Xiong et al. [87].

in a mixture of cosolvent H_2O/*i*-PrOH (1 : 3) at 60 °C, a set of optically pure vicinal stereocenters β-hydroxy sulfones were obtained in high yields (89–97%), with excellent enantioselectivities (92–99%) and high diastereomeric ratios (91 : 9–99 : 1), regardless of the electronic features or position of the substituents bonded to the aryl group (Scheme 5.77).

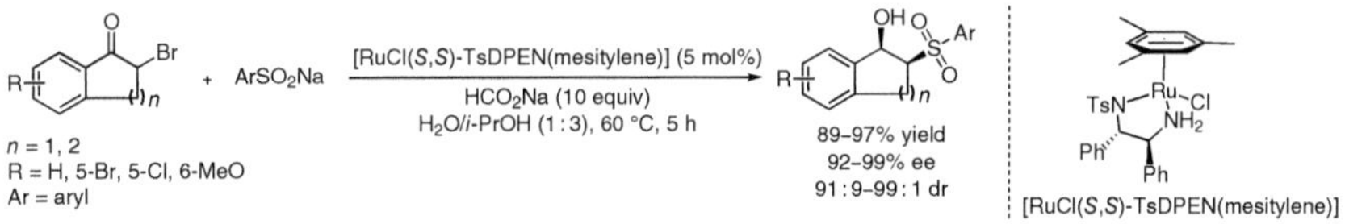

Scheme 5.77 Asymmetric synthesis of β-hydroxy sulfones from α-bromoindenones.

5.9 β-Substituted α-Keto Esters and Phosphonates

In 2013, Johnson and coworkers accomplished the first DKR of β-aryl α-keto esters via Ru-catalyzed ATH using a new α-naphthyl/diphenylbenzene sulfonamide catalyst (*S*,*S*)-**[Ru]**, prepared from $[RuCl_2(p\text{-cymene})]_2$ and a DPEN-based ligand (Scheme 5.78). The reaction generated three contiguous stereocenters through a reduction/lactonization sequence allowing the direct access to trisubstituted γ-butyrolactones in yields up to 94%, with remarkable diastereocontrol (diastereoselection >20 : 1) and high enantioinductions (up to 93% ee) [89].

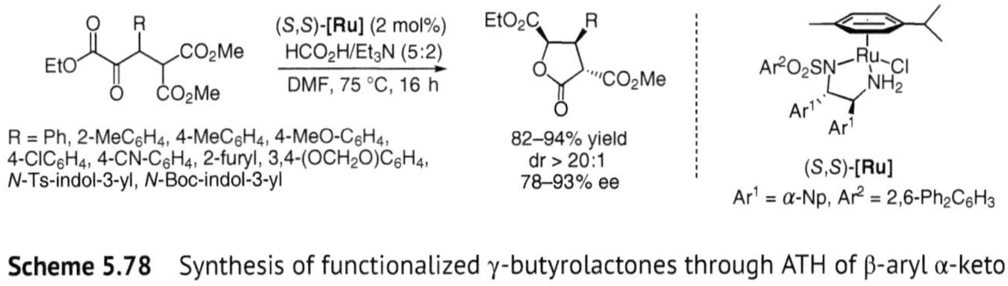

Scheme 5.78 Synthesis of functionalized γ-butyrolactones through ATH of β-aryl α-keto esters.

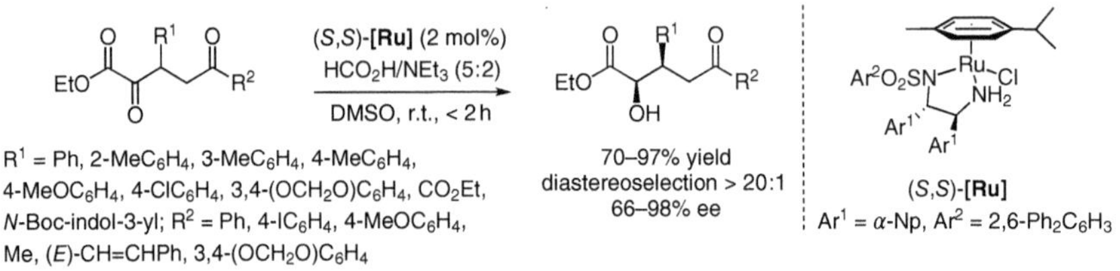

Scheme 5.79 Asymmetric synthesis of glycolic acid scaffolds.

The same group developed the chemoselective ATH/DKR of α,δ-diketo esters using privileged (arene)RuCl(monosulfonamide) complexes such as (*S*,*S*)-**[Ru]** (Scheme 5.79). The selective reduction of the α-keto ester in the presence of an aryl ketone provided the exclusive formation of the corresponding δ-keto α-hydroxy ester as a single diastereomer in yields ranging from 70% to 97% and ee values up to 98% ee for substrates having electron-donating, -withdrawing, and heteroaryl substituents at the β-position. Additionally, the ATH/DKR of β-chloro α-keto esters has been realized with a remarkable ligand-controlled inversion of the preference for the *syn*-selectivity to deliver the corresponding *anti*-chlorohydrins. Excellent levels of chemo-, enantio-, and diastereocontrol were achieved in the ATH/DKR of β-aryl β-chloro-α-ketoesters for a series of aliphatic and aromatic compounds (*anti*/*syn* up to >20 : 1, up to 98% ee) [90].

The same group studied the ATH/DKR of α,δ-diketo esters to give the *syn* products with high diastereoselectivities and good enantioselectivities. Interestingly, the reaction conditions allowed the chemoselective reduction of the α-keto ester in the presence of an aryl ketone. The authors used this strategy to achieve the asymmetric total synthesis of the α-benzylidene-γ-butyrolactone natural products megacerotonic acid and shimobashiric acid A (Scheme 5.80) [91].

This DKR/ATH approach was also used to achieve *anti*-chlorohydrins with high diastereo- and enantio-inductions starting from a wide range of

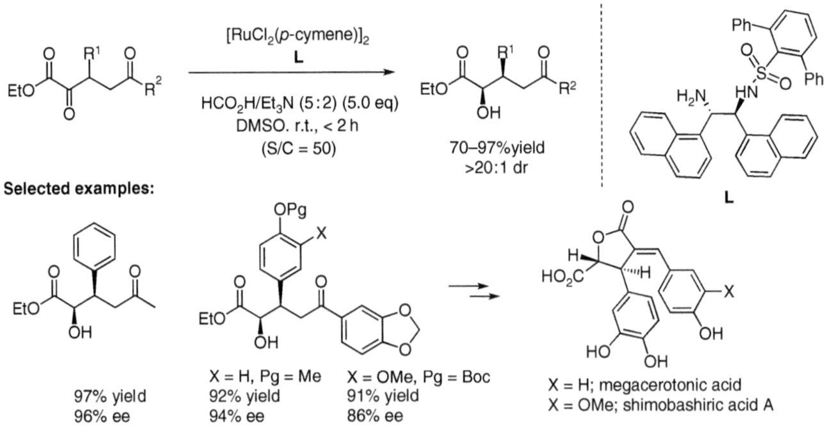

Scheme 5.80 Ru-catalyzed ATH/DKR of α,δ-diketo esters. Source: Steward and coworkers [91].

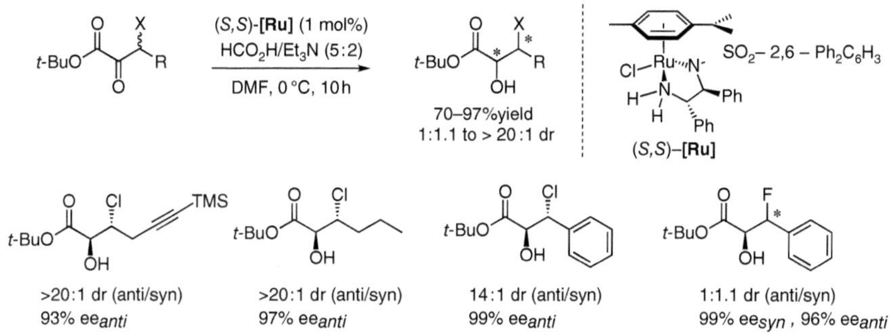

Scheme 5.81 Stereoselective synthesis of *anti*-chlorohydrins through DKR/ATH. Source: Based on Steward et al. [89].

β-chloro-α-ketoesters. However, low diastereoselectivities for β-fluoro-α-ketoesters were observed (Scheme 5.81) [91].

In 2013, Johnson and coworkers presented an approach to access enantioenriched *anti*-α-hydroxy-β-amino acid derivatives by enantioconvergent Ru(II)-catalyzed ATH of racemic α-keto esters readily prepared from the corresponding α-diazo esters by oxidation with oxone (Scheme 5.82). With the exception of aliphatic β-substituted substrates, yields ranging from 40% to 87% and routinely high levels of diastereo- and enantioselectivity were attained using (*S*,*S*)-**[Ru]** as a catalyst with heteroaromatic as well as electron-donating and -withdrawing aromatic systems (*anti*/*syn* up to >20 : 1, up to 98% ee) [92].

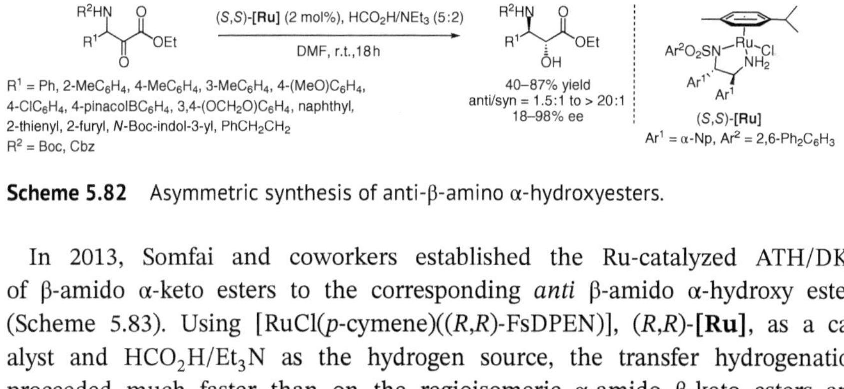

Scheme 5.82 Asymmetric synthesis of anti-β-amino α-hydroxyesters.

In 2013, Somfai and coworkers established the Ru-catalyzed ATH/DKR of β-amido α-keto esters to the corresponding *anti* β-amido α-hydroxy esters (Scheme 5.83). Using [RuCl(*p*-cymene)((*R*,*R*)-FsDPEN)], (*R*,*R*)-**[Ru]**, as a catalyst and HCO_2H/Et_3N as the hydrogen source, the transfer hydrogenation proceeded much faster than on the regioisomeric α-amido β-keto esters and

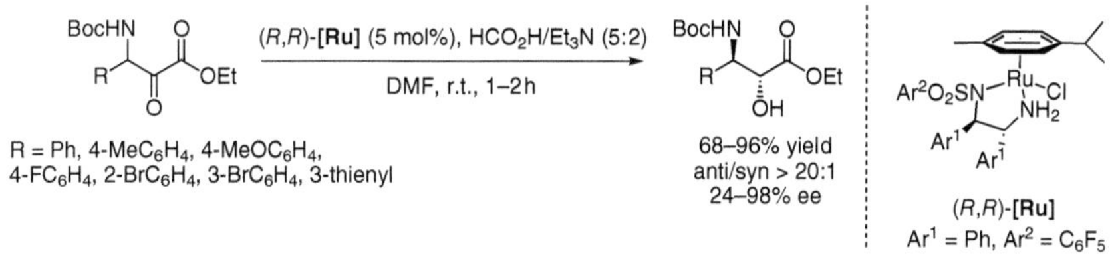

Scheme 5.83 Synthesis of *anti* β-amido α-hydroxy esters.

produced the *anti* aromatic and heteroaromatic compounds with good diastereomeric ratio and high enantioselectivities (up to 98% ee) as the only detectable products [93].

In 2013, Johnson and coworkers elaborated the first highly selective Ru-catalyzed ATH of acyl phosphonates using a DKR process with an unexpected reversal in facial selectivity as compared to the ATH of α-keto esters (Scheme 5.84). $[RuCl_2(p\text{-cymene})]_2$ combined with a bulky α-naphthyl ethylenediamine-derived ligand allowed the preparation of (*S*,*S*)-**[Ru]** as a catalyst to provide the corresponding reduced products with the highest diastereoselectivities (dr up to >20 : 1). The reaction accommodated various electron-donating and -withdrawing groups, including heteroaromatic, linear, and cyclic aliphatic substituents as well, yielding β-stereogenic α-hydroxy phosphonic acid derivatives in high yields and enantioselectivities up to 99% ee [94].

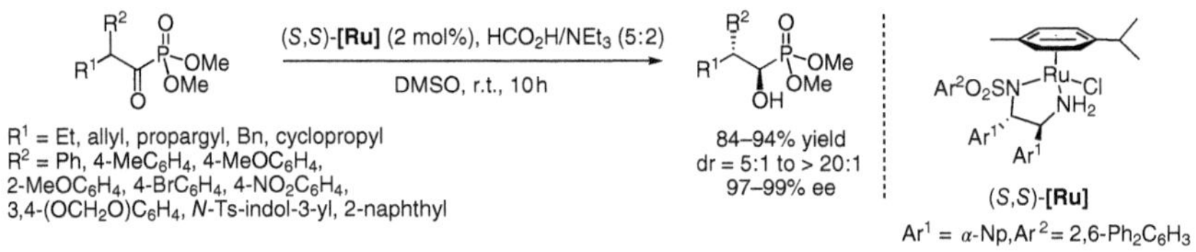

Scheme 5.84 Catalytic ATH of racemic acyl phosphonates.

5.10 β-Alkoxy Ketones

In 2015, Chen and coworkers demonstrated that a series of 3-(2-oxo-arylethyl) isobenzofuran-1(3H)-ones were efficiently reduced under mild conditions to the corresponding enantiomerically enriched phthalide derivatives bearing 1,3-diastereocenters using 0.2 mol% of [RuCl(mesitylene)((*S*,*S*)-TsDPEN)], (*S*,*S*)-**[Ru]**, as a catalyst and HCO_2H/Et_3N as the hydrogen donor in dichloromethane at 40 °C (Scheme 5.85). The 3-(2-hydroxy-2-arylethyl)isobenzoburan-1(3H)ones were produced in good yields (90–97%) and enantioselectivities up to 99% but moderate diastereomeric ratios ranging from 69 : 31 to 90 : 10. Phthalide frameworks are structural subunits found in a large number of natural products, exhibiting a range of biological activities [95].

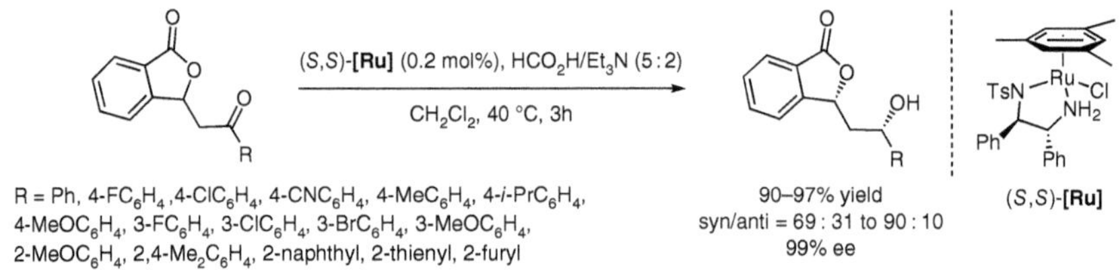

Scheme 5.85 Dynamic kinetic resolution of phthalide derivatives.

5.11 1,2-Diketones

Synthesis of *syn*-1,2-diols by ATH through DKR of symmetrical benzils using [RuCl(η^6-*p*-cymene)(*S*,*S*)-TsDPEN] in a mixture of HCO_2H/Et_3N was disclosed by Noyori, Ikariya et al. in 1999 (Scheme 5.86). High levels of stereoselectivity were achieved when the arene moiety was adorned in *para* position with either electron-donating substituents or electron-withdrawing groups (*dl*/*meso* = 94.2/5.8 to 98.6/1.4 and ee > 99%) in moderate to perfect yields (67–100%) [96].

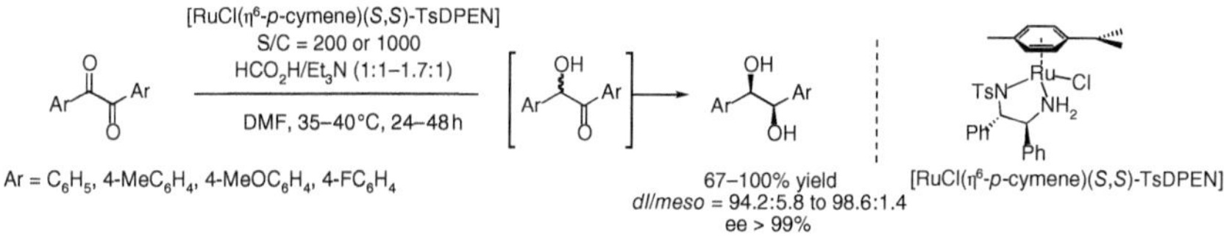

Scheme 5.86 A stereoselective synthesis of hydrobenzoins via ATH of benzils.

5.12 β-Substituted Ketones

Our group described in 2018 the first example of Ru-promoted ATH-DKR of β-amino ketones to give enantioenriched 1,3-aminoalcohols bearing a valuable tetrahydroisoquinoline scaffold, through epimerization of the β-stereocenter by an intermolecular elimination (retro-Mannich)/addition process [97]. Under optimal conditions, using [RuCl(S,S)-TsDPEN(benzene)] in dichloromethane in the presence of an azeotropic 5 : 2 mixture of HCO_2H and Et_3N as hydrogen source at 30 °C, various ring-substituted β-amino ketones that contained either electron-donating or -withdrawing substituents in *ortho*, *meta*, and *para* positions on the aryl groups R^1 and R^2 were reduced to their corresponding *syn* 2-(1,2,3,4-tetrahydro-1-isoquinolyl)ethanol derivatives in good yields (60–85%) with diastereomeric ratios up to 71 : 29 and excellent enantioselectivities (up to 99% ee) (Scheme 5.87).

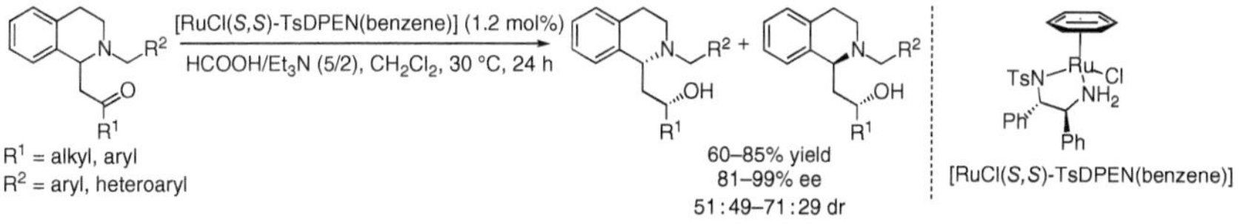

Scheme 5.87 Synthesis of *syn* 2-(1,2,3,4-tetrahydro-1-isoquinolyl)ethanol derivatives.

Wang and coworkers disclosed an elegant one-pot synthesis of biologically relevant 3,4-dihydro-2*H*-pyran-carbonitriles through Ru-catalyzed ATH of racemic α-aryl-γ-keto malononitriles coupled with a DKR process [98]. By employing the widely used chiral [RuCl(*S*,*S*)-TsDPEN(mesitylene)] in combination with HCO_2Na as a hydrogen source in a mixture of H_2O/*i*-PrOH (1 : 3) at 60 °C, diversely substituted 3,4-dihydro-2*H*-pyran-carbonitriles were obtained in high yields (71–99%), with excellent enantioselectivities (90–99%) and moderate to high diastereomeric ratios (80 : 20–98 : 2). From a mechanistic point of view, this method involved an Et_3N-mediated retro-Michael addition, followed by an ATH/DKR process and a subsequent spontaneous diastereoselective cyclization reaction of the hydrogenation products to deliver enantioenriched 3,4-dihydro-2*H*-pyran-carbonitriles (Scheme 5.88).

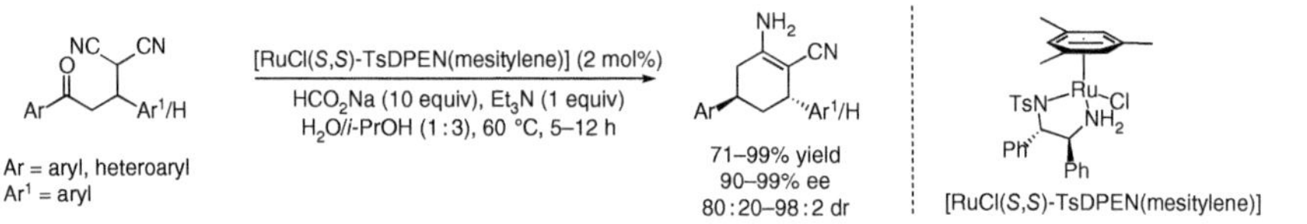

Scheme 5.88 Asymmetric synthesis of 3,4-dihydro-2*H*-pyran-carbonitriles.

5.13 α-Substituted Aldehydes

In 2009, Zhou and coworkers explored the AH of racemic α-aryl aldehydes upon DKR catalyzed by ruthenium complexes of chiral spirodiphosphines. Substrates bearing a small alkyl group (R = Me, Et, Bn) led to lower enantioselectivity (40–53% ee), while a bulkier alkyl group (R = *i*Pr, *t*Bu, *c*-pent, *c*-hex) gave higher enantioselectivity (71–81% ee). Furthermore, position and electronic properties of the substituents on the aryloxy moiety had no influence on the stereochemical outcome of the reaction (Scheme 5.89) [99].

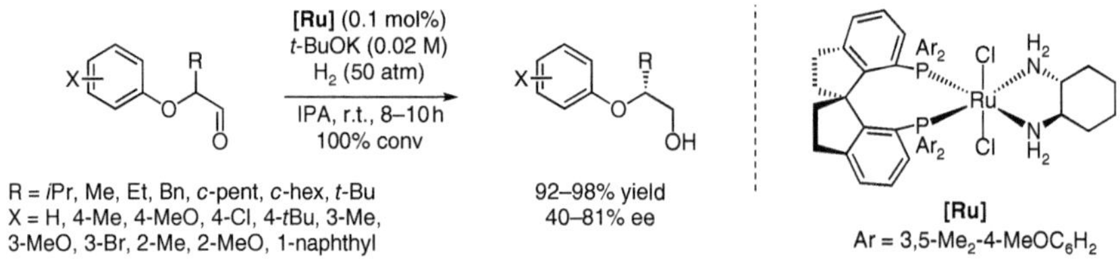

Scheme 5.89 Enantioselective synthesis of chiral β-aryloxy alcohols by AH via DKR. Source: Modified from Zhou et al. [99].

Interestingly, the group of Zhang showcased the use of ATH/DKR to synthesize optically pure β-sulfonyl primary alcohols (Scheme 5.90). Starting from racemic branched aldehydes, no new stereogenic center was created during the reaction, but, thanks to the DKR principle, a wide variety of enantiopure hydroxy sulfones were obtained with excellent yield (up to 92%) and good enantioselectivity (up to 90% ee) using [RuCl(η^6-*p*-cymene)(*S*,*S*)-TsDPEN] and HCO_2H/Et_3N (5 : 2) as the reducing agents [100].

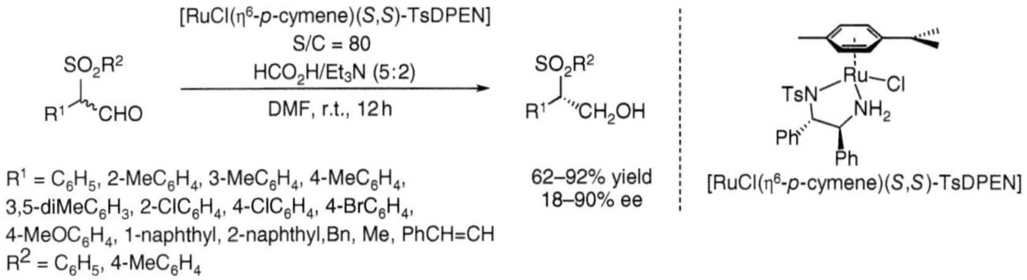

Scheme 5.90 DKR of racemic α-sulfonylaldehydes via ATH.

5.14 Summary and Conclusions

The homogeneous asymmetric reduction of ketone derivatives allows a straightforward and atom-economical access to enantiomerically enriched alcohols which are valuable intermediates for the manufacture of pharmaceuticals and fine chemicals. Transition-metal-mediated asymmetric hydrogenation and transfer hydrogenation are particularly efficient tools for this transformation, especially when combined with a DKR process. A range of ruthenium, rhodium, and iridium complexes were developed for the asymmetric (transfer) hydrogenation of substituted ketones. Valuable building blocks bearing up to three stereogenic centers were easily prepared that way in one single step with very high levels of enantio- and diastereoselectivity in good to excellent yields.

References

1 (a) Blacker, A.J. (2007). *Handbook of Homogeneous Hydrogenation*, vol. 1 (eds. J.G. de Vries and C.J. Elsevier). Weinheim: Wiley-VCH. (b) Andersson, P.G. and Munslow, I.J. (eds.) (2008). *Modern Reduction Methods*. Weinheim: Wiley-VCH. For comprehensive reviews and chapters on asymmetric hydrogenation (AH). see: (c)Knowles, W.S. (2002). *Angew. Chem. Int. Ed.* 41: 1998. (d) Noyori, R. (2002). *Angew. Chem. Int. Ed.* 41: 2008. (e) Genêt, J.-P. (2003). *Acc. Chem. Res.* 36: 908. (f) Blaser, H.-U., Malan, C., Pugin, B. et al. (2003). *Adv. Synth. Catal.* 345: 103. (g) Zhou, Y.-G. (2007). *Acc. Chem. Res.* 40: 1357. (h) Shang, G., Li, W., and Zhang, X. (2010). *Catalytic Asymmetric Synthesis*, 3e (ed. I. Ojima), 343. New York: Wiley. (i) Xie, J.H., Zhu, S.F., and Zhou, Q.L. (2011). *Chem. Rev.* 111: 1713. (j) Gopalaiah, K. and Kagan, H.B. (2011). *Chem. Rev.* 111: 4599. (k) Wang, D.-S., Chen, Q.-A., Lu, S.-M., and Zhou, Y.-G. (2012). *Chem. Rev.* 112: 2557. (l) Chen, Q.-A., Ye, Z.-S., Duan, Y., and Zhou, Y.-G. (2013). *Chem. Soc. Rev.* 42: 497. (m) Hamada, Y. (2014). *Chem. Rec.* 14: 235. (n) Li, Y.-Y., Yu, S.-L., Shen, W.-Y., and Gao, J.-X. (2015). *Acc. Chem. Res.* 48: 2587. (o) Xie, J.-H., Bao, D.-H., and Zhou, Q.-L. (2015). *Synthesis* 47: 460. (p) Zhang, Z., Butt, N.A., and Zhang, W. (2016). *Chem. Rev.* 116: 23. (q) Li, W., Lu, B., and Zhang, Z. (2016). *Chem.*

Rec. 16: 2506. (r) Mashima, K., Higashida, K., Iimuro, A. et al. (2016). *Chem. Rec.* 16: 2585. (s) Luo, Y.-E., He, Y.-M., and Fan, Q.-H. (2016). *Chem. Rec.* 16: 2697. (t) Ohkuma, T. and Arai, N. (2016). *Chem. Rec.* 16: 2801. (u) Imamoto, T. (2016). *Chem. Rec.* 16: 2659.

2 (a) For comprehensive reviews and chapters on asymmetric transfer hydrogenation (ATH), see: Zassinovich, G., Mestroni, G., and Gladiali, S. (1992). *Chem. Rev.* 92: 1051. (b) de Graauw, C.F., Peters, J.A., van Bekkum, H., and Huskens, J. (1994). *Synthesis*: 1007. (c) Noyori, R. and Hashiguchi, S. (1997). *Acc. Chem. Res.* 30: 97. (d) Palmer, M.J. and Wills, M. (1999). *Tetrahedron: Asymmetry* 10: 2045. (e) Pàmies, O. and Backvall, J.-E. (2001). *Chem. Eur. J.* 7: 5052. (f) Everaere, K., Mortreux, A., and Carpentier, J.-F. (2003). *Adv. Synth. Catal.* 345: 67. (g) Gladiali, S. and Alberico, E. (2006). *Chem. Soc. Rev.* 35: 226. (h) Joseph, S.M., Samec, J.S., Bäckvall, J.-E. et al. (2006). *Chem. Soc. Rev.* 35: 237. (i) Ikariya, T. and Blacker, A.J. (2007). *Acc. Chem. Res.* 40: 1300. (j) Wang, C., Wu, X., and Xiao, J. (2008). *Chem. Asian J.* 3: 1750. (k) Ikariya, T. (2011). *Bull. Chem. Soc. Jpn.* 84: 1. (l) Zheng, C. and You, S.-L. (2012). *Chem. Soc. Rev.* 41: 2498. (m) Bartoszewicz, A., Ahlsten, N., and Martín-Matute, B. (2013). *Chem. Eur. J.* 19: 7274. (n) Slagbrand, T., Lundberg, H., and Adolfsson, H. (2014). *Chem. Eur. J.* 20: 16102. (o) Štefane, B. and Požgan, F. (2014). *Catal. Rev.* 56: 82. (p) Ito, J.-I. and Nishiyama, H. (2014). *Tetrahedron Lett.* 55: 3133. (q) Wang, D. and Astruc, D. (2015). *Chem. Rev.* 115: 6621. (r) Foubelo, F., Nájera, C., and Yus, M. (2015). *Tetrahedron: Asymmetry* 26: 769. (s) Nedden, H.G., Zanotti-Gerosa, A., and Wills, M. (2016). *Chem. Rec.* 16: 2623. (t) Štefane, B. and Požgan, F. (2016). *Top. Curr. Chem.* 18: 374. (u) Ayad, T., Phansavath, P., and Ratovelomanana-Vidal, V. (2016). *Chem. Rec.* 16: 2754. (v) Matuška, O., Kindl, M., and Kačer, P. (2017). *New Advances in Hydrogenation Process* (ed. T.M. Ravanchi), 37. Croatia: InTech; Matsunami, A. and Kayaki, Y. (2018). *Tetrahedron Lett. 59*: 504. (w) Whittlesey, M.K. (2018). *N-Heterocyclic Carbenes in Catalytic Organic Synthesis 1* (ed. S.P. Nolan). Science and Synthesis, 285. Weinheim: Wiley-VCH. (x) Zhang, Z., Butt, N.A., Zhou, M. et al. (2018). *Chin. J. Chem.* 36: 443.

3 (a) Kitamura, M., Ohkuma, T., Inoue, S. et al. (1988). *J. Am. Chem. Soc.* 110: 629. (b) Mashima, K., Kusano, K., Ohta, T. et al. (1989). *J. Chem. Soc. Chem. Commun.*: 1208. (c) Ohkuma, T., Kitamura, M., and Noyori, R. (1990). *Tetrahedron Lett.* 31: 5509.

4 (a) Hashiguchi, S., Fujii, A., Takehara, J. et al. (1995). *J. Am. Chem. Soc.* 117: 7562. (b) Fujii, A., Hashiguchi, S., Uematsu, N. et al. (1996). *J. Am. Chem. Soc.* 118: 2521. (c) Haack, K.-J., Hashiguchi, S., Fujii, A. et al. (1997). *Angew. Chem. Int. Ed.* 36: 285. (d) Ikariya, T., Murata, K., and Noyori, R. (2006). *Org. Biomol. Chem.* 4: 393.

5 (a) Noyori, R., Ikeda, T., Ohkuma, T. et al. (1989). *J. Am. Chem. Soc.* 111: 9134. (b) Noyori, R., Tokunaga, M., and Kitamura, M. (1995). *Bull. Chem. Soc. Jpn.* 68: 36. (c) Kitamura, M., Tokunaga, M., and Noyori, R. (1993). *J. Am. Chem. Soc.* 115: 144. (d) Kitamura, M., Tokunaga, M., and Noyori, R. (1993). *Tetrahedron* 49: 1853.

6 Genêt, J.-P., Jugé, S., Mallart, S., and Laffitte, J.A. (1989). French Patent n°8911159.

7 For previous comprehensive reviews covering this topic, see: (a)Xie, J.-H. and Zhou, Q.-L. (2015). *Aldrichimica Acta* 48: 33. (b) Caddick, S. and Jenkins, K. (1996). *Chem. Soc. Rev.* 25: 447. (c) Ward, R.S. (1995). *Tetrahedron: Asymmetry* 6: 1475. (d) Sturmer, R. (1997). *Angew. Chem. Int. Ed.* 36: 1173. (e) El Gihani, M.T. and Williams, J.M.J. (1999). *Curr. Opin. Chem. Biol.* 3: 11. (f) Ratovelomanana-Vidal, V. and Genêt, J.-P. (2000). *Can. J. Chem.* 851: 846. (g) Huerta, F.F., Minidis, A.B.E., and Bäckvall, J.-E. (2001). *Chem. Soc. Rev.* 30: 321. (h) Faber, K. (2001). *Chem. Eur. J.* 7: 5005. (i) Pàmies, O. and Bäckvall, J.-E. (2003). *Chem. Rev.* 103: 3247. (j) Pellissier, H. (2003). *Tetrahedron* 59: 8291. (k) Turner, N.J. (2004). *Curr. Opin. Chem. Biol.* 8: 114. (l) Vedejs, E. and Jure, M. (2005). *Angew. Chem. Int. Ed.* 44: 3974. (m) Martín-Matute, B. and Bäckvall, J.-E. (2007). *Curr. Opin. Chem. Biol.* 11: 226. (n) Pellissier, H. (2008). *Tetrahedron* 64: 1563. (o) Pellissier, H. (2011). *Tetrahedron* 67: 3769. (p) Echeverria, P.-G., Ayad, T., Phansavath, P., and Ratovelomanana-Vidal, V. (2016). *Synthesis* 48: 2523.

8 Chen, C.-Y., Frey, L.F., Shultz, S. et al. (2007). *Org. Process Res. Dev.* 11: 616.

9 (a) Arai, N., Ooka, H., Azuma, K. et al. (2007). *Org. Lett.* 9: 939. (b) Ooka, H., Arai, N., Azuma, K. et al. (2008). *J. Org. Chem.* 73: 9084.

10 Xie, J.-H., Liu, S., Kong, W.-L. et al. (2009). *J. Am. Chem. Soc.* 131: 4222.

11 Hibino, T., Makino, K., Sugiyama, T., and Hamada, Y. (2009). *ChemCatChem* 1: 237.

12 Bai, W.-J., Xie, J.-H., Li, Y.-L. et al. (2010). *Adv. Synth. Catal.* 352: 81.

13 Chiwara, V.I., Haraguchi, N., and Itsuno, S. (2009). *J. Org. Chem.* 74: 1391.

14 Akashi, M., Arai, N., Inoue, T., and Ohkuma, T. (2011). *Adv. Synth. Catal.* 353: 1955.

15 Chung, J.Y.L., Steinhuebel, D., Krska, S.-W. et al. (2012). *Org. Process. Res. Dev.* 16: 1832.

16 Chen, C.-Y. and Weisel, M. (2013). *Synlett* 24: 189.

17 Wu, W., You, C., Yin, C. et al. (2017). *Org. Lett.* 19: 2548.

18 (a) Xu, F., Zacuto, M.J., Kohmura, Y. et al. (2014). *Org. Lett.* 16: 5422. (b) Chung, J.Y.L., Scott, J.P., Anderson, C. et al. (2015). *Org. Process. Res. Dev.* 19: 1760.

19 Liu, S., Xie, J.-H., Wang, L.-X., and Zhou, Q.-L. (2007). *Angew. Chem. Int. Ed.* 46: 7506.

20 (a) Cheng, L.-J., Xie, J.-H., Chen, Y. et al. (2012). *Adv. Synth. Catal.* 354: 1105. (b) Li, G., Xie, J.-H., Hou, J. et al. (2013). *Adv. Synth. Catal.* 355: 1597. (c) Cheng, L.-J., Xie, J.-H., Chen, Y. et al. (2013). *Org. Lett.* 15: 764.

21 Cheng, J.-Q., Xie, J.-H., Bao, D.-H., and Zhou, Q.-L. (2012). *Org. Lett.* 14: 2714.

22 Yin, C., Dong, X.-Q., and Zhang, X. (2018). *Adv. Synth. Catal.* 360: 4319.

23 Zatolochnaya, O.V., Rodriguez, S., Zhang, Y. et al. (2018). *Chem. Sci.* 9: 4505.

24 Fernández, R., Ros, A., Magriz, A. et al. (2007). *Tetrahedron* 63: 6755.

25 (a) Vyas, V.K. and Bhanage, B.M. (2016). *Org. Lett.* 18: 6436. (b) Vyas, V.K. and Bhanage, B.M. (2018). *Asian J. Org. Chem.* 7: 346.

26 Qin, T. and Metz, P. (2017). *Org. Lett.* 19: 2981.
27 Fang, L., Lyu, Q., Lu, C. et al. (2016). *Adv. Synth. Catal.* 358: 3196.
28 Alnafta, N., Schmidt, J.P., Nesbitt, C.L., and McErlean, C.S.P. (2016). *Org. Lett.* 18: 6520.
29 Fang, L., Liu, S., Han, L. et al. (2017). *Organometallics* 36: 1217.
30 Fang, L., Zhao, F., Hu, S. et al. (2018). *J. Org. Chem.* 83: 12213.
31 Zhang, Y.-M., Zhang, Q.-Y., Wang, D.-C. et al. (2019). *Org. Lett.* 21: 2988.
32 Bromhead, L.J., Visser, J., and McErlan, C.S.P. (2014). *J. Org. Chem.* 79: 1516.
33 Jeran, M., Cotman, A.E., Stephan, M., and Mohar, B. (2017). *Org. Lett.* 19: 2042.
34 He, B., Phansavath, P., and Ratovelomanana-Vidal, V. (2019). *Org. Lett.* 21: 3276.
35 Wu, Y., Geng, Z., Bai, J., and Zhang, Y. (2011). *Chin. J. Chem.* 29: 1467.
36 Cotman, A.E., Cahard, D., and Mohar, B. (2016). *Angew. Chem. Int. Ed.* 55: 5294.
37 Liu, C., Xie, J.-H., Li, Y.-L. et al. (2013). *Angew. Chem. Int. Ed.* 52: 593.
38 Lin, H., Xiao, L.-J., Zhou, M.-J. et al. (2016). *Org. Lett.* 18: 1434.
39 Ashley, E.R., Sherer, E.C., Pio, B. et al. (2017). *ACS Catal.* 7: 1446.
40 Cotman, A.E., Modec, B., and Mohar, B. (2018). *Org. Lett.* 20: 2921.
41 Makino, K., Goto, T., Hiroki, Y., and Hamada, Y. (2004). *Angew. Chem. Int. Ed.* 43: 882.
42 Mordant, C., Dünkelmann, P., Ratovelomanana-Vidal, V., and Genêt, J.-P. (2004). *Chem. Commun.*: 1296.
43 Makino, K., Goto, T., Hiroki, Y., and Hamada, Y. (2008). *Tetrahedron: Asymmetry* 19: 2816.
44 Makino, K., Hiroki, Y., and Hamada, Y. (2005). *J. Am. Chem. Soc.* 127: 5784.
45 Maeda, T., Makino, K., Iwasaki, M., and Hamada, Y. (2010). *Chem. Eur. J.* 16: 11954.
46 Hamada, Y., Koseki, Y., Fujii, T. et al. (2008). *Chem. Commun.*: 6206.
47 Tone, H., Buchotte, M., Mordant, C. et al. (2009). *Org. Lett.* 11: 1995.
48 Prévost, S., Ayad, T., Phansavath, P., and Ratovelomanana-Vidal, V. (2011). *Adv. Synth. Catal.* 353: 3213.
49 Echeverria, P.-G., Prévost, S., Cornil, J. et al. (2014). *Org. Lett.* 16: 2390.
50 Echeverria, P.G., Férard, C., Cornil, J. et al. (2014). *Synlett*: 2761.
51 Seashore-Ludlow, B., Villo, P., Häcker, C., and Somfai, P. (2010). *Org. Lett.* 12: 5274.
52 Liu, Z., Shultz, C.S., Sherwood, C.A. et al. (2011). *Tetrahedron Lett.* 52: 1685.
53 Seashore-Ludlow, B., Villo, P., and Somfai, P. (2012). *Chem. Eur. J.* 18: 7219.
54 Seashore-Ludlow, B., Saint-Dizier, F., and Somfai, P. (2012). *Org. Lett.* 14: 6334.
55 Perez, M., Echeverria, P.G., Martinez-Arripe, E. et al. (2015). *Eur. J. Org. Chem.* 2015: 5949.
56 Echeverria, P.G., Cornil, J., Férard, C. et al. (2015). *RSC Adv.* 5: 56815.
57 Wang, X., Xu, L., Yan, L. et al. (2016). *Tetrahedron* 72: 1787.
58 Sun, G., Zhou, Z., Luo, Z. et al. (2017). *Org. Lett.* 19: 4339.
59 Touge, T., Kuwana, M., Komatsuki, Y. et al. (2019). *Org. Process. Res. Dev.* 23: 452.

60 Prévost, S., Gauthier, S., Cano de Andrade, M.C. et al. (2010). *Tetrahedron: Asymmetry* 21: 1436.
61 Bai, J., Miao, S., Wu, Y., and Zhang, Y. (2011). *Chin. J. Chem.* 29: 2476.
62 Fang, Z. and Wills, M. (2013). *J. Org. Chem.* 78: 8594.
63 Cartigny, D., Püntener, K., Ayad, T. et al. (2010). *Org. Lett.* 12: 3788.
64 Echeverria, P.-G., Férard, C., Phansavath, P., and Ratovelomanana-Vidal, V. (2015). *Catal. Commun.* 62: 95.
65 Monnereau, L., Cartigny, D., Scalone, M. et al. (2015). *Chem. Eur. J.* 21: 11799.
66 He, B., Zheng, L.-S., Phansavath, P., and Ratovelomanana-Vidal, V. (2019). *ChemSusChem* 12: 3032.
67 Rast, S., Modec, B., Stephan, M., and Mohar, B. (2016). *Org. Biomol. Chem.* 14: 2112.
68 Plantan, I., Stephan, M., Urleb, U., and Mohar, B. (2009). *Tetrahedron Lett.* 50: 2676.
69 Mohar, B., Stephan, M., and Urleb, U. (2010). *Tetrahedron* 66: 4144.
70 Li, X., Tao, X., Ma, X. et al. (2013). *Tetrahedron* 69: 7152.
71 Hou, C.-J. and Hu, X.-P. (2016). *Org. Lett.* 18: 5592.
72 Liu, Y.-T., Chen, J.-Q., Li, L.-P. et al. (2017). *Org. Lett.* 19: 3231.
73 Magnus, N.A., Astleford, B.A., Laird, D.L.T. et al. (2013). *J. Org. Chem.* 78: 5768.
74 Lynch, D., Deasy, R.E., Clarke, L.-A. et al. (2016). *Org. Lett.* 18: 4978.
75 Bao, D.-H., Gu, X.-S., Xie, J.-H., and Zhou, Q.-L. (2017). *Org. Lett.* 19: 118.
76 Son, S.-M. and Lee, H.-K. (2013). *J. Org. Chem.* 78: 8396.
77 Kumaraswamy, G., Narayana Murthy, A., Narayanarao, V. et al. (2013). *Org. Biomol. Chem.* 11: 6751.
78 Kumaraswamy, G., Narayanarao, V., and Shanigaram, P. (2015). *Tetrahedron* 71: 8960.
79 Bisset, A.A., Dishington, A., Jones, T. et al. (2014). *Tetrahedron* 70: 7207.
80 Cotman, A.E., Lozinšek, M., Wang, B. et al. (2019). *Org. Lett.* 21: 3644.
81 Tao, X., Li, W., Ma, X. et al. (2012). *J. Org. Chem.* 77: 8401.
82 Tao, X., Li, W., Li, X. et al. (2013). *Org. Lett.* 15: 72.
83 Son, S.-M. and Lee, H.-K. (2014). *J. Org. Chem.* 79: 2666.
84 Ding, Z., Yang, J., Wang, T. et al. (2009). *Chem. Commun.*: 571.
85 Huang, X.-F., Zhang, S.-Y., Geng, Z.-C. et al. (2013). *Adv. Synth. Catal.* 355: 2860.
86 Geng, Z., Wu, Y., Miao, S. et al. (2011). *Tetrahedron Lett.* 52: 907.
87 Xiong, Z., Pei, C., Xue, P. et al. (2018). *Chem. Commun.* 54: 3883.
88 Hu, X., Zhang, K., Chang, F. et al. (2018). *Mol. Catal.* 452: 271.
89 Steward, K.M., Gentry, E.C., and Johnson, J.S. (2012). *J. Am. Chem. Soc.* 134: 7329. Corrigendum: (2015). *J. Am. Chem. Soc. 137*: 3715.
90 Steward, K.M., Corbett, M.T., Goodman, C.G., and Johnson, J.S. (2012). *J. Am. Chem. Soc.* 134: 20197. Corrigendum: (2015). *J. Am. Chem. Soc. 137*: 3991.
91 (a) Krabbe, S.W. and Johnson, J.S. (2015). *Org. Lett.* 17: 1188.
92 Goodman, C.G., Do, D.T., and Johnson, J.S. (2013). *Org. Lett.* 15: 2446.
93 Villacrez, M. and Somfai, P. (2013). *Tetrahedron Lett.* 54: 5266.
94 Corbett, M.T. and Johnson, J.S. (2013). *J. Am. Chem. Soc.* 135: 594.

95 Chen, T., Ye, Q., Zhao, Q., and Liu, G. (2015). *Org. Lett.* 17: 4972.

96 Murata, K., Okano, K., Miyagi, M. et al. (1999). *Org. Lett.* 1: 1119.

97 Zheng, L.-S., Phansavath, P., and Ratovelomanana-Vidal, V. (2018). *Org. Chem. Front.* 5: 1366.

98 Zheng, D., Zhao, Q., Hu, X. et al. (2017). *Chem. Commun.* 53: 6113.

99 Zhou, Z.-T., Xie, J.-H., and Zhou, Q.-L. (2009). *Adv. Synth. Catal.* 351: 363.

100 Wu, G., Zhu, J., Ding, Z. et al. (2009). *Tetrahedron Lett.* 50: 427.

6

Industrial Applications of Asymmetric (Transfer) Hydrogenation

Xumu Zhang[1] and Pan-Lin Shao[2]

[1] *Southern University of Science and Technology, Guangdong Provincial Key Laboratory of Catalysis, Department of Chemistry, 1088 Xueyuan Road, Shenzhen 518000, People's Republic of China*
[2] *Southern University of Science and Technology, College of Innovation and Entrepreneurship, 1088 Xueyuan Road, Shenzhen 518000, People's Republic of China*

6.1 Introduction

Since the seminal work in the development of chiral catalysts for asymmetric hydrogenation by Knowles [1] and Horner [2] in the late 1960s, and the first commercial application of asymmetric hydrogenation in the early 1970s by Monsanto in the production of the *anti*-Parkinsonian drug L-DOPA [3], asymmetric hydrogenation has developed into a powerful chemical transformation [4].

However, the development in the early 1980s was mainly focused on chiral Rh catalysts, and the substrate scope was limited to α-dehydroamino acids. One of the key milestones in the development of asymmetric hydrogenation was the discovery of BINAP [2,2′-bis(diphenylphosphino)-1,1′-binaphthyl] by Noyori and coworkers, which opened up opportunities to broaden the scope of utility of asymmetric hydrogenation [5]. A wide variety of prochiral olefin and ketone substrates were hydrogenated with excellent enantioselectivity [6]. For their pioneering and continuous excellent work, Knowles [3] and Noyori [7] were awarded the Nobel Prize in 2001.

Over the past few decades, thousands of efficient and powerful chiral phosphorus ligands with diverse structures have been developed for asymmetric hydrogenation; several representative chiral phosphorus ligands are shown in Figure 6.1 [4, 8]. Numerous unsaturated compounds, such as enamides [9], ketones [10], olefins [11], imines [12], and so on, can be hydrogenated in excellent enantioselectivities. Based on the privileges of high atom economy, high reactivity, mild reaction conditions, high cost efficiency, and environmental benign, asymmetric hydrogenation has been intensively employed in the pharmaceutical, agrochemical, fragrance, and flavor industries.

Several overviews of asymmetric hydrogenation in industrial processes have been published [13]. The examples below relate to applications that can be considered significant to the implementation of the methodology. The chapter is not intended to discuss what specific ligand or class of ligands are best for a specific substrate class,

Asymmetric Hydrogenation and Transfer Hydrogenation, First Edition.
Edited by Virginie Ratovelomanana-Vidal and Phannarath Phansavath.

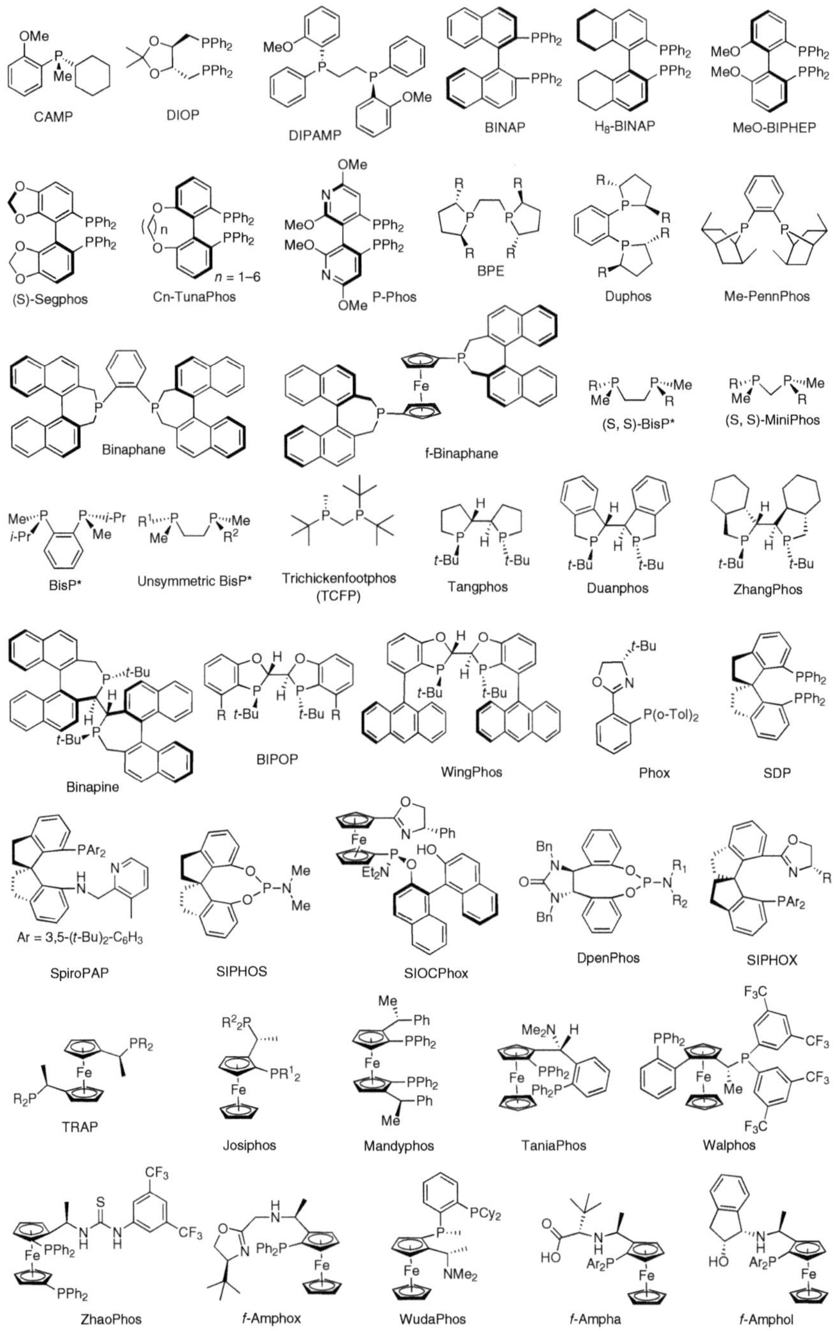

Figure 6.1 Representative chiral phosphorus ligands for asymmetric hydrogenation. Source: Tang and Zhang [4a], Zhou [4b], Roseblade and Pfaltz [4c], Xie et al. [4d], Corey and Kurti [4e], and Wu et al. [8a, b].

although some of the factors influencing catalyst choice are given in the illustrative examples. Additional examples of asymmetric hydrogenation used in industry are given throughout this chapter.

6.2 Industrial Applications of Asymmetric Hydrogenation

6.2.1 Asymmetric Hydrogenation of Enamide

6.2.1.1 L-DOPA

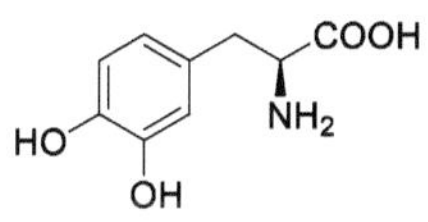

1, *L*-DOPA

L-DOPA (**1**) is the precursor to the neurotransmitters dopamine, norepinephrine, and epinephrine. As a drug, it is used in the clinical treatment of Parkinson's disease and dopamine-responsive dystonia [14].

Monsanto Corporation developed the practical synthesis of L-DOPA via asymmetric hydrogenation using DIPAMP as ligand, due to its high catalytic efficiency in Rh-catalyzed asymmetric hydrogenation of dehydroamino acids [15]. The success of the synthesis of L-DOPA constituted a milestone work, which has enlightened chemists to realize the power of asymmetric hydrogenation for the synthesis of chiral compounds. In the process of manufacture, the synthesis of the enamide **2** is a little long, and the nitrogen acyl group has to be removed after the reduction (Scheme 6.1). The asymmetric induction is not perfect; enantioenrichment of the reduction product **3** can be achieved through crystallization from the reaction mixture.

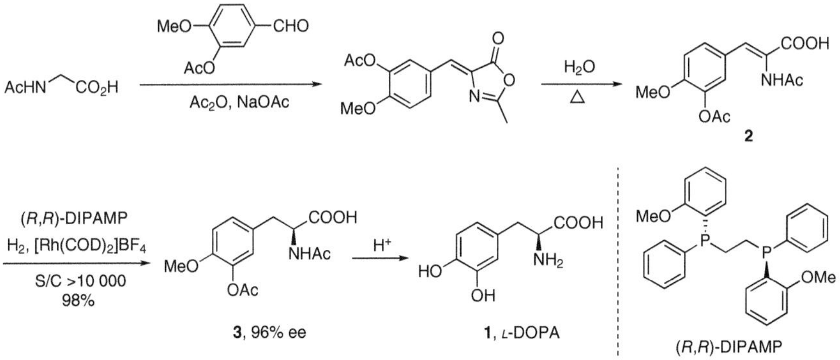

Scheme 6.1 Synthesis of L-DOPA.

6.2.1.2 Ramipril

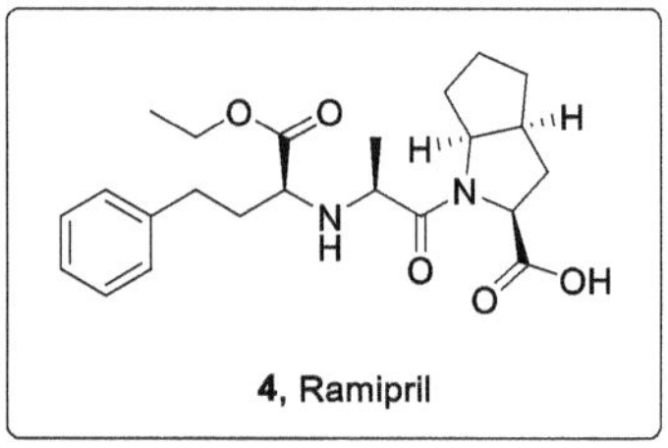

Ramipril (**4**) [16], patented in 1981 and approved for medical use in 1989, belongs to a class of angiotensin converting enzyme (ACE) inhibitors, which are used for treating high blood pressure and congestive heart failure, and also for preventing kidney failure due to high blood pressure and diabetes. The sales amounted to $940 million in 2017, consuming 54 tons of active pharmaceutical ingredients (APIs).

The traditional synthetic route is 10 steps to prepare the precursor of **5** from commercially available materials, followed by resolution using (*S*)-(+)-mandelic acid and crystallization (Scheme 6.2). The synthesis efficiency is rather low, and a large number of wastes will be produced, resulting in serious pollution. Every 40 tons of products produce 60 tons of waste [17].

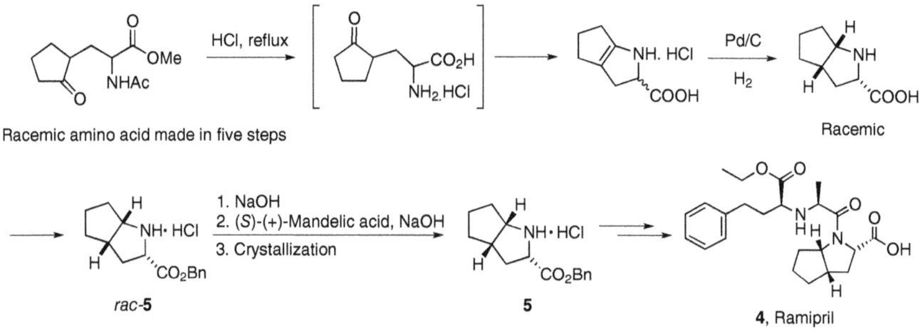

Scheme 6.2 Traditional synthetic route of Ramipril.

Zhang group cooperated with Chiral Quest Inc. and employed Rh/DuanPhos catalytic system for the asymmetric hydrogenation achieving the highly enantioselective synthesis of the key intermediate **5** (up to 80 000 turnover number (TON), >99% ee) on a ton scale (Scheme 6.3). One kilogram of catalyst can produce 36 tons of products. The process was patented by Chiral Quest Inc. [18]. To the best of our knowledge, this process represents one of the most concise and economical routes for the synthesis of Ramipril, and the TON is also the best one for hydrogenation of conjugated enamides for the synthesis of chiral γ,δ-unsaturated amino acid derivatives. This is a simple and more environmentally friendly route to the Ramipril intermediate. A significant cost reduction in the synthesis of Ramipril was realized. More than 30 tons of the Ramipril intermediate have been made by this route.

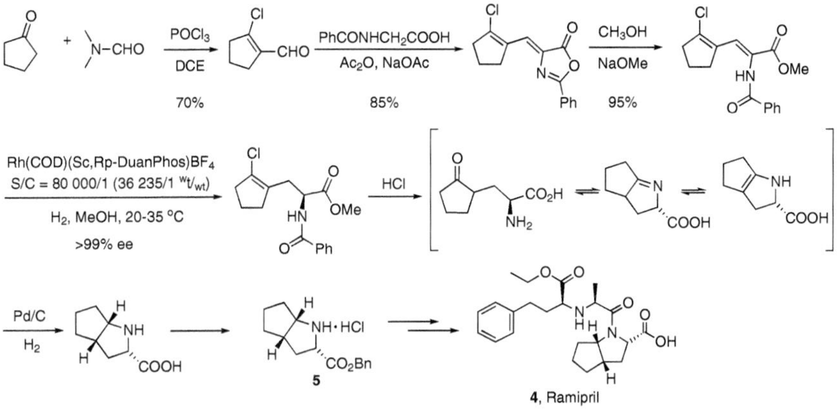

Scheme 6.3 Synthetic route of Ramipril developed by Chiral Quest Inc.

6.2.1.3 Sitagliptin

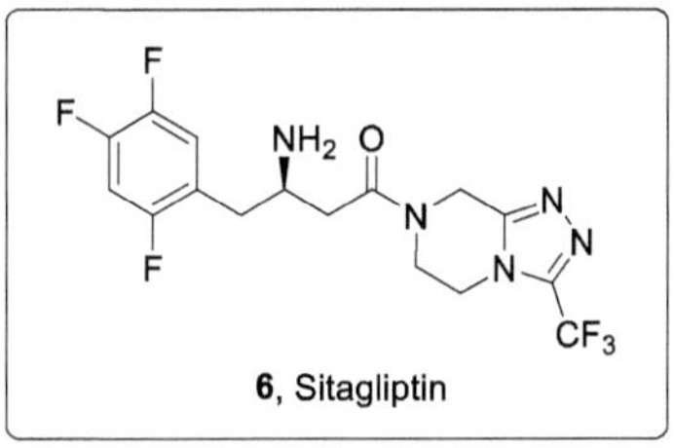

Sitagliptin, an oral antihyperglycemic of the dipeptidyl peptidase-4 (DPP-4) inhibitor, is the key component of the medications Januvia and Janumet [19]. The sales amounted to $10.5 billion in 2017, consuming 420 tons of APIs.

Merck developed the hydrogenation of the unprotected β-enamine amide **7** depicted as the final step for the production of Sitagliptin (Scheme 6.4). The β-enamine amide with the entire structure of sitagliptin **6** can be prepared in a

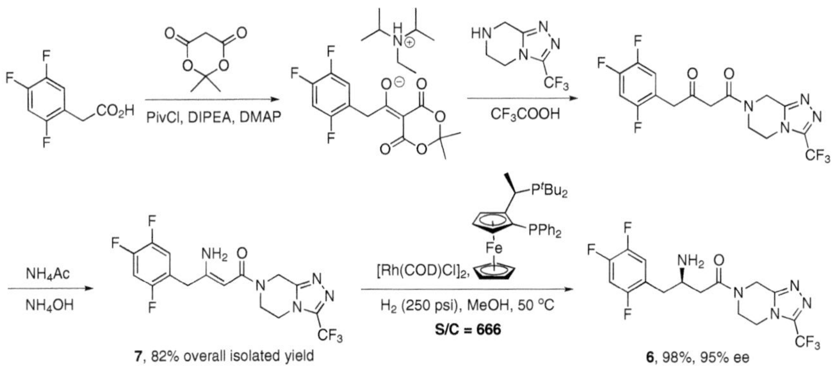

Scheme 6.4 Synthetic route for Sitagliptin developed by Merck.

one-pot process in 82% overall isolated yield. Hydrogenation of **7** in the presence of NH_4Cl (0.15 mol%), $[Rh(COD)Cl]_2$ (0.15 mol%), and ^{t}Bu-Josiphos (0.155 mol%) in MeOH under 250 psi of hydrogen at 50 °C for 16–18 hours was proved to be extremely robust and afforded **6** in 98% yield and 95% ee reproducibly (TON = 666) [20]. This manufacturing process represents an important breakthrough for the catalytic synthesis of chiral β-amino acids and received the Presidential Green Chemistry Challenge Award (2006) for alternative synthetic pathways and the IChemE Astra-Zeneca Award for excellence in Green Chemistry and Chemical Engineering (2005).

Mechanistic studies of this transformation indicate that the hydrogenation plausibly proceeded through the imine tautomer. Kinetic studies showed that the reaction rate slowed down over the course of the hydrogenation. The binding of the hydrogenated product **6** to the rhodium metal catalyst could have an adverse effect on the reaction rate and the asymmetric environment of the catalyst. Although the observed enantioselectivity did not change throughout the course of the hydrogenation, product inhibition of the catalyst was confirmed by a product-doping experiment unambiguously (Scheme 6.5).

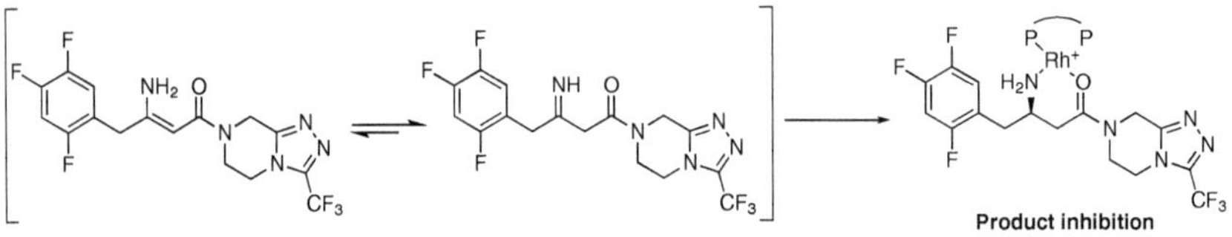

Scheme 6.5 Product inhibition.

In Merck's approach, some drawbacks cannot be ignored: low catalytic efficiency, product inhibition, low conversion number, high catalyst consumption, and low optical purity (95% ee). Chiral Quest Inc. developed a more efficient asymmetric hydrogenation route using Rh/DuanPhos complex (Scheme 6.6) [21]. Due to the potential amine inhibition of the catalyst, N-acylated β-enamine ester **8** was employed as the substrate. This strategy exhibits excellent activity and enantioselectivity (S/C = 10 000, 99% ee). One kilogram of catalyst can produce 2.4 tons of

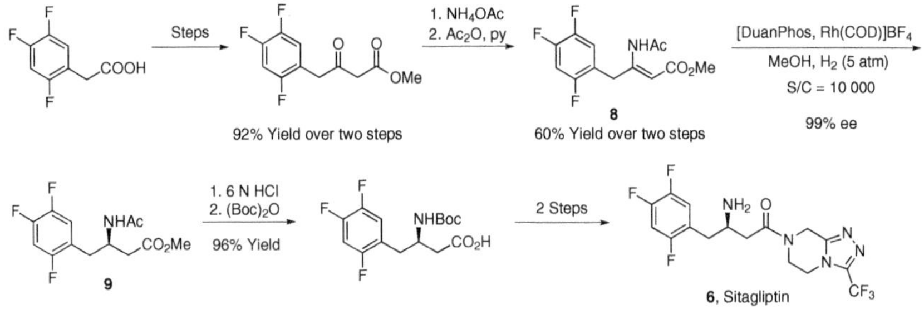

Scheme 6.6 Synthetic route for Sitagliptin developed by Chiral Quest Inc. Source: From Wu et al. [21].

the product. At present, the key intermediate **9** has achieved 50 tons of industrial synthesis, and the new process reduces the cost by 15%.

6.2.1.4 (*R*)-3-Amino-1-butanol

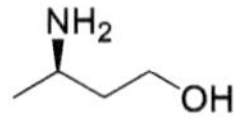

10
(*R*)-3-Amino-1-butanol

Enantiomerically pure amino alcohols and their derivatives are of great importance in synthetic and pharmaceutical chemistry. (*R*)-3-Amino-1-butanol (**10**) is the key intermediate to synthesize Dolutegravir [22]. Dolutegravir, designed and developed by GSK and Shionogi, is an integrase inhibitor for the treatment of the human immunodeficiency virus (HIV). Dolutegravir sodium (**11**) has been widely used worldwide since its launch into the market in 2013. The sales were $5.6 billion in 2017.

Zhang group and Chiral Quest Inc. developed an asymmetric hydrogenation route using Rh/DuanPhos complex to achieve the β-amino ester **13** firstly and then hydrolysis to form β-amino acid **14** (Scheme 6.7) [23]. N-acylated β-enamine ester **12** was employed as the substrate of the asymmetric hydrogenation, and this strategy exhibits excellent activity and enantioselectivity (S/C = 4000, 99% ee). One kilogram of catalyst can produce >1000 kg of product. At present, the key intermediate (hydrogenation product) has achieved 50 tons of industrial synthesis.

In addition, β-amino acids and their derivatives are important building blocks in chiral pharmaceuticals as shown in Figure 6.2. In 1999, Zhang group reported the first highly efficient asymmetric synthesis of β-amino acid derivatives via rhodium-catalyzed hydrogenation of β-(acylamino)acrylates using Rh-BICP and Rh-DuPhos catalytic systems [24].

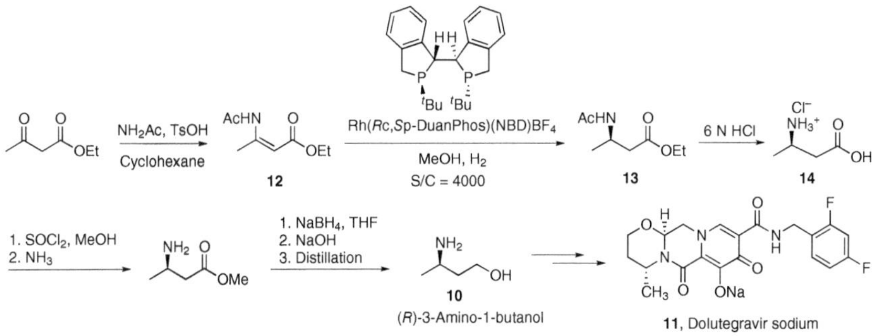

Scheme 6.7 Synthetic route for (*R*)-3-amino-1-butanol developed by Chiral Quest Inc. Source: Liu et al. [23a], Huang et al. [23b], Zhang [23c], Zhang and Tang [23d], and Wu et al. [23e].

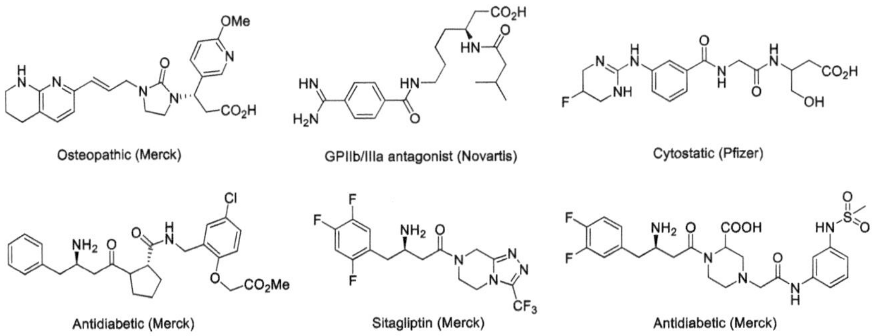

Figure 6.2 Pharmaceuticals with β-amino acids motifs.

6.2.1.5 (*S*)-2,6-Dimethyltyrosine

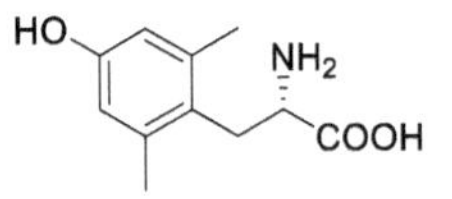

15, (*S*)-2,6-Dimethyltyrosine

(*S*)-2,6-Dimethyltyrosine (**15**) is the key intermediate to synthesize eluxadoline [25] and some clinical candidates, as shown in Figure 6.3 [26]. Although the structure of **15** is simple, its 2,6-dimethyl substituents make the assembly of this unnatural amino acid to be rather challenging. To date, two major approaches have been developed for preparing (*S*)-2,6-dimethyltyrosine and its derivatives. One is alkylation of protected (2,6-dimethyl-4-hydroxy)benzyl halides with

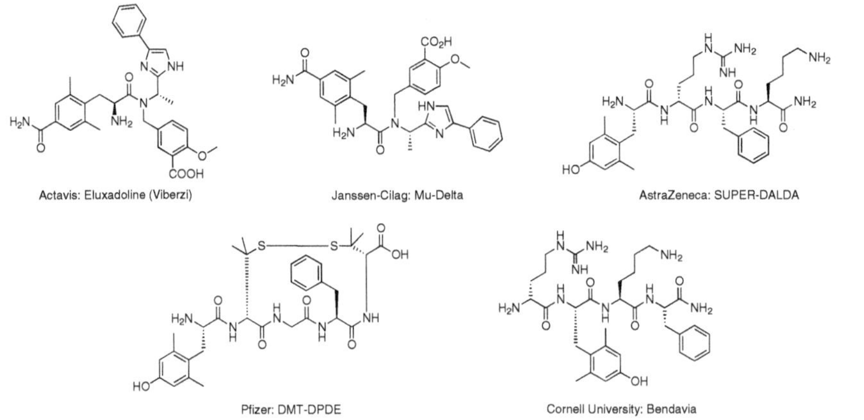

Figure 6.3 Clinical candidates bearing (*S*)-2,6-dimethyltyrosine motifs. Source: Wang et al. [26a] and Imamoto et al. [26b].

glycine-derived chiral synthons [27], while another one is asymmetric hydrogenation of (*Z*)-2-acetamido-3-(4-acetoxy-2,6-dimethyl-phenyl)-2-propenoate **16** to afford enantioenriched **17**, and the asymmetric hydrogenation approach allows for the synthesis of (*S*)-2,6-dimethyltyrosine (**15**) on a large scale [28].

The asymmetric hydrogenation route shown in Scheme 6.8 had been previously demonstrated on kilogram scale [29]. In this process, the key steps were a modified palladium-catalyzed Heck coupling of 4-iodo-3,5-dimethylphenyl acetate with methyl 2-acetamidoacrylate and hydrogenation of the resulting sterically hindered dehydroamino ester **16** using [Rh(1,5-COD)(*R*,*R*-DIPAMP)]BF_4 as catalyst under relatively forcing conditions (60 °C, 60 psi) and high catalyst loading (1 mol%) (Scheme 6.8). However, the hydrogenation proceeded slowly, producing the corresponding α-amino ester **17** in 87% yield with 92% ee. One kilogram of catalyst just produces only 40 kg of product. The hydrolysis of **17** with 12 N hydrochloric acid provided (*S*)-2,6-dimethyltyrosine hydrochloride **15·HCl** in 97% yield with no loss of optical purity.

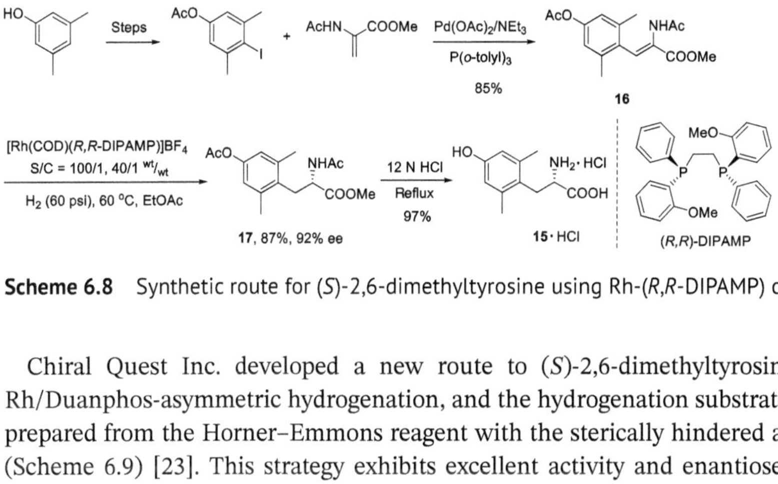

Scheme 6.8 Synthetic route for (*S*)-2,6-dimethyltyrosine using Rh-(*R*,*R*-DIPAMP) catalyst.

Chiral Quest Inc. developed a new route to (*S*)-2,6-dimethyltyrosine using Rh/Duanphos-asymmetric hydrogenation, and the hydrogenation substrate **16** was prepared from the Horner–Emmons reagent with the sterically hindered aldehyde (Scheme 6.9) [23]. This strategy exhibits excellent activity and enantioselectivity

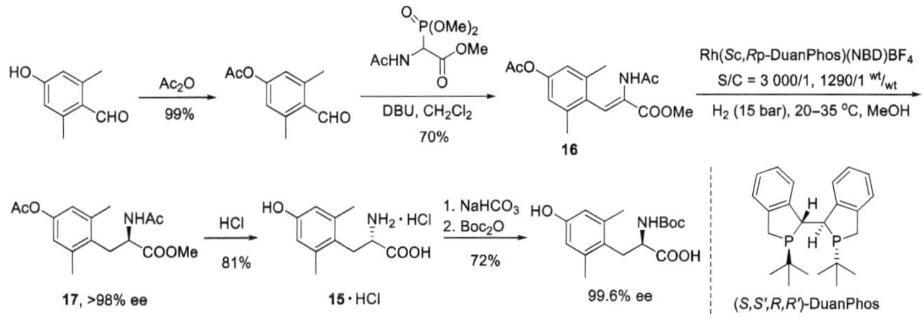

Scheme 6.9 Synthetic route for (*S*)-2,6-dimethyltyrosine developed by Chiral Quest Inc. Source: Hoge et al. [9a], Sun et al. [9b], Chen et al. [9c], Geng et al. [9d], Zhang et al. [9e], and Liu et al. [9f].

(S/C = 3000, >98% ee). This method is now routinely used for α-amino acid manufacture. The new process has high catalytic efficiency and can produce 1.3 tons of pharmaceutical intermediates with 1 kg of catalyst. The annual output of the key intermediates amounted to 60 tons. Chiral Quest Inc. has become the sole supplier of (*S*)-2,6-dimethyltyrosine.

6.2.1.6 Apremilast

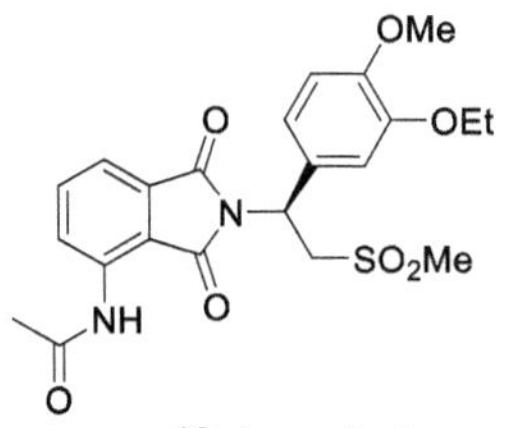

19, Apremilast

Apremilast (**19**) is a novel anti-inflammatory drug, acting as the first phosphodiesterase-4 (PDE4) inhibitor used in psoriasis (Ps) and psoriatic arthritis (PsA) treatment [30]. Apremilast's sales amounted to \$1.5 billion in 2017, consuming 1.3 tons of APIs.

In recent years, the development of technologies for the catalytic asymmetric hydrogenation of unprotected enamines has been reported [31], including the well-documented process for the synthesis of the diabetes drug Sitagliptin shown in Scheme 6.4. Celgene Corporation developed an analogous enantioselective enamine (**20**) reduction which comprises the most promising and concise route to the key aminosulfone intermediate **21**. It was established that 0.5 mol% catalyst with 1.0 mol% ligand under 90 psi hydrogen gas at 50 °C was sufficient enough to achieve >99.5% conversion within 18 hours. The product could be conveniently isolated by conversion into its *N*-Ac-L-leucine salt, resulting in a chiral upgrade to 99.2% ee and 80% isolated yield (Scheme 6.10).

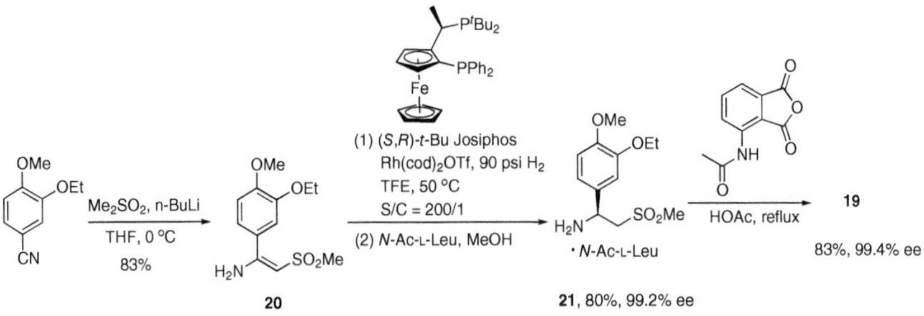

Scheme 6.10 Synthetic route for Apremilast developed by Celgene Corporation.

6.2.2 Asymmetric Hydrogenation of Ketone

6.2.2.1 Duloxetine

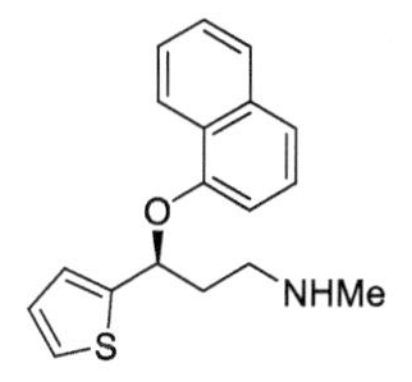

22, (*S*)-Duloxetine

(*S*)-Duloxetine (**22**) is an antidepressant drug targeting the presynaptic cell. It is a dual inhibitor preventing the reuptake of serotonin and norepinephrine. In addition to being an important pharmaceutical in the short-term treatment of major depressive disorders, it can be used to treat urinary incontinence and obsessive compulsive disorder [32]. Duloxetine's sales amounted to $1.5 billion in 2017, consuming 135 tons of APIs.

In addition to Resolution-Racemization-Recycle (*RRR*) synthesis [33], Duloxetine can be prepared by an asymmetric reduction of the ketone **23** (Scheme 6.11). The PhanePhos (**24**) variation of the Noyori method was used and resulted in 93.4% ee. The enantiopurity could be increased during downstream processes.

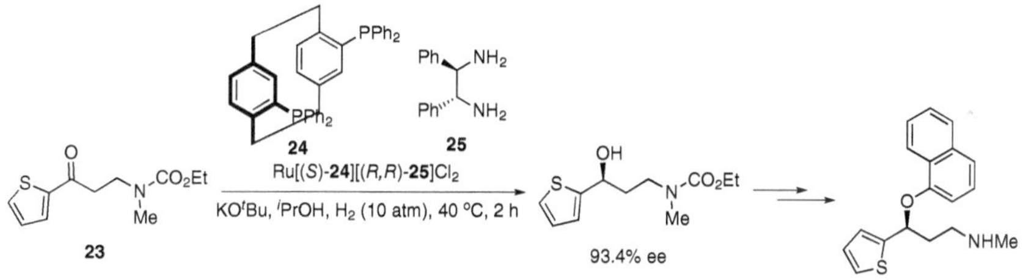

Scheme 6.11 Synthesis of duloxetine developed by Eli Lilly.

A more efficient approach would be the direct asymmetric hydrogenation of the secondary γ-secondary amino ketone. Zhang and Chiral Quest Inc. have solved this problem by the use of a rhodium catalyst with the ligand DuanPhos [34]. They explored the reduction of β-secondary-amino ketone with a low catalyst loading of the Rh–DuanPhos complex (S/C = 10 000, Scheme 6.12). The described catalytic system is highly efficient in terms of both enantioselectivity and reactivity (>99% conversion, >99% ee). The route provides one of the shortest (three steps overall) syntheses of Duloxetine, avoiding resolution and the harsh conditions in N-demethylation steps. Asymmetric hydrogenation reactions (1000 l) have been successfully completed and could be repeatable at a catalyst loading of S/C = 9000.

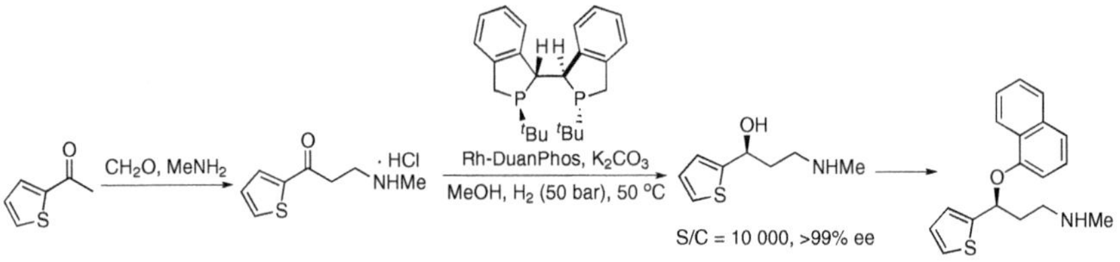

22, (*S*)-Duloxetine

Scheme 6.12 Synthesis of duloxetine developed by Chiral Quest Inc.

It has been put into operation in Jiangxi factories with an annual output of 28 tons. Compared with the traditional route, the total cost in this environment-friendly "green" process is reduced by 74%, and the amount of waste produced in the process was reduced by 82%.

6.2.2.2 Dorzolamide

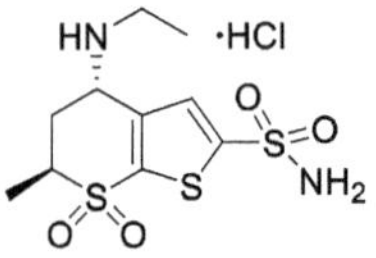

26, Dorzolamide

Dorzolamide (**26**) is a carbonic anhydrase inhibitor and used to lower intraocular pressure in open-angle glaucoma and ocular hypertension [35].

Chiral Quest Inc. developed the ruthenium-catalyzed asymmetric hydrogenation of methyl acetoacetate [36] for the enantioselective synthesis of methyl (*R*)-3-hydroxybutyrate, which is a very important chiral building block for the synthesis of Dorzolamide (Scheme 6.13). Up to 99% ee values (S/C >100 000) were achieved using chiral diphosphine C_3*-TunePhos as ligand. Asymmetric hydrogenation of methyl acetoacetate requires a Hastelloy reactor, and Chiral Quest Inc. has a 1000 l Hastelloy hydrogenation vessel. Manufacture of the *cis*-hydroxy sulfone intermediate for Dorzolamide is underway at Chiral Quest Inc., Suzhou.

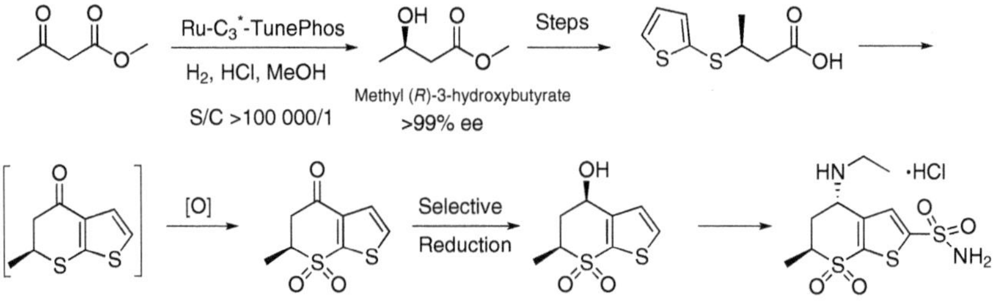

26, Dorzolamide

Scheme 6.13 Synthetic route for Dorzolamide developed by Chiral Quest Inc.

C_3*-TunePhos, a derivative of the C_n-TunePhos (n = 0–6) series, was developed by Zhang group in 2000 [37]. The procedures to synthesize these privileged ligands suffer from long synthetic routes and limited possibilities for derivatization [38]. An improved and divergent preparation to five C_3*-TunePhos derivatives is shown in Scheme 6.14 [39].

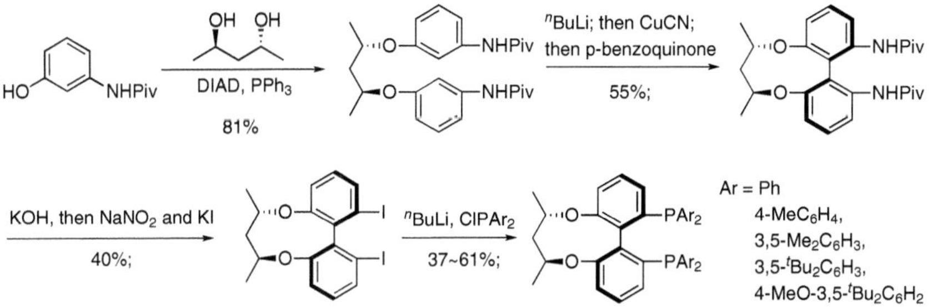

Scheme 6.14 Improved synthetic route to C_3*-TunePhos Source: Tan et al. [39a] and Zuo et al. [39b].

6.2.2.3 (*R*)-1-(3,5-Bis(trifluoromethyl)-phenyl)ethanol

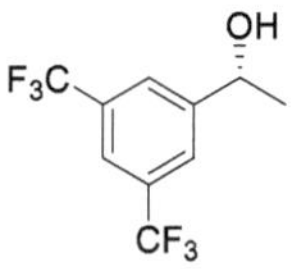

27, (*R*)-1-(3,5-Bis(trifluoromethyl)-phenyl)ethanol

(*R*)-1-(3,5-Bis(trifluoromethyl)-phenyl)ethanol (**27**) is used as a building block for NK-1 receptor antagonists, for example, aprepitant, fosaprepitant, and rolapitant (Figure 6.4). Different groups have reported syntheses of this compound. In addition to reduction with the CBS reagent or via enzymatic methodology, transfer hydrogenation and hydrogenation approaches have been developed [40, 41].

Merck has used an asymmetric hydrogen transfer reaction to prepare **27**, presumably due to its operational simplicity (Scheme 6.15) [40]. Although a series of patented monotosyl diamines proved to be the best ligands in terms of

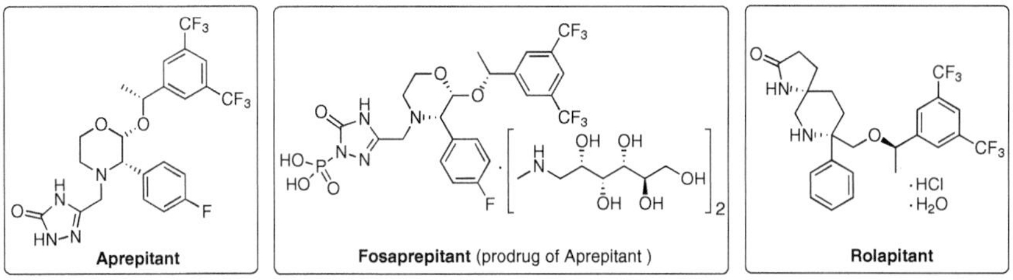

Figure 6.4 Aprepitant, fosaprepitant, and rolapitant.

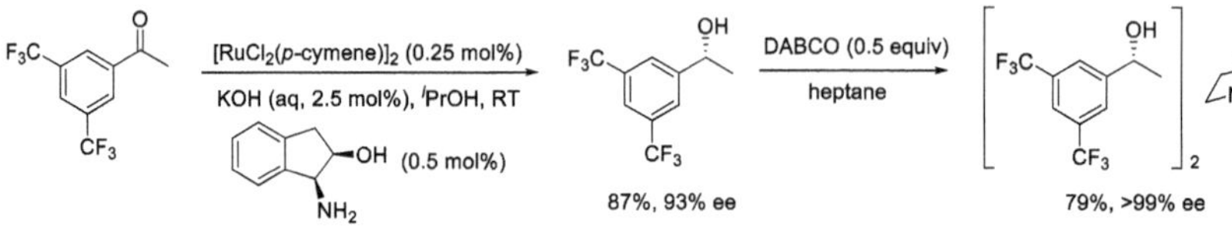

Scheme 6.15 Transfer hydrogenation for (*R*)-1-(3,5-bis(trifluoromethyl)phenyl)ethanol. Source: Hansen et al. [40a] and Brands et al. [40b].

enantioselectivity, the Merck group settled on *cis*-1-amino-2-indanol, because it was patent free. The enantioselectivity from the reduction was not, however, high enough to be used for the downstream reactions, but formation of a DABCO inclusion complex allowed for the necessary upgrade. The scale-up of the process was smoothly scaled to 40 kg batches.

Huang et al. described an efficient and practical pilot process for the synthesis of **27** using $RuCl_2$[(*R*)-Diop][(*R*)-D-Me-BIMAH] complex catalyst with an S/C ratio of 20 000 under 30 atm H_2 at 25 °C (Scheme 6.16). Scale-up to 5 kg scale was experimentally performed four times. All four batches proceeded smoothly in 99.9% conversion after 16-hour periods and afforded crude (*R*)-1-(3,5-bis(trifluoromethyl)phenyl)ethanol with 89.3–89.4% ee. The combined batches of crude product were recrystallized from heptane with 1,4-diazabicyclo[2.2.2]octane (DABCO) to afford the desired product (60–65% yield, 98–98.75% ee) in a reasonably acceptable cost according to the reported Hansen protocol [42].

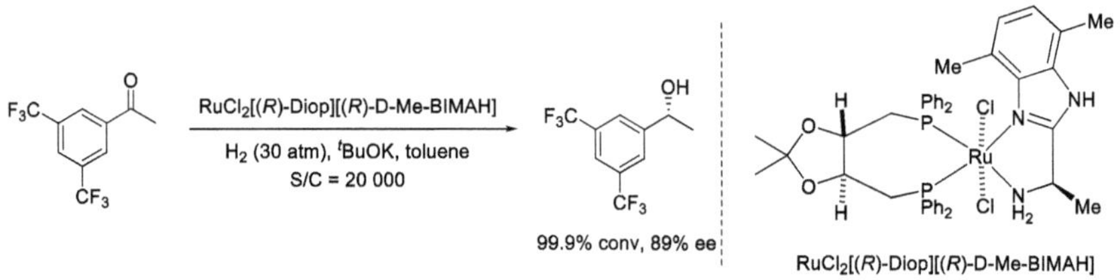

Scheme 6.16 Synthetic route for (*R*)-1-(3,5-bis(trifluoromethyl)phenyl)ethanol developed by Huang et al.

Zhang group reported the achievements in the practical asymmetric hydrogenations of a wide range of unfunctionalized ketones in ideal-approaching enantioselectivities (>99% ee) with only a ppm level of C_3*–TunePhos/diamine-Ru(II) catalysts (TON up to 1 000 000) (Scheme 6.17). It is noteworthy that the result (99.6% ee)

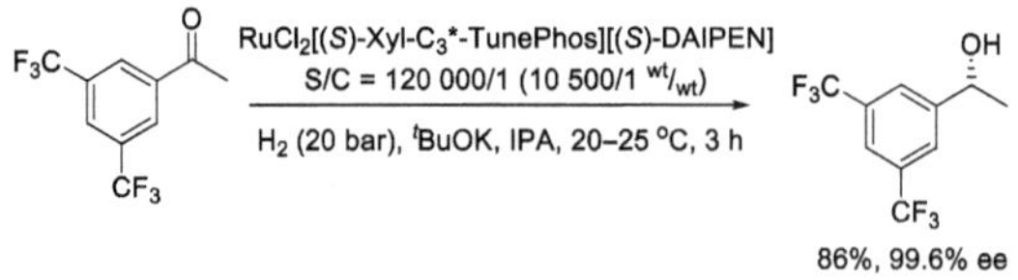

Scheme 6.17 Synthetic route for (*R*)-1-(3,5-bis(trifluoromethyl)phenyl)ethanol using C_3*-TunePhos/diamine-Ru(II) catalysts.

justified the potential utility of this Ru-C_3*–TunePhos complex in the practical synthesis of aprepitant, and 1000 kg of the intermediate has been made [43].

Recently, Zhang group successfully developed air-stable and high-active tridentate ferrocene-based aminophosphoxazoline (*f*-amphox) ligands [8], amino-phosphine-alcohol (*f*-amphol) ligands [8], and amino-phosphine acid (*f*-ampha) ligands [8] (Scheme 6.18). These highly modular ligands exhibited excellent performance in Ir-catalyzed asymmetric hydrogenation of simple ketones affording chiral alcohols (full conversions, up to >99% ee). These ligands are bound to play a strong role in the synthesis of (*R*)-1-(3,5-bis(trifluoromethyl)-phenyl)ethan-1-ol.

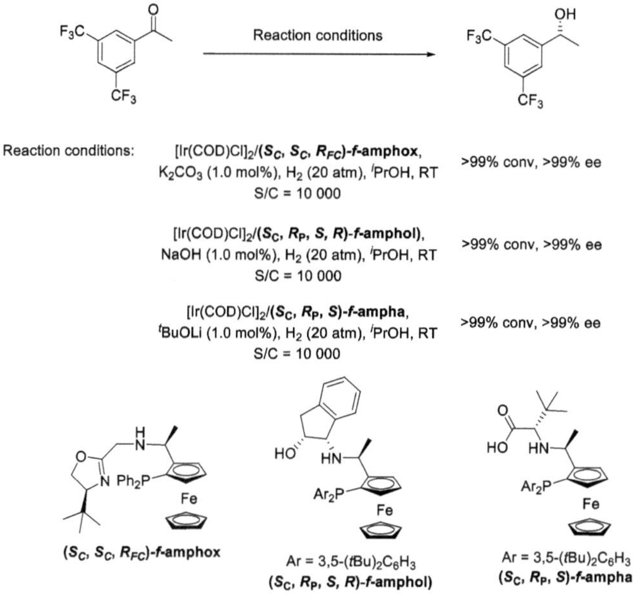

Scheme 6.18 Synthetic route for (*R*)-1-(3,5-bis(trifluoromethyl)phenyl)ethanol using *f*-amphox, *f*-amphol, and *f*-ampha ligands.

6.2.2.4 4-AA (Key Intermediate to Carbapenem Antibiotics)

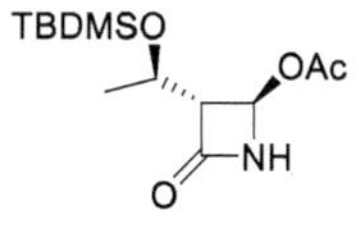

28, 4-AA

Diastereo- and enantioselective hydrogenation of 2-substituted 3-oxo carboxylic esters was reported in 1989 [44]. Takasago developed an asymmetric hydrogenation accompanied by dynamic kinetic resolution *en route* to a key intermediate to carbapenem antibiotics; both excellent enantio- and diastereoselectivities were obtained in hydrogenation of 2-benzamidomethyl-3-oxobutanoate **29** with a (−)-DTBM-Segphos-Ru catalyst [45]. The *cis*-(2*S*,3*R*)-methyl 2-benzamidomethyl-3-hydroxybutanoate **30** was obtained in 99.4% ee and in 98.6% de (Scheme 6.19). The product can be transformed into the key intermediate **28** of carbapenem antibiotics [46]. This procedure has been successfully applied to the industrial production of the key intermediate on a scale of 100 tons per year.

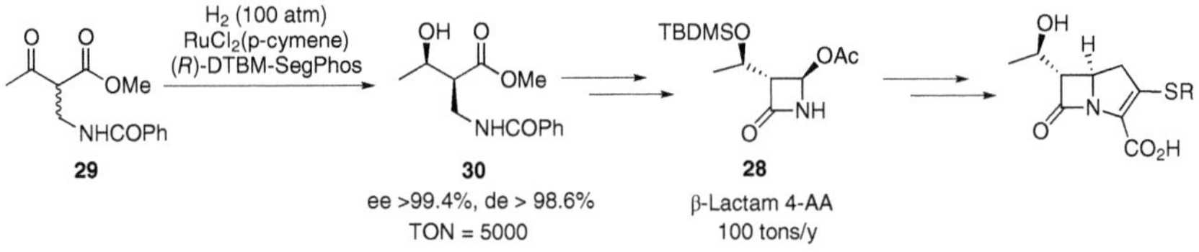

Scheme 6.19 Synthetic route for 4-AA developed by Takasago Inc.

6.2.2.5 Rivastigmine

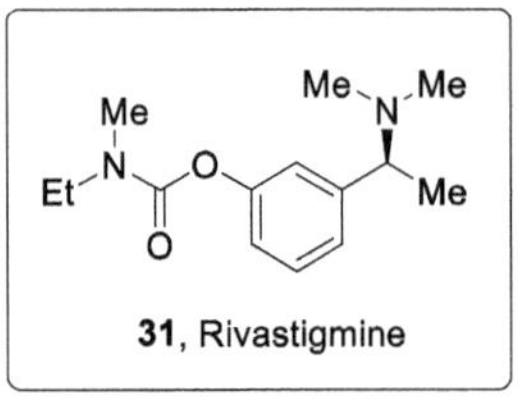

Rivastigmine (**31**), (*S*)-3-[1-(dimethylamino)ethyl]phenyl ethyl (methyl) carbamate, represents one of the most potent drugs for the treatment of mild-to-moderate dementia of the Alzheimer's type [47].

To date, several methods for the synthesis of enantiomerically pure Rivastigmine have been reported [48]. Transition metal-catalyzed asymmetric transfer hydrogenation has also been applied to the synthesis of Rivastigmine [49], but the catalyst loading was usually very high (10–0.05 mol%). Foulkes et al. at Novartis reported an improved process using an Ru-catalyzed enantioselective hydrogenation of *m*-hydroxyacetophenone to access optically pure alcohol intermediate **32** with 98% ee and 85% yield after crystallization at a ratio of substrate to catalyst (S/C) of 10 000 [50].

Jiuzhou Pharmaceutical Co., Ltd. developed two efficient processes for the synthesis of Rivastigmine [51]. Of particular note is the process used for the asymmetric hydrogenation by applying the highly efficient chiral spiro catalyst, Ir-SpiroPAP. The iridium catalysts with chiral tridentate spiro ligands (SpiroPAP) have proved to be extremely efficient for the asymmetric hydrogenation of ketones, especially for the acetophenone derivatives, TON up to 4 550 000 [10, 52]. The first route was easy to scale up in industry and provided the commercial intermediate (*S*)-3-(1-dimethylaminoethyl)phenol **33** (Scheme 6.20). Based on the demonstration in the lab, the process was carried out in pilot batches and provided a total of 51.7 kg of alcohol (*R*)-**32** in 91–96% yield with 96–97% ee. The ee value of (*R*)-**32** could be upgraded to >99% by crystallization from ethyl acetate/heptanes.

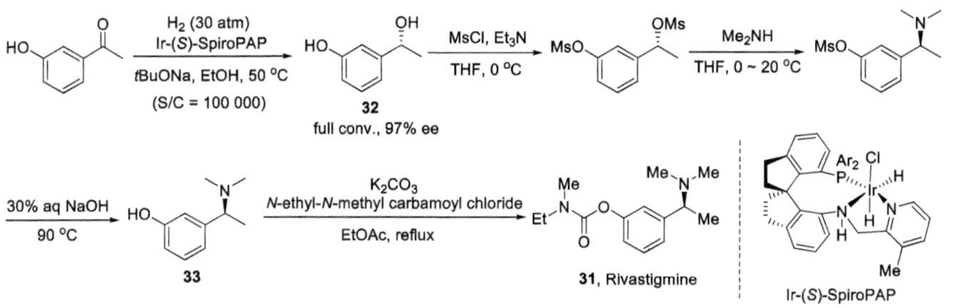

Scheme 6.20 Synthetic route for Rivastigmine developed by Jiuzhou Pharmaceutical Co., Ltd.

In order to avoid the use of a stoichiometric amount of ^{t}BuONa in the asymmetric hydrogenation step in the above route, the second route was performed to the synthesis of Rivastigmine in four steps and 84% overall yield, as shown in Scheme 6.21.

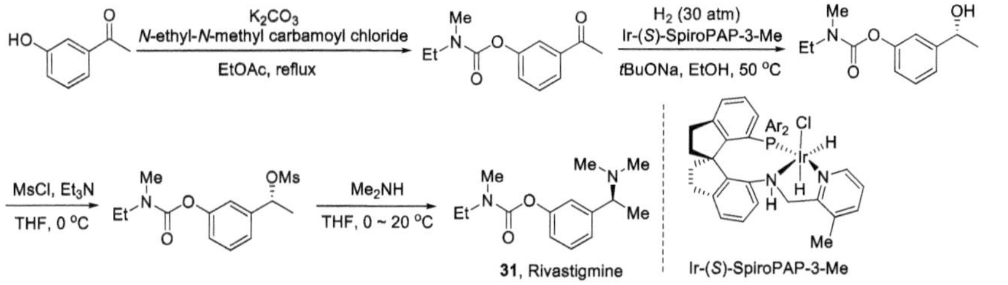

Scheme 6.21 Improved synthetic route for Rivastigmine developed by Jiuzhou Pharmaceutical Co., Ltd.

6.2.2.6 Montelukast

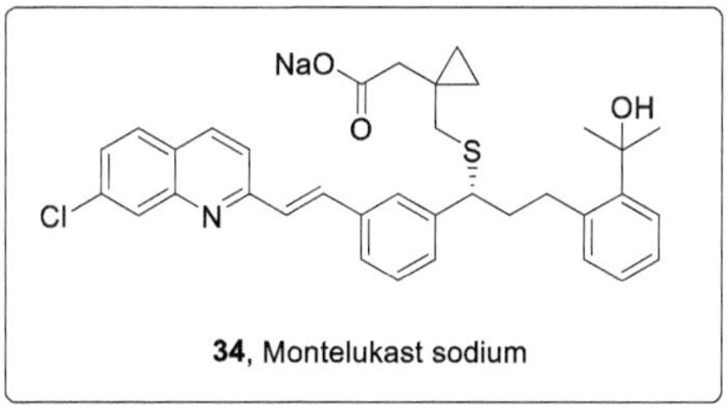

Montelukast sodium (**34**) is a leukotriene receptor antagonist (LTRA) developed by Merck, which is used for the maintenance treatment of asthma and to relieve symptoms of seasonal allergies [53]. The sales amounted to $2 billion in 2017, consuming 53.5 tons of APIs.

The conventional process for production of montelukast sodium (**34**) exploits commercially available alcohol **37** and thiol carboxylic acid **38** as starting materials (Scheme 6.22). Although its structure is complex, montelukast sodium has only a single stereocenter, which is usually installed via asymmetric reduction of the highly functionalized ketone **35**.

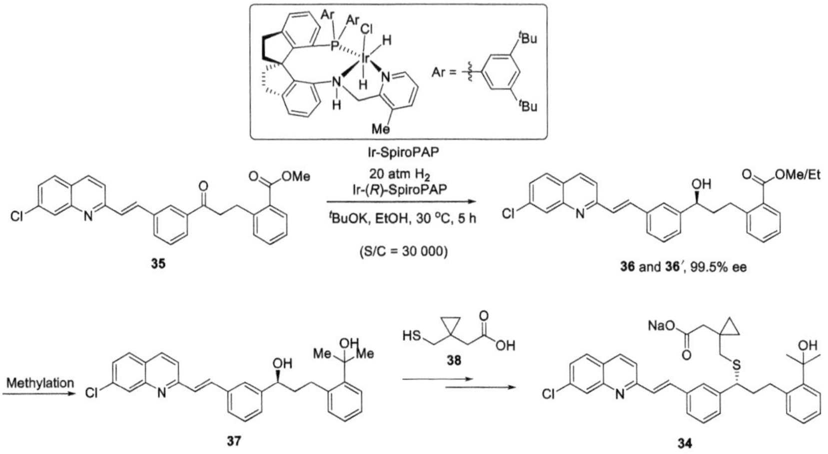

Scheme 6.22 Synthetic route for montelukast developed by Zhou and Jiuzhou Pharmaceutical Co., Ltd. Source: Modified from Zhu et al. [54].

Zhou and Jiuzhou Pharmaceutical Co., Ltd. developed an economical and robust process by applying the Ir/SpiroPAP catalyst (Scheme 6.22) [54]. The process was conducted at mild temperature (30 °C) under a hydrogen pressure of 20 atm with low catalyst loading (S/C = 30 000) and afforded the desired products **36** and **36′** in 99.5% ee. After crystallization from an aqueous ethanol solvent system, the monohydrates of **(*S*)-36** and **36′** were obtained in 90–95% yield, >99.9% ee. However, this process is sensitive to the purity of the substrate ketone **35**, especially depends on the remaining traces of palladium in ketone **35** from the former coupling reaction. And careful refinement of ketone was necessary to ensure reproducibility of the hydrogenation reaction. Therefore, the large-scale experiments were conducted at S/C = 30 000 to ensure that acceptable quality was achieved. The developed asymmetric hydrogenation conditions were applied on the large-scale production for multiple batches (from 25 to 75 kg), and all gave the same results as from the lab scale, and the quality was proved up to the USP 35 standard.

6.2.2.7 Crizotinib

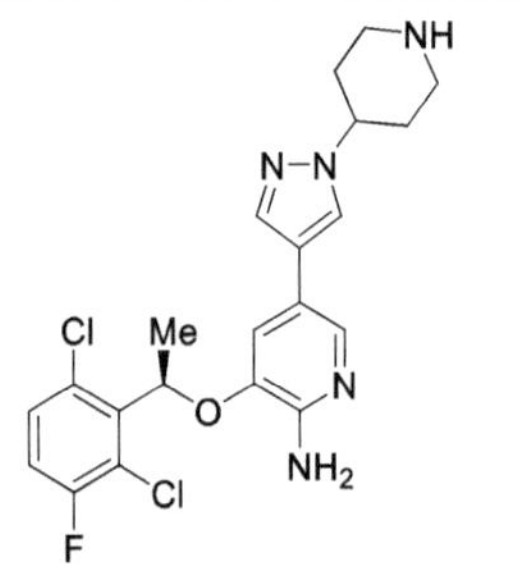

39, Crizotinib

Crizotinib (**39**) has been developed by Pfizer and was approved by FDA in 2011 for the treatment of locally advanced or metastatic non-small cell lung cancer [55]. The sales amounted to $550 million in 2017, consuming over 12 tons of APIs.

In the disclosed synthetic routes, (*S*)-1-(2,6-dichloro-3-fluoro-phenyl)ethanol (**41**) is a key intermediate, for which lots of efforts have been carried out to produce this alcohol with high enantioselectivity. However, asymmetric reduction with well-developed boran reagents such as BH_3·THF/(*R*)-methyl-CBS-oxazaborolidine (cat.), (−)-DIPCl, and $NaBH_4$/TMSCl/(*S*)-α,α-diphenylpyrrolidinemethanol all gave low optical purity of the alcohol, and in most of the cases gave low conversion of the reaction.

Zhou group and Jiuzhou Pharmaceutical Co. Ltd have demonstrated a new approach for the asymmetric synthesis of Crizotinib with high enantioselectivity. The key step of this approach is the asymmetric synthesis of chiral alcohol **41** through asymmetric hydrogenation of 2,6-dichloro-5-fluoroacetophenone precursor **40** using Ir[(*R*)-DTB-SpiroPAP-3-Me] as catalyst. The asymmetric hydrogenation was practiced in the pilot plant at 100-kg scale, and the chiral alcohol **41** could be obtained in 99.5% ee with TON of 100 000 (Scheme 6.23) [56].

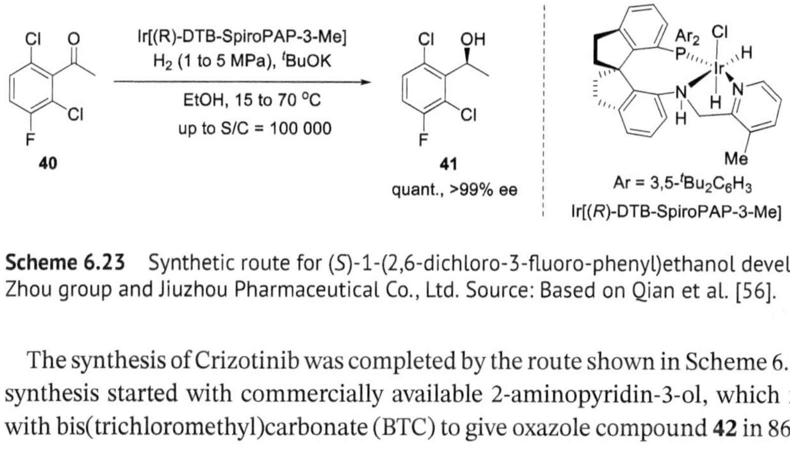

Scheme 6.23 Synthetic route for (*S*)-1-(2,6-dichloro-3-fluoro-phenyl)ethanol developed by Zhou group and Jiuzhou Pharmaceutical Co., Ltd. Source: Based on Qian et al. [56].

The synthesis of Crizotinib was completed by the route shown in Scheme 6.24. The synthesis started with commercially available 2-aminopyridin-3-ol, which reacted with bis(trichloromethyl)carbonate (BTC) to give oxazole compound **42** in 86% yield

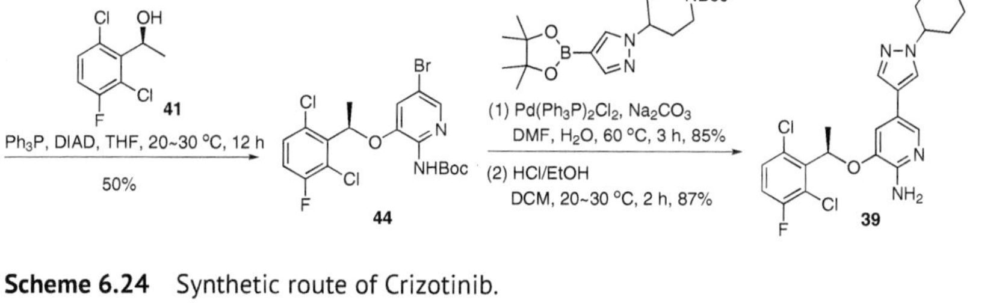

Scheme 6.24 Synthetic route of Crizotinib.

followed by bromination, hydrolysis, and protection of the amino group by the Boc group to afford **43** in good yield. After coupling of **43** and **41** under Mitsunobu condition, intermediate **44** could be isolated in 50% yield. Suzuki coupling of **44** with boronate, followed by the removal of Boc groups with HCl in ethanol, gave Crizotinib in good yield with 99.5% ee.

6.2.2.8 (*R*)-Phenylephrine

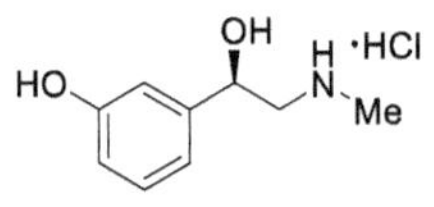

45, (*R*)-Phenylephrine

(*R*)-Phenylephrine (**45**), first marketed in 1936 by Boehringer Ingelheim, is a popular α-adrenergic receptor agonist used as a nasal decongestant [57]. The sales amounted to $647 million in 2017, consuming over 100 tons of APIs.

Various synthetic methods have been reported for the preparation of enantiopure (*R*)-Phenylephrine. However, only a few are of practical significance and impact from an economic perspective. The classical industrial synthetic route was developed in the late 1920s and early 1930s and published in a series of patents by Legerlotz [58]. These describe the synthesis of racemic Phenylephrine and give details of the resolution procedure using tartaric acid and the Walden inversion, by which the undesired enantiomer is transformed into the desired enantiomer. This route was subsequently investigated further and refined. Finally, two industrially used processes resulted, which differ only in the protecting group for the phenolic hydroxyl group. The synthetic route which uses the *O*-acetyl group has been performed at Boehringer Ingelheim for about 65 years (Scheme 6.25).

Starting from very cheap acetophenone, via nitration, reduction, diazotization/hydrolysis, and acetylation, 3-*O*-acetylacetophenone is obtained. Bromination followed by substitution with benzylmethylamine delivers, after treatment with hydrochloric acid during workup, *N*-benzylphenylephrone **46** as the hydrochloride salt. Catalytic hydrogenation cleaves off the *N*-benzyl group and reduces the carbonyl group to form *rac*-Phenylephrine hydrochloride **47**. This had to be resolved with L-tartaric acid in a relatively complicated way. Initially, the undesired enantiomer forms a crystalline L-tartrate had to be converted into the desired isomer by a Walden inversion. From the mother liquors of the L-tartrate formation, the desired isomer also had to be isolated and purified by repeated recrystallization of the free base. The combined crude (*R*)-Phenylephrine base finally is transformed into its hydrochloride salt **45**.

Takeda et al. developed the first asymmetric route to Phenylephrine in 1989 (Scheme 6.26) [59]. The reduction of the ketone **48** with Rh-MCCPM gave the corresponding alcohol **49** with 85% ee, but an upgrade could be realized by recrystallization. The protecting groups contribute to more than half of the formula

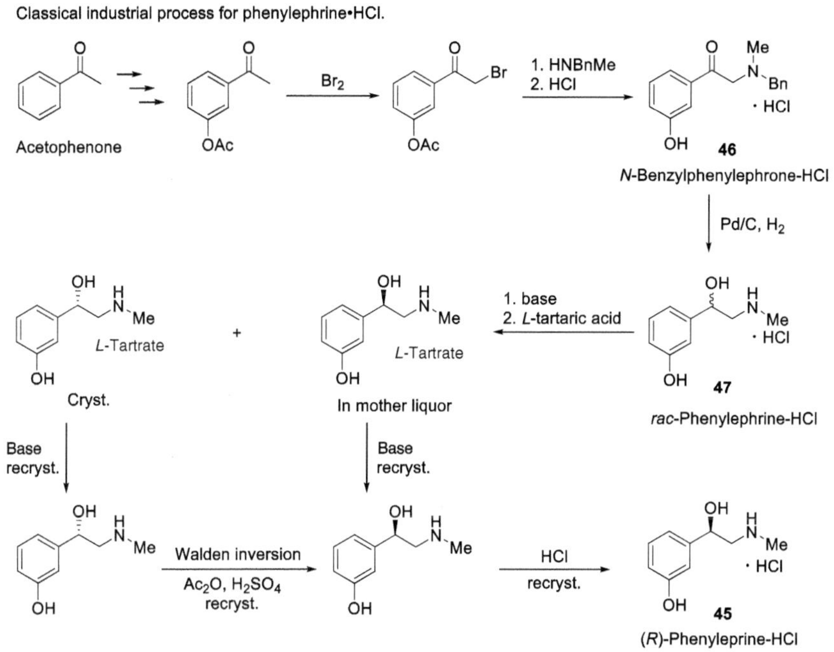

Scheme 6.25 Synthetic route for (*R*)-Phenylephrine performed at Boehringer Ingelheim.

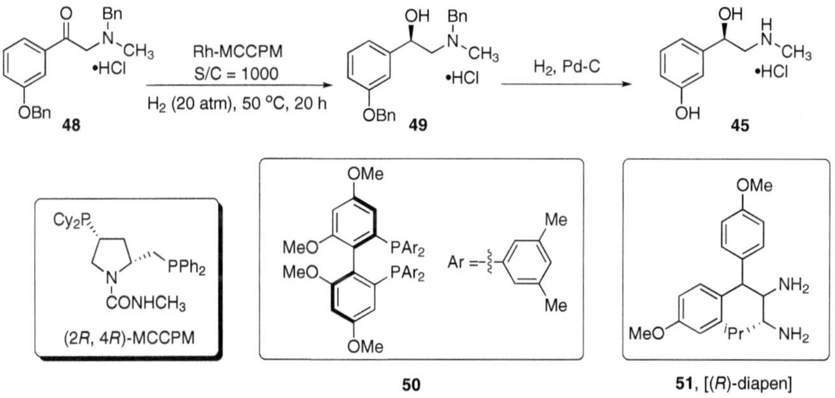

Scheme 6.26 First asymmetric synthesis of (*R*)-Phenylephrine. Source: Based on Takeda et al. [59].

weight of **48**. Optimization studies with Rh-MCCPM led to 92% ee, TON >40 000, and turnover frequency (TOF) >5000 h^{-1}. A further study showed that the use of [Ru(Xyl-PPhos)(DAIPEN)] [Ru(**50**)(**51**)] could completely circumvent the need for nitrogen protection while still giving high stereoselectivity and reactivity.

In principle, this route is industrially applicable, but it has several disadvantages. More than 50% of the molecular weight of the substrate for the asymmetric hydrogenation consists of protecting groups, which have to be removed in a later step. The

TON (TON = 2000) and the TOF (TOF = 100 h^{-1}) of the first hydrogenation step are relatively low. In addition, the optical purity of the crude product is only 85% ee, and the final product cannot be efficiently purified by recrystallization.

Zhou group developed a highly efficient iridium-catalyzed asymmetric hydrogenation of α-amino ketones for the synthesis of chiral 1,2-amino alcohols. With chiral spiro iridium catalyst Ir-(*R*)-SpiroPAP, a series of α-amino ketones were hydrogenated to chiral β-amino alcohols in excellent enantioselectivities (up to 99.9% ee) with TON up to 100 000. (*R*)-Phenylephrine was prepared in very high efficiency based on the efficient method (Scheme 6.27) [60].

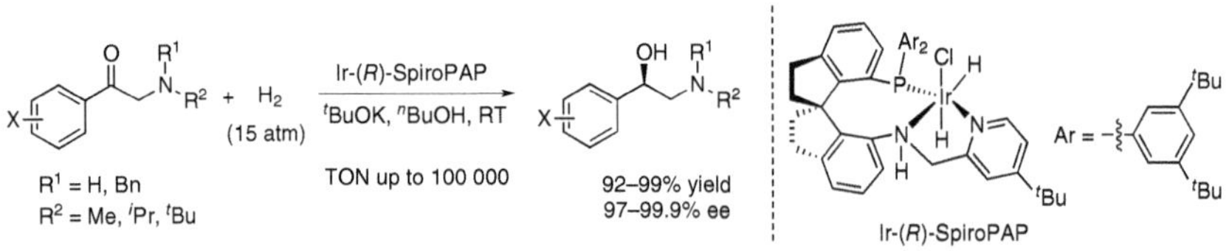

Scheme 6.27 Synthetic route for chiral 1,2-amino alcohols developed by Zhou group. Source: Based on Yuan et al. [60].

The substrate **46** containing an unprotected phenol group was hydrogenated on a gram scale with catalyst Ir-(*R*)-SpiroPAP (0.001 mol%) in ethanol in the presence of 3.0 equiv of potassium hydroxide as a base under 15 atm of hydrogen pressure at 30 °C for 12 hours to yield (*R*)-**52** in nearly quantitative yield with 99% ee. The (*R*)-Phenylephrine hydrochloride was obtained in 92% yield by removing the *N*-benzyl group by Pd/C-catalyzed hydrogenation, followed by acidification (Scheme 6.28).

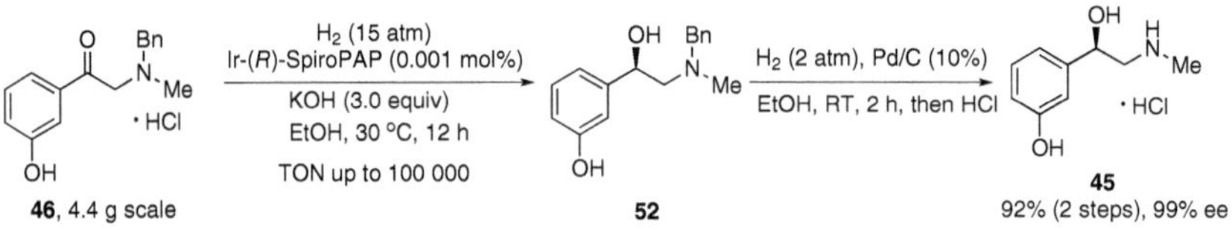

Scheme 6.28 Synthetic route for (*R*)-Phenylephrine developed by Zhou group.

In addition, (*R*)-clorprenaline (**53**), a β_2-agonist for the treatment of diverse disease states such as bronchitis and asthma, was synthesized with this highly efficient method. A one-step reaction, Pd/C-catalyzed hydrogenation to remove the *N*-benzyl group, gave (*R*)-clorprenaline in 85% yield (Scheme 6.29).

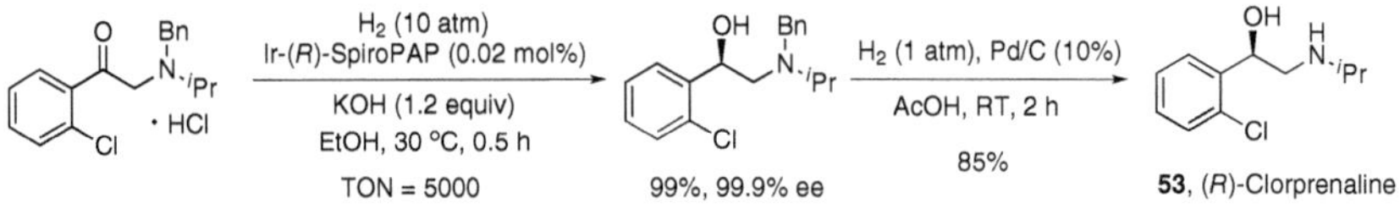

Scheme 6.29 Synthetic route for (*R*)-Clorprenaline.

6.2.2.9 Atorvastatin Calcium Salt

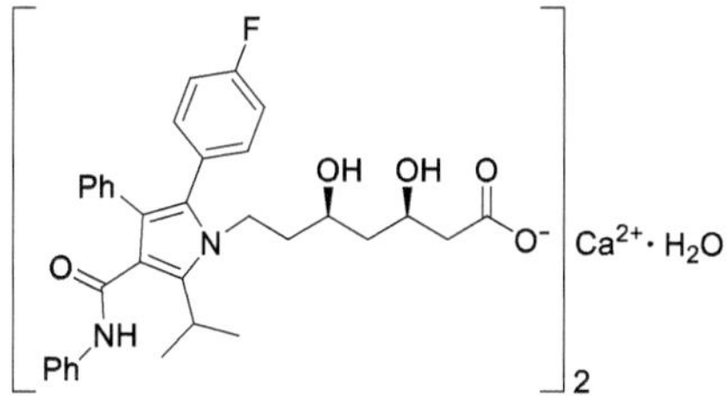

54, Atorvastatin calcium salt

Atorvastatin calcium (**54**) was first introduced to the market in 1997 by Pfizer as an effective HMG-CoA reductase inhibitor for the treatment of hypercholesterolemia and arteriosclerosis and now ranks at the top of the drugs best sold in the world. The sales amounted to $6 billion in 2017, consuming 554 tons of APIs. To date, joint synthetic development efforts led to a number of strategies that make use of chemoenzymatic resolution or chiral auxiliary at an appropriate stage, enantioselective catalysis, or starting from the chiral pool [61].

In the convergent synthesis of atorvastatin calcium, enantiomerically pure (*S*)-ethyl 4-chloro-3-hydroxybutanoate **56** is the key chiral intermediate, which can be prepared via Ru-catalyzed asymmetric hydrogenation of ethyl 4-chloroacetoacetate **55**. Using the Ru/C_3-TunePhos system, excellent enantioselectivity (>98% ee) has been achieved at 45 000 turnovers in complete conversion (Scheme 6.30) [62].

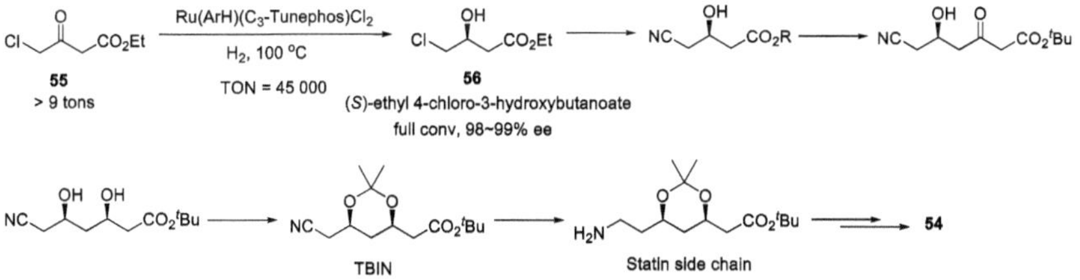

Scheme 6.30 Synthesis of (*S*)-ethyl 4-chloro-3-hydroxybutanoate using Ru/C_3-TunePhos as catalyst. Source: Baumann et al. [62a] and Zhang et al. [62b].

The β-hydroxy ester **56** has also been achieved with Ru-TMBTP from the reduction of the chloro compound **55** at lower pressures [63]. In addition to Ru-BINAP which requires high pressure to attain high enantioselectivity [64], {Ru(p-cymene)[(*S*)-TMBTP]} can be used at lower hydrogen pressure resulting in 97% ee (Scheme 6.31).

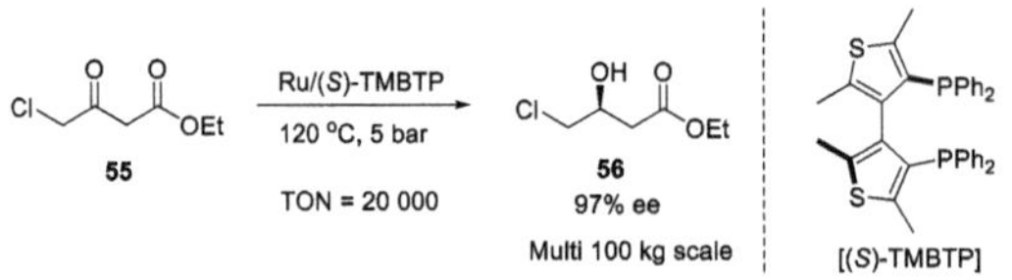

Scheme 6.31 Synthesis of (*S*)-ethyl 4-chloro-3-hydroxybutanoate using Ru/(*S*)-TMBTP as catalyst.

6.2.2.10 Orlistat

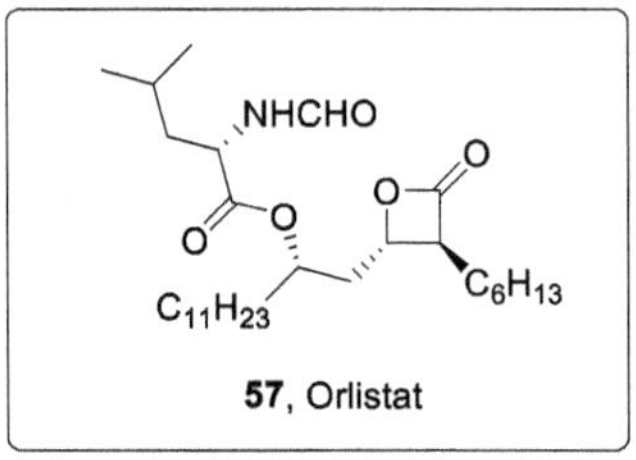

Orlistat (**57**), or tetrahydrolipstatin (THL), approved by FDA for the treatment of obesity in 1999, inhibits gastric and pancreatic lipases in the lumen of the gastrointestinal tract to decrease systemic absorption of dietary fat. The sales amounted to $390 million in 2017, consuming over 72 tons of APIs [65].

The first-generation synthesis of Orlistat (**57**) based on the optical resolution of the racemic key intermediate **58** is shown in Scheme 6.32. The overall yield of 18–22% is a remarkable achievement considering that the undesired diastereomer (*R*,*R*,*S*)-**58** could not be recycled [66].

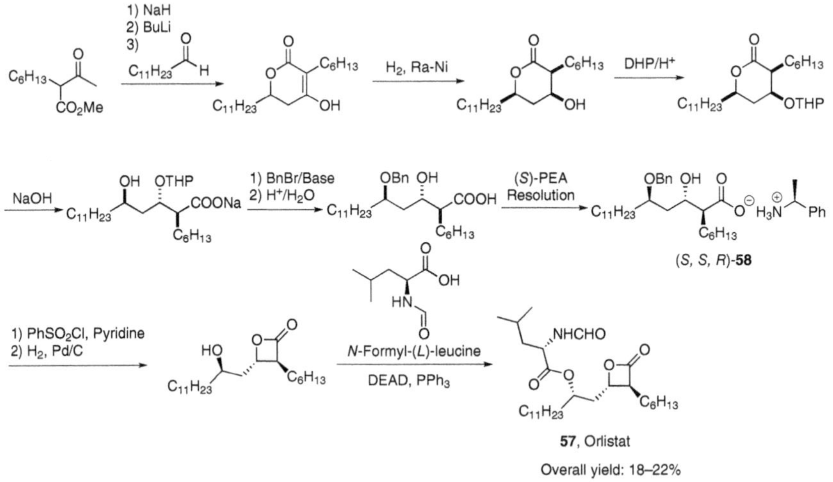

Scheme 6.32 First-generation, optical resolution-based synthesis of Orlistat.

The second-generation synthesis of Orlistat (**57**) was developed enantioselectively, which is suitable for large-scale preparation. The approach relies on an asymmetric reduction of the β-keto ester **59** using {$RuCl_n$[(*R*)-MeOBIPHEP]} as catalyst. The hydrogenation was run at 40 wt.% concentration at S/C 50 000 under 40 bar of hydrogen at 80 °C to achieve full conversion within four hours. The scale-up from a 2 l to a 2 m^3 autoclave proceeded uneventfully such that a total of 2.2 tons of (*R*)-**60** with 99% ee and 99% yield were produced in up to 240 kg batches. The strategy has been on the one hand to retain the final reaction sequence from (*S*,*S*,*R*)-**58**–**57** in order to affect the impurity profile of **57** as little as possible. On the other hand, however, the late-stage resolution of *rac*-**58** has been avoided by the use of enantiomerically pure intermediates rather than their racemates. The overall yield from the β-keto ester **59** amounts to 41–48% (Scheme 6.33) [66].

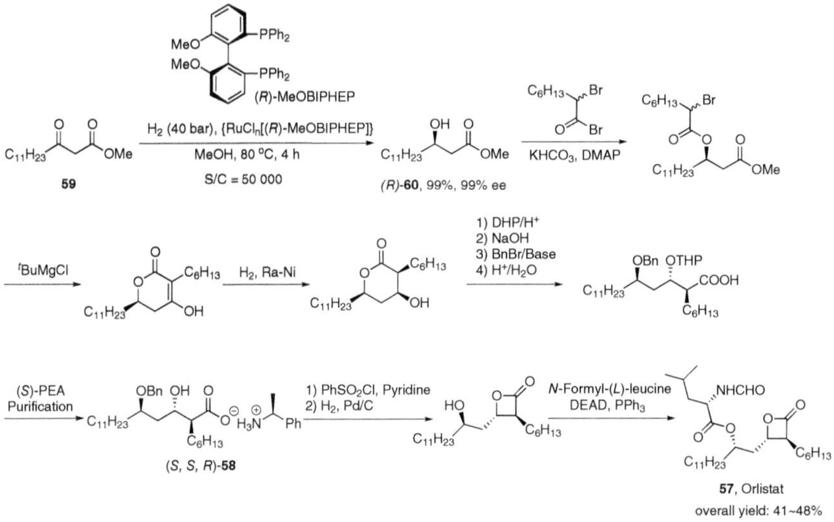

Scheme 6.33 Second-generation, new enantioselective synthesis of Orlistat. Source: Modified from Schwindt et al. [66].

Particularly, the asymmetric hydrogenation proved to be highly robust, the large-scale production of β-hydroxyester **60** with very high ee (>99%), and purity (>99%) was straightforward and practical. The second-generation processes utilized the same chemistry for the later steps as the first process. This improvement removed the late-stage resolution in the first synthesis and allowed doubling the throughput.

6.2.2.11 Ezetimibe

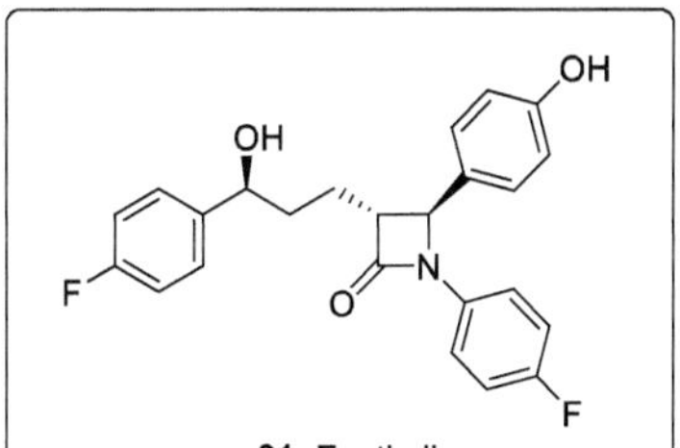

61, Ezetimibe

Ezetimibe (**61**) is the first lipid-lowering drug that inhibits intestinal uptake of dietary and biliary cholesterol without affecting the absorption of fat-soluble nutrients [67]. The sales amounted to $4.3 billion in 2017, consuming 26 tons of APIs. The novel structure of ezetimibe has prompted intense synthetic interest in the synthetic community, which led to the development of several syntheses for this molecule [68].

In the commonly adopted synthetic route of ezetimibe, as shown in Scheme 6.34, the key step is the CBS reduction, which was originally described by Corey, Bakshi, and Shibata. The stoichiometric reducing agent is usually the THF or DMS complex of BH_3, which is associated with serious storage and handling hazards [69]. Both from a safety and from an efficiency standpoint (the TON of these systems is low), the CBS reduction is not as convenient to run as methods based on molecular hydrogen or transfer hydrogenation. However, their versatility and often high enantioselectivity has made them the object of intense application in the API arena, even on very large scale.

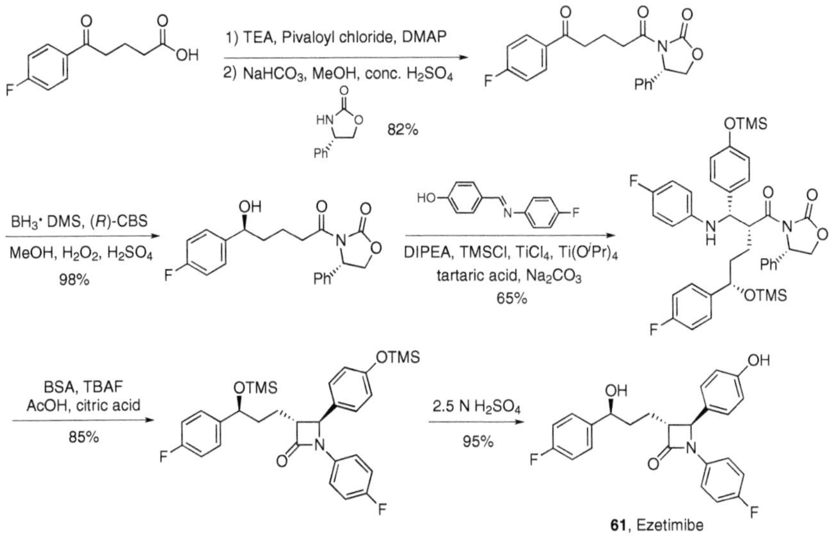

Scheme 6.34 Synthetic route of Ezetimibe via CBS reduction.

In 2018, Ding group developed a practical enantioselective synthesis of cyclohexyl-fused chiral spirobiindane derivatives, and a series of new chiral ligands bearing the backbone were easily accessed [70]. The utility of these ligands was demonstrated in the catalytic asymmetric hydrogenation of the key intermediate (3*R*,4*S*)-**62** for the synthesis of Ezetimibe, affording **63** in 95% yield and 99 : 1 diastereoselectivity under a catalyst loading of 0.01 mol% (Scheme 6.35).

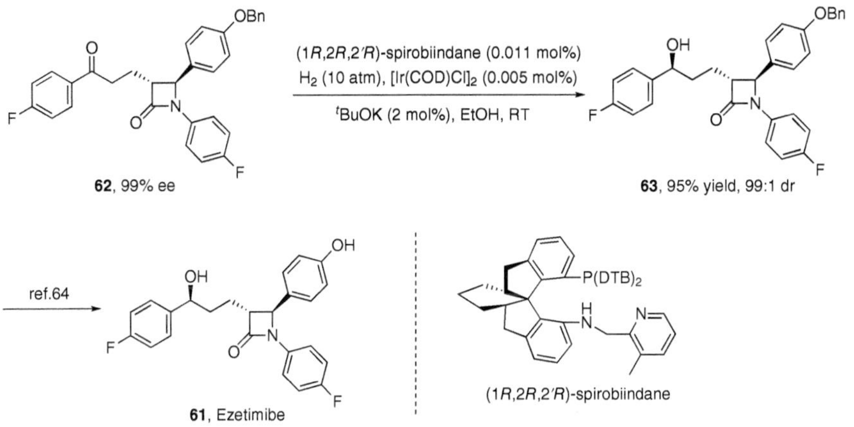

Scheme 6.35 Synthetic route of Ezetimibe developed by Ding group.

6.2.3 Asymmetric Hydrogenation of Olefin

6.2.3.1 L-Menthol

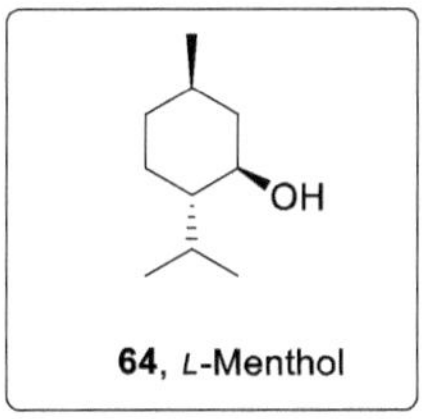

L-Menthol (**64**) is one of the most widely used compounds in the flavors and fragrances industry [71]. Roughly one-quarter of the world's L-menthol supply is already currently produced synthetically using two different routes.

Takasago uses a process with five steps starting from myrcene (**65**) (Scheme 6.36) [7]. The heart of this process is the homogeneously catalyzed asymmetric isomerization of an allylic amine to an enamine using Noyori's Rh-(*S*)-BINAP catalyst with

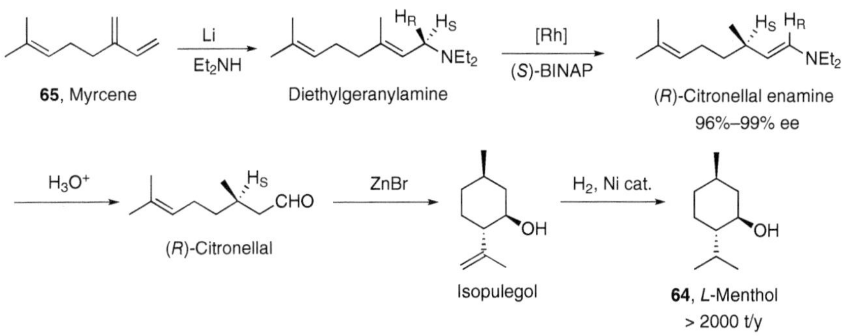

Scheme 6.36 Takasago's myrcene-based menthol synthesis. Source: Based on Noyori [7].

excellent enantioselectivities. The world market for L-menthol based on this method is exceeding 2000 tons per year.

BASF developed an efficient synthesis of L-menthol which would allow an extremely attractive synthesis with three steps based on citral (**66**) (Scheme 6.37) [72]. Central to this synthesis is a highly efficient (*R*)-citronellal synthesis. This is an unusual, selective hydrogenation of the double bond in conjugation with the aldehyde functionality that does not touch the other alkene or the aldehyde functionality. (*R*)-Citronellal was obtained in good yield with 86% ee using Rh-Chiraphos as catalyst. In the continuous process eventually developed, the hydrogenation takes place in the presence of small amounts of CO to maintain the integrity of the catalyst. The resultant (*R*)-citronellal is cyclized to L-isopulegol in a highly diastereoselective ene-reaction similar to the Takasago process; in the BASF process, an aluminum-based Lewis acid is used. The remaining double bond is hydrogenated off using a heterogeneous Ni-based catalyst to give L-menthol. After the final crystallization, the ee of the product **62** is >99%. The world market for L-menthol based on this method has exceeded 20 000 tons per year.

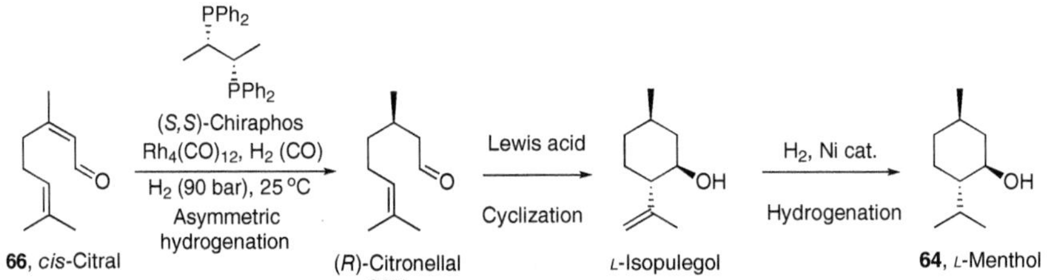

Scheme 6.37 BASF's citral-based menthol synthesis. Source: Lawrence [72a] and Heydrich et al. [72b].

6.2.3.2 Sacubitril

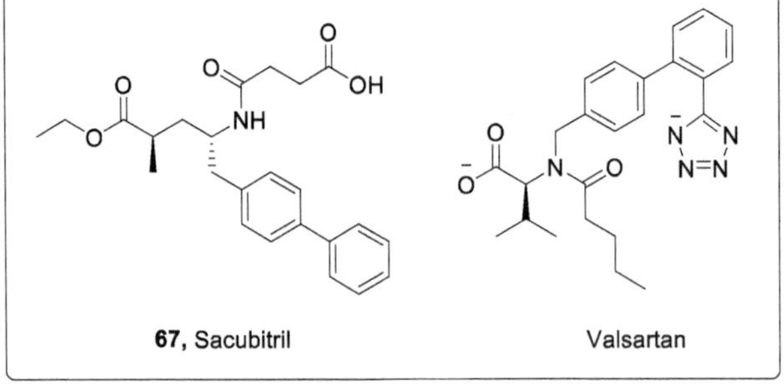

Sacubitril (**67**) is a neprilysin inhibitor prodrug developed by Novartis that was approved as part of an orally administered supramolecular sodium salt complex with the angiotensin receptor blocker valsartan in the United States and the European

Union in 2015. Sacubitril/valsartan (also known as LCZ-696) is a first-in-class dual angiotensin receptor blocker–neprilysin inhibitor marketed for the treatment of chronic heart failure with reduced ejection fraction [73].

The industrial scale synthesis of the key intermediate **68** for Sacubitril has been reported, and this route is described in Scheme 6.38 [74]. Interestingly, a separate patent disclosed the stereoselective hydrogenation of the trisubstituted olefin **68**, in which subjection of **68** to catalytic $[Ru(p\text{-cymene})I_2]_2$ and chiral phosphine ligand Mandyphos under 20 bar of hydrogen gas in warm ethanol delivered **69** in 99:1 dr before recrystallization. Subsequently, the esterification and removal of the Boc group revealed a primary amine which then reacted with succinic anhydride to ultimately deliver Sacubitril **67**.

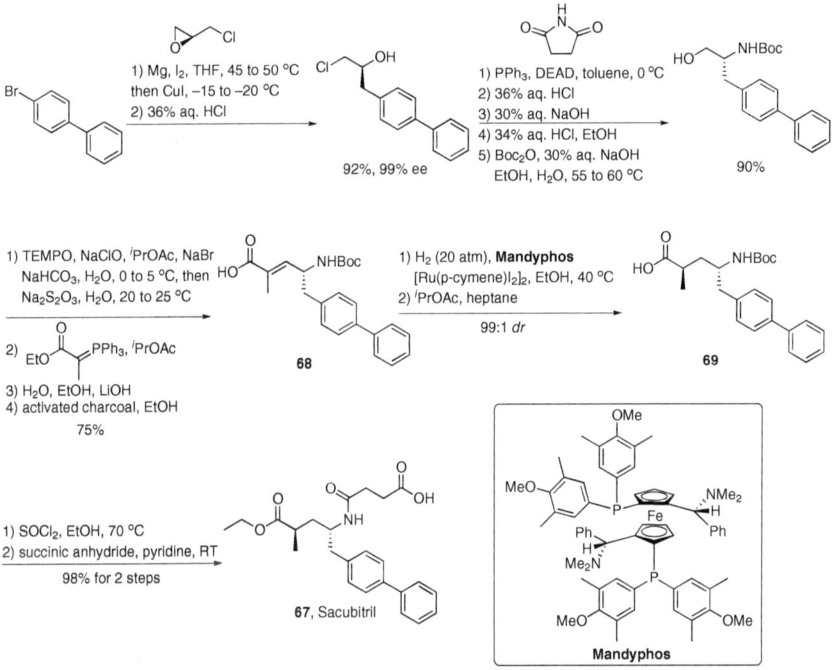

Scheme 6.38 Synthesis of Sacubitril. Source: Flick et al. [74a], Zhu et al. [74b], and Halama and Zapadlo [74c].

The freebase form of sacubitril does not readily crystallize; the isolation of a number of pharmaceutically acceptable salts of sacubitril via crystallization, most preferably the calcium salt or sodium salts, has been reported. Preparation of the sacubitril/valsartan complex (trisodium salt and hemihydrate) has been described on a kiloscale from sacubitril calcium salt via neutralization to the freebase and subsequent complexation with valsartan in ^{i}PrOAc/acetone. Addition of NaOH and crystallization then provided the desired trisodium salt hemihydrate.

6.2.3.3 Naproxen, Ibuprofen, and Flurbiprofen

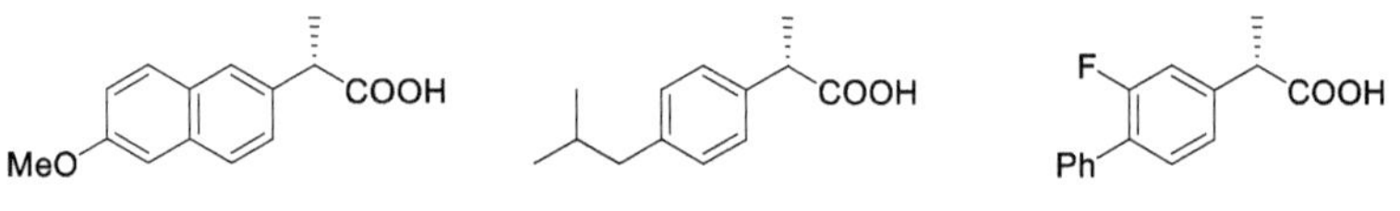

70, (*S*)-Naproxen **71**, (*S*)-Ibuprofen **72**, (*S*)-Flurbiprofen

Chiral carboxylic acids are important moieties in pharmaceuticals, agrochemicals, flavors, fragrances, and health supplements. For example, the α-aryl propionic acids Naproxen (**70**), Ibuprofen (**71**), and Flurbiprofen (**72**) are well-known nonsteroidal anti-inflammatory drugs [75].

The asymmetric hydrogenation of α-arylacrylic acids is highly useful for the synthesis of well-known nonsteroidal anti-inflammatory drugs such as Ibuprofen and Naproxen, and thus this reaction has been studied extensively [76]. However, chiral ruthenium catalysts generally require high hydrogen pressure for high enantioselectivity, while the rhodium catalysts show modest TONs or turnover frequencies. With catalyst SiPhox/Ir, various α-aryl- and α-alkylacrylic acids can be hydrogenated to the corresponding chiral carboxylic acids, including Ibuprofen, Naproxen, and Furbiprofen, in excellent yields (98–99%) and enantioselectivities (96–98% ee). The catalyst loading can be reduced to 0.01 mol% without diminishing the conversion or enantioselectivity of the reaction, although a longer reaction time is required (Figure 6.5).

Zhang group developed a new class of ferrocenyl chiral bisphosphorus ligand, Wudaphos, which exhibits excellent ee and activity (up to 99% ee, TON up to 20 000) for the asymmetric hydrogenation of both 2-aryl and 2-alkyl acrylic acids through ion-pair noncovalent interaction under base free and mild reaction conditions. Well-known anti-inflammatory drugs such as Naproxen and Ibuprofen and some bioactive compounds were also efficiently obtained with excellent ee (Scheme 6.39) [77].

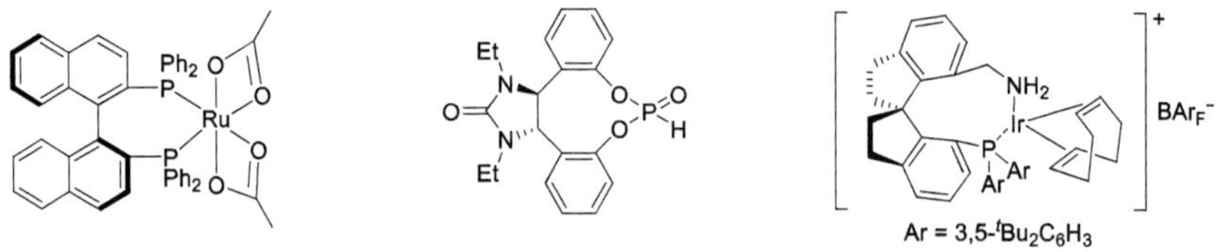

Ar = 3,5-$^tBu_2C_6H_3$

(*S*)-BINAP-Ru
S/C = 400, 11.2 MPa, 24 h
(*S*)-Naproxen, 97% ee

[Rh(m-Cl)(COD)]$_2$/L, NEt_3
S/C = 200, 5 MPa, 24 h
(*S*)-Naproxen, 95% ee
(*S*)-Ibuprofen, 93% ee

SiPhox/Ir
S/C = 1000, 0.6 MPa, 15 min
Cs_2CO_3 (0.5 equiv)
(*S*)-Naproxen, 97% ee
(*S*)-Ibuprofen, 98% ee
(*S*)-Flurbiprofen, 96% ee

Figure 6.5 Representative asymmetric hydrogenation catalytic systems for preparing Naproxen, Ibuprofen, and Flurbiprofen.

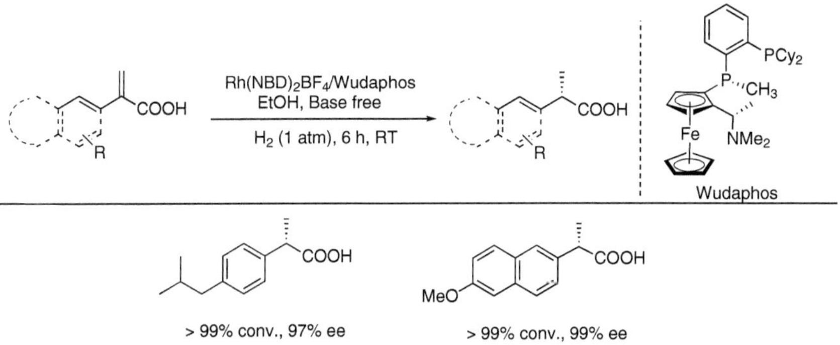

Scheme 6.39 Synthetic route for Naproxen and Ibuprofen developed by Zhang group. Source: Based on Chen et al. [77].

6.2.3.4 Ramelteon

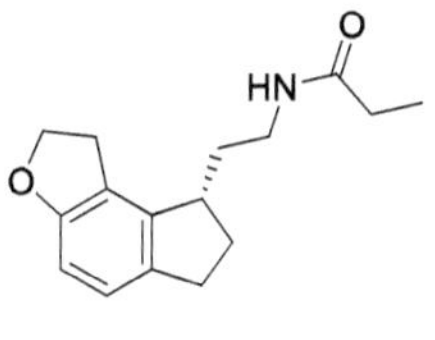

73, Ramelteon

Ramelteon (**73**) is well known as a selective melatonin MT1/MT2 receptor agonist which primarily has been used as a treatment of insomnia [78]. It was developed by Takeda Pharmaceuticals North America and approved by FDA in 2005.

Ramelteon has a simple but unique chemical scaffold containing a furan-fused tricycle ring and a stereogenic center at the benzylic position. Currently, most of the reported syntheses of Ramelteon involve catalytic asymmetric hydrogenation or chiral resolution of the corresponding racemic mixtures [79].

In 2005, Takeda Pharmaceutical Co., Ltd. developed synthetic approaches to Ramelteon, based on chelation-controlled asymmetric hydrogenation of bicyclic substrates using Ru-BINAP catalysts in 99% yield and 93% ee. The effectiveness and robustness of the reaction were demonstrated on a 700-g scale (Scheme 6.40) [80]. In 2009, they developed the direct enantioselective hydrogenation of the three-fused-ring allylamine catalyzed by Rh-Josiphos (Scheme 6.40). The procedure is suitable for the preparation of Ramelteon owing to the excellent enantiomeric excess and low hydrogen pressure [81].

6.2.3.5 Aliskiren

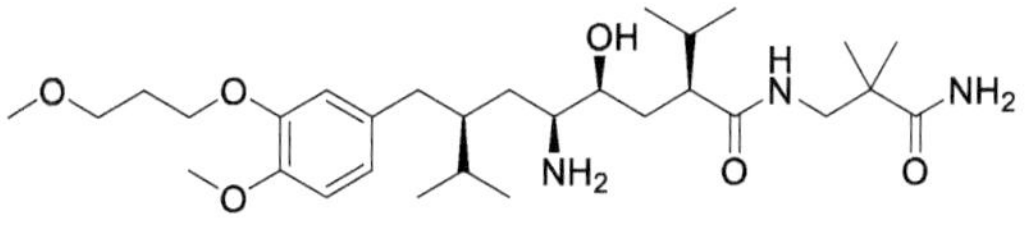

74, Aliskiren

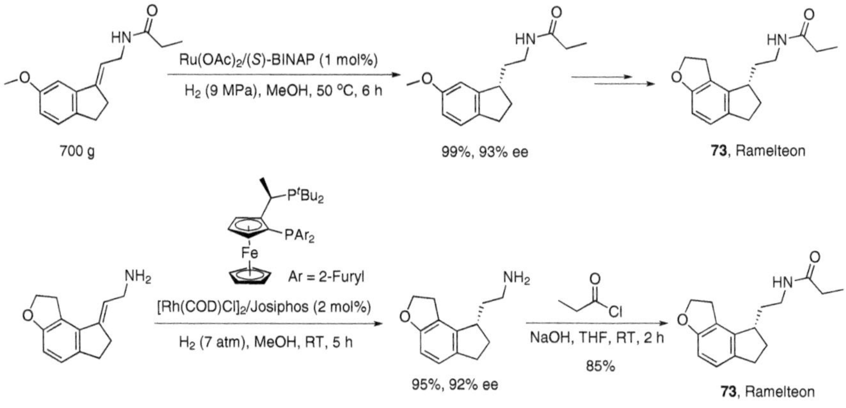

Scheme 6.40 Synthetic route for Ramelteon. Source: Uchikawa et al. [80a] and Yamano et al. [80b].

Aliskiren (**74**), Novartis' blood pressure-lowering agent, is the first of a new class of orally effective direct renin inhibitors, which was approved by FDA in 2007 [82]. Its current licensed indication is essential (primary) hypertension. In December 2011, Novartis halted a clinical trial of the drug after discovering increased incidence of nonfatal stroke, kidney complications, high blood potassium, and low blood pressure in people with diabetes and kidney impairment. However, it is still a successful example in drug synthesis, especially in asymmetric hydrogenation (Scheme 6.41).

In 2002, Spindler and Weissensteiner developed the route to synthon **77** with the key asymmetric hydrogenation step [85]. The application of an Rh catalyst generated *in situ* from $[Rh(NBD)_2]BF_4$ and Walphos afforded the saturated acid **76** with 95% ee at 50 bar hydrogen and ambient temperature. At bench scale under technical reaction conditions, 12 g of **75** could be hydrogenated with an S/C = 5700, yielding intermediate **76** with 95% ee (Scheme 6.41).

In 2007, an asymmetric hydrogenation process for the α-isopropyl dihydrocinnamic acid en route to Aliskiren was developed using a rhodium catalyst ligated with a chiral monodentate phosphoramidite and a nonchiral phosphine. While the catalyst based on 2 equiv of MonoPhos to the $[Rh(COD)_2]BF_4$ gave promising results, the rate of hydrogenation and ee of the product could be improved spectacularly by the addition of monodentate nonchiral triarylphosphine (PPh_3) to the catalyst. This remarkable mixed-ligand catalyst has been identified using high-throughput experimentation. At a larger scale (autoclaves between 150 and 450 mL) using scalable conditions (S/C = 1000, 80 bar H_2, 55 °C), the product was produced with full conversion and 90% ee (Scheme 6.41) [83].

In 2008, Zhou group carried out the hydrogenation of α-isopropylcinnamic acid derivative **75** to prepare the acid (*R*)-**76**, which is the key intermediate for the synthesis of Aliskiren [86]. The compound **75** was successfully hydrogenated by catalyst Ir-SIPHOX. The carboxylic acid (*R*)-**76** was attained in high yield with 98% and 95% ee at catalyst loadings of 0.017 mol% (S/C = 6000) and 0.01 mol% (S/C = 10 000), respectively (Scheme 6.41). This primary result was superior to those reported previously.

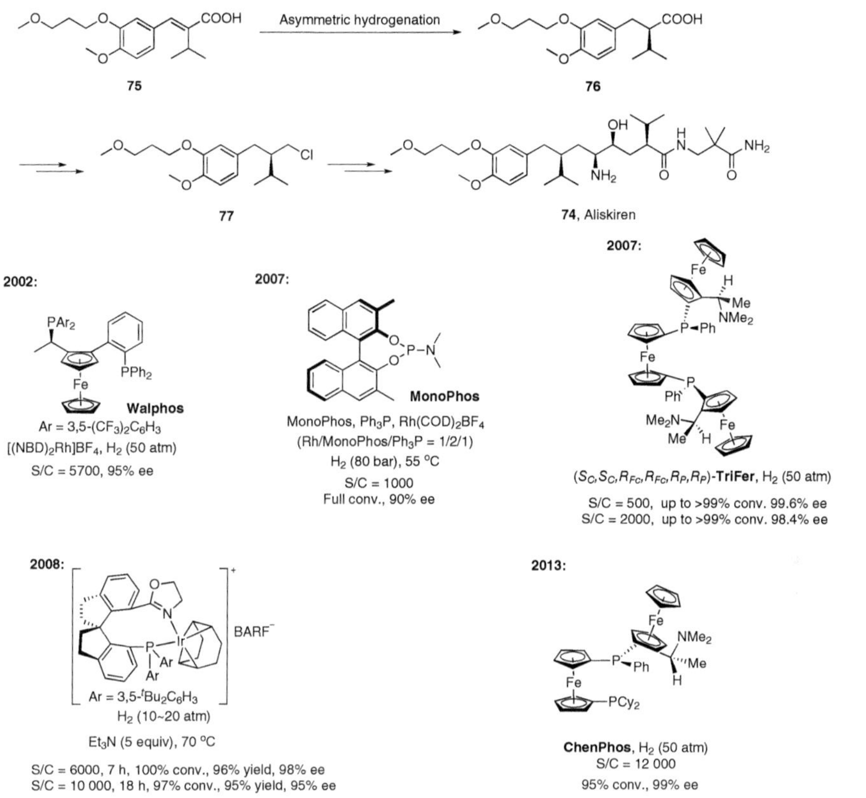

Scheme 6.41 Synthesis of the chiral drug Aliskiren via asymmetric hydrogenation. Source: Boogers et al. [83], Chen et al. [84a,b].

In 2007 and 2013, Chen group developed ferrocene-based P-stereogenic phosphines TriFer and ChenPhos ligands, respectively, which showed high enantioselectivity, activity, and productivity for the rhodium-catalyzed asymmetric hydrogenation of α-isopropylcinnamic acid derivative **75** (Scheme 6.41) [84].

6.2.3.6 (+)-*cis*-Methyl Dihydrojasmonate

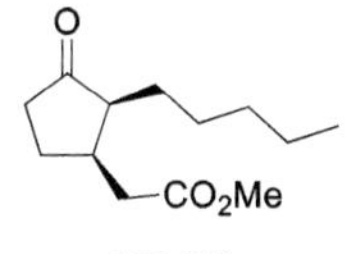

(1*R*,2*S*)

78, (+)-*cis*-Methyl dihydrojasmonate

Methyl dihydrojasmonate is an important, unnatural perfumery chemical (jasminelike odor) and is used in many fragrance mixtures. The worldwide volume of use for methyl dihydrojasmonate is in the region of greater than 1000 metric tons per year [87]. Since the early 1970s, the success of the industrial synthesis of a near-equilibrium mixture of the racemic *trans/cis* isomers of methyl dihydrojasmonate was commercialized as *Hedione* (*trans/cis* 90/10) and allowed replacement of the natural extract. Among the four isomers of methyl dihydrojasmonate, only the (1*R*,2*S*)-*cis* isomer **78** was found to have an intense and distinct odor, while the other isomers are substantially weaker or almost odorless [88]. As a matter of fact, an increasing concern for human health, environmental preservation, and cost control has favored the trend to employ the single enantiomer of odor-active chemicals in the field of fragrance chemistry [89].

Enantioselective syntheses of methyl dihydrojasmonate have been disclosed relying on enamine alkylation of a chiral precursor [90], asymmetric Michael addition [91], catalytic asymmetric hydrogenation [88], or a variant of the Claisen rearrangement [92]. The industrial-scale process to (+)-*cis*-methyl dihydrojasmonate **78** developed by Firmenich-Genêt was the homogeneous enantioselective hydrogenation of what is known as diDeHydroHedione (DHH) [93] **79** bearing electron-deficient tetrasubstituted C=C bond [94], with excellent *cis/trans* selectivities (≥99 : 1) and good enantioselectivities (up to 88% ee). This accomplishment was the result of the discovery of a cationic, more electronic, and highly unsaturated ruthenium catalyst with a chiral diphosphine ligand [Ru(P*P)(H)(η^6-cot]BF_4 (cot = 1,3,5-cyclooctatriene) [95]. A good enantioselectivity (64% ee) was obtained with (2*R*,5*R*)-Me-DuPhos as chiral ligand, and after optimization of the reaction conditions using (*R*,S_{Fc})-Josiphos as chiral ligand, a higher enantioselectivity (88% ee) was reached (Scheme 6.42). The Firmenich-Genêt procedure has the advantage of allowing the straightforward synthesis of the Ru complexes with a

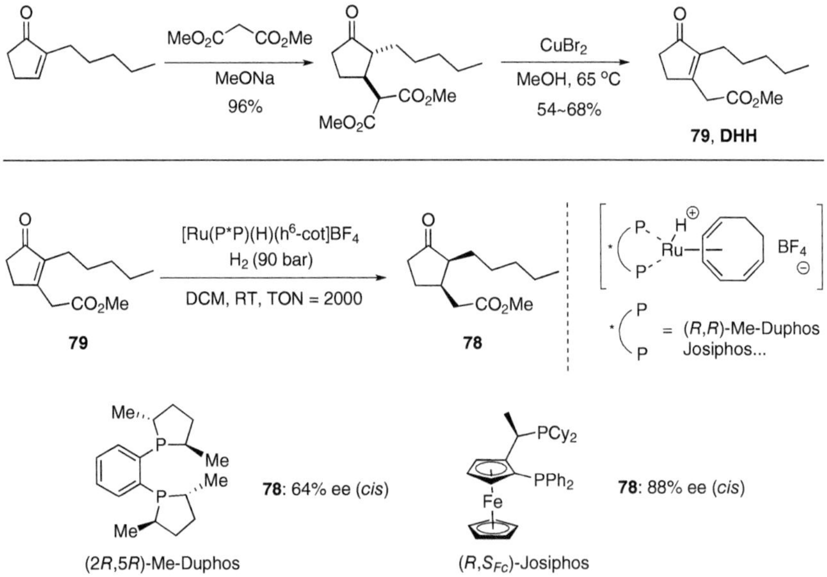

Scheme 6.42 The industrial-scale process to (+)-*cis*-methyl dihydrojasmonate.

number of different diphosphane ligands [96], and the neutral conditions might avoid epimerization at C(2) of the *cis* isomer.

6.2.4 Asymmetric Hydrogenation of Imine

6.2.4.1 Solifenacin

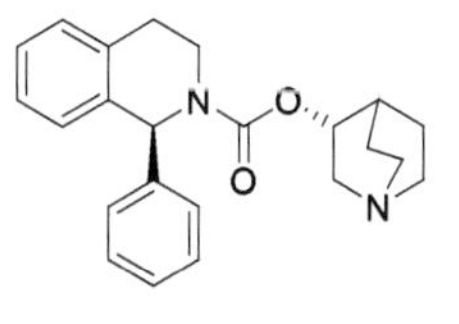

80, Solifenacin

Solifenacin (**80**) is a urinary antispasmodic, launched in 2004 for the treatment of overactive bladder. The common feature of approaches to solifenacin is the use of (*S*)-1-phenyl-1,2,3,4-tetrahydroisoquinoline **82** as the key intermediate [97].

The chiral intermediate can be manufactured using a classical resolution [98]. The strategy used on industrial scale for the synthesis of solifenacin involves a Bischler–Napieralski cyclization followed by reduction of the resulting imine **81** (Scheme 6.43). Commonly, diastereomeric resolution of the tartrate salt of racemic **82** provides (*S*)-**82**, which then reacts with a (halo)alkyl haloformate to yield the desired **83**.

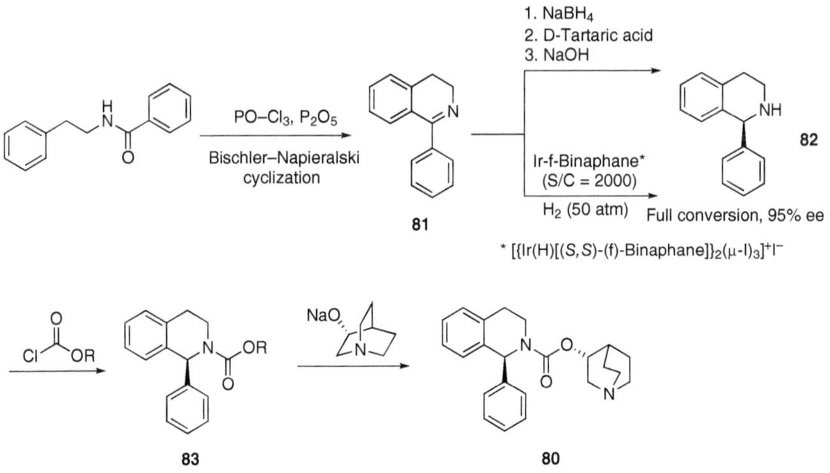

Scheme 6.43 Synthetic route for solifenacin. Source: Based on Chang et al. [99].

It is widely accepted that asymmetric catalysis in industrial processes can provide superior yield and reduce the waste streams during the synthesis of pharmaceutical intermediates. In this contribution, Zhang group and Chiral Quest Inc. developed the synthesis of enantiomerically enriched (*S*)-**82** (Scheme 43) [99]. The highly effective iodine-bridged dimeric [{Ir(H)[(*S*,*S*)-(f)-binaphane]}$_2$(μ-I)3]$^+$I$^-$ complex has been applied in the asymmetric hydrogenation with excellent enantioselectivity (S/C = 2000, full conversion, 95% ee). The use of I_2 as an additive enhanced the performance of this catalyst.

6.2.4.2 (*S*)-Metolachlor

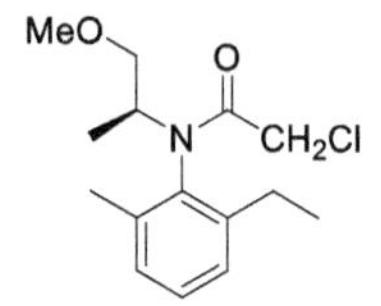

84, (*S*)-Metolachlor

Metolachlor (**84**) is a pesticide sold under the trade names DUAL MAGNUM™ and DUAL GOLD™. Up to now, only a few industrial applications of C=N bond hydrogenations have been reported. The (*S*)-metolachlor process carried out by Ciba-Geigy/Syngenta with a volume of >20 000 tons per year is not only the largest scale asymmetric hydrogenation production process, but the Ir/Xyliphos complex is also one of the most active and productive enantioselective catalysts known, as shown in Scheme 6.44 [100].

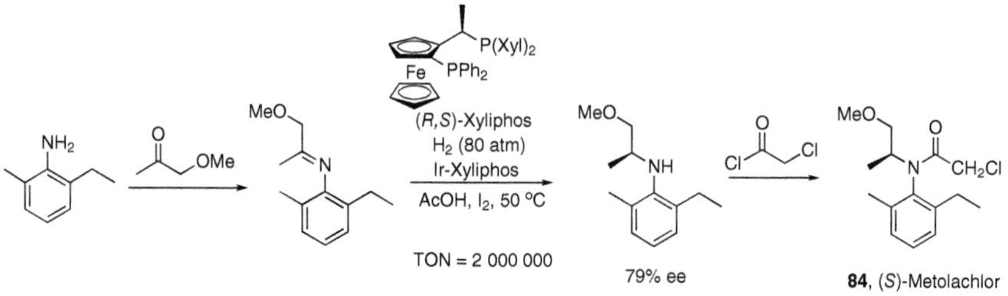

Scheme 6.44 Synthesis of pesticide (*S*)-metolachlor. Source: Blaser [100a] and Blaser et al. [100b].

Although the enantioselectivity of the asymmetric reduction is only 79%, it is sufficient to provide an economic advantage over the use of the racemic mixture. The reduction employs a catalyst prepared *in situ* from [Ir(COD)Cl]$_2$ and the (*R*,*S*)-Xyliphos ligand. The addition of both iodide anion and acetic acid additives is critical to the catalytic activity. The high catalyst cost could be offset by high catalytic activity (turnovers of >600 000 h^{-1}) and low catalyst usage (S/C up to 2 000 000). The first production batch was run with overwhelming success: about 10 tons of *N*-alkylated aniline (NAA) with 79% ee were produced using about 34 g of Ir complex, ca. 70 g of xyliphos ligand and some acid and iodide; reaction time was two hours and conversion 99.6%.

6.2.5 Asymmetric Transfer Hydrogenation

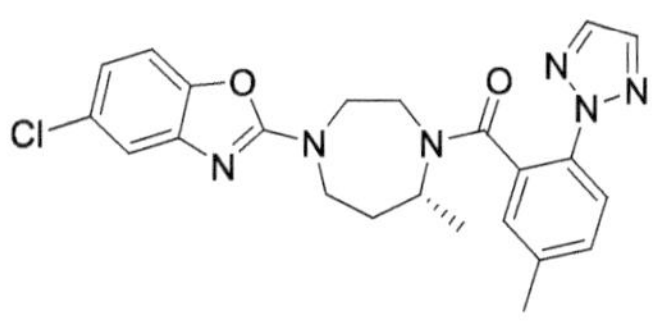

85, Suvorexant

Suvorexant (**85**) is a selective dual orexin antagonist that promotes sleep by reducing wakefulness and arousal [101]. It has been approved for the treatment of insomnia by FDA in 2018.

In 2011, Merck & Co. Inc. reported the first multikilogram synthesis of **85**. The key structural features of **83** include a 1,4-diazepane ring system with a methyl substituent in the 7-position, with the ring nitrogen atoms bearing benzoxazole and aromatic amide moieties. The preparation of the chiral diazepane ring moiety was achieved through a racemic reductive amination and classical resolution sequence to afford a 27% overall yield with 96% ee (Scheme 6.45) [102].

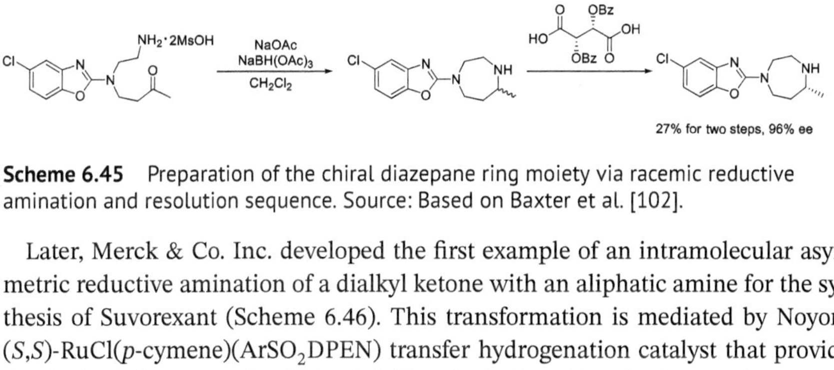

Scheme 6.45 Preparation of the chiral diazepane ring moiety via racemic reductive amination and resolution sequence. Source: Based on Baxter et al. [102].

Later, Merck & Co. Inc. developed the first example of an intramolecular asymmetric reductive amination of a dialkyl ketone with an aliphatic amine for the synthesis of Suvorexant (Scheme 6.46). This transformation is mediated by Noyori's (*S*,*S*)-RuCl(*p*-cymene)($ArSO_2$DPEN) transfer hydrogenation catalyst that provides the desired diazepane ring in 97% yield and 94.5% ee. Despite the requirements of low temperature, high pH, high dilution, and bulky catalyst needed to provide high yield and ee, a reduction of the catalyst loading to 2 mol% proved feasible. This process has been run on >100 kg scale [103].

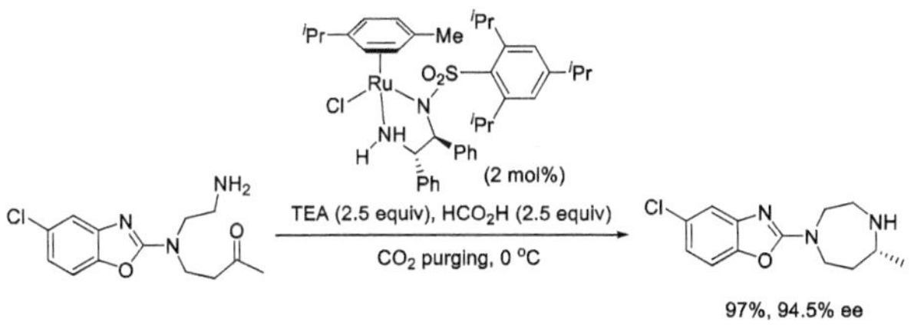

Scheme 6.46 Ruthenium-catalyzed intramolecular asymmetric reductive amination en route to Suvorexant.

6.3 Summary and Conclusions

Catalytic asymmetric hydrogenation is one of the powerful methods to prepare chiral molecules. In practice, asymmetric hydrogenation offers an economical method for the large-scale preparation of chiral products, such as the famous Monsanto's L-DOPA synthesis, Takasago's carbapenem synthesis, and Syngenta's (*S*)-metolachlor synthesis. Thus far, we have been able to synthesize many pharmaceutically important chiral building blocks, including amino acids, amines, alcohols, esters, acids, and amino alcohols, in a highly efficient way.

However, chiral ligands which can be applied in the manufacturing are still rare, as in the manufacturing scale not only high ee but also a high TON and TOF are required, as the costs of catalysts are normally key issues in industry scale. Most of the reported catalysts have TONs lower than 1000 and cause the high cost of the catalysts which are unsuitable for the application in industry. We are convinced that in the near future the industrial application of enantioselective hydrogenation technology will accelerate further [104].

References

1 Knowles, W.S. and Sabacky, M.J. (1968). *Chem. Commun.*: 1445–1446.

2 Horner, L., Siegel, H., and Büthe, H. (1968). *Angew. Chem. Int. Ed.* 7: 942–942.

3 Knowles, W.S. (2002). *Angew. Chem. Int. Ed.* 41: 1998–2007.

4 (a) Tang, W. and Zhang, X. (2003). *Chem. Rev.* 103: 3029–3070. (b) Zhou, Y.-G. (2007). *Acc. Chem. Res.* 40: 1357–1366. (c) Roseblade, S.J. and Pfaltz, A. (2007). *Acc. Chem. Res.* 40: 1402–1411. (d) Xie, J.-H., Zhu, S.-F., and Zhou, Q.-L. (2010). *Chem. Rev.* 111: 1713–1760. (e) Corey, E.J. and Kurti, L. (2010). *Enantioselective Chemical Synthesis.* Dallas: Direct Book Publishing. (f) Zhou, Q.-L. (2011). *Privileged Chiral Ligands and Catalysts.* Weinheim: Wiley-VCH. (g) Karamé, I. (2012). *Asymmetric Hydrogenation.* London, UK: InTech. (h) Xie, J.-H. and Zhou, Q.-L. (2012). *Acta Chim. Sin.* 70: 1427–1438. (i) Liu, Y., Wang, Z., and Ding, K. (2012). *Acta Chim. Sin.* 70: 1464–1470. (j) Kamer, P.C.J. and Van Leeuwen, P.W.N.M. (2013). *Phosphorus (III) Ligands in Homogeneous Catalysis.* Chichester: Wiley.

5 (a) Miyashita, A., Yasuda, A., Takaya, H. et al. (1980). *J. Am. Chem. Soc.* 102: 7932–7934. (b) Miyashita, A., Takaya, H., Souchi, T., and Noyori, R. (1984). *Tetrahedron* 40: 1245–1253. (c) Noyori, R., Ohta, M., Hsiao, Y. et al. (1986). *J. Am. Chem. Soc.* 108: 7117–7119. (d) Hitamura, M., Hsiao, Y., Noyori, R., and Takaya, H. (1987). *Tetrahedron Lett.* 28: 4829–4832. (e) Takaya, H., Ohta, T., Sayo, N. et al. (1987). *J. Am. Chem. Soc.* 109: 1596–1597. (f) Ohta, T., Takaya, H., Kitamura, M. et al. (1987). *J. Org. Chem.* 52: 3174–3176.

6 (a) Noyori, R., Ohkuma, T., Kitamura, M. et al. (1987). *J. Am. Chem. Soc.* 109: 5856–5858. (b) Kitamura, M., Ohkuma, T., Inoue, S. et al. (1988). *J. Am. Chem. Soc.* 110: 629–631. (c) Kitamura, M., Kasahara, I., Manabe, K. et al. (1988). *J. Org. Chem.* 53: 708–710. (d) Ohkuma, T., Ooka, H., Hashiguchi, S. et al. (1995).

J. Am. Chem. Soc. 117: 2675–2676. (e) Ohkuma, T., Ooka, H., Ikariya, T., and Noyori, R. (1995). *J. Am. Chem. Soc.* 117: 10417–10418.

7 Noyori, R. (2002). *Angew. Chem. Int. Ed.* 41: 2008–2022.

8 (a) Wu, W., Liu, S., Duan, M. et al. (2016). *Org. Lett.* 18: 2938–2941. (b) Wu, W., Xie, Y., Li, P. et al. (2017). *Org. Chem. Front.* 4: 555–559. (c) Wu, W., You, C., Yin, C. et al. (2017). *Org. Lett.* 19: 2548–2551. (d) Zhao, Q., Li, S., Huang, K. et al. (2013). *Org. Lett.* 15: 4014–4017. (e) Zhao, Q., Wen, J., Tan, R. et al. (2014). *Angew. Chem. Int. Ed.* 53: 8467–8470. (f) Yu, J., Long, J., Yang, Y. et al. (2017). *Org. Lett.* 19: 690–693. (g) Yu, J., Duan, M., Wu, W. et al. (2017). *Chem.-Eur. J.* 23: 970–975. (h) Chen, C., Zhang, Z., Jin, S. et al. (2017). *Angew. Chem. Int. Ed.* 56: 6808–6812. (i) Dong, X.-Q., Zhao, Q., Li, P. et al. (2015). *Org. Chem. Front.* 2: 1425–1431. (j) Chen, C., Wen, S., Geng, M. et al. (2017). *Chem. Commun.* 53: 9785–9788. (k) Huang, K., Emge, T.J., and Zhang, X. (2014). *Heteroat. Chem.* 25: 131–134. (l) Chen, C., Dong, X.-Q., and Zhang, X. (2016). *Chem. Rec.* 16: 2674–2686.

9 (a) Hoge, G., Wu, H., Kissel, W.S. et al. (2004). *J. Am. Chem. Soc.* 126: 5966–5967. (b) Sun, X., Zhou, L., Wang, C., and Zhang, X. (2007). *Angew. Chem. Int. Ed.* 46: 2623–2626. (c) Chen, J., Zhang, W., Geng, H. et al. (2009). *Angew. Chem. Int. Ed.* 48: 800–802. (d) Geng, H., Zhang, W., Chen, J. et al. (2009). *Angew. Chem. Int. Ed.* 48: 6052–6054. (e) Zhang, X., Huang, K., Hou, G. et al. (2010). *Angew. Chem. Int. Ed.* 49: 6421–6424. (f) Liu, T.-L., Wang, C.-J., and Zhang, X. (2013). *Angew. Chem. Int. Ed.* 52: 8416–8419. (g) Wang, Q., Huang, W., Yuan, H. et al. (2014). *J. Am. Chem. Soc.* 136: 16120–16123. (h) Molinaro, C., Scott, J.P., Shevlin, M. et al. (2015). *J. Am. Chem. Soc.* 137: 999–1006. (i) Li, P., Zhou, M., Zhao, Q. et al. (2016). *Org. Lett.* 18: 40–43. (j) Gao, W., Lv, H., and Zhang, X. (2017). *Org. Lett.* 19: 2877–2880. (k) Gao, W., Lv, H., Zhang, T. et al. (2017). *Chem. Sci.* 8: 6419–6422.

10 (a) Doucet, H., Ohkuma, T., Murata, K. et al. (1998). *Angew. Chem. Int. Ed.* 37: 1703–1707. (b) Jiang, Y., Jiang, Q., and Zhang, X. (1998). *J. Am. Chem. Soc.* 120: 3817–3818. (c) Ohkuma, T., Koizumi, M., Doucet, H. et al. (1998). *J. Am. Chem. Soc.* 120: 13529–13530. (d) Li, W., Hou, G., Wang, C. et al. (2010). *Chem. Commun.* 46: 3979–3981. (e) Xie, J.-H., Liu, X.-Y., Xie, J.-B. et al. (2011). *Angew. Chem. Int. Ed.* 50: 7329–7332. (f) Gu, G., Lu, J., Yu, O. et al. (2018). *Org. Lett.* 20: 1888–1892.

11 (a) Kanazawa, Y., Tsuchiya, Y., Kobayashi, K. et al. (2006). *Chem.- Eur. J.* 12: 63–71. (b) von Matt, P. and Pfaltz, A. (1991). *Tetrahedron: Asymmetry* 2: 691–700. (c) Shang, J., Han, Z., Li, Y. et al. (2012). *Chem. Commun.* 48: 5172–5174. (d) Shi, L., Wei, B., Yin, X. et al. (2017). *Org. Lett.* 19: 1024–1027. (e) Yin, X., Chen, C., Dong, X.-Q., and Zhang, X. (2017). *Org. Lett.* 19: 2678–2681. (f) Yin, X., Chen, C., Li, X. et al. (2017). *Org. Lett.* 19: 4375–4378. (g) Han, Z., Li, P., Zhang, Z. et al. (2016). *ACS Catal.* 6: 6214–6218. (h) Li, S., Huang, H., and Zhang, X. (2014). *Chem. Commun.* 50: 8878–8881. (i) Li, P., Hu, X., Dong, X.-Q., and Zhang, X. (2016). *Chem. Commun.* 52: 11677–11680. (j) Li, S., Huang, K., Cao, B. et al. (2012). *Angew. Chem. Int. Ed.* 51: 8573–8576. (k) Kraft, S., Ryan, K., and Kargbo, R.B. (2017). *J. Am. Chem. Soc.* 139: 11630–11641.

12 (a) Willoughby, C.A. and Buchwald, S.L. (1994). *J. Am. Chem. Soc.* 116: 8952–8965. (b) Morimoto, T. and Achiwa, K. (1995). *Tetrahedron: Asymmetry* 6: 2661–2664. (c) Zhu, G. and Zhang, X. (1998). *Tetrahedron: Asymmetry* 9: 2415–2418. (d) Cobley, C.J. and Henschke, J.P. (2003). *Adv. Synth. Catal.* 345: 195–201. (e) Guiu, E., Claver, C., Benet-Buchholz, J., and Castillon, S. (2004). *Tetrahedron: Asymmetry* 15: 3365–3373. (f) Jackson, M. and Lennon, I.C. (2007). *Tetrahedron Lett.* 48: 1831–1834. (g) Li, C. and Xiao, J. (2008). *J. Am. Chem. Soc.* 130: 13208–13209. (h) Gao, K., Yu, C.-B., Li, W. et al. (2011). *Chem. Commun.* 47: 7845–7847. (i) Li, P., Huang, Y., Hu, X. et al. (2017). *Org. Lett.* 19: 3855–3858. (j) Wen, J., Tan, R., Liu, S. et al. (2016). *Chem. Sci.* 7: 3047–3051.

13 (a) Johnson, N.B., Lennon, I.C., Moran, P.H., and Ramsen, J.A. (2007). *Acc. Chem. Res.* 40: 1291–1299. (b) Blaser, H.-U. and Federsel, H.-J. (2010). *Asymmetric Catalysis on Industrial Scale: Challenges, Approaches, and Solutions*, 2e. Weinheim: Wiley-VCH. (c) Chen, C., Dong, X.-Q., and Zhang, X. (2017). *Chin. J. Pharm.* 48: 943–964. (d) Ager, D.J., de Vriesb, A.H.M., and de Vries, J.G. (2012). *Chem. Soc. Rev.* 41: 3340–3380.

14 Fahn, S. (1996). *Adv. Neurol.* 69: 477–486.

15 (a) Vineyard, B.D., Knowles, W.S., Sabacky, M.J. et al. (1977). *J. Am. Chem. Soc.* 99: 5946–5952. (b) Knowles, W.S. (1983). *Acc. Chem. Res.* 16: 106–112. (c) Knowles, W.S. (1986). *J. Chem. Educ.* 63: 222–225.

16 Malakondaiah, G.C., Gurav, V.M., Reddy, L.A. et al. (2008). *Synthetic Commun.* 38: 1737–1744.

17 Kondaiah, G.C.M., Vivekanandareddy, M., Reddy, L.A. et al. (2011). *Synthetic Commun.* 41: 1186–1191.

18 Gao, M., Meng, J.-J., Lv, H., and Zhang, X. (2015). *Angew. Chem. Int. Ed.* 54: 1885–1887.

19 Zhou, G., Grosser, S., Sun, L. et al. (2016). *Org. Process Res. Dev.* 20: 653–660.

20 Hansen, K.B., Hsiao, Y., Xu, F. et al. (2009). *J. Am. Chem. Soc.* 131: 8798–8799.

21 Wu, S., Yu, B., Wang, Y. et al. (2010). Process and intermediates for the preparation of N-acylated-4-aryl beta-amino acid derivatives. PCT Int. Appl., WO 2010078440 A1.

22 (a) Sankareswaran, S., Mannam, M., Chakka, V. et al. (2016). *Org. Process Res. Dev.* 20: 1461–1468. (b) Aoyama, Y., Hakogi, T., Fukui, Y. et al. (2019). *Org. Process Res. Dev.* 23: 558–564. (c) Yasukata, T., Masui, M., Ikarashi, F. et al. (2019). *Org. Process Res. Dev.* 23: 565–570. (d) Hughes, D.L. (2019). *Org. Process Res. Dev.* 23: 716–729.

23 (a) Liu, D. and Zhang, X. (2005). *Eur. J. Org. Chem.*: 646–649. (b)Huang, K., Zhang, X., Emge, T.J. et al. (2010). *Chem. Commun.* 46: 8555–8557. (c) Zhang, X. (1999). Catalytic asymmetric hydrogenation, hydroformylation, and hydrovinylation via transition metal catalysts with phosphines and phosphites. PCT, WO9959721. (d) Zhang, X. and Tang, W. (2005). P-chiral phospholanes and phosphocyclic compounds and their use in asymmetric catalytic reactions. PCT, WO2005117907. (e) Wu, S., Tian, Z., Hu, S. et al. (2015). Preparation method for optical activity active 3-amino butanol and optical activity 3-amino butyric acid. CN104370755.

24 Zhu, G., Chen, Z., and Zhang, X. (1999). *J. Org. Chem.* 64: 6907–6910.

25 (a) Garnock-Jones, K.P. (2015). *Drugs* 75: 1305–1310. (b) Davenport, J.M., Covington, P., Bonifacio, L. et al. (2015). *J. Clin. Pharmacol.* 55: 534–542.

26 (a) Wang, X., Niu, S., Xu, L. et al. (2017). *Org. Lett.* 19: 246–249. (b) Imamoto, T., Tamura, K., Zhang, Z. et al. (2012). *J. Am. Chem. Soc.* 134: 1754–1769.

27 (a) Soloshonok, V.A., Tang, X., and Hruby, V.J. (2001). *Tetrahedron* 57: 6375–6382. (b) Balducci, D., Contaldi, S., Lazzari, I., and Porzi, G. (2009). *Tetrahedron: Asymmetry* 20: 1398–1401.

28 Praquin, C.F.B., de Koning, P.D., Peach, P.J. et al. (2011). *Org. Process Res. Dev.* 15: 1124–1129.

29 (a)Dygos, J.H., Yonan, E.E., Scaros, M.G. et al. (1992). *Synthesis*: 741–743. (b) Getman, D.P., Periana, R.A., and Riley, D.P. (1987). US Patent 4879398.

30 Dudek, M.K., Kostrzewa, M., Paluch, P., and Potrzebowski, M.J. (2018). *Cryst. Growth Des.* 18: 3959–3970.

31 (a) Schafer, P. (2012). *Biochem. Pharmacol.* 83: 1583–1590. (b) Ruchelman, A.L. and Connolly, T.J. (2015). *Tetrahedron: Asymmetry* 26: 553–559.

32 (a) Wong, D.T. (1998). *Expert Opin. Invest. Drugs* 7: 1691–1699. (b) Norman, T.R. and Olver, J.S. (2010). *Drug Des. Dev. Ther.* 4: 19–31. (c) Bymaster, F.P., Beedle, E.E., Findlay, J. et al. (2003). *Bioorg. Med. Chem. Lett.* 13: 4477–4480. (d) Cashman, J.R. and Ghirmai, S. (2009). *Bioorg. Med. Chem.* 17: 6890–6897. (e) Zhou, J.-N., Fang, Q., Hu, Y.-H. et al. (2014). *Org. Biomol. Chem.* 12: 1009–1017.

33 (a) Fujima, Y., Ikunaka, M., Inoue, T., and Matsumoto, J. (2006). *Org. Process Res. Dev.* 10: 905–913. (b) Träff, A., Lihammar, R., and Bäckvall, J.-E. (2011). *J. Org. Chem.* 76: 3917–3921.

34 (a) Liu, D., Gao, W., Wang, C., and Zhang, X. (2005). *Angew. Chem. Int. Ed.* 44: 1687–1689. (b) Ratovelomanana-Vidal, V., Girard, C., Touati, R. et al. (2003). *Adv. Synth. Catal.* 345: 261–274.

35 (a) Blacklock, T.J., Sohar, P., Butcher, J.W. et al. (1993). *J. Org. Chem.* 58: 1672–1679. (b) Fu, J., Sun, F., Liu, W. et al. (2016). *Mol. Pharm.* 13: 2987–2995. (c) Huang, Q., Rui, E.Y., Cobbs, M. et al. (2015). *J. Med. Chem.* 58: 2821–2833.

36 Jeulin, S., de Paule, S.D., Ratovelomanana-Vidal, V. et al. (2004). *Proc. Natl. Acad. Sci. U. S. A.* 101: 5799–5804.

37 Zhang, Z., Qian, H., Longmire, J., and Zhang, X. (2000). *J. Org. Chem.* 65: 6223–6226.

38 (a) Qiu, L., Kwong, F.Y., Wu, J. et al. (2006). *J. Am. Chem. Soc.* 128: 5955–5965. (b) Sun, X., Zhou, L., Li, W., and Zhang, X. (2008). *J. Org. Chem.* 73: 1143–1146. (c) Sun, X., Li, W., Hou, G. et al. (2009). *Adv. Synth. Catal.* 351: 2553–2557. (d) Zou, Y., Geng, H., Zhang, W. et al. (2009). *Tetrahedron Lett.* 50: 5777–5779.

39 (a) Tan, X., Gao, S., Zeng, W. et al. (2018). *J. Am. Chem. Soc.* 140: 2024–2027. (b) Zuo, W., Lough, A.J., Li, Y.F., and Morris, R.H. (2013). *Science* 342: 1080–1083.

40 (a) Hansen, K.B., Chilenski, J.R., Desmond, R. et al. (2003). *Tetrahedron: Asymmetry* 14: 3581–3587. (b) Brands, K.M.J., Payack, J.F., Rosen, J.D. et al. (2003). *J. Am. Chem. Soc.* 125: 2129–2135.

41 (a) Liu, W.-P., Yuan, M.-L., Yang, X.-H. et al. (2015). *Chem. Commun.* 51: 6123–6125. (b) Cettolin, M., Puylaert, P., Pignataro, L. et al. (2017). *ChemCatChem* 9: 3125–3130. (c) Wang, Y., Yang, G., Xie, F., and Zhang, W. (2018). *Org. Lett.* 20: 6135–6139. (d) Zhang, L., Tang, Y., Han, Z., and Ding, K. (2019). *Angew. Chem. Int. Ed.* 58: 4973–4977.
42 Xu, L., Huang, Z.-H., Sandoval, C.A. et al. (2014). *Org. Process Res. Dev.* 18: 1137–1141.
43 Li, W., Sun, X., Zhou, L. et al. (2009). *J. Org. Chem.* 74: 1397–1399.
44 (a) Noyori, R., Ikeda, T., Ohkuma, T. et al. (1989). *J. Am. Chem. Soc.* 111: 9134–9135. (b) Ohkuma, T., Kitamura, M., and Noyori, R. (2000). *Asymmetric hydrogenation*. In: *Catalytic Asymmetric Synthesis*, 2e, 1. New York: Wiley.
45 (a) Kumobayashi, H., Miura, T., Sayo, N. et al. (2001). *Synlett*: 1055–1064. (b) Farina, V., Reeves, J.T., Senanayake, C.H., and Song, J.J. (2006). *Chem. Rev.* 106: 2734–2793.
46 Shimizu, H., Nagasaki, I., Matsumura, K. et al. (2007). *Acc. Chem. Res.* 40: 1385–1393.
47 (a) Polinsky, R.J. (1998). *Clin. Ther.* 20: 634–647. (b) Rosler, M., Anand, R., Cicin-Sain, A. et al. (1999). *Br. Med. J.* 318: 633–638. (c) Farlow, M.R. and Cummings, J.L. (2007). *Am. J. Med.* 120: 388–397.
48 (a) Boezio, A.A., Pytkowicz, J., Côté, A., and Charette, A.B. (2003). *J. Am. Chem. Soc.* 125: 14260–14261. (b) Mangas-Sánchez, J., Rodríguez-Mata, M., Busto, E. et al. (2009). *J. Org. Chem.* 74: 5304–5310. (c) Han, K., Kim, C., Park, J., and Feng, M.-J. (2010). *J. Org. Chem.* 75: 3105–3108. (d) Fuchs, M., Koszelewski, D., Tauber, K. et al. (2010). *Chem. Commun.* 46: 5500–5502. (e) Rao, R., Shewalkar, M.P., Nandipati, R. et al. (2012). *Synth. Commun.* 42: 589–598.
49 (a) Fieldhouse, R. (2005). PCT, WO2005058804. (b) Guijarro, D., Pablo, O., and Yus, M. (2010). *J. Org. Chem.* 75: 5265–5270. (c) Mathes, C., Foulkes, M., and Kesselgruber, M. (2010). PCT, WO2010072798.
50 Foulkes, M., Mathes, C., Spindler, F. et al. (2011). Process for the preparation of optically active compounds using pressure hydrogenation. WO2011073362.
51 Yan, P.-C., Zhu, G.-L., Xie, J.-H. et al. (2013). *Org. Process. Res. Dev.* 17: 307–312.
52 (a) Xie, J.-H., Liu, X.-Y., Yang, X.-H. et al. (2012). *Angew. Chem. Int. Ed.* 51: 201–203. (b) Yang, X.-H., Xie, J.-H., and Zhou, Q.-L. (2014). *Org. Chem. Front.* 1: 190–193.
53 (a) Labelle, M., Belley, M., Gareau, Y. et al. (1995). *Bioorg. Med. Chem. Lett.* 5: 283–288. (b) Schoors, D.F., De Smet, M., Reiss, T. et al. (1995). *Br. J. Clin. Pharmacol.* 40: 277–280. (c) Markham, A. and Faulds, D. (1998). *Drugs* 56: 251–256.
54 Zhu, G.-L., Zhang, X.-D., Yang, L.-J. et al. (2016). *Org. Process Res. Dev.* 20: 81–85.
55 Hallberg, B. and Palmer, R.H. (2010). *New Engl. J. Med.* 363: 1760–1762.
56 Qian, J.-Q., Yan, P.-C., Che, D.-Q. et al. (2014). *Tetrahedron Lett.* 55: 1528–1531.

57 (a) The Merck Index (2006). *An Encyclopaedia of Chemicals, Drugs and Biologicals*, 14e. Whitehouse Station: Merck and Co. Inc. (b) McGarrity, J.F. and Zanotti-Gerosa, A. (2010). *Tetrahedron: Asymmetry* 21: 2479–2486.

58 (a) Legerlotz, H. (1929). Verfahren zur herstellung optisch aktiver monooxyphenylalkamine. German Patent DE543529. (b) Legerlotz, H. (1932). Verfahren zur darstellung von aromatischen monooxyaminoalkoholen und deren derivaten. German Patent DE566578. (c) Legerlotz, H. (1933). Verfahren zur umkehrung des drehungssinnes von optisch aktiven phenylalkaminen. German Patent DE585164.

59 Takeda, H., Tachinami, T., Aburatani, M. et al. (1989). *Tetrahedron Lett.* 30: 367–370.

60 Yuan, M.-L., Xie, J.-H., Yang, X.-H., and Zhou, Q.-L. (2014). *Synthesis* 46: 2910–2916.

61 (a) Chen, X., Xiong, F., Chen, W. et al. (2014). *J. Org. Chem.* 79: 2723–2728. (b) Pan, J., Zheng, G.-W., Ye, Q., and Xu, J.-H. (2014). *Org. Process Res. Dev.* 18: 739–743. (c) Gong, X.-M., Zheng, G.-W., Liu, Y.-Y., and Xu, J.-H. (2017). *Org. Process Res. Dev.* 21: 1349–1354.

62 (a) Baumann, K.L., Butler, D.E., Deering, C.F. et al. (1992). *Tetrahedron Lett.* 33: 2283–2284. (b) Zhang, W., Chi, Y., and Zhang, X. (2007). *Acc. Chem. Res.* 40: 1278–1290.

63 Blaser, H.-U., Spindler, F., and Studer, M. (2001). *Appl. Catal. A-Gen.* 221: 119–143.

64 Ager, D.J. and Laneman, S.A. (1997). *Tetrahedron: Asymmetry* 8: 3327–3355.

65 Heck, A.M., Yanovski, J.A., and Calis, K.A. (2000). *Pharmacotherapy* 20: 270–279.

66 Schwindt, A.A., Fleming, M.P., Han, Y.-K. et al. (2007). *Org. Process Res. Dev.* 11: 524–533.

67 Rosenblum, S.B. (2007). *The Art of Drug Synthesis* (eds. D.J. Johnson and J.J. Li), 183–196. New York, NJ: Wiley-Intersience.

68 (a) Busscher, G.F., Lefort, L., Cremers, J.G.O. et al. (2010). *Tetrahedron: Asymmetry* 21: 1709–1714. (b) Sniezek, M., Stecko, S., Panfil, I. et al. (2013). *J. Org. Chem.* 78: 7048–7057.

69 (a)Sasikala, C.H.V.A., Padi, P.R., Sunkara, V. et al. (2009). *Org. Process Res. Dev.* 13: 907–910. (b) Mannam, M.R., Sankareswaran, S., Gaddam, V.R. et al. (2019). *Org. Process Res. Dev.* 23: 919–925.

70 Zheng, Z., Cao, Y., Chong, Q. et al. (2018). *J. Am. Chem. Soc.* 140: 10374–10381.

71 (a) Schäfer, B. (2013). *Chem. Unserer Zeit* 47: 174–182. (b) Plößer, J., Lucas, M., Wärnå, J. et al. (2016). *Org. Process Res. Dev.* 20: 1647–1653.

72 (a) Lawrence, B.M. (2006). *Mint: The Genus Mentha*. New York: CRC Press. (b) Heydrich, G., Gralla, G., Rauls, M. et al. (2009). Method for producing optically active, racemic menthol. WO2009068444.

73 (a) McCormack, P.L. (2016). *Drugs* 76: 387–396. (b) Chen, A.Y., Adamek, R.N., Dick, B.L. et al. (2019). *Chem. Rev.* 119: 1323–1455.

74 (a) Flick, A.C., Ding, H.X., Leverett, C.A. Jr., et al. (2017). *J. Med. Chem.* 60: 6480–6515. (b) Zhu, G., Ye, W., Zheng, H. et al. (2014). New process. WO

2014032627. (c) Halama, A. and Zapadlo, M. (2019). *Org. Process Res. Dev.* 23: 102–107.

75 (a) Landoni, M.F. and Soraci, A. (2001). *Curr. Drug Metab.* 2: 37–51. (b) Harrington, P.J. and Lodewijk, E. (1997). *Org. Process Res. Dev.* 1: 72–76.

76 (a) Zhu, S.-F., Yu, Y., Li, S. et al. (2001). *Angew. Chem. Int. Ed.* 51: 8872–8875. (b) Dong, K., Li, Y., Wang, Z., and Ding, K. (2014). *Org. Chem. Front.* 1: 155–160. (c) Li, J., Shen, J., Xia, C. et al. (2016). *Org. Lett.* 18: 2122–2125. (d) Zhu, S.-F. and Zhou, Q.-L. (2017). *Acc. Chem. Res.* 50: 988–1001.

77 Chen, C., Wang, H., Zhang, Z. et al. (2016). *Chem. Sci.* 7: 6669–6673.

78 (a) Koike, T., Hoashi, Y., Takai, T. et al. (2011). *J. Med. Chem.* 54: 3436–3444. (b) Zlotos, D.P., Jocker, R., Cecon, E. et al. (2014). *J. Med. Chem.* 57: 3161–3185.

79 (a) Fu, X., Guo, X., Li, X. et al. (2013). *Tetrahedron: Asymmetry* 24: 827–832. (b) Fukatsu, K., Uchikawa, O., Kawada, M. et al. (2002). *J. Med. Chem.* 45: 4212–4221. (c) Xiao, S., Chen, C., Li, H. et al. (2015). *Org. Process Res. Dev.* 19: 373–377.

80 (a) Uchikawa, O., Fukatsu, K., Tokunoh, R. et al. (2002). *J. Med. Chem.* 45: 4222–4239. (b) Yamano, T., Yamashita, M., Adachi, M. et al. (2006). *Tetrahedron: Asymmetry* 17: 184–190.

81 Yamashita, M. and Yamano, T. (2009). *Chem. Lett.* 38: 100–101.

82 (a) Maibaum, J. and Feldman, D.L. (2009). *Annu. Rep. Med. Chem.* 44: 105–127. (b) Jensen, C., Herold, P., and Brunner, H.R. (2008). *Nat. Rev. Drug Discov.* 7: 399–410. (c) Siragy, H.M., Kar, S., and Kirkpatrick, P. (2007). *Nat. Rev. Drug. Discov.* 6: 779–780.

83 Boogers, J.A.F., Felfer, U., Kotthaus, M. et al. (2007). *Org. Process Res. Dev.* 11: 585–591.

84 (a) Chen, W., Spindler, F., Pugin, B., and Nettekoven, U. (2013). *Angew. Chem. Int. Ed.* 52: 8652–8656. (b) Chen, W., McCormack, P.J., Mohammed, K. et al. (2007). *Angew. Chem. Int. Ed.* 46: 4141–4144.

85 Sturm, T., Weissensteiner, W., and Spindler, F. (2003). *Adv. Synth. Catal.* 345: 160–164.

86 Li, S., Zhu, S.-F., Zhang, C.-M. et al. (2008). *J. Am. Chem. Soc.* 130: 8584–8585.

87 Scognamiglio, J., Jones, L., Letizia, C.S., and Api, A.M. (2012). *Food Chem. Toxicol.* 50: 562–571.

88 Dobbs, D.A., Vanhessche, K.P.M., Brazi, E. et al. (2000). *Angew. Chem. Int. Ed.* 39: 1992–1995.

89 Abate, A., Brenna, E., Fuganti, C. et al. (2005). *J. Org. Chem.* 70: 1281–1290.

90 Hill, R.K. and Edwards, A.G. (1965). *Tetrahedron* 21: 1501–1507.

91 (a) Perrard, T., Plaquevent, J.-C., Desmurs, J.-R., and Hébrault, D. (2000). *Org. Lett.* 2: 2959–2962. (b) Maruoka, K. (2008). *Org. Process Res. Dev.* 12: 679–697.

92 Fehr, C. and Galindo, J. (2000). *Angew. Chem. Int. Ed.* 39: 569–573.

93 Chapuis, C. and Richard, C.-A. (2018). *Helv. Chim. Acta* 101: e1800063.

94 Crawford, K., Rautenstrauch, V., and Uijttewaal, A. (2001). *Synlett*: 1127.

95 Wiles, J.A., Bergens, S.H., Vanhessche, K.P.M. et al. (2001). *Angew Chem. Int. Ed.* 40: 914–919.

96 Genêt, J.P. (2003). *Acc. Chem. Res.* 36: 908–918.

97 (a) Jiang, T., Chen, W.-W., and Xu, M.-H. (2017). *Org. Lett.* 19: 2138–2141. (b) Naito, R., Yonetoku, Y., Okamoto, Y. et al. (2005). *J. Med. Chem.* 48: 6597–6606.

98 (a) Perlman, N. and Nidam, T. (2007). Processes for preparing solifenacin. WO2007076116A2, (b) Puig, J., Sanchez, L., Masllorens, E. et al. (2008). An improved process for the synthesis of solifenacin. WO2008062282. (c) Dave, M.G., Pandey, B., Kothari, H.M., and Patel, P.R. (2009). Process for preparing chemically and chirally pure solifenacin base and its salts. WO2009087664. (d) Bolchi, C., Pallavicini, M., Fumagalli, L. et al. (2013). *Org. Process Res. Dev.* 17: 432–437. (e) Trinadhachari, G.N., Kamat, A.G., Balaji, B.V. et al. (2014). *Org. Process Res. Dev.* 18: 934–940.

99 Chang, M., Li, W., and Zhang, X. (2011). *Angew. Chem. Int. Ed.* 50: 10679–10681.

100 (a) Blaser, H.-U. (2002). *Adv. Synth. Catal.* 344: 17–31. (b) Blaser, H.-U., Buser, H.-P., Coers, K. et al. (1999). *Chimia* 53: 275–280.

101 Cox, C.D., Breslin, M.J., Whitman, D.B. et al. (2010). *J. Med. Chem.* 53: 5320–5332.

102 Baxter, C.A., Cleator, D., Brands, K.M.J. et al. (2011). *Org. Process Res. Dev.* 15: 367–375.

103 Strotman, N.A., Baxter, C.A., Brands, K.M.J. et al. (2011). *J. Am. Chem. Soc.* 133: 8362–8371.

104 Knowles, W.S. and Noyori, R. (2007). *Acc. Chem. Res.* 40: 1238–1239.

7

Tethered Ruthenium(II) Catalysts in Asymmetric Transfer Hydrogenation

Vijyesh K. Vyas, Richard C. Knighton and Martin Wills

Department of Chemistry, The University of Warwick, Coventry, CV4 7AL, UK

7.1 Introduction: The Rationale Behind Tethered Catalysts Design

The [(η^6-arene)Ru(TsDPEN)Cl] (where TsDPEN = *N*-tosyl-1,2-diphenylethylene-1, 2-diamine) class of half-sandwich complex first introduced by Noyori et al. in 1995 represented a major advance in asymmetric transfer hydrogenation (ATH) [1]. These highly practical reagents have a broad substrate range, and their mechanism has now been extensively studied and is well understood [2]. The sense of enantioselective reduction can also be predicted for many substrates, notably acetophenone derivatives and propargyl ketones bearing a triple bond adjacent to the ketone (Figure 7.1).

The selectivity arises through the combination [3] of (i) a ruthenium hydride formed in predominantly one diastereoisomeric form, thus creating a predictable chiral center at the metal, (ii) a hydrogen bond from the N–H on the basic amine to the carbonyl group of the substrate, (iii) an attractive electrostatic interaction between the positive charge on the hydrogen atoms around the η^6-arene ring and either the aryl or alkyne ring of the substrate, and (iv) a disfavorable interaction between the SO_2 of the TsDPEN group and the substrate aryl ring in the *minor* diastereomer, i.e. leading to the minor product (not illustrated). This is a simplified analysis; however, the model in Figure 7.1 represents a guide for the prediction of enantioselectivity in asymmetric ketone reductions. A more complex pattern of results emerges for enones [4].

Since their initial reports, many derivatives of the Noyori catalysts have been reported, including those commonly termed "tethered" derivatives [5], containing a covalent link between the η^6-arene ring and the diamine ligand. Such catalysts are often highly active, while retaining the same enantioselective directing effects. This additional activity is likely to be the result of a combination of (i) extra stability due to the three-point attachment of the ligand to the metal [6] and (ii) subtle changes to the conformation of the complex which improve the alignment of groups toward

Asymmetric Hydrogenation and Transfer Hydrogenation, First Edition.
Edited by Virginie Ratovelomanana-Vidal and Phannarath Phansavath.

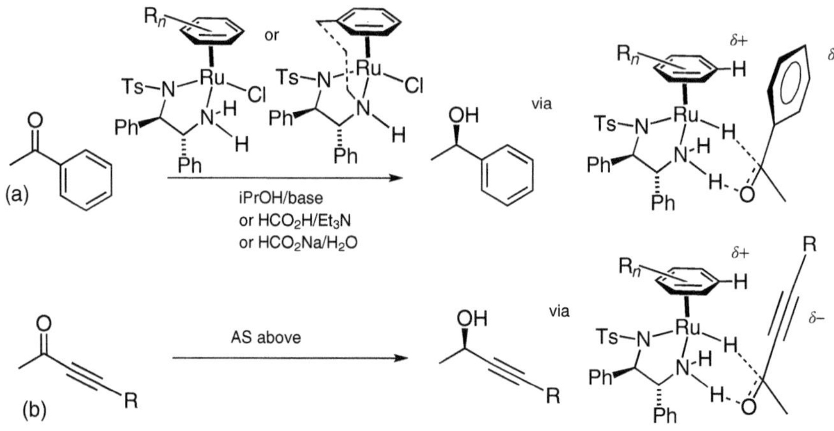

Figure 7.1 Sense of asymmetric reduction of acetophenone (a) and propargylic ketones (b) by [(arene)Ru((*R*,*R*)-TsDPEN)Cl] catalysts and tethered derivatives.

the hydrogen-transfer transition state. The improved activity frequently allows the catalysts to be used at lower loadings than is typical for Ru(II)-based ATH catalysts, and the tethering group also prevents free rotation of the η^6-arene. In several cases, the syntheses of the catalysts have been optimized and scaled up by industrial suppliers, and several tethered catalysts are now commercially available.

7.2 Tethered Ru(II) Catalysts and Their Syntheses

A selection of tethered [(arene)Ru(TsDPEN)X] catalysts which have been described to date are shown in Figures 7.2 and 7.3 [7–29]. Although tethered amino alcohol-containing catalysts were among the first to be reported [7], tethered TsDPEN derivatives proved to be much more efficient, and since the first reports of this type of derivative, the "3C-tethered-TsDPEN" derivatives **B** and **C3** [9, 10] and the majority of the subsequent derivatives have featured a sulfonated 1,2-diphenylethylene-1,2-diamine (DPEN) ligand as the homochiral entity. In most cases the tether is linked to the η^6-arene through the "basic" nitrogen atom of the TsDPEN; however, in some cases (e.g. **B** and **J**) the link is through the sulfonate group. Wills et al. prepared sulfonate-linked **B** first and reported this in 2004 [9]; however, it was a less-active catalyst than **C3**, which we reported in the following year and which has become a popular choice in catalyst screening [10].

In 2011, Ikariya et al., in collaboration with Takasago, reported a modified catalyst **HT** (TsDENEB®) and its derivatives) which contained an ether linkage between the η^6-arene and the TsDPEN component [11]. Wills et al. disclosed their own work on this catalyst in 2012 [27]. A related series of tethered catalysts, **J** and **K**, was also reported by Mohar et al., who are a prolific team of researchers in asymmetric catalysis [12–16]. These catalysts are characterized by the inclusion of a sulfamoylamino (NSO_2N) function, either through the link to the DPEN ligand itself or on the DPEN nitrogen opposite the position of tethering.

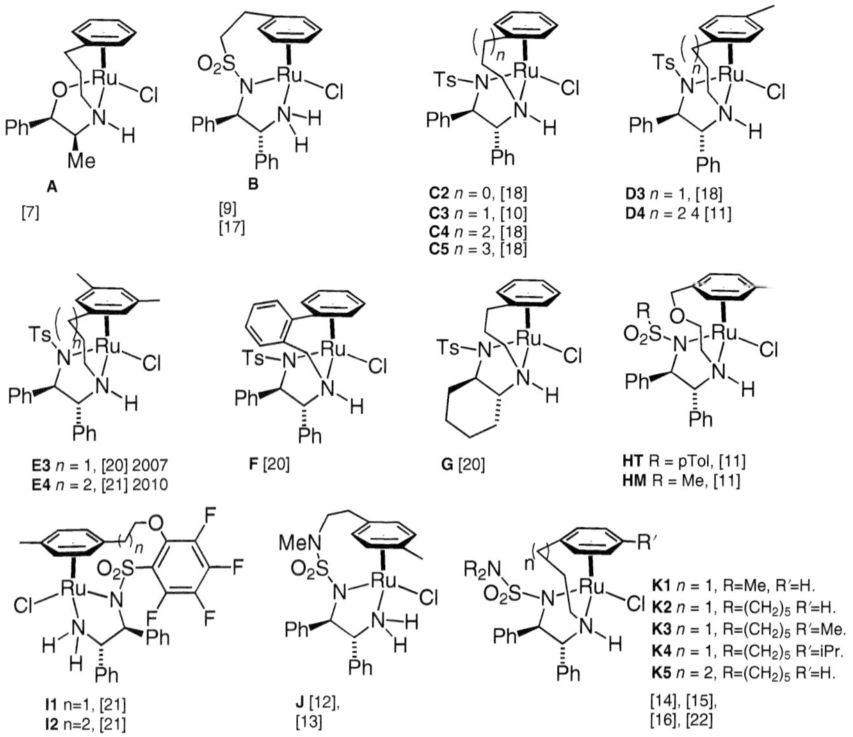

Figure 7.2 Selected tethered catalysts, part 1: principle classes. Source: Cheung et al. [7]; Knighton et al. [8].

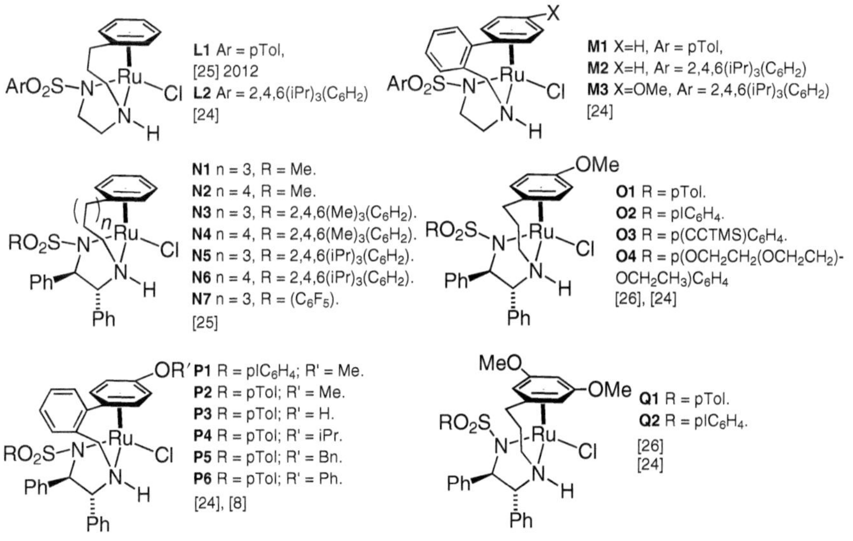

Figure 7.3 Selected tethered catalysts, part 2: racemic and η^6-arene electron rich . Source: Cheung et al. [7]; Knighton et al. [8].

In tethered catalyst derivatives, the linkage is typically three or four atoms in length, and our experience is that a longer or shorter link is less efficient, and the link can include an aromatic ring or a linear chain. The sulfonate group can be varied significantly without detriment to catalytic activity, and the range of groups includes the smaller Ms group, electron-deficient Tf (or otherwise fluorinated) and hindered groups such as Mes (2,4,6 trimethylphenyl) and *Tris* (2,4,6-triisopropylphenyl), and iodotosyl. The range of reported substitution on the η^6-arene ring includes methyl and methoxy groups. However, this remains an area for significant optimization of the catalysts. In some cases the screening of a range of catalysts reveals one derivative which outperforms the other, either in terms of activity or selectivity, or both; however, few strong correlations have yet emerged between the catalyst structure and the substrate class. Tethered catalysts containing iodide in place of the chloride on the metal have been reported [28]. An alternative design of catalysts tethered to the center of the ligand was reported; however, they gave products in moderate ees, up to 72% ee in the best cases (not illustrated) [29].

It is noteworthy that closely related Rh(III)-tethered catalysts have also been reported, indeed predating the Ru(II) versions [30], and tested in ATH and asymmetric hydrogenation (AH), and have given excellent results in many cases [5, 31]. A solid-supported tethered Rh(III) catalyst has been reported [32]. A screen of reduction catalysts is advised to include examples of tethered Rh(III) catalysts. However, these are outside the remit of this chapter and will not be covered in detail here.

7.2.1 Synthetic Approaches to Tethered Catalysts

An outline of the synthetic approaches to tethered catalysts is illustrated in Figure 7.4. In the majority of cases, tethered catalysts are prepared by the attachment of a cyclohexadiene-containing group to a ligand and then reacted with $RuCl_3$ to form an Ru(II) dimer. This dimer can be used directly in ATH applications as it will form the tethered catalyst *in situ* [1]. More commonly, however, the tethered complex is prepared and isolated before use, through treatment of the dimer with a mild base. In the case of DENEB®, the Ikariya/Takasago route involves a substitution reaction of an η^6-arene containing a bromomethyl group [11, 33]. Racemic versions of tethered Ru(II) catalysts have been reported and are valuable for general reductions of otherwise challenging substrates. The ligand precursors of the catalysts can be prepared via amide intermediates followed by reduction (not illustrated), and this approach can be valuable in cases where competing dialkylation products are formed, as in the case of the less sterically hindered monotosylated 1,2-diamino ethane [34, 35].

In some cases, catalysts containing electron-rich η^6-arene groups can be prepared through an "arene-substitution" route [26, 36] in which the arene group on the ligand directly displaces a less electron-rich one in an Ru(II) dimer precursor [8, 24]. This route provides a more rapid access to some catalysts, for example those linked

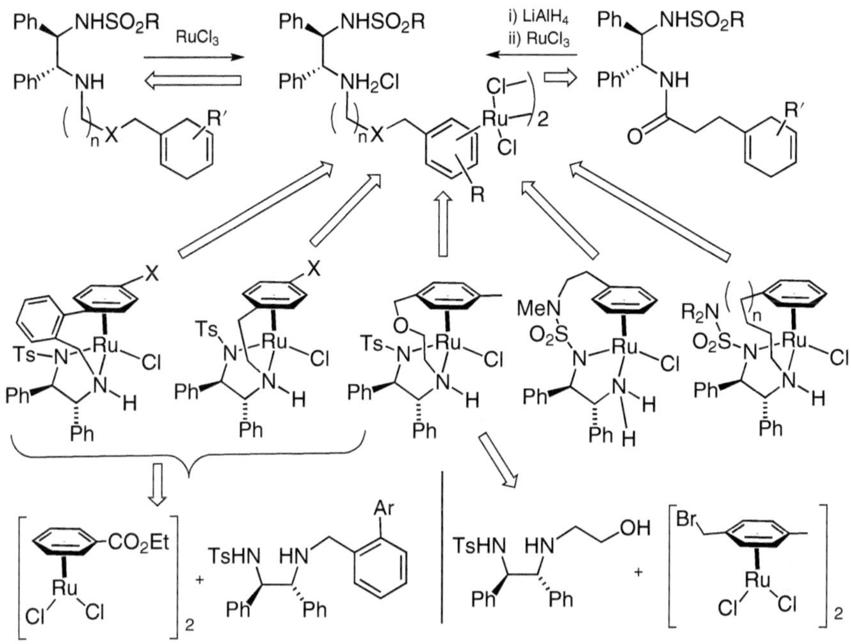

Figure 7.4 Summary of common approaches to the synthesis of tethered catalysts.

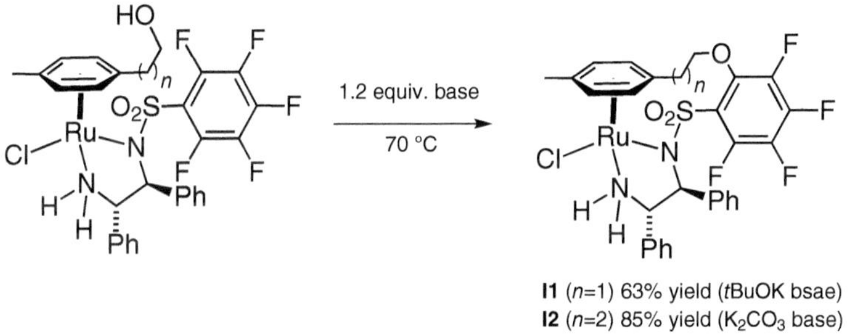

Figure 7.5 Synthesis of tetrafluorophenyl-bridged tethered catalysts **I1** and **I2**. Source: Modified from Matsunami et al. [21].

through a benzyl group, which are otherwise difficult to prepare through the cyclohexadiene route. The recently reported "tetrafluorophenyl-linked" complexes **I1** and **I2** were prepared through selective substitution of a fluoride with the alcohol on a chain attached to the η^6-arene (Figure 7.5) [21]. This catalyst was more active than the nontethered equivalent.

The commercial significance of several of the tethered catalysts is reflected in patenting activity relating to both the synthesis of the catalysts and their applications to specific target molecules [33, 36–39]. Cationic versions of tethered complexes have been reported [40], as has a silica-supported version of the tethered catalysts [41] and other supported versions [42–44].

7.3 Applications to Asymmetric Reductions of Ketones and Imines

7.3.1 Reductions of Acetophenone Derivatives

In the majority of reductions, the reducing agent is an azeotropic 5/2 combination of formic acid and trimethylamine (FA/TEA); examples are given in Figures 7.6–7.10. In some cases, an additional solvent (typically DCM, MeCN, MeOH, or EtOAc) can be used if the substrate is insoluble in FA/TEA. A different ratio of FA/TEA is

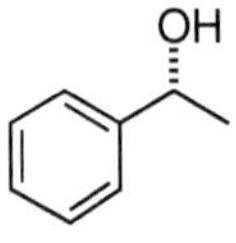

Cat	Loading/%	T/°C	t/h	Conv	Yield	ee	Ref
C3	0.5	40	3	100	-	96	[10]
C4	0.2	40	8	100	-	96	[18]
F	0.5	40	24	100	-	95	[20]
HT	0.1	60	3	-	>99	97	[11]
I1	0.1	60	15	-	97	96	[21]
I1	0.1	60	15	-	99	94	[21]
J	0.5	60	2	>99	-	94	[13]
K1	0.1	60	8.5	100	-	96.2	[14]
N1	0.2	40	5	100	-	96	[25]
N5	0.2	40	24	85	-	96	[25]
O1	0.1	60	2	100	-	96	[26]
O2	1.0	40	7	>99	-	97	[24]
O3	1.0	40	22	>99	-	96	[24]
O4	1.0	40	4	>99	-	97	[24]
O4	1.0	60	1	99	-	97	[24]
O4	1.0	60	2	99	-	97	[24]
P4	1.0	40	24	>99	-	97	[8]
P5	0.5	40	24	100	-	98	[24]
Q1	0.1	60	4	100	-	89	[26]

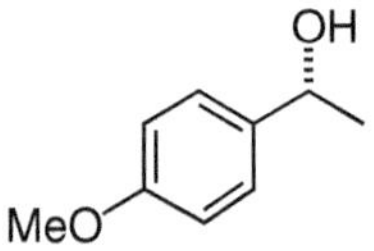

Cat	Loading/%	T/°C	t/h	Conv	Yield	ee	Rref
C3	0.5	40	1.7	100	-	94	[45]
C4	0.2	40	7	96	-	95	[18]
J	1	40	20	95	-	93	[13]
N1	0.2	40	24	100	-	95	[25]
N3	0.2	60	24	91	-	92	[25]
P5	0.25	40	72	100	82	99	[24]
O2	1.0	40	72	94	72	96	[24]

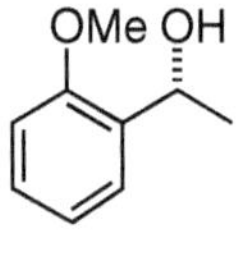

Cat	Loading/%	T/°C	t/h	Conv	Yield	ee	Ref
C3	0.5	40	1.3	100	-	70	[45]
C4	0.2	40	5	96	-	88	[18]
HT	0.1	60	24	-	>99	93	[11]
I1	0.33	40	24	-	85	86	[13]
J	1.0	40	15	100	-	93	[13]
O1	0.1	60	1.5	100	-	96	[26]
P4	1.0	40	23	97	65	87	[8]
P5	0.25	40	72	100	73	87	[24]

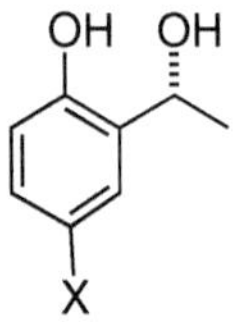

X	Cat	Loading/%	T/°C	t/h	Conv	Yield	ee	Ref
Cl	**C3**	0.5	40	4.8	99	72	90	[47]
Br	**C3**	0.5	40	4	>99	65	91	[47]
OMe	**C3**	0.5	40	5.5	92	24	93	[47]
H	**J**	0.5	40	7	100	-	97	[13]
H	**K**	0.1	60	2	100	-	94.5	[15]
H	**O1**	0.1	60	3	100	-	99	[26]
Cl	**O1**	0.5	40	23	97	94	93	[47]
Br	**O1**	0.5	40	5.5	98	42	94	[47]
OMe	**O1**	0.5	40	6.5	91	84	93	[47]
H	**Q1**	0.1	60	7	100	-	96	[26]

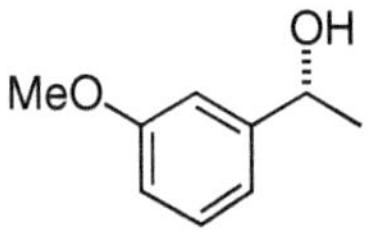

Cat	loading/%	T/°C	t/h	conv	Yield	ee	ref
C3	0.02	40	20	100	-	94	[45]
J	0.5	60	2	100	-	98	[13]

Figure 7.6 First series of ketone reduction products using ATH in FA/TEA.

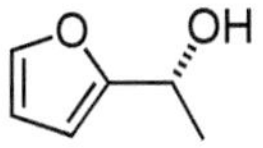

Cat	Loading/%	T/°C	t/h	Conv	Yield	ee	Ref
C3	0.5	40	3	100	-	98	18
O1	0.1	60	2	100	-	98	26
HT	0.1	60	5	-	>99	98	11
K2	0.1	40	15	100	-	98.3	14
Q1	0.1	60	3	99	-	96	26
P2	0.25	45	72	100	59	99	24

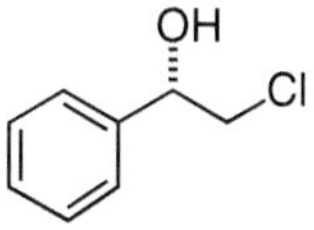

Cat	Loading/%	T/°C	t/h	Conv	Yield	ee	Ref
C3	0.5	40	1.5	100	-	97	[45]
C4	0.2	40	1	96	-	96	[18]
F	0.5	28	24	100	-	92	[20]
HT	0.1	60	5	-	>99	97	[11]
J	0.2	60	24	-	100	96	[21]
N1	0.2	40	3	72	-	>95	[25]
N5	0.2	40	1	100	-	93	[25]
P2	0.25	45	72	100	79	97	[24]
O1	0.1	60	1	>99	-	97	[26]
O2	1.0	45	25	>99	61	88	[24]
Q1	0.1	60	2	99	-	94	[26]

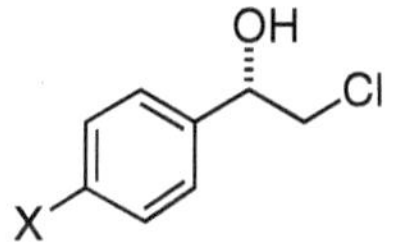

X	Cat	Loading/%	T/°C	t/h	Conv	Yield	ee	Ref
OMe	**C3**	0.5	rt	3	100	-	95	[18]
Cl	**C3**	0.5	rt	3	100	-	93	[18]
OMe	**C4**	0.5	rt	1	100	-	96	[18]
Cl	**C4**	0.5	rt	1	100	-	94	[18]
Cl	**HT**	0.5	28	18	>99	-	96	[27]
OMe	**HT**	0.5	28	18	>99	-	97	[27]

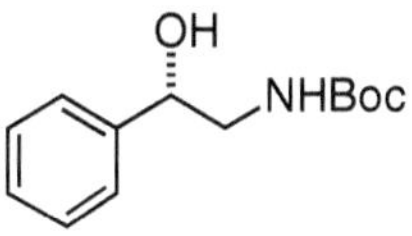

Cat	Loading/%	T/°C	t/h	Conv	Yield	ee	Ref
HT	0.5	28	18	99	-	>99	[27]

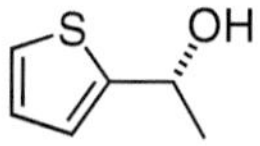

Cat	Loading/%	T/°C	t/h	Conv	Yield	ee	Ref
C3	0.5	40	1.5	100	-	97	45
K2	0.1	40	15	100	-	96.8	13

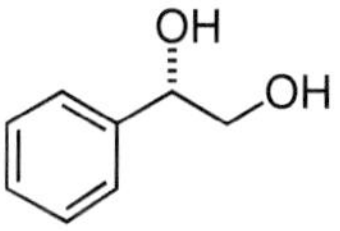

Cat	Loading/%	T/°C	t/h	Conv	Yield	ee	Ref
C4	0.2	40	5	100	-	>95	[18]
HT	0.1	60	5	-	98	96	[11]
J1	0.1	30	24	-	79	90	[21]
J2	0.1	30	24	-	88	90	[21]
N1	0.2	40	6	100	-	>95	[25]
N5	0.2	40	5	99	-	96	[25]
O1	0.1	60	2	100	-	98	[26]
Q1	0.1	60	2	100	-	95	[26]

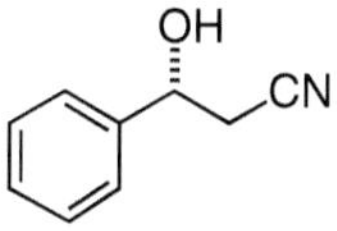

Cat	Loading/%	T/°C	t/h	Conv	Yield	ee	Ref
HT	0.1	60	5	-	>99	95	[11]
J	0.5	60	1	100	-	96	[12]
J1	0.2	30	24	-	96	96	[21]
O1	0.1	60	1	100	-	98	[26]
Q1	0.1	60	2	99	-	93	[26]

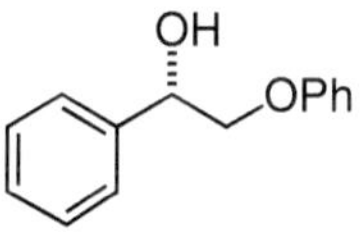

Cat	Loading/%	T/°C	t/h	Conv	Yield	ee	Ref
C3	0.5	40	-	100	-	95	[45]
F	0.5	40	5	99.5	-	91	[20]
HT	0.5	rt	18	>99	-	95	[27]
P2	1.0	45	72	100	75	93	[24]

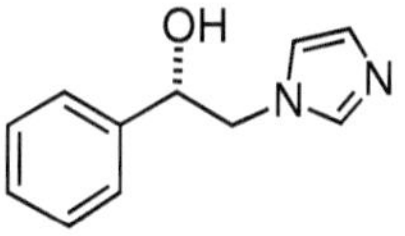

Cat	Loading/%	T/°C	t/h	Conv	Yield	ee	Ref
C3	0.5	40	18	100	-	99	[45]

Figure 7.7 Second series of ketone reduction products using ATH in FA/TEA.

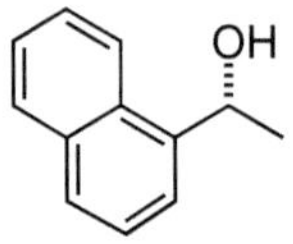

Cat	Loading/%	T/°C	t/h	Conv	Yield	ee	Ref
HM	0.1	60	24	-	96	97	[11]
HT	0.1	60	24	-	94	84	[11]
J	0.5	40	5	-	100	99.9	[12]
O1	0.1	60	3	100	-	99	[26]

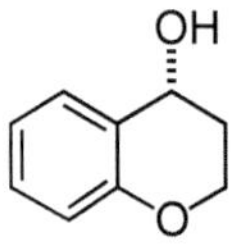

Cat	Loading/%	T/°C	t/h	Conv	Yield	ee	Ref
HT	0.1	40	5	>99	-	>99	[11]
K1	0.1	60	2	100	-	94.4	[14]
O1	0.1	60	3	100	-	99	[26]
P1	0.25	45	72	100	78	99	[8]
Q1	0.1	60	3	-	100	99	[26]

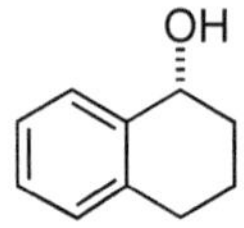

Cat	Loading/%	T/°C	t/h	Conv	Yield	ee	Ref
C4	0.2	40	5	100	-	95	[25]
HT	0.1	60	24	-	85	99	[11]
J	0.5	40	20	100	-	99.9	[13]
K5	0.1	60	3	100	-	99.9	[14]
N1	0.2	40	24	95	-	98	[25]
N5	0.2	40	24	82	-	99	[25]
O1	0.2	60	5	99	-	99	[26]
P1	0.25	-	72	100	86	99	[8]
Q1	0.1	60	5	100	-	99	[26]

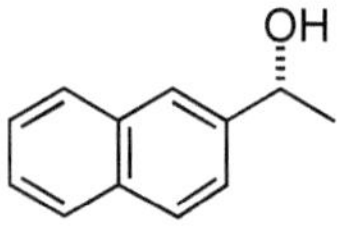

Cat	Loading/%	T/°C	t/h	Conv	Yield	ee	Ref
J	0.5	60	2	100	-	96	[13]
O1	0.1	60	2	100	-	94	[26]

Figure 7.8 Third series of ketone reduction products using ATH in FA/TEA.

sometimes used. Alternative amines, such as DABCO, can be used in place of TEA. An advantage of the use of FA as the hydrogen source for the reduction is that the reaction is driven by the formation of carbon dioxide gas, which makes the reaction largely irreversible. However, some carbon dioxide will remain in solution, and therefore some reversibility, particularly on a large scale, cannot be ruled out.

In some cases, an alcohol, usually isopropanol, is used as both the solvent and the reducing agent, following the precedent of the Meerwein–Ponndorf–Verley reduction reaction. However, the formation of acetone as the side product creates reversibility in this case, and to drive the reaction, high dilutions are required. A diol has been reported as a hydrogen source for reductions: Williams et al. obtained excellent results using tethered catalyst C3 with *cis*-1,4-butenediol as the source of hydrogen for the reduction of acetophenone derivatives at low temperatures and with excellent enantioselectivities [46]. The driving force for this is the formation of a stable lactone following initial oxidation to 4-hydroxybutanal. Tethered catalysts can be used in aqueous solution, in which case sodium formate, or another formate salt, is used as the reducing agent in ATH. In some cases a water-miscible cosolvent (e.g. MeOH) can be used in the reaction to improve substrate solubility.

7.3.1.1 Asymmetric Transfer Hydrogenation Using Formic Acid

As a result of the high activity of the catalysts, they can often be used in reductions of substrates that have proven challenging to untethered versions. An early example of this difference can be found in the reduction of acetophenone using catalyst C3, which can be conducted at an S/C as high as 10 000 [10]. The same catalyst, under the same conditions, was found to be capable of the ATH of the very hindered ketone

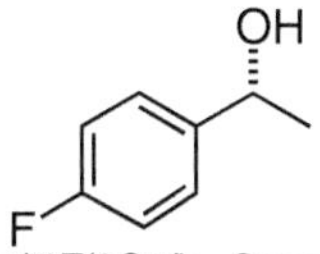

Cat	Loading/%	T/°C	t/h	Conv	Yield	ee	Ref
N4	0.2	40	24	99	-	96	[25]

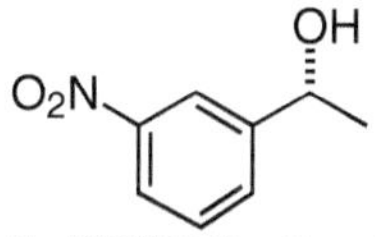

Cat	Loading/%	T/°C	t/h	Conv	Yield	ee	Ref
C3	0.2	40	5	100	-	79	[25]
N2	0.2	40	5	100	-	79	[25]

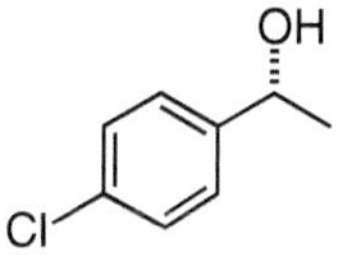

Cat	Loading/%	T/°C	t/h	Conv	Yield	ee	Ref
C3	0.2	40	5	100	-	93	[25]
F	0.5	28	24	100	-	89	[20]
HT	0.5	30	18	>99	-	96	[27]
P1	1.0	45	21.5	99	63	94	[24]
P2	0.25	45	72	100	67	87	[24]

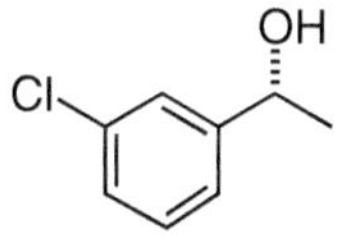

Cat	Roading/%	T/°C	t/h	Conv	Yield	ee	Ref
F	0.5	28	24	97	-	88	[20]
HT	0.5	30	-	>99	-	94	[27]

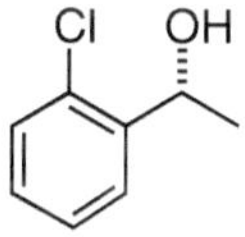

Cat	Loading/%	T/°C	t/h	Conv	Yield	ee	Ref
F	0.5	28	16	100	-	68	[20]
HT	0.5	30	36	100	-	87	[27]
J	0.33	40	24	-	100	86	[21]
N2	0.2	40	5	100	-	86	[25]
P1	1.0	45	23	97	65	87	[24]
P2	0.25	45	72	100	87	90	[24]

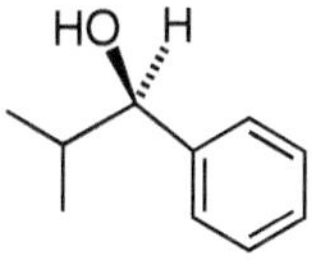

Cat	Loading/%	T/°C	t/h	Conv	Yield	ee	Ref
C3	0.5	40	24	92	-	95	[45]
HT	0.5	30	36	11	-	27	[27]

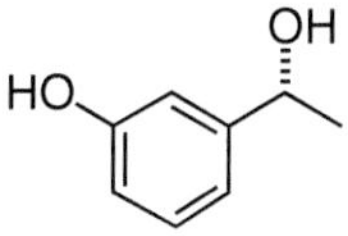

Cat	Loading/%	T/°C	t/h	Conv	Yield	ee	Ref
C3	0.2	40	5	98	-	99	[25]
K2	0.1	60	5	100	-	95.7	[14]
N1	0.2	40	7	100	-	98	[25]
N2	0.2	40	5	100	-	95	[25]
N6	0.2	40	5	96	-	98	[25]

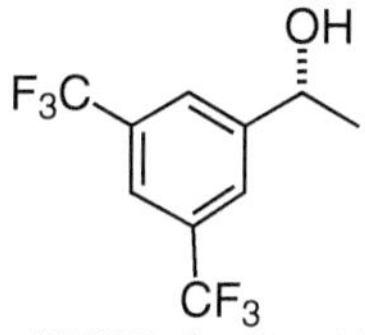

Cat	Loading/%	T/°C	t/h	Conv	Yield	ee	Ref
HT	0.5	30	18	>99	-	60	[27]
N6	0.2	40	5	100	-	81	[25]

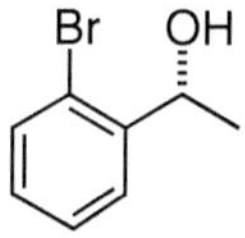

Cat	Loading/%	T/°C	t/h	Conv	Yield	ee	Ref
I1	0.33	40	24	-	97	88	[21]
I2	0.33	40	24	-	90	90	[21]

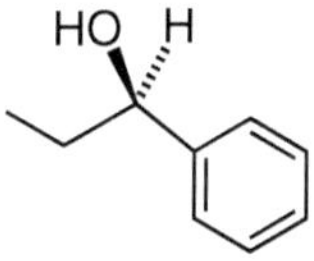

Cat	Loading/%	T/°C	t/h	Conv	Yield	ee	Ref
C3	0.2	40	24	100	-	94	[25]
HT	0.1	60	24	-	>99	94	[11]
N7	0.2	40	24	100	-	95	[25]
O1	0.1	60	3.5	100	-	97	[26]
P1	1.0	45	26	99	51	94	[24]
P2	0.25	45	72	100	86	94	[24]
Q1	0.1	60	5	-	100	88	[26]

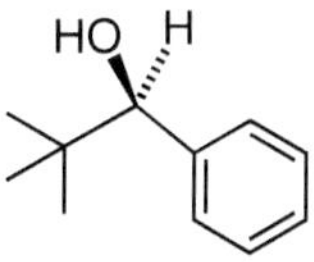

Cat	Loading/%	T/°C	t/h	Conv	Yield	ee	Ref
C3	0.5	40	24	92	-	77	[45]
HT	0.5	30	36	6	-	14	[27]

Figure 7.9 Fourth series of ketone reduction products using ATH in FA/TEA.

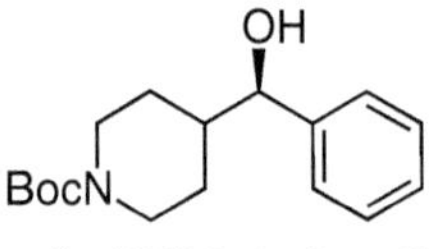

Cat	Loading/%	T/°C	t/h	Conv	Yield	ee	Ref
C3	0.5	40	16	100	-	93	[45]

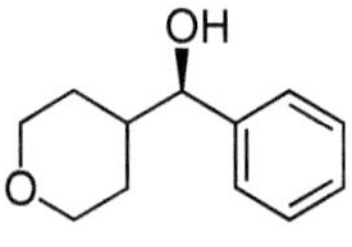

Cat	Loading/%	T/°C	t/h	Conv	Yield	ee	Ref
C3	0.5	40	6	100	-	92	[45]

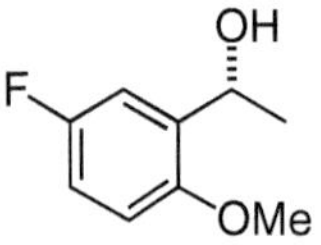

Cat	Loading/%	T/°C	t/h	Conv	Yield	ee	Ref
N2	0.5	40	3	100	-	67	[25]

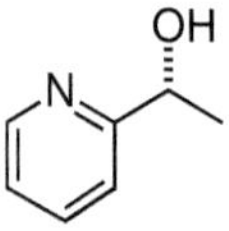

Cat	Loading/%	T/°C	t/h	Conv	Yield	ee	Ref
C4	0.5	rt	2	100	-	91	[18]

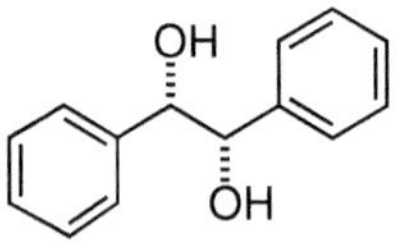

Cat	Loading/%	T/°C	t/h	Conv	Yield	dr	ee	ref
HT	0.1	60	5	-	-	97.2:2.8	>99	[11]
K1	0.1	60	6	100	-	96:4	99	[15]

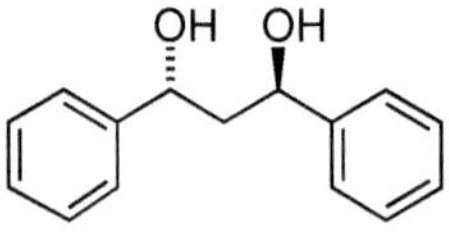

Cat	Loading/%	T/°C	t/h	Conv	Yield	dr	ee	Ref
K1	0.2	60	4	100	98	>99.9:0.1	99.9	[15]

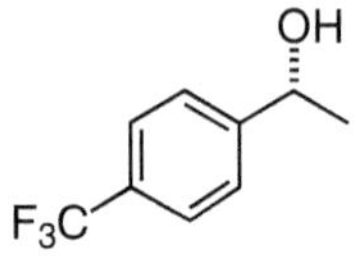

Cat	loading/%	T/°C	t/h	Conv	Yield	ee	Ref
P2	0.25	45	72	100	79	95	[26]

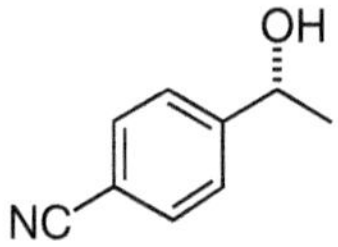

Cat	Loading/%	T/°C	t/h	Conv	Yield	ee	Ref
O1	0.1	60	1.5	99	-	88	[26]

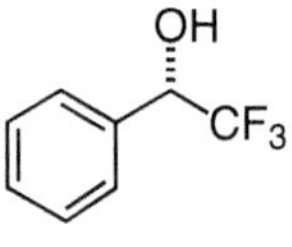

Cat	Loading/%	T/°C	t/h	Conv	Yield	ee	Ref
C4	0.5	40	5	100	-	56	[25]
J1	0.1	60	3	-	100	45	[21]
J1	2.0	0	24	-	100	60	[21]

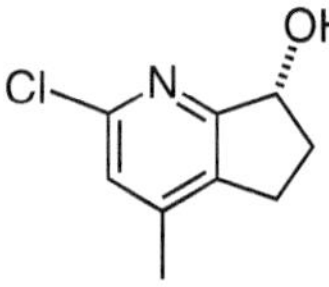

Cat	Loading/%	T/°C	t/h	Conv	Yield	ee	Ref
C3	0.5	rt	3	100	-	95	[18]
C4	0.5	rt	1	100	-	96	[18]

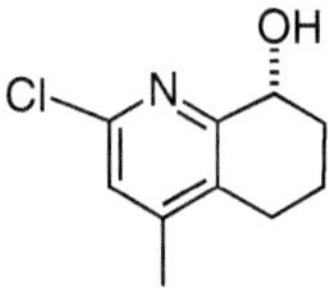

Cat	Loading/%	T/°C	t/h	Conv	Yield	ee	Ref
C3	0.5	rt	3	100	-	94	[18]
C4	0.5	rt	1	100	-	97	[18]

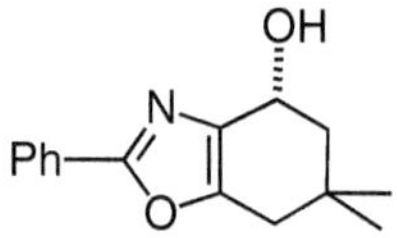

Cat	Loading/%	T/°C	t/h	Conv	Yield	ee	Ref
C3	0.5	rt	3	100	-	94	[18]
C4	0.5	rt	1	100	-	94	[18]

Figure 7.10 Fifth series of ketone reduction products using ATH in FA/TEA.

tBuCOPh in 95% conversion and 77% ee after 32 h, whereas essentially no reduction was observed using an untethered catalyst [10].

The largest group of substrates which have been reduced using Ru(II)-tethered catalysts are acetophenone derivatives, i.e. in which an aromatic ring and a relatively unhindered alkyl group flank the ketone, respectively. A comprehensive set of these is illustrated in Figures 7.6–7.10. *Ortho*-substituted aromatic rings in the substrate can be problematic for reasons of steric hindrance, however; more detail on this observation which can be turned to be an advantage will follow in the section on benzophenone reductions. For acetophenone derivatives, the configuration of the major product follows that outlined in Figure 7.1 for a given configuration of catalyst. The alkyl group can be one of a very wide range of functionalities including *O*- and *N*-containing saturated heterocyclic groups and heteroaromatic groups separated by a saturated carbon atom. Acetophenones containing a range of α-substituents, including halide and heterocyclic groups, are also well tolerated.

Some trends can be observed; catalysts containing 3,5-dimethyl or 3,5-dimethoxy substituents on the η^6-arene ring generally give products in lower ees [24, 26].Most other catalysts work well provided they contain a three to four atom tether. Following screening, some catalysts emerged as the most suitable ones for certain substrates; for example the Mohar catalysts **J** and **K** appear to be well suited for reduction of naphthyl ketones and trifluoromethyl-substituted ketones [12, 15]. However, it is difficult to predict an optimal catalyst for a new substrate, so full screening in any application is still recommended.

In some cases, some unexpected results have been observed. For example, 2-methoxyacetophenone is reduced in just 68% ee using catalyst **C3**, whereas this increases to a remarkable 96% ee using the p-methoxy catalyst **O1** for reasons that are not clear (Figure 7.6) [26]. In contrast, both catalysts reduce 2-hydroxy acetophenones in very high ees. Understanding of such subtle effects helps to inform the design of improved catalysts in the future.

7.3.1.2 Reduction Under Aqueous Conditions

As is the case for the nontethered parent complexes, tethered [(arene) Ru(TsDPEN)Cl] complexes can be used under aqueous conditions, or in combinations of water and a compatible solvent such as methanol, provided that the ketone substrate benefits from sufficient solubility [47]. The use of the water/methanol solvent combination, with sodium formate as the hydrogen source, appears to be particularly valuable for the reduction of electron-rich acetophenone derivatives, which are otherwise very slowly reduced substrates under FA/TEA conditions (Figure 7.11).

Ikariya et al. have demonstrated the efficient reduction of α-halo and α-oxo ketones using TsDENEB®, achieving this through use of HCO_2H/ HCO_2K as the hydrogen source, in place of HCO_2H/Et_3N [48]. The authors synthesized a range of derivatives, of up to 97% ee, which can be used to access a range of synthetically valuable 2-amino-1-arylethanols, key intermediates in β-adrenergic receptor agonists.

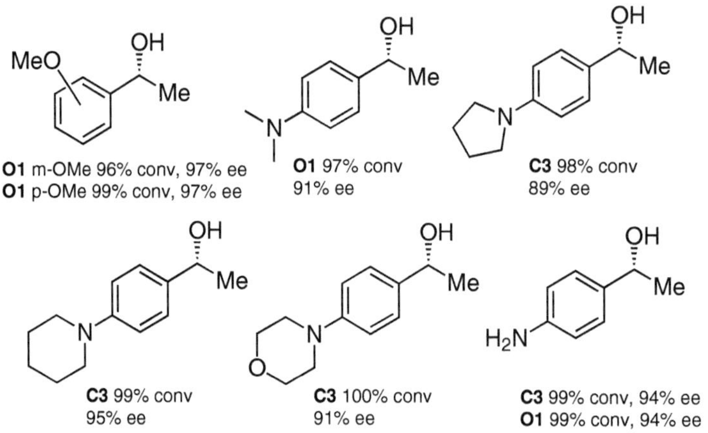

Figure 7.11 Products of ATH of electron-rich acetophenones under aqueous conditions with tethered catalysts (using 1 mol% catalyst **C3** or **O1**, 60 °C, MeOH/H_2O, and sodium formate).

7.3.1.3 Hydrogenation with Hydrogen Gas

AH reactions, usually carried out under neutral conditions in a solvent such as methanol, can be achieved using tethered complexes. Although not illustrated, examples have been reported with the 3C-tethered complex **C3** [23], its OMe derivatives **O1** and **Q1** [26], and DENEB® **HT** [11]. An interesting transformation of alkynes into chiral alcohols via TfOH-catalyzed hydration followed by Ru-catalyzed tandem AH was reported by Zhou et al. [49, 50]. (*S,S*)-TsDENEB® **HT** was tested in this application and gave good results although the untethered catalysts were taken forward for detailed studies in this instance.

7.3.1.4 Racemic Catalysts for Reductions

Racemic tethered Ru(II) catalysts can be used productively in ketone and aldehyde reductions. The 3C-tethered catalyst **L1** reduces several substrates in the presence of functional groups such as nitro, cyano, and aromatic halides, which are not themselves reduced [23, 34]. The recently reported BnTsEN, BnTrisEN, and BnTrisMeOEN complexes (**L1, L2,** and **M1–M3**) [24] are capable of the racemic reduction of acetophenone. Using FA/TEA, reductions were complete at 60 °C at 0.2 mol% loading in four to five hours and 99–100% conversion. Reductions could also be carried out in water using sodium formate as the hydrogen source within 1.5–2.5 h at 60 °C. The two Tris catalysts also served to achieve hydrogenation in methanol at 60 °C within 16 h under 30 bar hydrogen, with full conversions achieved using just 0.2 mol% catalyst.

7.3.1.5 Specific Applications to Complex Acetophenone Derivatives

In a work reported by Teijin Pharma [51], a range of tethered catalysts were used for reductions of hetereoaromatic ketones containing alpha Br, Cl, or amino groups for the preparation of β_2-adrenergic receptor agonists with the amino alcohol structure. Products were formed in high ees, typically 90% ee or above, and good conversions.

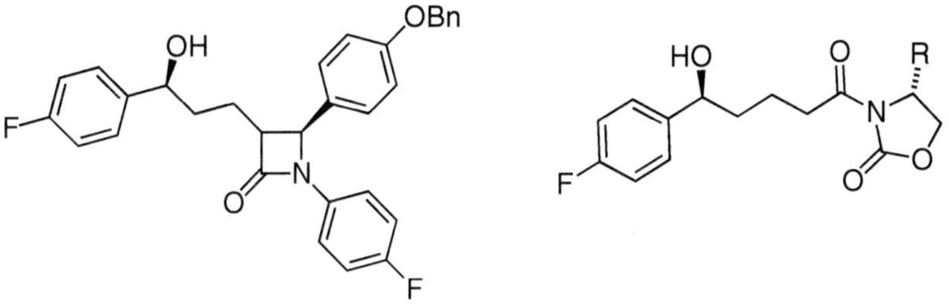

Figure 7.12 Intermediates formed in ATH in two routes used in the synthesis of Ezetimibe. Source: Zou et al. [52]; Zhang et al. [53].

Tethered catalysts have been applied to the synthesis of pharmaceutically active Ezetimibe, an alternative to statins for lowering cholesterol. Two alternative routes have been developed, both of which make use of ruthenium-catalyzed ATH with tethered catalysts (Figure 7.12) [52, 53].

Tethered catalysts, alongside other ATH catalysts, were screened in a synthesis of Lorlatinib, a drug for the treatment of non-small cell lung cancer, by Pfizer [54] and gave excellent results; 99.9% conversion and 99.7% ee in the best case (Figure 7.13).

Several applications of reductions using tethered catalyst to specific targets have been reported, notably to pharmaceutical targets. Catalyst **C3** was employed in a late-state reduction in the synthesis of bronchodilators for chromic obstructive pulmonary disease (COPD) [55]. The reduction occurs in the penultimate step, confirming the catalysts selectivity and tolerance to a range of functional groups (Figure 7.14).

The synthesis of Montelukast (singulair) – a drug for treating asthma and rhinitis – has been reported by Synthon BV, utilizing the 3C-tethered catalyst to

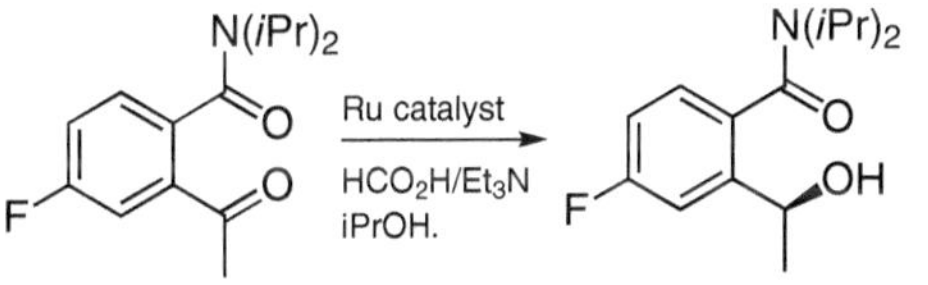

Catalyst	Conv./%	ee/%
[(p-cymene)Ru(*S*,*S*)-TsDPEN)Cl]	100	98.5
C4-(*S*,*S*)-teth-TrisDPEN **N6**	100	99.0
C4-(*S*,*S*)-teth TsDPEN **C4**	99.9	99.7
C3-(*S*,*S*)-teth MtsDPEN **N3**	100	96.4
C4-(*S*,*S*)-teth MsDPEN **N2**	100	86.5

Figure 7.13 Use of tethered catalyst alongside nontethered ones in ketone ATH.

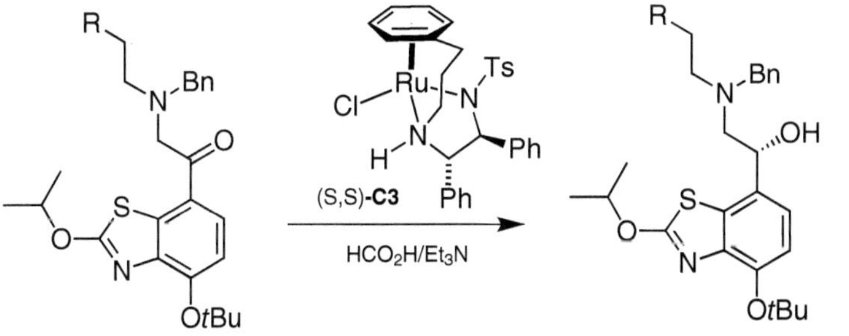

Figure 7.14 ATH step in the synthesis of a bronchodilator drug by AstraZeneca.

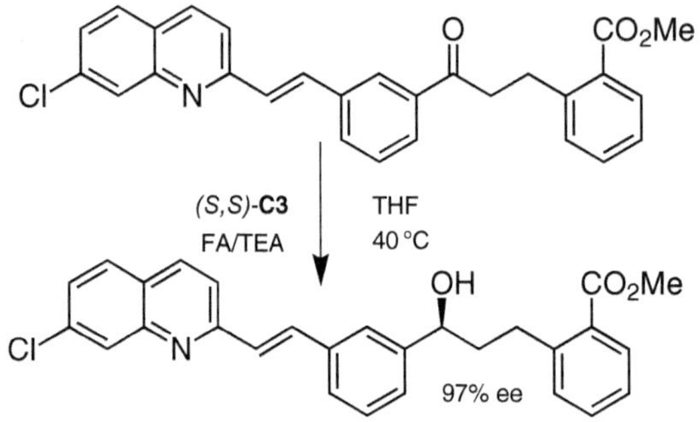

Figure 7.15 ATH using tethered catalyst in the formation of an intermediate to Montelukast. Source: Based on Vinent and Thijs [56].

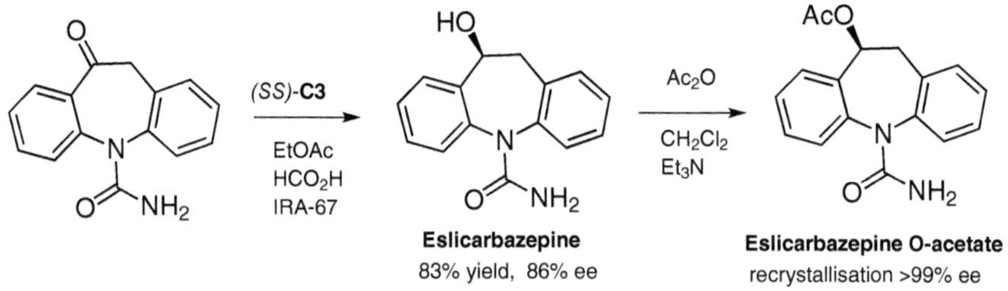

Figure 7.16 ATH to produce eslicarbazepine.

reduce a ketone to obtain the corresponding enantioenriched alcohol precursor (Figure 7.15) [56].

Archimica has reported the synthesis of eslicarbazepine acetate (an antiepileptic drug) using ATH at a critical step to create the required chiral center in the product. The use of (*S,S*)-**C3** and a weak base yielded the desired eslicarbazepine in good enantiomeric excess [57], which could be acetylated and recrystallized to >99% ee (Figure 7.16).

The synthesis of eslicarbazepine was also reported using TsDENEB® **HT** as the tethered catalyst [58]. Ranbaxy scientists [59, 60] also reported an analogous synthesis using (*S,S*)-**C3**. Boehringer Ingelheim GmbH (Germany) [61] has reported the utilization of the 3C-tethered catalyst for the production of key intermediates in the synthesis of a family of CCR2 antagonists. A representative example is provided (Figure 7.17 left), whereby the (*S,S*)-**C3** catalyst reduced a 4-trifluoromethylacetophenone derivative on a 40-g scale with excellent catalyst loadings (S/C (w/w) = 2000 : 1, 20 mg required for the reduction).

In the synthesis of adrenergic receptor agonist (Figure 7.17 right), MsDENEB® **HM** proved to be more enantioselective in the reduction than TsDENEB® (95%

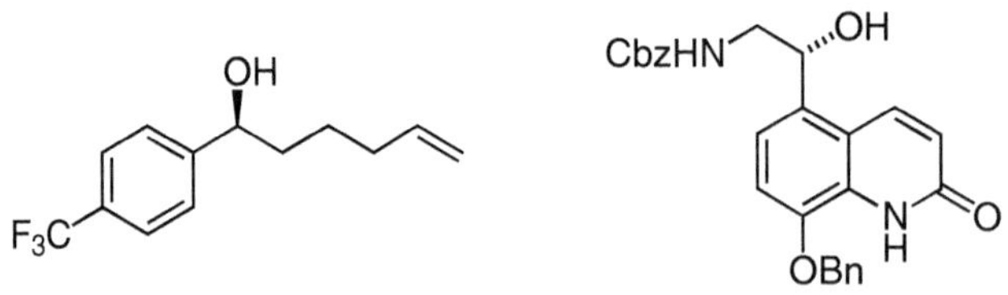

Figure 7.17 Pharmaceutical intermediates prepared by ATH of ketones.

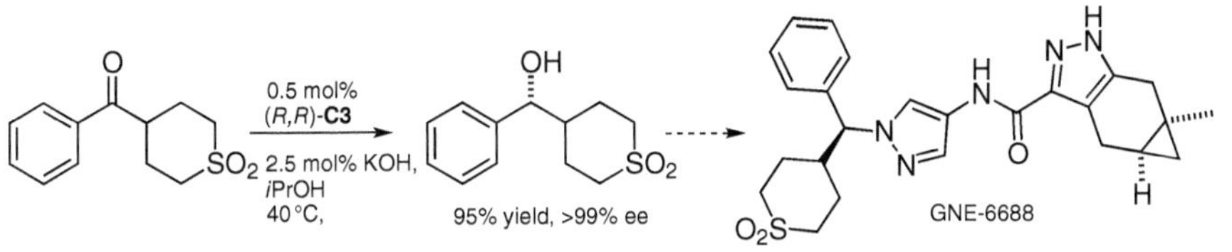

Figure 7.18 Synthesis of intermediate to an ITK inhibitor using ATH.

ee vs. 75% ee) [62]. In a recent study [63], Genentech and Pharmaron Beijing Ltd reported the synthesis of an inhibitor of interleukin-2-inducible T-cell kinase (ITK) (Figure 7.18). Key to the synthesis was the ATH of a heterocyclic ketone with catalyst **C3**, which gave a product of 99% ee in 95% yield. This was elaborated to the product through a substitution with inversion of configuration.

7.3.2 Reductions of Acetylenic Ketones

Examples of products of asymmetric reduction of acetylenic ketones formed using tethered complexes include diketones and β-, γ-, and δ-ketoesters [64, 65]. Substrates containing two conjugated triple bonds adjacent to the ketone are very sensitive, and a relatively high catalyst loading (10 mol%) is required for their reduction, but a successful concise application to the synthesis of panaxjapyne has been reported [66].

7.3.3 Reductions of Benzophenone Ketones

Several reports have been published on the applications of ATH with tethered catalysts to the asymmetric reduction of unsymmetrical benzophenone derivatives and related compounds (including ketones flanked by heteroaromatic groups). An early example [67] was a racemic reduction in which high catalytic activity was required. Eli Lilly, in collaboration with Johnson Matthey, described the reduction using just 0.01 mol% catalyst **L1** [23]. A separate catalytic hydrogenation completed the selective removal of the ketone group to give the required product. In a patented application, by Ambit Biosciences, of the asymmetric reduction of a challenging heterocyclic/aryl ketone substrate precursor, tethered catalysts were screened alongside a range of other catalysts, giving products of up to 64% ee [68]. The 3C-tethered catalyst **C3** has been screened for ATH of challenging class of tricyclic ketones, i.e. *N*-substituted dibenzo[b,e]azepine-6,11-dione having an amide linkage in the skeleton (Figure 7.19) [69]. Although the arene ring in the skeleton appears to offer little scope for differentiation, the corresponding chiral alcohols were produced in up to 94% ee in the example shown and in >99% ee in the best case for a derivative.

Ikariya et al. demonstrated the power of electronic and steric discrimination in biaryl ketone reduction by substitution at the *ortho* position at one arene ring, producing products of excellent ee (>99% in the best cases) (Figure 7.20) [70]. Reagents containing electron-withdrawing substituents including −F, −Cl, −Br, and $-CF_3$ at the *ortho* position of one ring gave products of at least 90% ee in all cases, whereas those with an −OH at the *ortho* position could only produce products

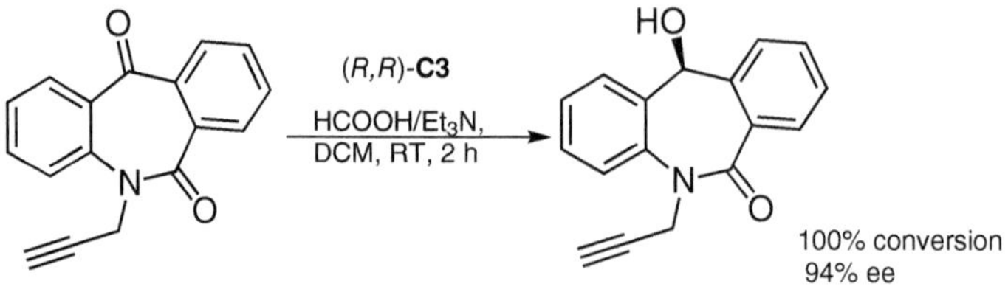

Figure 7.19 ATH of *N*-substituted dibenzo[b,e]azepine-6,11-dione.

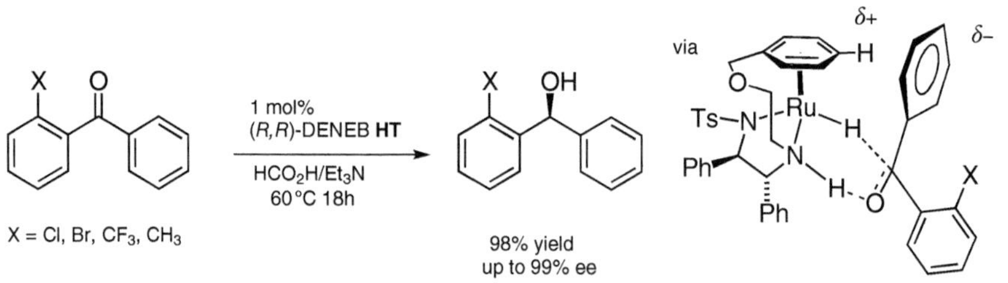

Figure 7.20 Reduction of unsymmetrical benzophenones and control of reduction selectivity by a combination of steric and electronic factors. Source: Based on Touge et al. [70].

in 77% ee. This synthetic method gave access to easy synthesis of optically active 6-phenyl-6*H*benzo[c]chromenes which are otherwise difficult to prepare and was tested with benzoylferrocene and 3-nitrophenyl 2-thienyl ketone to good effect. The origin of the enantioselectivity is believed to arise from an out-of-plane twisting of the *ortho*-substituted ring which reduces the propensity for it to engage in the established edge/face η^6H/aromatic ring interaction (Figure 7.20).

Tethered catalyst **HT** was also applied to the ATH of aryl *N*-heteroaryl ketones using sodium formate as a hydrogen source [71–73]. The stereochemical control produced products in excellent enantioselectivities (up to 99.9%) (Figure 7.21). 2-Pyridines gave the best results when the opposing ring contained an *ortho*-substituent. However, an arene ring without *ortho* substituents gave chiral products when the pyridine ring was converted to the *N*-oxide (up to 97.8% ee).

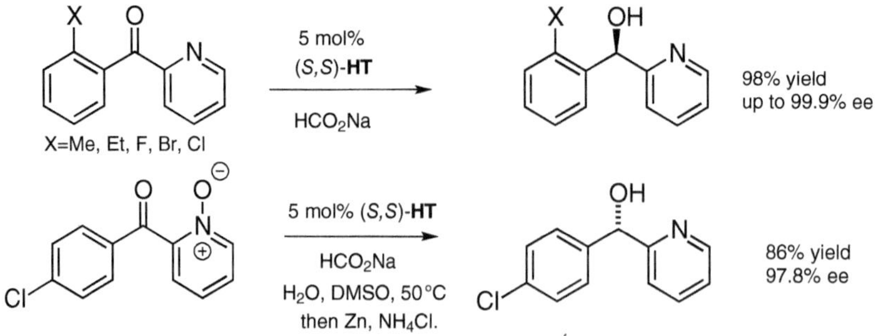

Figure 7.21 ATH of pyridyl ketones and pyridyl *N*-oxide ketones.

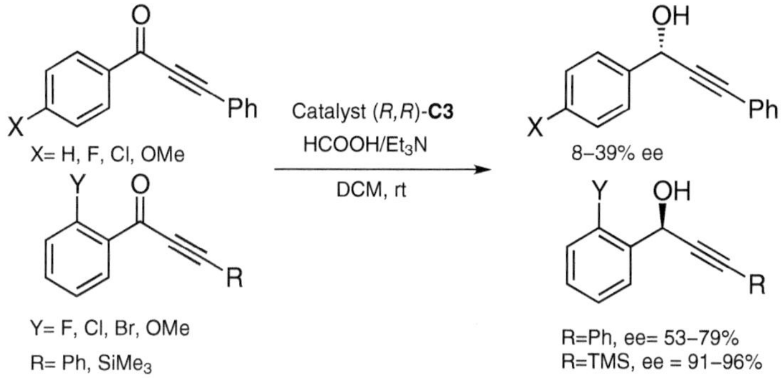

Figure 7.22 Optimized substrate structure for asymmetric reductions. Source: Based on Vyas et al. [74].

Wills et al. reported the synthesis of propargylic alcohols in high enantiomeric excess by fine-tuning the electronic and steric factors of the substrates (Figure 7.22) [74]. The introduction of a substituent at the *ortho* position of the substrate resulted in reversal of the enantioselectivity, whereas replacing the phenyl with trimethylsilyl group produced a dramatic rise in ee. This method was applied to the synthesis of an allocolchicine analogue in high ee.

7.3.4 Reductions of Diverse Ketones

Dialkyl ketones are challenging substrates for reduction by [(arene)Ru(II)Cl (TsDPEN)] catalysts, presumably because of lack of appropriate electron-rich regions to interact with the η^6-arene group of the catalyst. There are some examples in which improved selectivities have been obtained. The 3C-tethered catalyst **C3** will reduce acetyl cyclohexanone in 69% ee, whereas this is improved to 89% ee when the derivative with two Me groups on the η^6-arene ring, i.e. **E3**, is used [18]. The S product is formed when the (*R*,*R*) catalyst is employed. Likewise, complex **Q1**, bearing two methoxy groups, gives product of higher ee than the p-OMe catalyst **O1**, again suggesting the operation of a steric effect [26]. However, very little systematic work has been carried out on ATH of dialkyl ketones using the [(arene)Ru(TsDPEN)Cl] class of catalyst, either tethered or untethered, and this remains an important challenge for future research work. While α-tetralones and related compounds are excellent substrates for ATH, the situation for β-tetralones, with the aromatic ring set further back from the ketone, is less obvious. Beta-tetralone itself is reduced in just 88% ee. However, and surprisingly, substrates with additional substituents between the ketone and the aromatic ring are reduced in higher ee (Figure 7.23) [75].

Closely related to β-tetralones is an interesting example of a ketone ATH by Shipman et al. promoted by tethered 3C catalyst **C3** (Figure 7.24a) [76]. This challenging substrate requires not only regioselectivity in terms of the ketone which is reduced but also enantioselectivity.

The asymmetric reduction of 2,2-dimethyl-6-(2-oxoalkyl/oxoaryl)-1,3-dioxin-4-ones has been achieved using a tethered catalyst (Figure 7.24b) [65]. The products

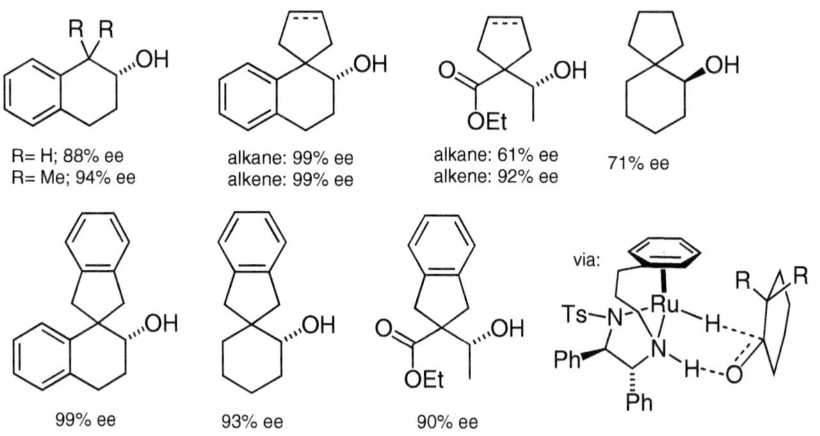

Figure 7.23 Products of ATH of hindered β-tetralone derivatives. Source: Based on Soni et al. [75].

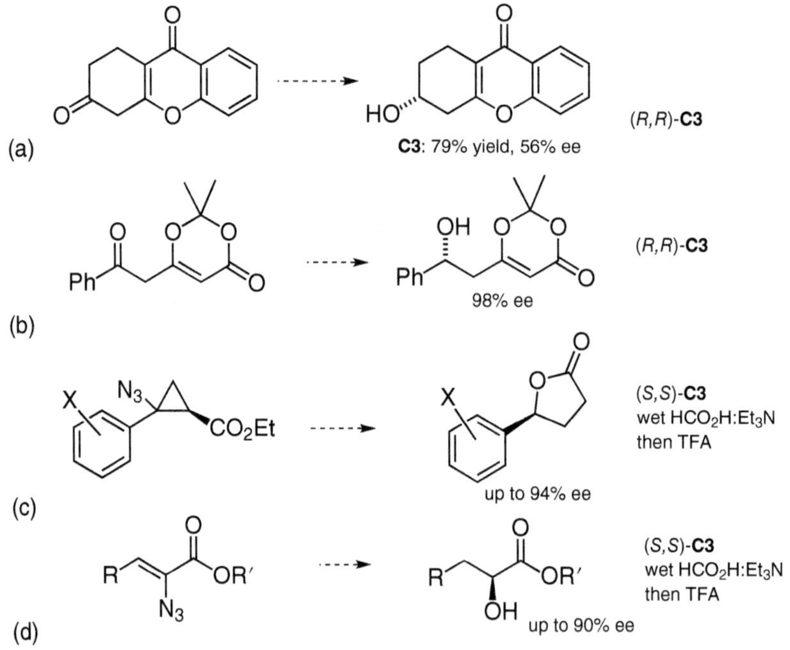

Figure 7.24 (a) Selective ATH reduction in the synthesis of kigamycin A. Source: Based on Turner et al. [76]. (b) Reduction of 2,2-dimethyl-6-(2-oxoalkyl/oxoaryl)-1,3-dioxin-4-ones. Source: Based on Fang et al. [65]. (c) Reductive cyclopropane opening and reduction. Source: Based on Su et al. [77]. (d) Reductive conversion of vinyl azide to a chiral alcohol. Source: Based on Ji et al. [78].

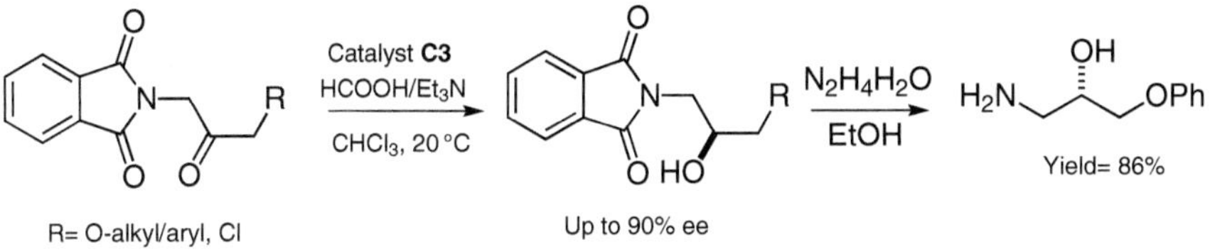

Figure 7.25 ATH of ketones flanked by alkoxy and phthalimide groups. Source: Based on Chew and Wills [80].

can be converted into 5-hydroxy-3-ketoesters which are commonly found in statin-type pharmaceuticals, although in the report on this work the final targets were enantiomerically enriched diols and triols. A combination of azide reduction and imine hydrolysis followed by ATH of the resulting ketone using tethered catalyst **C3** has been used to create a chiral lactone (Figure 7.24c) [77] and α-hydroxy esters (Figure 7.24d) [78]. Trichloromethyl ketones make remarkably good substrates for ATH using both tethered and untethered complexes (not illustrated), with products of typically 96–99% ee from the alkyl/trichloromethyl substrates [79]. Aromatic/trifluoromethyl ketones, such as $PhCO(CF_3)$, however, remain a challenging class of substrate which has been reduced in only 56% ee to date using a tethered catalyst [25].

The ATH of aminomethyl/aryloxy/alkoxymethyl ketones to produce β-amino alcohols is challenging due to lack of differential electronic or steric densities immediately flanking the carbonyl group. Wills et al. demonstrated the use of tethered catalyst **C3** for the synthesis of optically active β-amino alcohols by masking the amino functionality with the phthalimide moiety and reducing the electron density from nitrogen – giving rise to the formation of one enantiomer predominantly (Figure 7.25) [80].

The ATH of 1,3-alkoxy/aryloxy propanones represents an even more challenging target, given the extremely close structural relationship between the groups flanking the ketone. Using tethered arene/Ru(II)/TsDPEN complexes such as **C3** to investigate the effect of more distal electronic and steric effects within the substrate revealed some subtle directing effects (Figure 7.26) [81]. A variety of substrates with combinations of aryloxy and alkoxy substituents were subjected to ATH, and products of low to moderate ees were observed when only electronic factors were taken into account. However, improved ees (up to 68%) were observed in cases where electronic and steric effects are matched to the catalyst.

Chen, Wills et al. reported the ATH of racemic 1,2-(α-ketotetramethylene)ferrocene via a remarkably selective and controlled kinetic resolution method using tethered catalyst **C3** (the untethered catalyst also worked in this application) [82]. The corresponding ketone and alcohol were both produced with high ee in the kinetic resolution (Figure 7.27).

Ratovelomanana-Vidal and Phansavath et al. screened the tethered catalyst for the synthesis of α,α-dichloro- and α,α-difluoro-β-hydroxy esters and amides from their corresponding ketones [83]. Tethered catalysts were applied to the ATH of benzyl 2,2-dichloro-3-oxobutanoate and gave products of up to 99% ee, although an

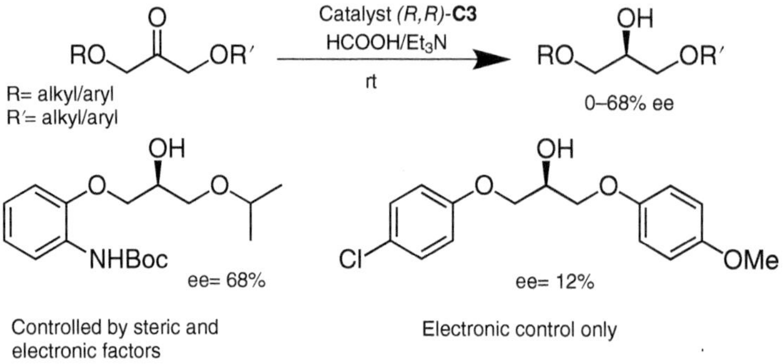

Figure 7.26 ATH of challenging 1,3-dialkoxy propanones can be improved by matching electron and steric directing effects. Source: Based on Forshaw et al. [81].

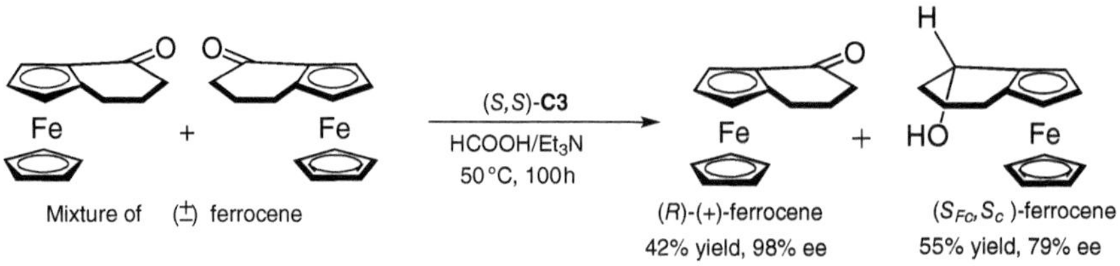

Figure 7.27 Kinetic resolution of ferrocenyl ketones.

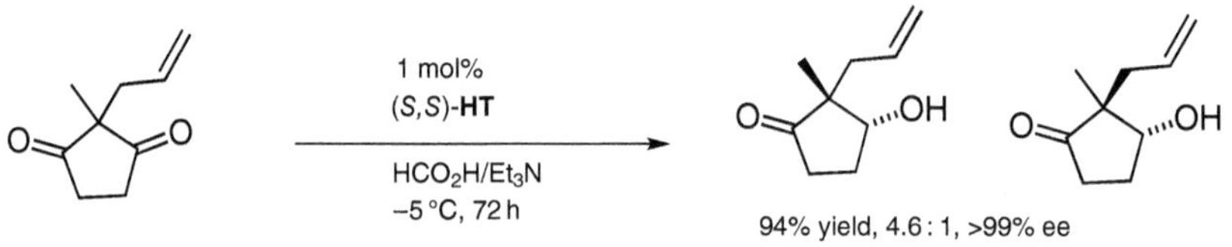

Figure 7.28 Selective ATH of a 1,3-cyclopentadione.

untethered catalyst was applied to study substrate scope. A clever application [84] of the selective reduction of one ketone of a 1,3-cyclopentadione, containing a quaternary center to avoid enolate formation between the carbonyls, used TsDENEB® **HT**. The process creates two chiral centers as a result, the ee being very high, with the diastereomeric ratio c. 4.6 : 1 (Figure 7.28).

In another clever application of a selective reduction which creates a new chiral center at the alcohol and also at the hetereocyclic rings, tethered catalyst **C3** was applied to the selective reduction of one ketone in a symmetrical diketone (Figure 7.29) [85].

A robust and effective enantioselective synthesis of the potent anti-HIV nucleoside EFdA was demonstrated by McLaughlin and coworkers [86, 87]. A key structural feature of EFdA includes an alkyne-bearing stereocenter at the 4-position along with the synthetic challenge of introducing the 3-hydroxy-bearing stereocenter. ATH was used at a key step to produce the 3-hydroxy stereocenter;

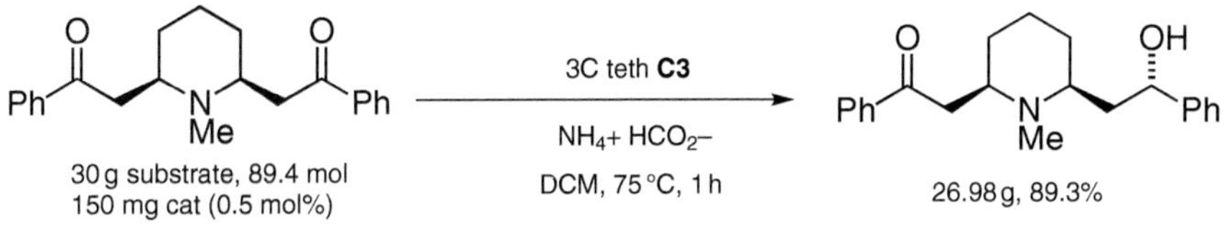

Figure 7.29 Selective reduction of a diketone which creates several chiral centers. Source: Based on Wensen [85].

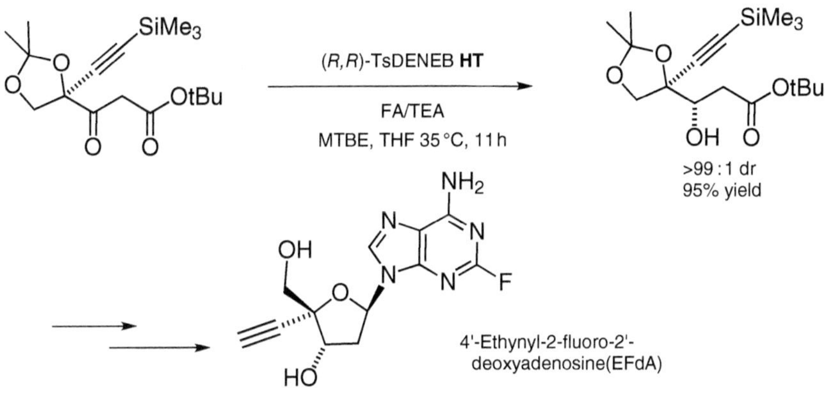

Figure 7.30 ATH of β-keto ester precursor of EfdA.

(R,R)-TsDENEB® **HT** catalyst was in this case used in FA/TEA to obtain the desired isomer of β-hydroxyester with a dr of >99 : 1 (Figure 7.30).

Lianyungang Duxiang Chemicals Co. Ltd., China, reported the preparation of an intermediate for a product that features use of the 3C-tethered Ru(II) catalyst **C3**. The target is prezista (darunavir) that is used as an anti-HIV therapeutic (Figure 7.31) [88].

Syngenta disclosed a clever asymmetric reduction/kinetic resolution (Figure 7.32) en route to the synthesis of a fungicide [89]. TsDENEB® **HT** was used and is listed alongside other reagents including Ru/diphosphine catalysts.

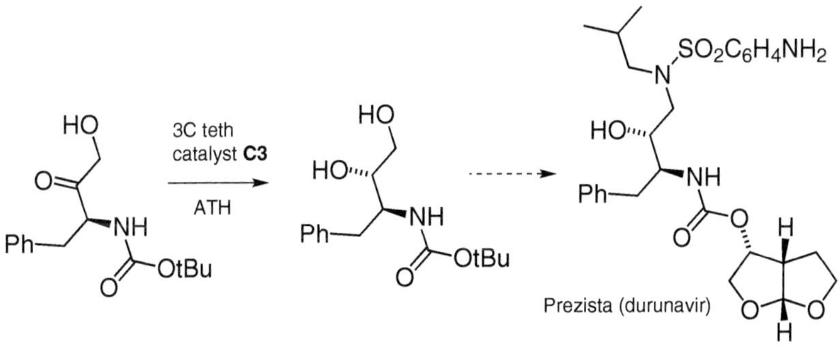

Figure 7.31 Asymmetric reduction of α-amino ketone. Source: Based on Huang et al. [88].

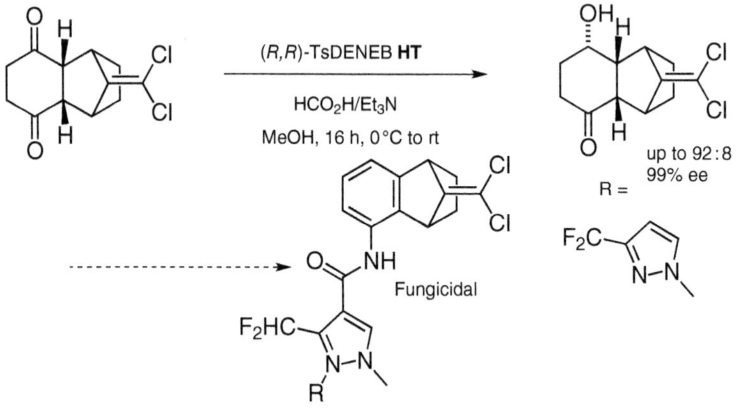

Figure 7.32 Asymmetric reduction of a symmetrical diketone en route to a fungicidal target.

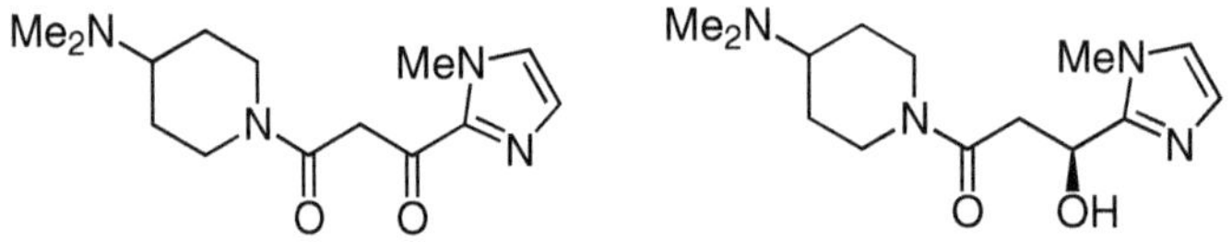

Figure 7.33 β-keto amide and its DENEB® reduction product. Source: Based on Takahashi et al. [95].

Eli Lilly used TsDENEB® **HT** for the efficient and selective ATH of a ketone in a nucleoside-related pharmaceutical precursor molecule [90, 91]. Cyclopropyl-substituted allyl alcohols have been prepared through ATH of ArCH—CH—CO-cyclopropyl ketones [92, 93]. TsDENEB® **HT** was the only tethered catalyst screened (not illustrated). BASF reported a specific reduction of a cyclic ketone containing multiple conjugated alkenes without concomitant reduction of the double bonds using ATH [94]. An unusual reduction of a ketone attached to an imidazole was reported by Toray Industries (Figure 7.33) [95]. 1-(4-(Dimethylamino) piperidin-1-yl)-3-(1-methyl-1*H*-imidazol-2-yl) propane-1,3-dione was reduced by ATH in isopropyl alcohol using 2.5 mol% TsDENEB® **HT** (18 h, 80 °C) to give 2.45 g of the alcohol product (81%). The product was crystallized and characterized.

7.3.5 Dynamic Kinetic Resolutions

Dynamic kinetic resolution (DKR) reactions can create multiple chiral centers in one step but require a careful match of racemization to reduction rates. An early example of the DKR–ATH of a 2-substituted cyclic ketone (tetralone derivative) was reported by LEK Pharmaceuticals. The report documents the ability of 3-C-tethered catalyst **C3** to produce two adjacent stereocenters in one ATH step [96]. Through careful control of the reaction conditions, scientists at Merck and Co. optimized the DKR of a Boc-protected a-amino ketone using TsDENEB® **HT** in a key step to form an alcohol intermediate in the synthesis of the type 2 diabetes drug omarigliptin (Figure 7.34)

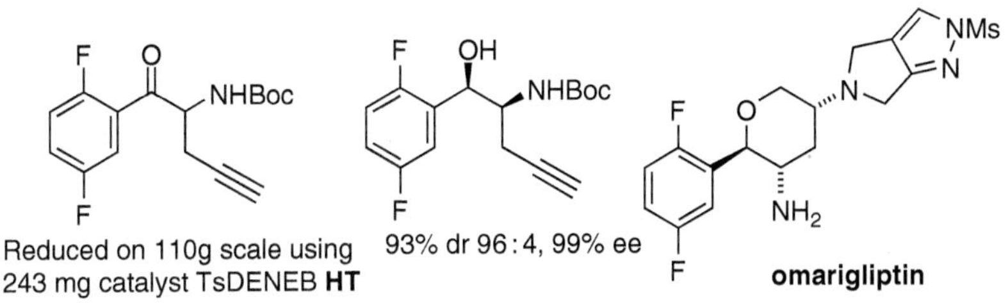

Figure 7.34 α-Aminoketone substrate, reduction product, and pharmaceutical target. Source: Chung et al. [97]; Xu et al. [98].

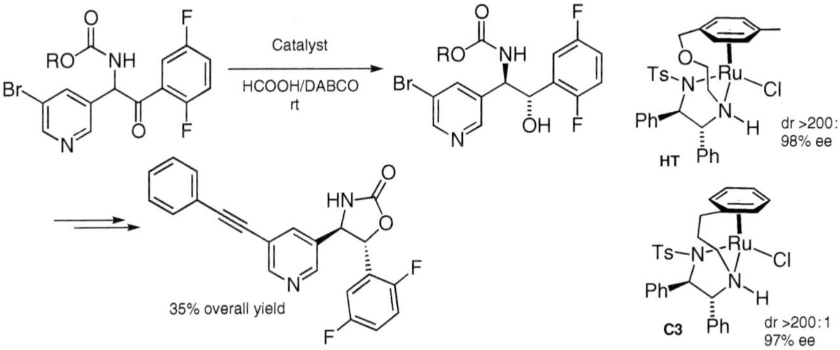

Figure 7.35 ATH–DKR of an α-amino ketone toward an mGluR5 receptor modulator. Source: Based on Gonzalez-Bobes et al. [99].

[97]. TsDENEB® **HT** gave an improved result over an untethered catalyst which had earlier been studied for this application [98].

ATH–DKR proved to be successful in a key step in the total synthesis of the metabotropic glutamate receptor 5 (mGluR5) (Figure 7.35). Tethered and nontethered complexes were screened [99], and all the complexes were effective at low catalyst loadings, giving products in high conversions. Both tethered catalyst **C3** and TsDENEB® **HT** gave the required product in excellent enantioselectivity and a high diastereomeric ratio.

Hu et al. have explored the use of tethered catalysts in a DKR–ATH one-pot sulfoarylation and reduction (not illustrated) [100]. Bhanage et al. reported the enantioselective synthesis of *cis*-β-heteroaryl amino cycloalkanols with two contiguous chiral centers using tethered catalyst **C3** (Figure 7.36) [101]. In this example, *cis*-amino cycloalkanols can be produced which would otherwise be difficult to generate, e.g. by ring-opening reactions. The corresponding chiral alcohols were obtained with excellent diastereoselectivities (up to 99/1) and enantioselectivities (up to 99% ee). The developed method has shown its potential for the synthesis of antitubercular drugs and chiral ionic liquids.

Mohar et al. recently reported the application of their ansa-Ru(II)-type tethered catalyst **K2** to the ATH–DKR of racemic 2,3-disubstituted 1-indanones possessing two stereolabile centers into the corresponding chiral 2,3-disubstituted-1-indanols

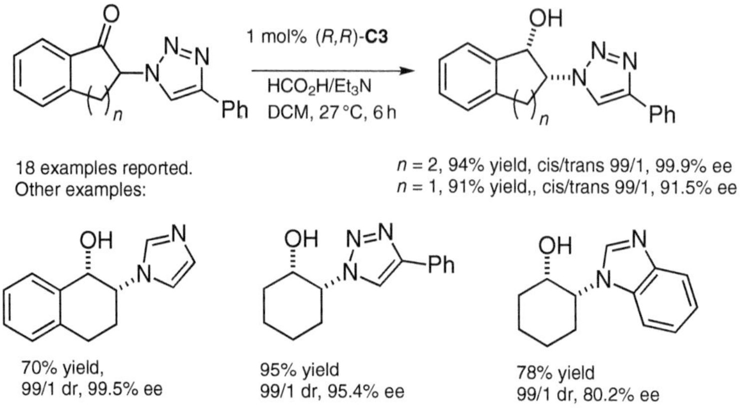

Figure 7.36 ATH to produce *cis*-β-heteroaryl amino cycloalkanols.

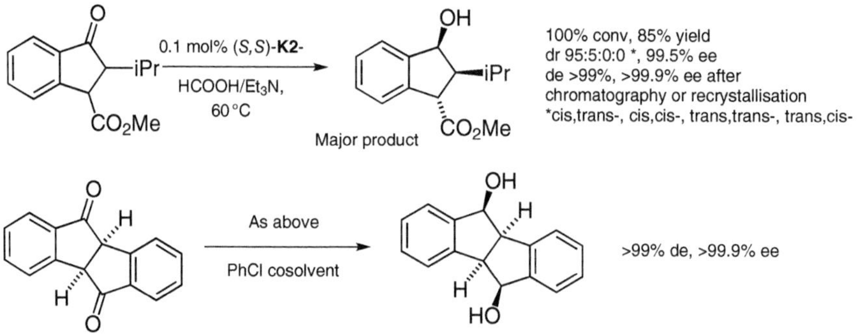

Figure 7.37 DKR–ATH of racemic 2,3-disubstituted 1-indanones. Source: Based on Cotman et al. [22].

with excellent diastereo- (>99/1) and enantioselectivities (>99.9 ee) (Figure 7.37) [22]. This methodology gives access to chiral alcohols with multiple chiral centers. The method has been applied to the synthesis of Pallidol analogues via a diol intermediate in high selectivity.

β-Keto esters and amides are excellent substrates for DKR reactions due to the ready racemization of the α-chiral center [51, 64, 102]. Tethered catalysts with sulfamoylamino (NSO_2N) groups were demonstrated by the group of Mohar as effective catalysts for the ATH of α-CF_3(CO) indanones via ATH–DKR (Figure 7.38). Both ketones were reduced in this application, with the trifluoromethyl ketone being reduced before the aromatic one.

α-Amino–β-keto esters containing primary amines are excellent DKR substrates for tethered catalysts and have been employed productively in recent research work. Reduction in up to 83/17 anti/syn and 98% ee and then subsequent acylation gave the amide products (Figure 7.39) [103].

A report [104] by Ishida et al. documents the reduction of the α-N-acylamino-β-keto esters using various (*R,R*)-TsDENEB® **HT** derivatives where $ArSO_2$ = Ts, Mes,

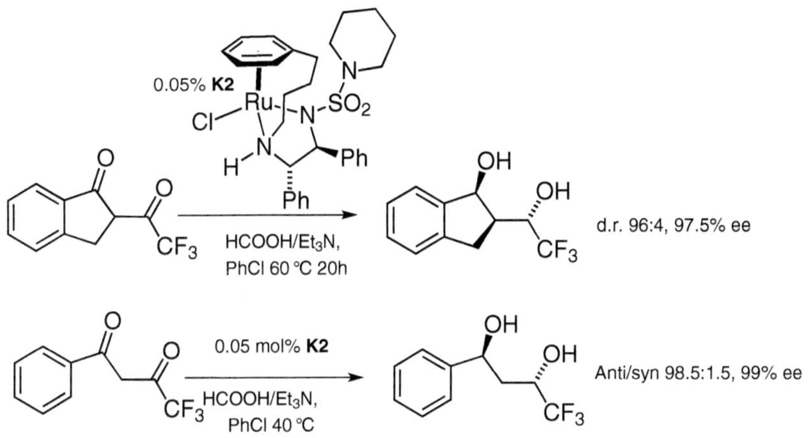

Figure 7.38 ATH–DKR of fluorinated substrates. Source: Based on Cotman et al. [15].

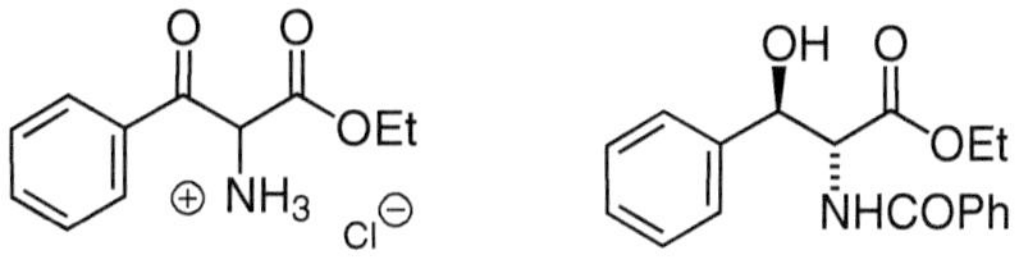

Figure 7.39 ATH–DKR of unprotected α-amino ketones gives amides after acylation. Source: Based on Echeverria et al. [103].

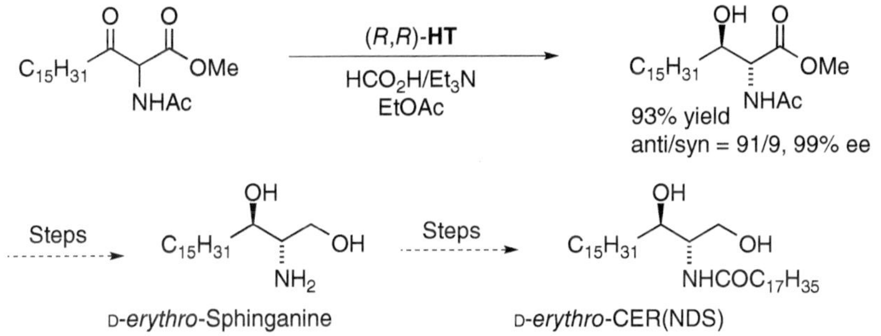

Figure 7.40 ATH of α-*N*-acylamino-β-keto esters. Source: Tanaka et al. [104]; Tanaka et al. [105].

2-$CF_3(C_6H_4)SO_2$, 2,3,5-$iPr_3(C_6H_4)SO_2$, as catalysts, and the subsequent conversion of the products to sphinganine derivatives (Figure 7.40) [104–106].

In a follow-up work [107], a flow process for the reaction in Figure 7.40 was described. A pipe-in-series flow reactor was able to process material on >50 kg scale, giving products of reduction in 86.2% yield at a low catalyst loading of just 0.1 mol%, a greater conversion and enantioselectivity compared to the nontethered complexes. Wang et al. described an efficient alternative DKR–ATH by demonstrating the synthesis of enantiomeric pure syn-aryl β-hydroxy α-dibenzylamino esters [108]. The tertiary nitrogen atom is well tolerated by (*R,R*)-TsDENEB® **HT**, producing products in excellent diastereo- (>20 : 1) and enantioselectivities (>99%) (Figure 7.41). The method was found to be useful for the synthesis of droxidopa on

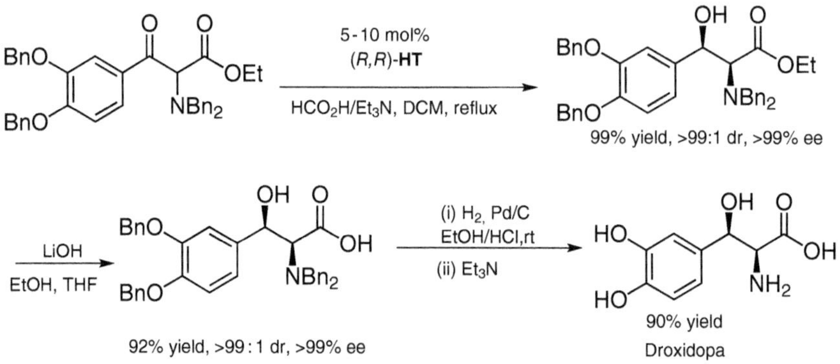

Figure 7.41 ATH/DKR of dibenzylamino β-ketoesters.

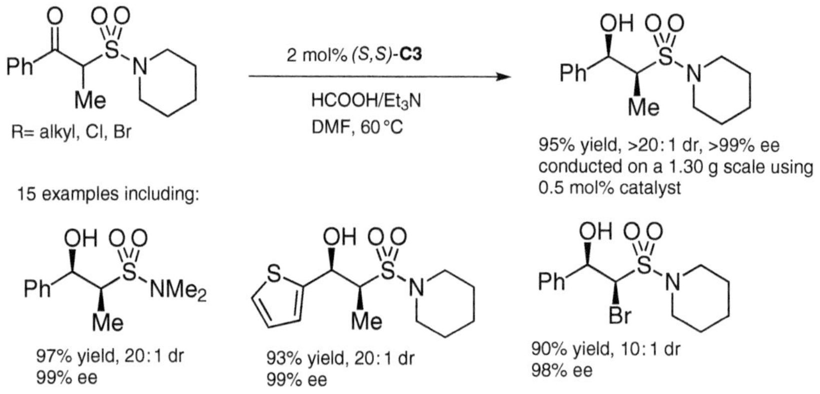

Figure 7.42 ATH/DKR of sulfonamides using 3C-tethered catalyst **C3**. Source: Based on Zhang et al. [110].

a gram scale by hydrolysis of the ester followed by a palladium-catalyzed *N*-alkyl cleavage without any detrimental effect on the diastereo- or enantioselectivity.

The first example of the production of a chiral indoline by ATH–DKR was reported by Wang and coworkers. Optically active *cis*-2-substituted-*N*-acetyl-3-hydroxyindolines were prepared in excellent diastereoselectivities (up to >99/1) and enantioselectivities (>99.9% in the best cases) [109]. Zhang et al. applied tethered catalysts to the ATH of racemic α-substituted β-keto sulfonamides to generate products with multiple chiral centers via ATH–DKR (Figure 7.42) [110]. In this work, with a sensitive functional group in the substrate, tethered catalysts showed good functional group tolerance, and biologically important β-hydroxy sulfonamides were produced in excellent diastereoselectivities (up to 20 : 1) and enantioselectivities (up to 99%).

The oxo-tethered catalyst TsDENEB® **HT** was screened for DKR–ATH of β-substituted chromanones to produce synthetically useful chromanols with good enantiomeric excess [111]. The stereoselectivity in this work is controlled by torsional strain and CH—O sulfonamide hydrogen bonds to produce 1,3-optically active chromanols. In this study, TsDENEB® **HT** and a TsDPEN-mesitylene catalyst

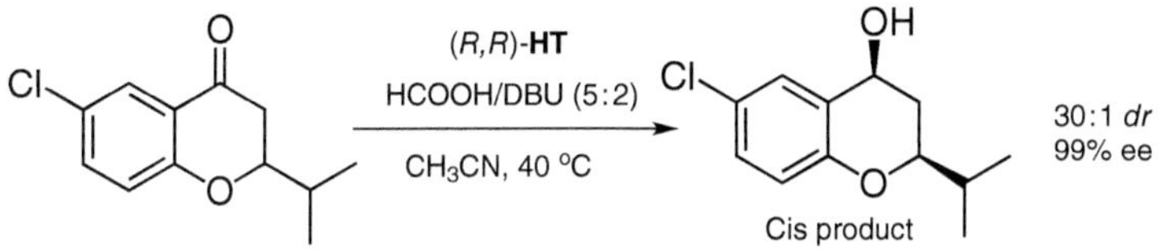

Figure 7.43 ATH of racemic chromanones.

gave best results, producing products in up to 99% ee, whereas the tethered catalyst was better in producing the *cis* isomer (30 : 1 dr) (Figure 7.43).

A similar concept was used in the asymmetric reduction of a racemic tetrahydroisoquinoline containing an external ketone (not illustrated) [112]. An interesting reduction of a six-membered enone containing an exo-carboxylic acid with DKR followed by cyclization of the new OH group onto the acid to form a five-membered lactone was reported (not illustrated) [113]. DENEB® (1 mol%) was used, and an example on an 80-g scale gave a product of 80% ee which could be recrystallized to > 99% ee.

7.3.6 Reductions of Imines

The ATH of imines remains a challenging area. Scientists at Merck published an excellent example of a practical application of tethered catalyst **C3** to the asymmetric reduction of a primary imine to give a key intermediate in the synthesis of a pharmaceutical target (Figure 7.44) [114, 115]. The use of an *ortho*-hydroxy imine is very important to the stability of the imine substrate.

Tethered catalyst **C3** was also demonstrated in chemoselective ATH of isoxazolium salts to produce optically active Δ^4-isoxazolines (Figure 7.45) [116]. The reduction was achieved with moderate to good yields and ees with no subsequent ring opening. The lower ee in the case of 3,5-diphenyl substrate reinforces our understanding of the workings of these Noyori-class catalysts.

The Ru-tethered catalyzed ATH of quinoline was demonstrated by Wills and coworkers [19]. This challenging class of substrate was subjected to ATH with tethered ruthenium and rhodium catalysts in $HCOOH/Et_3N$. However, a tethered

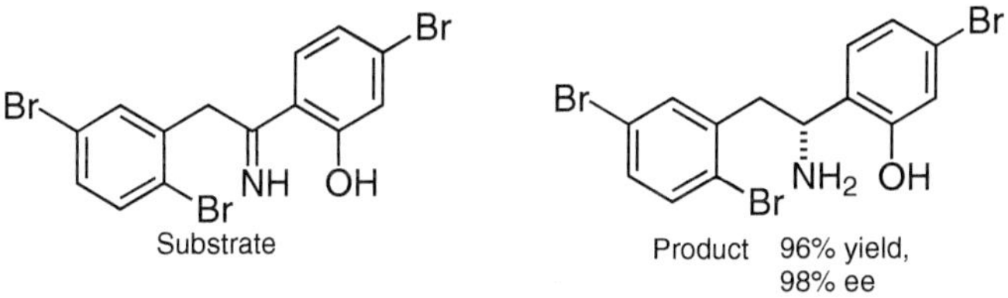

Figure 7.44 Formation of a primary amine by ATH from an imine was achieved using a tethered catalyst. Source: Mangion et al. [114]; Dirocco et al. [115].

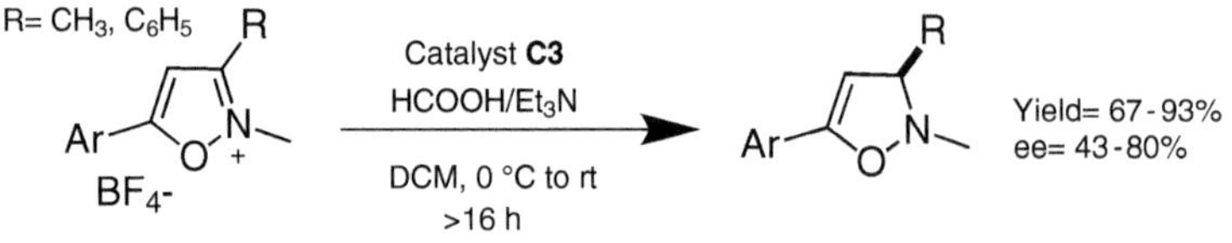

Figure 7.45 ATH of an iminium salt by a tethered catalyst. Source: Based on Chew and Wills [116].

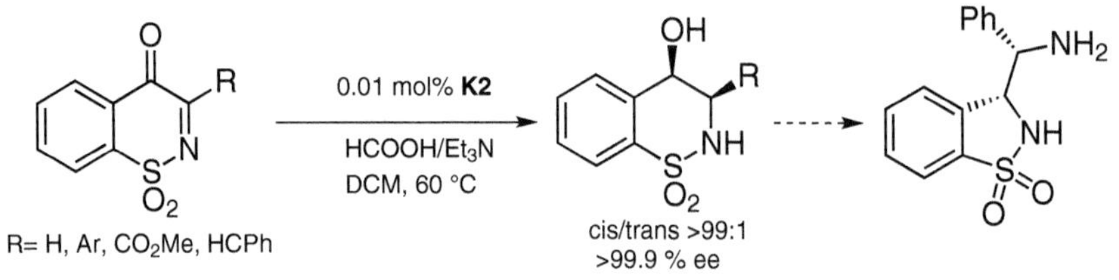

Figure 7.46 Enantioselective reduction of α-(*N*-sulfonylimino) ketones. Source: Based on Jeran et al. [16].

rhodium complex proved to be more effective for producing good ee (up to 94%) over ruthenium complex **C3** which generated the product in an ee of up to 73% in similar reaction conditions (not illustrated). Wills et al. also reported asymmetric reduction of a cyclic imine (dimethoxy-1-methyl-dihydroquinoline) using TsDENEB® **HT** in 100% conversion and 87% ee [27]. Kačer et al. have reported the results of a detailed study of the asymmetric *hydrogenation* (15 bar hydrogen gas) of a range of cyclic imines using untethered, tethered, and novel derivatives [117]. The reactions were carried out in MeOH with 1 mol% TFA, and 1 mol% of catalyst in each case at 40 °C, and a new hydroxyl-substituted catalyst was reported to show high activity and proved to be competitive with some of the best catalysts reported for this application.

The *ansa* catalyst **K2** was applied to the reduction of α-(*N*-sulfonylimino) and α-(*N*-sulfonylamino) aryl ketones to produce 4-hydroxy-benzo-δ- and 3-(α-hydroxy-arylmethyl)-benzo-γ-sultams, respectively (Figure 7.46) [16]. This DKR–ATH proceeds smoothly at very low catalyst loadings (S/C = 10 000) to afford a useful class of product in excellent diastereo- (>99 : 1 in the best cases) and enantioselectivity (>99.9 in the best cases).

7.4 Conclusions and Outlook

In conclusion, tethered Ru(II) catalysts are now established catalysts for use in the ATH of ketones and imines. Following their first report in the mid-2000s, they have grown rapidly in importance and have found many valuable applications in asymmetric catalysis and most recently in process chemistry applications. We recommend routine screening of a range of tethered Ru(II) catalysts in future reductions where a combination of practicality, versatility, high activity, and enantioselectivity is required.

References

1 (a) Hashiguchi, S., Fujii, A., Takehara, J. et al. (1995). *J. Am. Chem. Soc.* 117: 7562–7563. (b) Fujii, A., Hashiguchi, S., Uematsu, N. et al. (1996). *J. Am. Chem. Soc.* 118: 2521–2522.

2 (a) Haack, K.J., Hashiguchi, S., Fujii, A. et al. (1997). *Angew. Chem.* 109: 297–300; *Angew. Chem. Int. Ed.*, *36*, 285–288. (b) Brandt, P., Roth, P., and Andersson, P.G. (2004). *J. Org. Chem.* 69: 4885–4890.

3 (a) Dub, P.A. and Gordon, J.C. (2016). *Dalton Trans.* 45: 6756–6781. (b) Dub, P.A. and Gordon, J.C. (2017). *ACS Catal.* 7: 6635–6655.

4 (a) Xue, D., Chen, Y.-C., Cui, X. et al. (2005). *J. Org. Chem.* 70: 3584–3591. (b) Peach, P., Cross, D.J., Kenny, J.A. et al. (2006). *Tetrahedron* 62: 1864–1876.

5 Nedden, H.G., Zanotti-Gerosa, A., and Wills, M. (2016). *Chem. Rec.* 16: 2623–2643.

6 (a) Hall, A.M.R., Dong, P., Codina, A. et al. (2019). *ACS Catal.* 9: 2079–2090. (b) Toubiana, J., Medina, L., and Sasson, Y. (2014). *Mod. Res. Catal.* 3: 68–88.

7 Cheung, F.K., Hayes, A.M., Hannedouche, J. et al. (2005). *J. Org. Chem.* 70: 3188–3197.

8 Knighton, R.C., Vyas, V.K., Mailey, L.H. et al. (2018). *J. Organomet. Chem.* 875: 72–79.

9 Hannedouche, J., Clarkson, G., and Wills, M. (2004). *J. Am. Chem. Soc.* 126: 986–987.

10 Hayes, A.M., Morris, D.J., Clarkson, G.J., and Wills, M. (2005). *J. Am. Chem. Soc.* 127: 7318–7319.

11 Touge, T., Hakamata, T., Nara, H. et al. (2011). *J. Am. Chem. Soc.* 133: 14960–14963.

12 Kišić, A., Stephan, M., and Mohar, B. (2013). *Org. Lett.* 15: 1614–1617.

13 Kišić, A., Stephan, M., and Mohar, B. (2014). *Adv. Synth. Catal.* 356: 3193–3198.

14 Kišić, A., Stephan, M., and Mohar, B. (2015). *Adv. Synth. Catal.* 357: 2540–2546.

15 Cotman, A.E., Cahard, D., and Mohar, B. (2016). *Angew. Chem. Int. Ed.* 55: 5294–5298.

16 Jeran, M., Cotman, A.E., Stephan, M., and Mohar, B. (2017). *Org. Lett.* 19: 2042–2045.

17 Cheung, F.K., Hayes, A.M., Morris, D.J., and Wills, M. (2007). *Org. Biomol. Chem.* 5: 1093–1103.

18 Cheung, F.K., Lin, C., Minissi, F. et al. (2007). *Org. Lett.* 9: 4659–4662.

19 Parekh, V., Ramsden, J.A., and Wills, M. (2010). *Tetrahedron: Asymmetry* 21: 1549–1556.

20 Martins, J.E.D., Morris, D.J., Tripathi, B., and Wills, M. (2008). *J. Organomet. Chem.* 693: 3527–3532.

21 Matsunami, A., Ikeda, M., Nakamura, H. et al. (2018). *Org. Lett.* 20: 5213–5218.

22 Cotman, A.E., Modec, B., and Mohar, B. (2018). *Org. Lett.* 20: 2921–2924.

23 Jolley, K.E., Zanotti-Gerosa, A., Hancock, F. et al. (2012). *Adv. Synth. Catal.* 354: 2545–2555.

24 Soni, R., Jolley, K.E., Gosiewska, S. et al. (2018). *Organometallics* 37: 48–64.

25 Hodgkinson, R., Jurčik, V., Zanotti-Gerosa, A. et al. (2014). *Organometallics* 33: 5517–5524.

26 Soni, R., Jolley, K.E., Clarkson, G.J., and Wills, M. (2013). *Org. Lett.* 15: 5110–5113.

27 Parekh, V., Ramsden, J.A., and Wills, M. (2012). *Catal. Sci. Technol.* 2: 406–414.

28 Jolley, K.E., Clarkson, G.J., and Wills, M. (2015). *J. Organomet. Chem.* 776: 157–162.

29 Cheung, F.K., Graham, M.A., Minissi, F., and Wills, M. (2007). *Organometallics* 26: 5346–5351.

30 Cross, D.J., Houson, I., Kawamoto, A.M., and Wills, M. (2004). *Tetrahedron Lett.* 45: 843–846.

31 (a) Blacker, J. and Martin, J. (2004). *Asymmetric Catalysis on an Industrial Scale: Challenges, Approaches and Solutions* (eds. H.U. Blaser and E. Schmidt), 201–220. Weinheim: Wiley-VCH. (b) Matharu, D.S., Morris, D.J., Kawamoto, A.M. et al. (2005). *Org. Lett.* 7: 5489–5491. (c) He, B., Phansavath, P., and Ratovelomanana-Vidal, V. (2019). *Org. Lett.* 21: 3276–3280. (d) Echeverria, P.-G., Férard, C., Phansavath, P., and Ratovelomanana-Vidal, V. (2015). *Catal. Comm.* 62: 95–99. (e) Zheng, L.-S., Llopis, Q., Echeverria, P.-G. et al. (2017). *J. Org. Chem.* 82: 5607–5615. (f) Zheng, L.-S., Férard, C., Phansavath, P., and Ratovelomanana-Vidal, V. (2018). *Chem. Commun.* 54: 283–286.

32 Dimroth, J., Schedler, U., Keilitz, J. et al. (2011). *Adv. Synth. Catal.* 353: 1335–1344.

33 Hakamada, T., Nara, H., and Touge, T. (2011). Ruthenium-diamine complexes and method for producing optically active compounds. WO Patent 2012/026201A1, filed 17 June 2011 and issued 1 March 2012.

34 Hodgkinson, R., Jurčík, V., Nedden, H. et al. (2018). *Tetrahedron Lett.* 59: 930–933.

35 Nedden, H., Jurcik, V., Wills, M., and Hodgkinson R. (2014). Complexes and methods for their preparation. WO Patent 2016/042298A1, filed 11 September 2015 and issued 24 March 2016.

36 Wills, M., Jolley, K., and Soni, R. (2016). Catalyst and process for synthesising the same. US Patent 9,321,045B2, filed 1 May 2015 and issued 26 April 2016.

37 Dyke, A., Grainger, D. M., Medlock, J. A. et al. (2010). Process for hydrogenating ketones in the presence of Ru (II) catalysts. WO Patent 2010/106364A2, filed 17 March 2010 and issued 23 September 2010.

38 Touge, T., Hakamata, T., and Nara, H. (2012). Method for producing diamine compound. WO Patent 2012/147944, filed 27 April 2012.

39 Touge, T., Hakamata, T., and Nara, H. (2012). Ruthenium - production method of diamine complex and an optically active compound. WO Patent 2012/153684A1, filed 2 May 2012 and published 31 July 2014.

40 Abdur-Rashid, K., Amoroso, D., Sui-Seng, C. et al. (2010). Cationic transition-metal arene catalysts. WO Patent 2009/132443A1, filed 1 May 2009 and issued 5 November 2009.

41 Touge, T. and Nara, H. (2015). Solid-supported ruthenium-diamine complex, and method for manufacturing optically active compound. WO Patent 2016/056669, filed 13 October 2015 and published 14 April 2016.
42 Tan, Q., Xu, Y., Gao, X., and Xiao, H. (2015). Transition metal composite oxide catalyst. CN Patent 106925265, filed 30 December 2015 and published 7 July 2017.
43 Sanchez, A., Itov, G., Khandelwal, M. et al. (2018). N-H free and Si-rich perhydridopolysilzane compositions, their synthesis, and applications. WO Patent 2018/107155A8, filed 11 September 2017 and issued 14 June 2018.
44 Sanchez, A., Itov, G., Khandelwal, M. et al. (2018). N-H free and Si-rich per-hydridopolysilzane compositions, their synthesis, and applications. US Patent 2018/0072571A1, filed 27 July 2017 and issued 15 March 2018.
45 Morris, D.J., Hayes, A.M., and Wills, M. (2006). *J. Org. Chem.* 71: 7035–7044.
46 Wakeham, R.J., Morris, J.A., and Williams, J.M.J. (2015). *ChemCatChem* 7: 4039–4041.
47 Soni, R., Hall, T.H., Mitchell, B.P. et al. (2015). *J. Org. Chem.* 80: 6784–6793.
48 Yuki, Y., Touge, T., Nara, H. et al. (2018). *Adv. Synth. Catal.* 360: 568–574.
49 Liu, S., Liu, H., Zhou, H. et al. (2018). *Org. Lett.* 20: 1110–1113.
50 Zhou, H., Liu, S., Liu, H., and Liu, Q. (2018). Method for synthesizing chiral alcohol by hydration/asymmetric hydrogenation of alkyne in series. CN Patent 108101741A, filed 5 December 2017 and issued 1 June 2018.
51 Komiyama, M. and Ito, T. (2013). Process for the preparation of amino alcohol derivatives having b2 adrenergic receptor agonist activity and their intermediates. JP Patent 2015/017075, filed 12 July 2013 and published 29 January 2015.
52 Zou, Q., Shi, J., and Xiong, S. (2015). Synthesis method of ezetimibe. CN Patent 105503686, filed 31 December 2015 and issued 20 April 2016.
53 Zhang, Q., Huang, P., Xu, G., and Chen, B. (2013). Method for synthesising ezetimibe intermediate. CN Patent 102978253A, filed 28 November 2012 and issued 20 March 2013.
54 Li, B., Dugger, R.W., Conway, B. et al. (2017). *Org. Process Res. Dev.* 21: 1340–1348.
55 Barnwell, N., Cornwall, P., Gill, D.M. et al. (2012). Process. WO Patent 2012/156693A1, filed 11 May 2012 and issued 22 November 2012.
56 Vinent, H.T. and Thijs, L. (2009). Process for preparing montelukast intermediates. WO Patent 2009/130056A1, filed 23 April 2009 and issued 29 October 2009.
57 Wisdom, R., Jung, J., and Meudt, A. (2011). Process for the asymmetric transher hydrogenation of ketones. WO Patent 2011/131315A1, filed 14 April 2011 and issued 27 October 2011.
58 Zaramella, S., Rossi, E., De Lucchi, O., and Serafini, S. (2016). Improved process for the preparation of eslicarbazepine and eslicarbazepine acetate. EU Patent 3064490A1, filed 19 November 2015 and issued 7 September 2016.
59 Hirpara, K., Jesunadh, K., Sharma, M.K., and Khanduri, C.H. (2014). Process for the preparation of oxcarbazepine and its use as intermediate in the

preparation of eslicarbazepine acetate. WO Patent 2014/049550A1, filed 26 September 2013 and issued 3 April 2014.

60 Hirpara, K., Jesunadh, K., Sharma, M.K., and Khanduri, C.H. (2015). Process for the preparation of oxcarbazepine and its use as intermediate in the preparation of eslicarbazepine acetate. US Patent 2015/0232426A1, filed 26 September 2013 and issued 20 August 2015.

61 Ebel, H., Frattini, S., Giovannini, R. et al. (2013). New ccr2 antagonists. US Patent 2013/0217728A1, filed 26 May 2011 and issued 22 August 2013.

62 Komiyama, M., Itoh, T., and Takeyasu, T. (2015). *Org. Process Res. Dev.* 19: 315–319.

63 Moore, J., Lau, K., Xu, X. et al. (2019). *Tetrahedron Lett.* 60: 785–788.

64 Fang, Z. and Wills, M. (2013). *J. Org. Chem.* 78: 8594–8605.

65 Fang, Z., Clarkson, G.J., and Wills, M. (2013). *Tetrahedron Lett.* 54: 6834–6837.

66 Fang, Z. and Wills, M. (2014). *Org. Lett.* 16: 374–377.

67 Grainger, D., Zanotti-Gerosa, A., Mitchell, D. et al. (2013). *ChemCatChem* 5: 1205–1210.

68 Holladay, M.W. and Setti, E. (2012). Optically active pyrazolylaminoquinazoline, and pharmaceutical compositions and methods of use thereof. US Patent 2012/0053193A1, filed 31 August 2011 and issued 1 March 2012.

69 Vyas, V.K. and Bhanage, B.M. (2016). *Org. Chem. Front.* 3: 614–619.

70 Touge, T., Nara, H., Fujiwharam, M. et al. (2016). *J. Am. Chem. Soc.* 138: 10084–10087.

71 Zhou, H., Wang, B., Liu, Q., and Lu, G. (2017). Optical activity di(heteto)aryl methanol and asymmetric synthesis method thereof. CN Patent 106831550A, filed 17 January 2017 and issued 13 July 2017.

72 Zhou, H., Wang, B., Liu, Q., and Jiang, X. (2017). A kind of method of asymmetric synthesis of Claritin carbinoxamine. CN Patent 106831549B, filed 17 January 2017 and issued 13 June 2017.

73 Wang, B., Zhou, H., Lu, G. et al. (2017). *Org. Lett.* 19: 2094–2097.

74 Vyas, V.K., Knighton, R.C., Bhanage, B.M., and Wills, M. (2018). *Org. Lett.* 20: 975–978.

75 Soni, R., Collinson, J.-M., Clarkson, G.C., and Wills, M. (2011). *Org. Lett.* 13: 4304–4307.

76 Turner, P.A., Samiullah, Whatmore, J.L., and Shipman, M. (2013). *Tetrahedron Lett.* 54: 6538–6540.

77 Su, Y., Tu, Y.-Q., and Gu, P. (2014). *Org. Lett.* 16: 4204–4207.

78 Ji, Y., Xue, P., Ma, D.-D. et al. (2015). *Tetrahedron Lett.* 56: 192–194.

79 Perryman, M.S., Harris, M.E., Foster, J.L. et al. (2013). *Chem. Commun.* 49: 10022–10024.

80 Chew, R.J. and Wills, M. (2018). *J. Catal.* 361: 40–44.

81 Forshaw, S., Matthews, A.J., Brown, T.J. et al. (2017). *Org. Lett.* 17: 2789–2792.

82 Liu, R., Zhou, G., Hall, T.H. et al. (2015). *Adv. Synth. Catal.* 357: 3453–3457.

83 Zheng, L.-S., Phansavath, P., and Ratovelomanana-Vidal, V. (2018). *Org. Lett.* 20: 5107–5111.

84 Kuang, L., Liu, L.L., and Chiu, P. (2015). *Chem-Eur. J.* 21: 14287–14291.

85 Wensen, L. (2017). The synthetic method of 2- [(2R, 6S)-6-[(2S)-2-hydroxyl-2-phenethyl]-1-methylpiperidine]-1-acetophenone. CN Patent 106496099B, filed 12 October 2016 and issued 15 March 2017.
86 McLaughlin, M., Kong, J., Belyk, K.M. et al. (2017). *Org. Lett.* 19: 926–929.
87 Girijavallabhan, V.M., McLaughlin, M., Cleator, E. et al. (2017). 4'-substituted nucleoside reverse transcriptase inhibitors and preparations thereof. WO Patent 2017/053216A2, filed 19 September 2016 and issued 30 March 2017.
88 Huang, P., Zheng, X., and Wang, Y. (2016). One method of synthesis of darunavir key intermediate. CN Patent 106083656B, filed 20 June 2016 and issued 9 November 2016.
89 (a) Smejkal, T. (2013). Process for the stereoselective preparation of a pyrazole carboxamide. US Patent 9120759B2, filed 13 February 2012 and issued 11 December 2013; (b) Smejkal, T. (2013). Process for the stereoselective preparation of a pyrazole carboxamide. WO Patent 2013/120860A1, filed 13 February 2012 and issued 22 August 2013.
90 Bonday, Z.Q., Cortez, G.S., Dahnke, K.R. et al. (2016). 5'-substituted nucleoside analogs. CA Patent 2981097A1, filed 27 April 2015 and issued 10 November 2016.
91 Bonday, Z.Q., Cortez, G.S., Grogan, M.J. et al. (2018). *ACS Med. Chem. Lett.* 9: 612–617.
92 Zhou, H., Liu, S., Liu, Q., and Wang, C. (2017). Cyclopropyl substituted allyl alcohol and asymmetric synthesis method thereof. CN Patent 107473941A, filed 18 September 2017 and issued 15 December 2017.
93 Liu, S., Cui, P., Wang, J. et al. (2019). *Org. Biomol. Chem.* 17: 264–267.
94 Schaefer, B. and Siegel, W. (2018). Improved process for preparing astacene. WO Patent 2018/095863A1, filed 20 November 2017 and issued 31 May 2018.
95 Takahashi, H., Baba, Y., Morita, Y. et al. (2018). Crystals of cyclic amine derivative and pharmaceutical use thereof. WO Patent 2018/038255A1, filed 25 August 2017 and issued 1 March 2018.
96 Synthetic route for the preparation of substituted 2-phenyl-1,2,3,4-tetrahydronaphthalene-1-ols. EU Patent 2644603A1, filed 30 March 2012 and issued 2 October 2013.
97 Chung, J.Y.L., Scott, J.P., Andersson, C. et al. (2015). *Org. Process Res. Dev.* 19: 1760–1768.
98 Xu, F., Zacuto, M.J., Kohmura, Y. et al. (2014). *Org. Lett.* 16: 5422–5425.
99 Gonzalez-Bobes, G., Hanson, R., Strotman, N. et al. (2016). *Adv. Synth. Catal.* 358: 2077–2082.
100 Hu, X., Zhang, K., Chang, F. et al. (2018). *Mol. Catal.* 452: 271–276.
101 Vyas, V.K. and Bhanage, B.M. (2016). *Org. Lett.* 18: 8436–6439.
102 Monnereau, L., Cartigny, D., Scalone, M. et al. (2015). *Chem. Eur. J.* 21: 11799–11806.
103 Echeverria, P.-G., Cornil, J., Férard, C. et al. (2015). *RSC Adv.* 5: 56815–56819.
104 Tanaka, S., Touge, T., Nara, H., and Ishida, K. (2013). Method for producing optically active β-hydroxy-α-aminocarboxylic acid ester. WO Patent 2013/065867A1, filed 31 October 2011 and issued 10 May 2013.

105 Tanaka, S., Touge, T., Nara, H., and Ishida, K. (2014). Method for producing optically active beta-hydroxy-alpha-aminocarboxylic acid ester. US Patent 2014/0296562A1, filed 31 October 2012 and issued 2 October 2014.
106 Ishida, K., Obata, Y., Akagi, C. et al. (2014). *J. Drug Del. Sci. Tech.* 24: 689–693.
107 Touge, T., Kuwana, M., Komatsuki, Y. et al. (2019). *Org. Process Res.Dev.* 23: 452–461.
108 (a) Sun, G., Zhou, Z., Luo, Z. et al. (2017). *Org. Lett.* 19: 4339–4342. (b) Sun, G., Wang, Z., Wang, H. et al. (2018). Method for preparing droxidopa and intermediate thereof. WO Patent 2018/090670A1, filed 2 August 2017 and issued 24 May 2018.
109 Luo, Z., Sun, G., Zhou, Z. et al. (2018). *Chem. Commun.* 54: 13503–13506.
110 Zhang, X., Xiong, Z., Pei, C. et al. (2018). *Chem. Commun.* 54: 3883–3886.
111 Ashley, E.R., Sherer, E.C., Pio, B. et al. (2017). *ACS Catal.* 7: 1446–1451.
112 Zheng, L.-S., Phansavath, P., and Ratovelomanana-Vidal, V. (2018). *Org. Chem. Front.* 5: 1366–1379.
113 Yagi, K., Komatsuki, Y., Ujihara, H., and Ishida, K. (2012). Method for producing wine lactone. WO Patent 2012/165164A1, filed 17 May 2012 and issued 6 December 2012.
114 Mangion, I.K., Chen, C.-y., Li, H. et al. (2014). *Org. Lett.* 16: 2310–2313.
115 Dirocco, D.A., Davies, I., Feng, P. et al. (2016). Process for preparing substituted indole compounds. WO Patent 2016/073659A1, filed 5 November 2015 and issued 12 May 2016.
116 Chew, R.J. and Wills, M. (2018). *J. Org. Chem.* 83: 2980–2985.
117 Vihanová, B., Václavik, J., Šot, P. et al. (2016). *Chem. Commun.* 52: 362–365.

8

Homogeneous Asymmetric Hydrogenation of Heteroaromatic Compounds Catalyzed by Transition Metal Complexes

Qing-Hua Fan, Yan-Mei He and Fa-Ju Li

Beijing National Laboratory for Molecular Sciences, CAS Key Laboratory of Molecular Recognition and Function, Institute of Chemistry, Chinese Academy of Sciences (CAS), No. 2 First North Street, Zhongguancun, Beijing, 100190, P. R. China

8.1 Introduction

Optically active heterocyclic structures are ubiquitous in naturally occurring and artificial biologically active molecules. Catalytic asymmetric hydrogenation of readily available heteroaromatic compounds is a straightforward and atom-economical way to acquire a wide range of chiral heterocyclic compounds [1]. But the asymmetric hydrogenation of heteroaromatic compounds is still challenging due to the high chemical stability of heteroaromatics and the possible poisoning and/or deactivation of catalysts caused by the strong coordination between the heteroatoms and the transition metal centers. The past decade has been witnessing the significant progresses in this research field, which is highlighted by the evolution of effective chiral catalysts [2] and substrate activation strategies [3].

Developing more effective and enantioselective catalysts is the main issue to make breakthroughs in this research field. Since the pioneering work of Zhou [2a] on the asymmetric hydrogenation of quinoline derivatives in 2003, several catalytic systems have been demonstrated to be successful, including transition metal complexes with chiral phosphorus [2a, b], diamine [2c], and *N*-heterocyclic carbene (NHC) [2d] ligands, and organocatalysts like frustrated Lewis pairs (FLPs) [2f]. For some catalytic systems [2a], additives are needed to activate the catalyst and *in situ* generate the real catalytic species. The chiral ligands/catalysts appearing in this chapter are given in Scheme 8.1.

To address the problems caused by the inherent chemical properties of the heteroaromatic compounds, substrate activation strategies [3] have been developed. For the nitrogen-containing six-membered-ring substrates, quaternization of nitrogen groups is a general approach, no matter the positively charged substrates are presynthesized or *in situ* formed. Thus, the related problems of substrate aromaticity and catalyst poisoning are solved. On the other hand, these electron-deficient compounds facilitate the hydride addition from the catalytically active species.

Asymmetric Hydrogenation and Transfer Hydrogenation, First Edition.
Edited by Virginie Ratovelomanana-Vidal and Phannarath Phansavath.

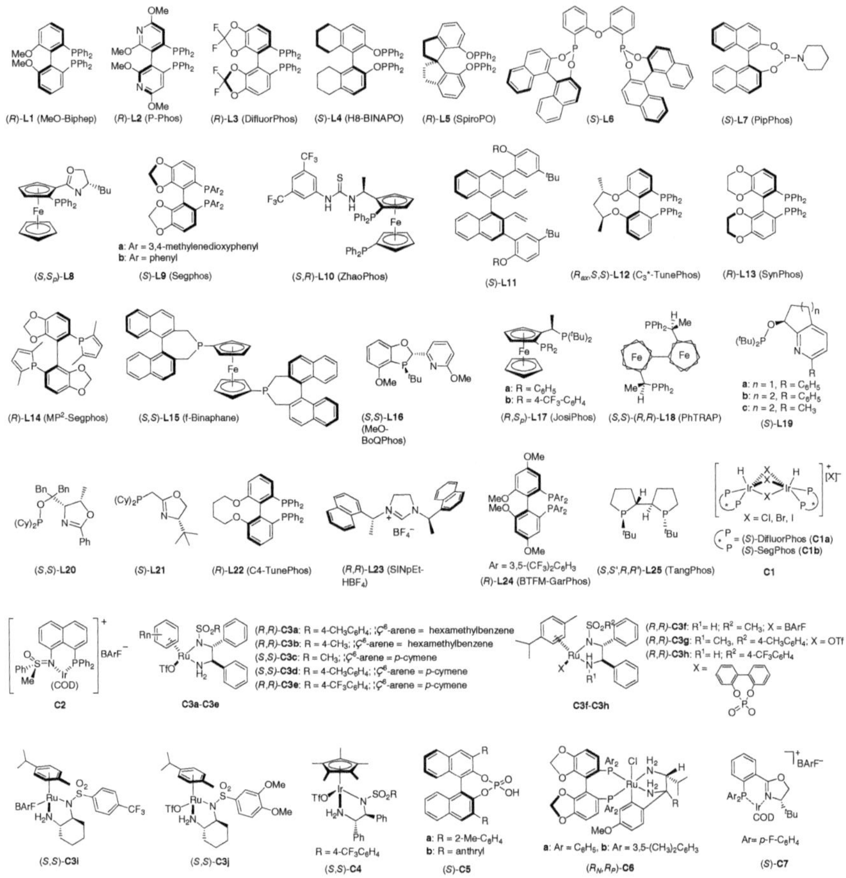

Scheme 8.1 Chiral ligands and catalysts for the asymmetric hydrogenation of heteroaromatic compounds included in Chapter 8.

Despite great advances achieved, the asymmetric hydrogenation of heteroaromatic compounds is less investigated with comparison to the well-established studies on alkenes, ketones, and imines. For most heteroaromatic compounds, the hydrogenation always includes sequential reduction of multiple and different types of unsaturated double bonds. Good controls in chemo-, regio-, and stereoselectivity are the key to obtain high yield and high optical purity of the chiral products. Meanwhile, the diverse types and complex chemical structures of heteroaromatic compounds decide that continuing and persistent efforts focused on seeking highly efficient and selective catalysts are still needed in future studies.

This chapter summarizes the mainline developments and provides representative examples, trying to give an overall perspective for the homogeneous asymmetric hydrogenation of heteroaromatic compounds catalyzed by transition metal complexes. Other related studies on heterogeneous asymmetric hydrogenation [4] and asymmetric transfer hydrogenation [2e] catalyzed by chiral organocatalysts will not be included in this chapter.

8.2 Asymmetric Hydrogenation of Quinolines

Quinolines are the most studied substrates for the catalytic asymmetric hydrogenation of heteroaromatic compounds. The partially reduced products, chiral 1,2,3,4-tetrahydroquinolines, are important motifs found in naturally occurring alkaloids, pharmaceuticals, and agrochemicals [5].

In 2003, Zhou and coworkers described the first highly enantioselective hydrogenation of quinolines catalyzed by chiral iridium/diphosphine MeO–BiPhep (**L1**, Scheme 8.1) with iodine as activator [6]. A series of 2-substituted tetrahydroquinolines were obtained in good yields with excellent enantioselectivities. Later in 2009, the same group extended the substrate scope to 2-benzyl, 2-functionalized, and 2,3-disubstituted quinolines [7]. Mechanistic studies disclosed that an Ir(III) complex, *in situ* formed by iodine oxidation from Ir(I) precatalyst, may be the catalytic active species, and the hydrogenation pathway involves a 1,4-hydride addition, isomerization, and 1,2-hydride addition.

Thereafter, inspired by Zhou's pioneering work, a variety of transition metal complexes of chiral phosphorus-containing ligands have proven to be successful in the asymmetric hydrogenation of a wide range of quinoline derivatives. For comparing their catalytic activity and stereoselectivity, the data obtained in the hydrogenation of 2-methyl and 2-phenyl quinolines are summarized and listed in Table 8.1.

Chan, Fan, and Xu et al. developed a highly effective and air-stable Ir/diphosphine P-Phos (**L2**) catalyst for the asymmetric hydrogenation of 2-substituted quinolines [8]. The hydrogenation carried out at a substrate to catalyst (S/C) ratio of 2000–50 000 could provide good enantioselectivities with turnover frequency (TOF) up to 4000 h^{-1} and turnover number (TON) up to 43 000 [8b]. And catalyst recover and reuse were achieved using liquid poly(ethylene glycol) dimethyl ether (DMPEG) as reaction medium.

Electron-deficient diphosphine ligand DifluorPhos (**L3**) was found to be highly efficient and enantioselective by Xu and coworkers in the Ir-catalyzed asymmetric hydrogenation of 2-substituted quinolines with quite low catalyst loading (S/C = 50 000) [9]. And Zhou and coworkers disclosed the asymmetric hydrogenation of *N*-phthaloyl-protected aromatic quinoline-3-amines with the same catalyst system; >20 : 1 dr and excellent ee values were obtained, providing a facile access to chiral exocyclic amines [16]. Ohshima, Ratovelomanana-Vidal, and Mashima et al. synthesized a cationic triply halogen-bridged dinuclear Ir(III) complex (**C1a**) with Difluorphos. In the asymmetric hydrogenation of 2-phenyl and 2-alkyl-substituted quinolinium salts, good to excellent enantioselectivities were obtained, although the conversions for 2-alkyl-substituted substrates are not gratifying [10].

Chiral diphosphinite ligands (**L4** and **L5**) derived from H8-1,1′-bi-2-naphthol (BINOL) and 1,1′-spiro-biindane-7,7′-diol, BINOL-derived diphosphonite (**L6**), and monodentate phosphoramidite PipPhos (**L7**) ligands have also been demonstrated to be highly effective and enantioselective for the Ir-catalyzed asymmetric hydrogenation of 2-substituted quinolines [11].

Zhou group and Bolm group utilized ferrocenyloxazoline- and sulfoximine-derived *P,N*-ligand/catalyst (**L8** and **C2**) for the study of Ir-catalyzed asymmetric

Table 8.1 Comparison of catalytic results with representative catalytic systems for the asymmetric hydrogenation of quinolines.

Entry	Ligand	Additive	Catalytic system	S/C	Yield (%)	Ee (%)	Ref.
1	(*R*)-**L1**	I_2	$[Ir(COD)Cl]_2$	200	**2a**: 94	**2a**: 94 (*R*)	[6]
					2b: 95	**2b**: 72 (*R*)	
2	(*R*)-**L2**	I_2	$[Ir(COD)Cl]_2$	10 000	**2a**: 99	**2a**: 91 (*R*)	[8b]
					2b: 98	**2b**: 60 (*R*)	
3	(*R*)-**L3**	I_2	$[Ir(COD)Cl]_2$	10 000	**2a**: 99	**2a**: 94 (*R*)	[9]
4[a)]	(*S*)-**L3**	—	(*S*)-**C1**	50	**2a**: 81	**2a**: 94 (*S*)	[10]
					2b: >95	**2b**: 91 (*S*)	
5	(*R*)-**L4**	I_2	$[Ir(COD)Cl]_2$	100	**2a**: 98	**2a**: 97 (*R*)	[11a]
					2b: 98	**2b**: 87(*R*)	
6	(*R*)-**L5**	I_2	$[Ir(COD)Cl]_2$	2000	**2a**: 100	**2a**: 92(*R*)	[11b]
					2b: 100	**2b**: 65(*R*)	
7	(*S*)-**L6**	I_2	$[Ir(COD)Cl]_2$	50	**2a**: >96	**2a**: 96(*S*)	[11c]
					2b: 42	**2b**: 73(*S*)	
8	(*S*)-**L7**	Piperidine hydrochloride	$[Ir(COD)Cl]_2$/ phosphine	100	**2a**: 100	**2a**: 89(*S*)	[11d]
					2b: 88	**2b**: 88(*S*)	
9	(*S,Sp*)-**L8**	I_2	$[Ir(COD)Cl]_2$	200	**2a**: 95	**2a**: 90 (*R*)	[12a]
					2b: 45	**2b**: 3 (*S*)	
10	*P,N*-ligand	—	(*R*)-**C2**	100	**2a**: >95	**2a**: 87(*S*)	[12b]
11	(*S,R*)-**L10**	—	$[Rh(COD)Cl]_2$	200	**2a**: 99	**2a**: 99(*S*)	[13]
12	Diamine	—	**1a**: (*R,R*)-**C3a**	500	**2a**: 99	**2a**: 99(*R*)	[14a]
			1b: (*S,S*)-**C3c**	100	**2b**: 94	**2b**: 92(*R*)	
13	Diamine	TFA	(*S,S*)-**C4**	500	**2a**: 99	**2a**: 98(*S*)	[14f]
					2b: 90	**2b**: 79(*R*)	
14	—	Phenanthridine	$[Ru(p\text{-cymene})I_2]_2$	200	**2b**: 95	**2b**: 91(*S*)	[15]
			(*S*)-**C5a**	25			

a) 2-Phenylquinolinium salt was used as substrate.

hydrogenation of 2-substituted quinolines, respectively [12]. The reactivity and enantioselectivity obtained with **L8** are higher than those with **C2**.

Zhou and coworkers reported the first example of Pd/diphosphine (**L9a**)-catalyzed asymmetric hydrogenation of 2-alkyl 3-phthalimido-substituted quinolines with the addition of a Brønsted acid to activate the substrates [17]. Zhang et al. described an Rh/thiourea chiral diphosphine ligand (**L10**) complex for the asymmetric hydrogenation of 2-substituted quinolium salts [13]. The anion binding between the catalyst and substrate is believed to be important to achieve high reactivity and enantioselectivity.

Phosphine-free chiral diamine ligands are found to be highly effective and enantioselective in quinoline hydrogenation [14]. Fan and coworkers established the first successful example of Ru/N-monosulfonylated diamine complexes (**C3**)-catalyzed asymmetric hydrogenation of quinolines without activation of either catalyst or substrates. A wide range of 2-alkyl, 2-aryl, 2-functionalized, and 2,3-disubstituted 1,2,3,4-tetrahydroquinolines were obtained with excellent reactivity and enantioselectivity in organic solvents [14a]. The reaction was proved to be also effective and enantioselective in ionic liquids [14b, c], undegassed water [14d], or even under solvent-free conditions [14e]. Interestingly, an unusual chemoselectivity was observed in ionic liquids [14c]. The hydrogenation of quinolines bearing a carbonyl group was more selective for the C=N bonds over the C=O bonds, and catalyst recycling was also attained. Meanwhile, Ir/diamine catalyst (**C4**) was demonstrated to be successful for the asymmetric hydrogenation of quinolines [14f]. The reaction proceeded well at a 1000/1 S/C ratio in undegassed methanol without inert gas protection. The mechanistic studies reveal that quinoline is reduced via an ionic and cascade reaction pathway as aforementioned for the Ir/diphosphine catalytic system, and the hydrogen addition undergoes a stepwise H^+/H^- transfer process outside the coordination sphere. Moreover, density functional theory (DFT) calculations indicate that a TfO^- anion-involved 10-membered-ring transition state and CH$\cdots\pi$ attraction between the η^6-arene ligand and the fused phenyl ring of dihydroquinoline intermediate determined the absolute configuration of 1,2,3,4-tetrahydroquinoline product (Scheme 8.2).

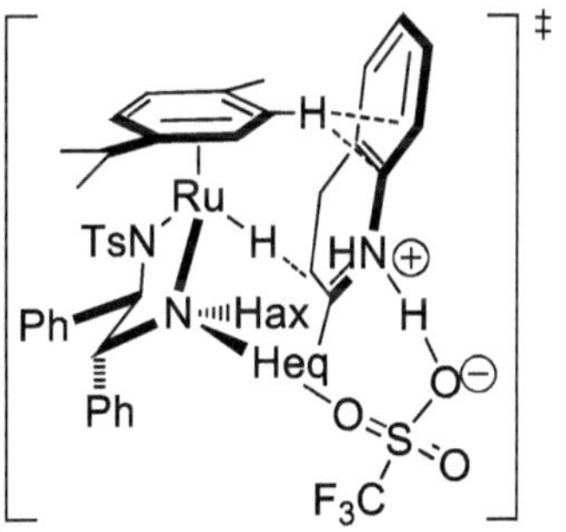

Scheme 8.2 Preferred transition structure for the hydride transfer to 3,4-dihydroquinoline. Source: Fan et al. [14a].

In recent years, FLPs have shown great potentials in metal-free hydrogenation. In 2015, Du and coworkers reported the first asymmetric hydrogenation of 2,3- and 2,4-disubstituted and 2,3,4-trisubstituted 2-arylquinolines using the chiral borane catalysts derived *in situ* from chiral diene (**L11**) [18]. *Cis*-products were mainly acquired with excellent enantioselectivities except for 2,3-disubstituted quinolines for which moderate to good ee values were observed.

Biomimetic asymmetric hydrogenation of 2-arylquinolines using catalytic amount of phenanthridine as additive was accomplished by Zhou and coworkers [15].

A combination of achiral $[Ru(p\text{-cymene})I_2]_2$ and chiral phosphoric acid (**C5a**) was used as catalyst. The hydrogenated dihydrophenanthridine was believed to act as a reductant in the hydrogenation of quinolines. It is noteworthy that an unexpected reversal of enantioselectivity was observed between the reactions promoted by different nicotinamide adenine dinucleotide (phosphate) (NAD(P)H) models.

8.3 Asymmetric Hydrogenation of Quinoxalines

Quinoxalines are the earliest studied aromatic compounds in asymmetric hydrogenation [19]. At that time, results obtained are not satisfying; the highest ee value (90%) was achieved only under harsh reaction conditions [20]. Inspired by the breakthrough made in the asymmetric hydrogenation of quinolines, quinoxalines have become the research objects. The most successful catalytic systems used for the asymmetric hydrogenation of quinoxalines are the same as those for quinolines hydrogenation. The main results are summarized in Table 8.2.

In 2009, Xu, Fan, and Chan et al. described a highly efficient asymmetric hydrogenation of quinoxalines using Ir/diphosphinite (**L4**)/I_2 catalytic system at a low catalyst loading (up to S/C = 20 000) [21]. A series of 2-substituted 1,2,3,4-terahydroquinoxalines were acquired with full conversions and excellent enantioselectivities.

Monodentate phosphoramidite ligand (*S*)-PipPhos was demonstrated to be highly effective and enantioselective in the Ir-catalyzed asymmetric hydrogenation of 2-substituted quinoxalines by Minnaard, Feringa, and de Vries et al. with the use of piperidine hydrochloride as additive [22].

Ohshima, Mashima, and Ratovelomanana-Vidal et al. utilized the triply halogen-bridged dinuclear Ir(III) complex (**C1a**) with DifluorPhos as chiral ligand in the asymmetric hydrogenation of 2-substituted quinoxalines, and excellent results were also obtained [23]. They found that the addition of an achiral amine could further improve the reactivity and enantioselectivity for the asymmetric hydrogenation of 2-arylquinoxalines [23c].

The Ru/diphosphine/diamine complexes (**C6**) were applied in the asymmetric hydrogenation of 2-substituted quinoxalines by Ohkuma and coworkers [24]. Excellent reactivity and enantioselectivity were observed for both 2-phenyl and 2-alkylquinoxalines substrates.

Fan and coworkers demonstrated that the cationic Ru/diamine complexes (**C3**) were highly effective and highly enantioselective for the asymmetric hydrogenation of 2-alkyl-, 2-aryl-, and 2,3-dialkyl-substituted quinoxalines, although the diastereoselectivity is not high [25]. The catalyst counteranion was found to obviously affect the enantioselectivity and/or diastereoselectivity.

FLP-catalyzed asymmetric hydrogenation of a variety of 2,3-disubstituted quinoxalines was successfully carried out by Du and coworkers with chiral alkene ((*S*)-**L11**)/Piers' borane as catalyst [28a]. Excellent *cis*-selectivity and high enantioselectivity were furnished. The same group also developed a chiral

Table 8.2 Comparison of catalytic results with representative catalytic systems for the asymmetric hydrogenation of quinoxalines.

Entry	Ligand	Additive	Catalytic system	S/C	Yield (%)	Ee (%)	Ref.
1	(*R*)-**L4**	I_2	$[Ir(COD)Cl]_2$	5000	**4a**: >99	**4a**: 93 (*S*)	[21]
					4b: >99	**4b**: 85 (*S*)	
2	(*S*)-**L7**	Piperidine hydrochloride	$[Ir(COD)Cl]_2$	100	**4a**: 85	**4a**: 96 (*S*)	[22]
					4b: 92	**4b**: 86 (*S*)	
3	(*S*)-**L3**	—	(*S*)-**C1**	100	**4a**: 99	**4a**: 94 (*S*)	[23a]
					4b: 98	**4b**: 89 (*S*)	[23b]
4	(*S*)-**L3**	MPA[a]	(*S*)-**C1**	200	**4b**: >99	**4b**: 93 (*S*)	[23c]
5	—	—	(R_N,R_P)-**C6b**	1000	**4a**: >99	**4a**: >99 (*S*)	[24]
			(R_N,R_P)-**C6a**	100	**4b**: >99	**4b**: 96 (*R*)	
6	Diamine	—	(*R*,*R*)-**C3d**	100	**4a**: 95	**4a**: 98 (*R*)	[25]
			(*S*,*S*)-**C3i**	100	**4b**: 90	**4b**: 94 (*S*)	
7	—	—	[Ru(*p*-cymene) $I_2]_2$	200	**4b**: 96	**4b**: 90 (*R*)	[26]
			(*S*)-**C5b**	83			
8	—	—	Fe–H complex	20	**4b**: 90	**4b**: 88 (*S*)	[27]
			(*R*)-**C5b**	100			
9	—	Phenanthridine	[Ru(*p*-cymene) $I_2]_2$	200	**4b**: 99	**4b**: 90 (*R*)	[15]
			(*S*)-**C5a**	100			

a) MPA, *N*-methyl-*p*-anisidine.

sulfinamide/Piers' borane-catalyzed asymmetric transfer hydrogenation of 2,3-disubstituted quinoxalines with ammonia borane as reductant [28b].

Zhou and Fan et al. developed a metal/Brønsted acid relay catalysis for the enantioselective reduction of quinoxalines using a combination of achiral metal complex ([Ru(*p*-cymene)$I_2]_2$) and chiral phosphoric acid **C5b** and hydrogen gas as reductant [26]. A range of 2-aryl-1,2,3,4-tetrahydroquinoxalines were obtained with high yield and enantioselectivity via a convergent asymmetric disproportionation process. Subsequently, Beller and coworkers described a similar catalytic system consisting of an achiral iron complex and a chiral phosphoric acid **C5b** for the asymmetric hydrogenation of 2-substituted quinoxalines [27]. Good results were obtained with 2-arylquinoxalines, while the enantioselectivities for 2-alkylquinoxalines were moderate.

The catalytic system [Ru(*p*-cymene)$I_2]_2$/**C5** for the biomimetic asymmetric hydrogenation of quinolines was demonstrated to be also effective for

2-arylquinoxalines [15]; excellent reactivity and enantioselectivity were achieved under mild conditions.

8.4 Asymmetric Hydrogenation of Isoquinolines

Comparing with quinolines, the asymmetric hydrogenation of isoquinolines is more challenging due to the reasons listed below: First, the reduced 1,2,3,4-tetrahydroisoquinoline products are alkyl amines, which are stronger coordinating agents with transition metals to deactivate the catalyst. Second, the catalytic mechanism for the asymmetric hydrogenation of different isoquinoline derivatives is more complicated than that for quinolines, no matter the reaction pathway or the unsaturated double bond reduced in the final step may be varied for different catalytic systems.

Zhou and coworkers developed a substrate-activating strategy for the Ir/diphosphine ((*S*)-**L9b**)-catalyzed asymmetric hydrogenation of isoquinolines using chloroformates as activating reagents (Scheme 8.3a) [29a]. Chiral dihydroisoquinolines were obtained with moderate to high yields and ee values. To improve the reaction, *N*-benzyl isoquinolinium salts were prepared and subjected to the hydrogenation using diphosphine (R_{ax},*S*,*S*)-**L12** as ligand (Scheme 8.3b) [29b]. Excellent yields and ee values were observed for 1-substituted isoquinolines, albeit the enantioselectivity for 3-substituted substrates was not so good.

Isoquinolinium salts were also demonstrated to be ideal activated substrates for asymmetric hydrogenation. In 2013, Mashima and coworkers applied their halogen-bridged dinuclear iridium(III) complexes (**C1a**) to the asymmetric hydrogenation of 1- and 3-substituted and 1,3-disubstituted isoquinolinium hydrochlorides (Scheme 8.3c) [30]. Excellent results were achieved

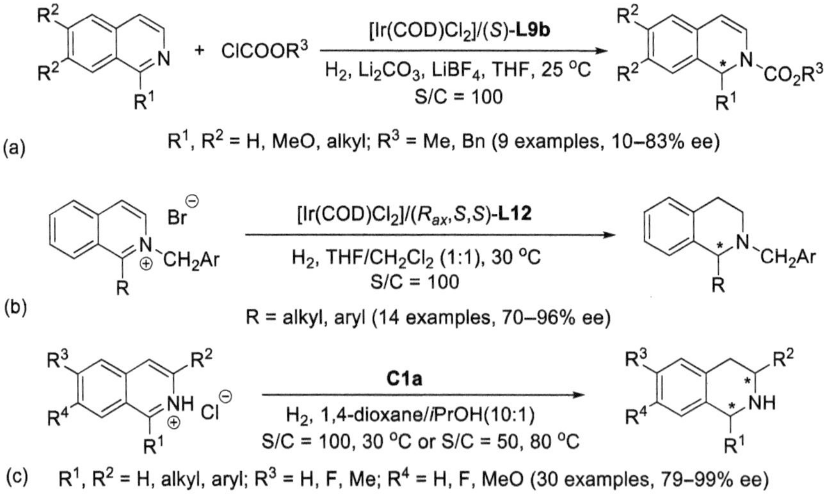

Scheme 8.3 Asymmetric hydrogenation of isoquinolines. Source: (a) Based on Lu et al. [29a]. (b) Based on Ye et al. [29b]. (c) Based on Iimuro et al. [30a] and Kita et al. [30b].

for all 1,2,3,4-tetrahydroisoquinolines products, and a Cl···H=N hydrogen bonding-involved six-membered outer-sphere ring was proposed for the transition state [30b].

Recently, Zhou and coworkers reported that an Ir/diphosphine ((*R*)-**L9b**)-catalyzed asymmetric hydrogenation of isoquinolines proceeded well by employing traceless trichloroisocyanuric acid (TCCA) as substrate-activating agent [31], affording tetrahydroisoquinolines with good to excellent catalytic activity and enantioselectivity. The mechanistic studies indicated that isoquinolinium chlorides *in situ* formed were the true substrates for hydrogenation.

Meanwhile, using thiourea-containing chiral diphosphine (*R*,*S*)-**L10** as ligand, an Rh-catalyzed asymmetric hydrogenation of isoquinolinium hydrochlorides was also realized by Zhao and Zhang et al. [13]. The anion binding between the ligand and substrate was implied to play an important role in this transformation.

The asymmetric hydrogenation of 3,4-disubstituted isoquinolines was conducted with Ir/diphosphine ((*R*)-**L13**) by Zhou and coworkers [32a]. With 1-bromo-3-chloro-5,5-dimethyl-hydantoin as halogen source to activate the catalyst, 3,4-disubstituted tetrahydroisoquinolines were produced with good to excellent yields and ee values. Subsequently, the same group described an efficient synthesis of 4-fluoro-3-alkyl tetrahydroisoquinolines by utilizing the same catalytic system starting from isoquinolinium hydrochlorides [32b].

8.5 Asymmetric Hydrogenation of Pyridines and Pyrazines

Monocyclic six-membered *N*-heteroarenes, such as pyridines and pyrazines, are more challenging substrates for asymmetric hydrogenation in comparison to their benzofused counterparts (quinolines, isoquinolines, and quinoxalines) due to their higher stability and stronger coordinating ability to transition metal centers. Therefore, to gain high catalytic activity and selectivity in asymmetric hydrogenation, the most used strategy is substrate activation [3, 33].

In 2005, Charette and coworkers developed an Ir/*P*,*N* catalyst ((*S*)-**C7**) for the asymmetric hydrogenation of *N*-iminopyridinium ylides, affording piperidines with moderate to high yields and enantioselectivities for a range of 2-alkyl-substituted substrates (Scheme 8.4a) [34].

From the activated *N*-benzyl-pyridium bromides, a variety of chiral 2-arylpiperidines were obtained with excellent activities and enantioselectivities through Ir/diphosphine ((*R*)-**L13**)-catalyzed asymmetric hydrogenation by Zhou and coworkers (Scheme 8.4b) [35]. However, the results achieved for the 2-alkylpyridium salts were less satisfactory.

Chen and Zhang et al. also realized the Ir-catalyzed asymmetric hydrogenation of *N*-benzyl-2-substituted pyridinium salts with similarly high yields and ee values as those obtained by Zhou's group [36]. An unusual steric-hindered diphosphine ligand ((*R*)-**L14**) proved to be critical for this enantioselective hydrogenation.

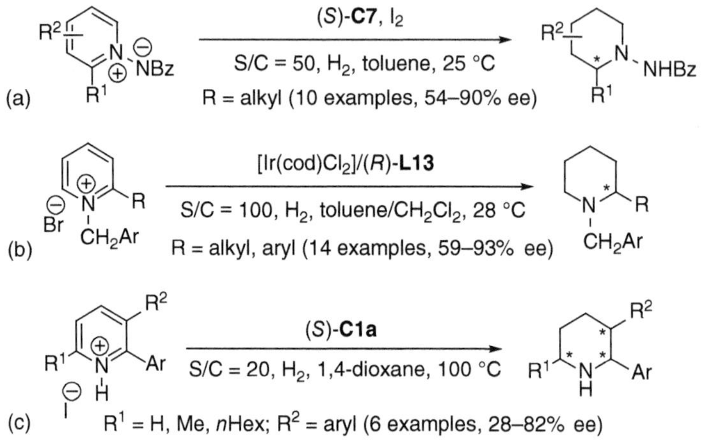

Scheme 8.4 Asymmetric hydrogenation of pyridines. Source: (a) Based on Legault and Charette [34]. (b) Based on Ye et al. [35]. (c) Based on Kita et al. [40a].

Ir/(*S*,*S*)-f-Binaphane ((*S*,*S*)-**L15**) was applied to the asymmetric hydrogenation of 6-substituted 3-hydroxypyridinium salts by Zhou and coworkers [37]. The fully reduced *trans* 6-substituted piperidin-3-ols were mainly obtained with excellent reactivity and enantioselectivity, albeit the *trans* to *cis* ratios were less than 9 : 1. Direct Swern oxidation of piperidin-3-ols led to chiral piperidin-3-ones with excellent enantioselectivity, and *cis* piperidines were also achieved via the reduction of piperidin-3-ones with K-selectride.

Qu and Kozlowski et al. described the highly enantioselective hydrogenation of 2-alkyl and 2-arylpyridinium salts using Ir/*P*,*N*-ligand MeO–BoPhos ((*S*,*S*)-**L16**) complex [38]. It is noteworthy that the reaction could tolerate a range of *O*- and/or *S*-containing aryl and heteroaryl substituents. The DFT calculations suggest an outer-sphere dissociative mechanism, and the final C2-hydride transfer to the iminium salt decided the absolute configuration of the product.

The asymmetric hydrogenation of N-benzylated 3-substituted pyridinium salts was described by Lefort and coworkers using Rh/diphosphine JosiPhos J002-2 (**L17a**) as catalyst [39]. It was noticed that the addition of an equivalent of organic base remarkably increased the reactivity and enantioselectivity.

Pyridinium halogen salts are another type of activated substrates for asymmetric hydrogenation. Mashima and coworkers employed their triply halogen-bridged dinuclear Ir(III) complex ((*S*)-**C1a**) for the highly diastereoselective asymmetric hydrogenation of multisubstituted pyridinium halogen salts to acquire *cis* piperidine derivatives (Scheme 8.4c) [40a]. But the enantiomeric excesses are between 28% and 82%. Later, the same group realized the asymmetric hydrogenation of 3-amido-2-arylpyridinium hydrochlorides using the same catalyst system [40b]. High ee values were observed only with the addition of an equivalent of Brønsted acid.

Zhou and coworkers realized the highly efficient and enantioselective hydrogenation of trisubstituted pyridinium hydrochlorides with Ir/(*R*)-DifluorPhos ((*R*)-**L3**) catalyst, affording *cis* 3-trifluoromethyl-substituted piperidines with excellent diastereoselectivity [41]. Similar catalyst system with C_3*TunePhos ((R_{ax},*S*,*S*)-**L12**) was also effective for the asymmetric hydrogenation of 2,6-disubstituted substrates.

Recently, activated pyrazines have proven to be smoothly hydrogenated catalyzed by chiral iridium catalysts. In 2016, Zhou and coworkers reported the asymmetric hydrogenation of 3-substituted and 3,5-disubstituted *N*-benzyl pyrazinium salts using Ir/diphosphine catalysts (Scheme 8.5a) [42]. In the meantime, Mashima and coworkers applied the triply halogen-bridged dinuclear Ir(III) complex ((*S*)-**C1a**) for the asymmetric hydrogenation of tosylamido-substituted pyrazines (Scheme 8.5b) [43]. The addition of an equivalent of *N,N*-dimethylanilinium bromide is essential to enhance both catalytic activity and enantioselectivity. The corresponding partially reduced chiral tetrahydropyrazines with an amidine skeleton were obtained in high yields with high ee values, which could be further reduced to provide chiral *tert*-butyl 3-phenylpiperazine-1-carboxylate.

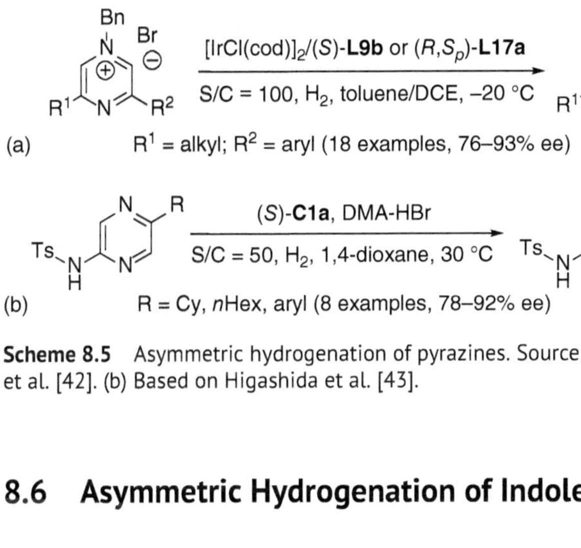

Scheme 8.5 Asymmetric hydrogenation of pyrazines. Source: (a) Based on Huang et al. [42]. (b) Based on Higashida et al. [43].

8.6 Asymmetric Hydrogenation of Indoles and Pyrroles

Five-membered heteroaromatics are less stabilized by the resonance energy compared to the six-membered ones. The asymmetric hydrogenation of N-protected indoles and pyrroles as well as the N-unprotected ones has been successfully accomplished using different chiral transition metal complexes.

Early in 2000, Kuwano and Ito et al. reported the first example of highly enantioselective hydrogenation of five-membered heteroarenes catalyzed with Rh/*trans*-chelating diphosphine PhTRAP ((*S*,*S*)-(*R*,*R*)-**L-18**) complex (Scheme 8.6a) [44]. A variety of chiral *N*-acetyl-protected 2-substituted indolines were obtained in high yields with high enantioselectivities.

Subsequently, the asymmetric hydrogenation of a series of N-protected indoles and pyrroles with the similar catalytic system was realized in which PhTRAP was used. Various *N*-tosyl 3-substituted indoles and *N*-Boc-protected 2- and 3-substituted indoles were successfully hydrogenated with excellent enantioselectivities using Rh or Ru catalyst, respectively (Scheme 8.6b,c) [45a, b]. The Ru-catalyzed asymmetric

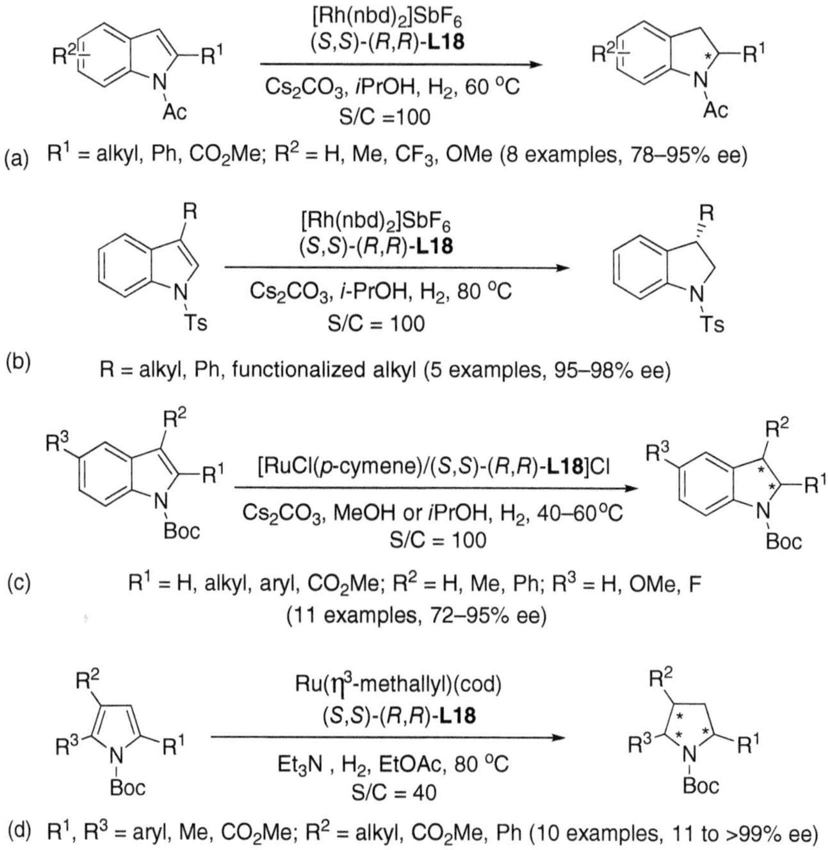

Scheme 8.6 Asymmetric hydrogenation of indoles and pyrroles catalyzed by Rh/TRAP complexes. Source: (a) Based on Kuwano et al. [44]. (b) Based on Kuwano et al. [45a]. (c) Based on Kuwano and Kashiwabara [45b] and (d) Based on Kuwano et al. [45c].

hydrogenation of *N*-Boc-2,3,5-trisubstituted pyrroles was also carried out, but only partially hydrogenated dihydropyrroles were observed in quantitative yields with excellent ee values for triphenyl-substituted substrates (Scheme 8.6d) [45c].

Pfaltz and coworkers demonstrated that Ir/*P*,*N*-ligand (**L19a**, **L20**, and **L21**) complexes were efficient in the asymmetric hydrogenation of *N*-Boc-, *N*-acetyl-, and *N*-tosyl-indoles [46]. It was found that the right combination of catalyst and protecting group was critical to achieve high catalytic activity and excellent enantioselectivity for various 2- and 3-substituted indoles.

Pd/diphosphine ((*R*)-**L4**) catalysts were developed for the asymmetric hydrogenation of simple N-unprotected indoles by Zhou and Zhang et al. (Scheme 8.7a) [47]. High yields and good to excellent enantioselectivities could be acquired with the addition of an equivalent of strong Brønsted acid for a wide range of 2-substituted and 2,3-disubstituted indoles. The mechanistic study revealed that the reaction underwent a stepwise outer-sphere and ionic mechanism, and the high

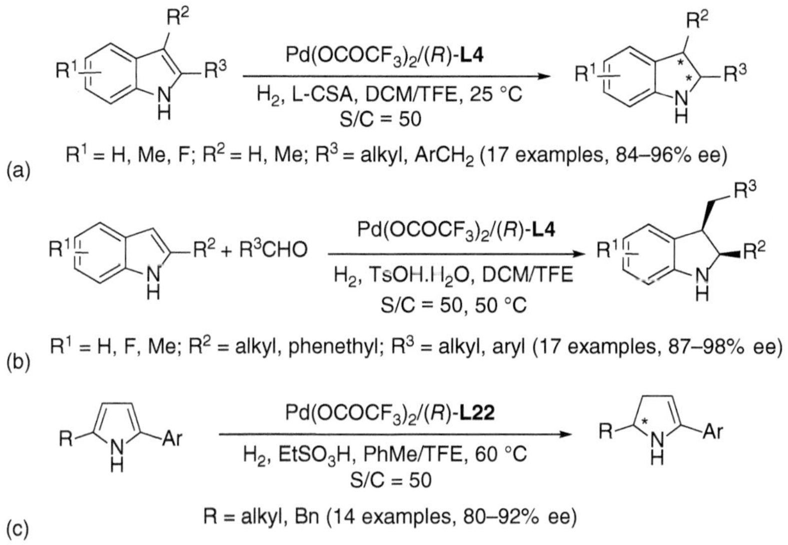

Scheme 8.7 Pd-catalyzed asymmetric hydrogenation of unprotected indoles and pyrroles. Source: (a) Based on Wang et al. [47]. (b) Based on Duan et al. [48b]. (c) Based on Wang et al. [49].

enantioselectivity arose from a hydrogen bonding between N—H of the iminium salt and oxygen of the coordinated trifluoroacetate in an eight-membered-ring transition state for the final hydride transfer. Afterward, Zhou and coworkers developed a rapid and divergent approach for the synthesis of chiral 2,3-disubstituted indolines through a Brønsted acid/Pd-complex-promoted cascade Friedel–Crafts reaction/asymmetric hydrogenation (Scheme 8.7b) [48].

Pd/diphosphine (*R*)-C4-TunePhos ((*R*)-**L22**) complex was proven to be effective for the partial hydrogenation of simple pyrroles with a Brønsted acid additive by Zhou, Fan, and coworkers (Scheme 8.7c) [49]. A series of 2-alkyl-5-arylpyrroles were converted to chiral 1-pyrrolines in good yield with up to 92% ee.

A cocatalysis of transition metal and anion binding was suggested by Chung and Zhang et al. using a chiral thiourea-containing diphosphine ZhaoPhos ((*S*,*R*)-**L10**) ligand for the Rh-catalyzed highly enantioselective asymmetric hydrogenation of 2-substituted and 2,3-disubstituted N-unprotected indoles [50]. Two equivalents of HCl were needed to form the iminium chloride intermediates. And the noncovalent interaction between the thiourea motif and the chloride anion was demonstrated to play a crucial role in the reaction.

Recently, Fan and coworkers applied the Ru/diamine complexes ((*R*,*R*)-**C3e** and (*R*,*R*)-**C3h**) for the asymmetric hydrogenation of 2-substituted and 2,3-disubstituted N-unprotected indoles [51]. Using hexafluoroisopropanol (HFIP) as reaction medium, a variety of chiral indolines were obtained with excellent reactivities, diastereoselectivities, and enantioselectivities under very mild conditions. Almost at the same time, similar work was reported by Touge and Arai [52].

8.7 Asymmetric Hydrogenation of Heteroarenes with Multi-*N*-Heterocycles

Most heteroarenes with multi-*N*-heterocycles usually contain two and more nitrogen atoms which could more strongly coordinate to transition metal centers to deactivate the catalysts [53]. Furthermore, the regio-, chemo-, and/or ring selectivity of the asymmetric hydrogenation may become more complicated, leading to various chiral products difficult for purification.

Chiral Ru/NHC (**L23**) complex was applied by Glorius and coworkers to the asymmetric hydrogenation of indolizines, 1,2,3-triazolo-[1,5-*a*]pyridines, and imidazo[1,2-*a*]pyridines in which one of the nitrogen atoms is in the bridgehead position of the fused heteroarenes units (Scheme 8.8) [54]. High yield, high enantioselectivity, and complete regioselectivity for the six-membered pyridine ring were achieved for all substrates investigated. The hydrogenated products could be further derivatized to other functional molecules.

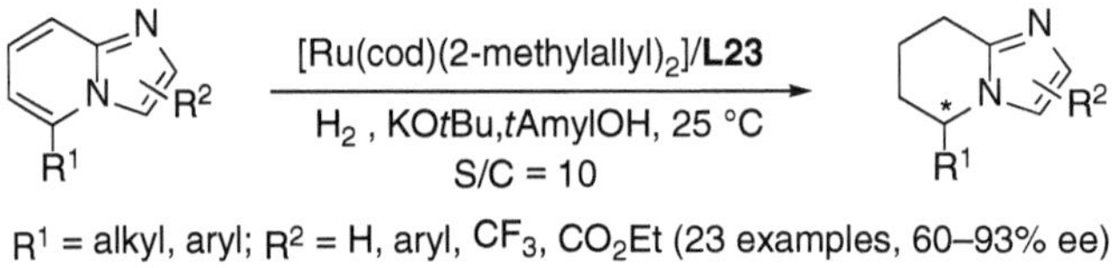

Scheme 8.8 Enantioselective hydrogenation of imidazo[1,2-*a*]pyridines by Ru/NHC complexes. Source: Based on Schlepphorst et al. [54b].

Kuwano and coworkers disclosed the asymmetric hydrogenation of *N*-Boc-protected azaindoles with Ru/diphosphine PhTRAP ((*S*,*S*)-(*R*,*R*)-**L-18**) catalyst; the five-membered ring was selectively hydrogenated (Scheme 8.9) [55]. A series of 2-substituted azaindoles, including 4-, 5-, 6-, and 7-azaindoles, were reduced to afford chiral azaindolines with moderate to high yields and enantioselectivities. Moreover, the azaindolines were further hydrogenated to chiral *cis*-octahydroazaindoles without the formation of other diastereomers or loss of enantioselectivities.

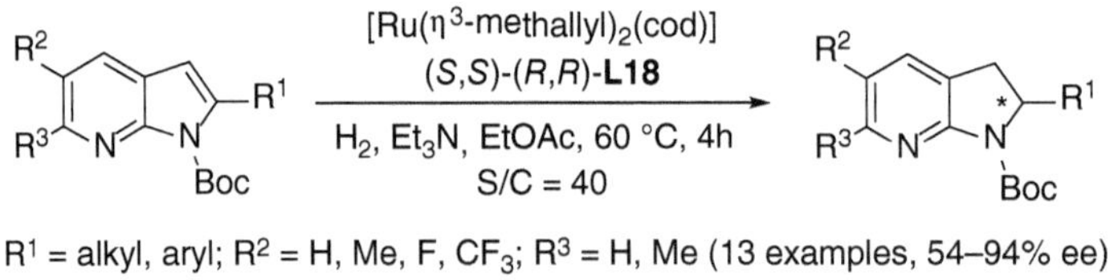

Scheme 8.9 Asymmetric hydrogenation of 7-azaindoles by Ru/TRAP complexes. Source: Based on Makida et al. [55].

By employing Ir/diphosphine complexes, Zhou and coworkers realized the asymmetric hydrogenation of *N*-benzyl-protected pyrrole[1,2-*a*]pyrazinium salts and unprotected pyrrole[1,2-*a*]pyrazines with diphosphine (R_{ax},*S*,*S*)-**L12** and (*R*)-**L24**

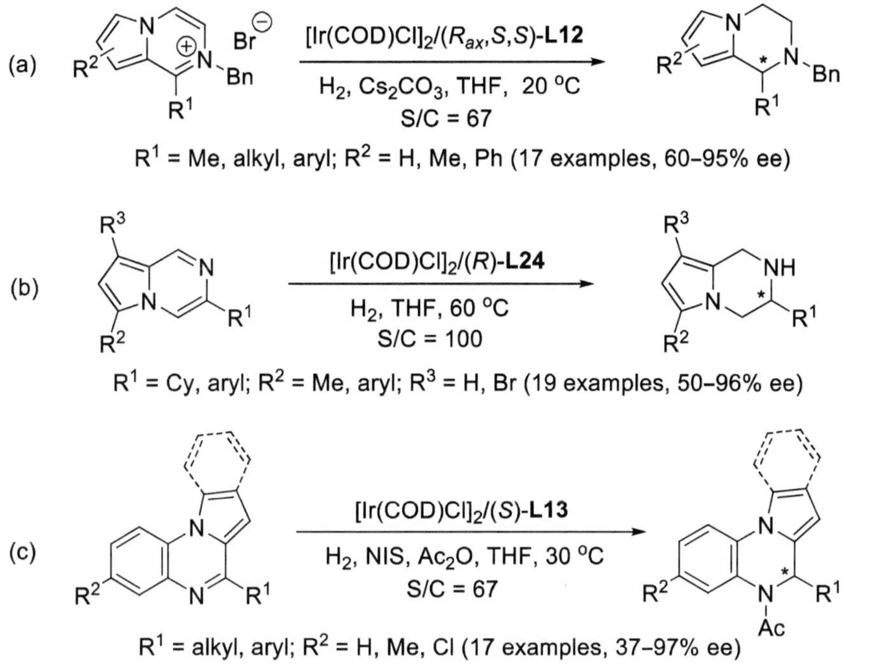

Scheme 8.10 Asymmetric hydrogenation of *N*-benzyl-protected pyrrole[1,2-*a*]pyrazinium salts, unprotected pyrrole[1,2-*a*]pyrazines, and pyrrolo/indolo[1,2-*a*]quinoxalines by Zhou and coworkers. Source: (a) Based on Huang et al. [56a]. (b) Based on Hu et al. [56b]. (c) Based on Hu et al. [56c].

as ligand, respectively (Scheme 8.10a,b) [56a, b]. The six-membered ring was selectively reduced, furnishing the chiral 1,2,3,4-tetrahydropyrrolo[1,2-*a*]pyrazines with high yields and ee values. More recently, the same group disclosed the asymmetric hydrogenation of 4-substituted pyrrolo/indolo[1,2-*a*]quinoxalines and 6-substituted phenanthridines with Ir/diphosphine ((*S*)-**L13**) complex (Scheme 8.10c) [56c]. The six-membered *N*-heteroarene unit was selectively reduced with excellent enantioselectivity. It is worth mentioning that 4 equiv. of acetic anhydride was needed for the *in situ* protection of hydrogenation products, inhibiting potential rearomatization of products and catalyst poisoning. And catalytic amount of *N*-iodosuccinimide (NIS) was also needed as additive to generate more active catalytic species and then improve the catalytic reactivity.

1,10-Phenanthrolines are versatile bidentate ligands for transition metal catalysis and are rarely treated as substrates of hydrogenation. Fan and coworkers developed the first highly enantioselective and diastereoselective hydrogenation of substituted 1,10-phenanthrolines using Ru/chiral diamine complexes (**C3d**) as catalysts (Scheme 8.11) [57]. The partially reduced 1,2,3,4-tetrahydro-1,10-phenanthrolines could be easily obtained with a combination of hydrogenation/dehydrogenation process. In addition, the fully hydrogenated 1,2,3,4,7,8,9,10-octahydro-1,10-phenanthrolines were prepared in high yield with excellent enantioselectivities and diastereoselectivities under high catalyst loading (2.0 mol% **C3d**).

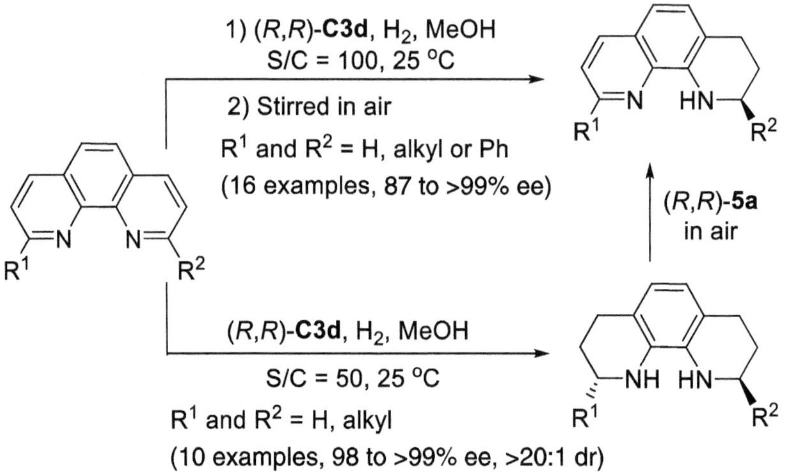

Scheme 8.11 Ru/diamine complexes-catalyzed asymmetric hydrogenation of 1,10-phenanthrolines by Fan and coworkers. Source: Based on Wang et al. [57].

Later, the same group established the first asymmetric hydrogenation of a broad range of 2,6-disubstituted and 2,3,6-trisubstituted 1,5-naphthyridines and 2,7-disubstituted 1,8-naphthyridines catalyzed by Ru/chiral cationic diamine complexes (**C3**) (Scheme 8.12a,b) [58]. Full conversions and up to 99% ee values were achieved for the tetrahydronaphthyridine products. It was observed that the hydrogenation was more selective for the electron-rich pyridine ring, and the level of ring selectivity was determined by the difference of electronic properties between the two substituents.

Recently, the substrate scope of Ru/chiral diamine ((*R*,*R*)-**C3d** and (*R*,*R*)-**C3g**)-catalyzed asymmetric hydrogenation of *N*-heteroarenes was expanded to 2,2′-bisquinolines, 2,2′-bisquinoxalines, and bis(quinoline-2-yl)methanes by Fan and coworkers (Scheme 8.12c) [59]. Full conversions and an unprecedented level of enantioselectivity were achieved for all the substrates investigated. These works provide a practical and facile approach to optically pure endocyclic diamines, which are potential novel chiral ligands for asymmetric catalysis. Particularly, these chiral diamines were easily derived to five- and six-membered NHCs which are difficult to access through other methods.

8.8 Asymmetric Hydrogenation of Other *N*-Heteroarenes

The asymmetric hydrogenation of heteroaromatic compounds with two or more *N* or *O* heteroatoms in a single ring has been investigated. In 2011, Kuwano and coworkers disclosed the first successful catalytic asymmetric reduction of five-membered aromatic rings containing two or more heteroatoms via asymmetric hydrogenation [60]. With the use of Ru/diphosphine PhTRAP ((*S*,*S*)-(*R*,*R*)-**L18**)

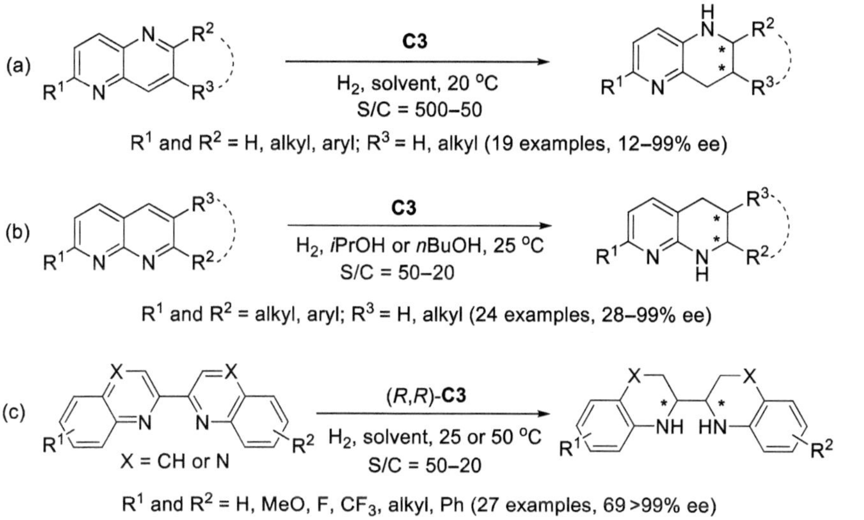

Scheme 8.12 Ru/diamine complexes-catalyzed asymmetric hydrogenation of 1,5-naphthyridines, 1,8-naphthyridines, and 2,2′-bisquinolines by Fan and coworkers. Source: (a) Based on Zhang et al. [58a]. (b) Based on Ma et al. [58b]. (c) Based on Ma et al. [59a].

complex, a series of *N*-Boc-4-alkyl-2-phenylimidazoles and 4- and 5-substituted 2-phenyloxazoles were hydrogenated to give partially reduced dihydro products in high yields and good to excellent enantioselectivities (Scheme 8.13a). These C=N double bond-containing unsaturated products could be further transformed to important chiral 1,2-diamines and β-amido alcohols.

Later, the same group realized the asymmetric hydrogenation of six-membered 2,4-disubstituted pyrimidines with Ir/diphosphine Josiphos (**L17a**)/I_2 catalytic system, providing partially reduced 1,4,5,6-tetrahydropyrimidines in high yields

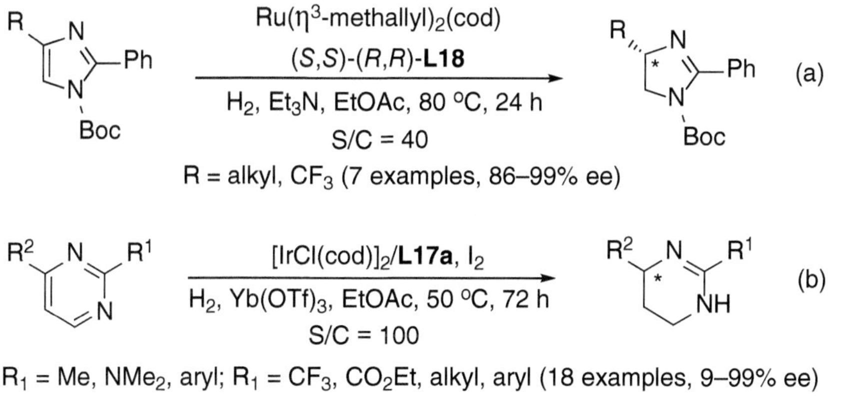

Scheme 8.13 Asymmetric hydrogenation of *N*-Boc-imidazoles and pyrimidines by Kuwano and coworkers. Source: (a) Based on Kuwano et al. [60]. (b) Based on Kuwano et al. [61]

and excellent enantioselectivities (Scheme 8.13b) [61]. More importantly, $Yb(OTf)_3$ additive was demonstrated to remarkably improve both the activity and stereoselectivity of the reaction.

Recently, Shi and Zhou et al. have made continuous contributions in the asymmetric hydrogenation of heteroarenes with two heteroatoms. By the application of Pd/diphosphine Josiphos ((*R*,*Sp*)-**L17**)/Brønsted acid catalytic system, the asymmetric hydrogenation of mono-, di-, and tri-substituted 2-hydroxypyrimidines was carried out to offer chiral cyclic ureas with good yields and enantioselectivities (Scheme 8.14a) [62a]. Subsequently, a variety of 4,6-disubstituted 2-hydroxypyrimidines were subjected to the asymmetric hydrogenation with Ir/diphosphine f-Binaphane ((*S*,*S*)-**L15**)/1,3,5-TCCA catalytic system [62b]. Excellent *cis* diastereoselectivity and enantioselectivity were obtained. The cyclic urea products could be readily transformed to chiral 1,3-diamines and thiourea derivatives without loss of optical purity. It was also found that the equilibrium of the lactame–lactime tautomerism of 2-hydroxypyrimidine was more toward the oxo form with lower aromaticity in the presence of the *in situ* generated hydrogen halide, which effectively enhanced reactivity.

The similar Pd/diphosphine MeO-Biphep ((*S*)-**L1**) or TangPhos ((*S*,*S*′,*R*,*R*′)-**L25**)/Brønsted acid catalytic system was demonstrated to be successful by the same group in the asymmetric hydrogenation of five-membered 3-perfluoroalkyl-substituted pyrazol-5-ols (Scheme 8.14b) [63]. A wide range of 2,5-disubstituted and 2,4,5-trisubstituted pyrazolidinones were thus prepared with excellent yield and enantioselectivity. The Brønsted acid-promoted tautomerization of pyrazol-5-ols is crucial for this asymmetric hydrogenation.

Mashima and coworkers realized the asymmetric hydrogenation of several quinazolinium salts by the application of triply halogen-bridged dinuclear Ir(III) complex ((*S*)-**C1b**) (Scheme 8.15) [64]. The selectivity for the fully hydrogenated

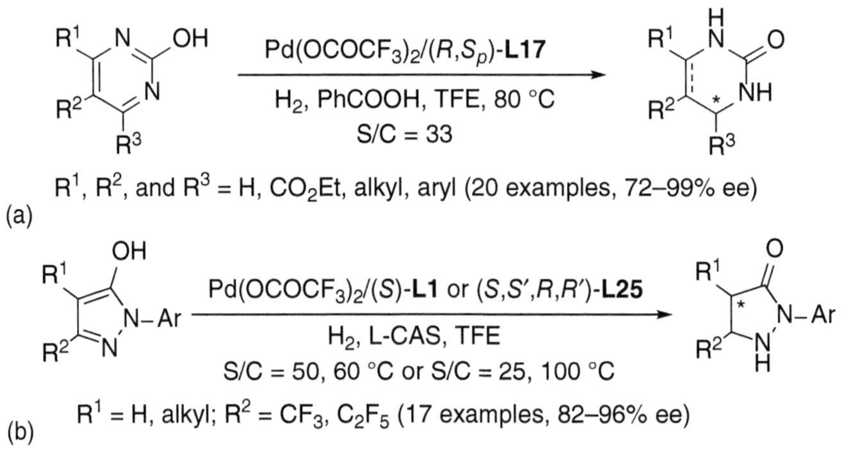

Scheme 8.14 Asymmetric hydrogenation of 2-hydroxypyrimidines and pyrazol-5-ols by Zhou and coworkers. Source: (a) Based on Feng et al. [62a]. (b) Based on Chen et al. [63].

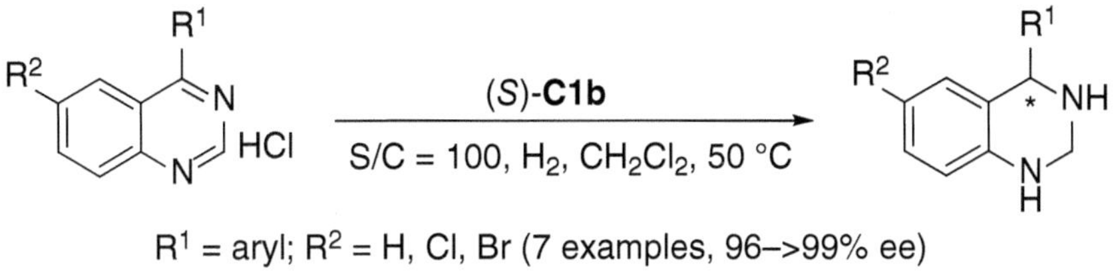

Scheme 8.15 Asymmetric hydrogenation of quinazolinium salts by Mashima and coworkers. Source: Based on Kita et al. [64].

1,2,3,4-tetrahydroquinazolines was not satisfactory, while the enantioselectivity was excellent.

Fan and coworkers reported the first asymmetric hydrogenation of phenanthridines catalyzed by Ru/diamine complex (*R*,*R*)-**C3h** (Scheme 8.16) [65]. A series of optically active 5,6-dihydrophenanthridines were acquired with good yields and high ee values. The catalyst counteranion was found to be critically important to achieve high enantioselectivity. An achiral Brønsted acid-catalyzed transfer hydrogenation of 3-phenyl-2*H*-1,4-benzoxazine with (*R*)-6-methyl-5,6-dihydrophenanthridine as a chiral hydride proceeded well, affording the reduced product with 92% yield and 91% ee.

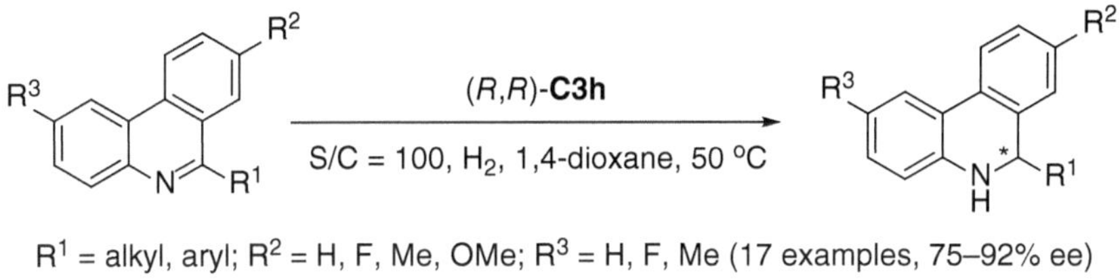

Scheme 8.16 Asymmetric hydrogenation of phenanthridines by Fan and coworkers. Source: Modified from Yang et al. [65].

8.9 Asymmetric Hydrogenation of *O*- and *S*-Heteroarenes

The asymmetric hydrogenation of *O*- and *S*-heteroarenes is different from that of *N*-heteroarenes due to the divergence of chemical properties between *N* and *O*, *S* atoms. On one hand, the substrate activation strategies for *N*-heteroarenes are no more suitable for *O*- and *S*-heteroarenes. Introducing electron-withdrawing functional groups into the substrates is sometimes used to weaken the aromaticity. On the other hand, the catalytic mechanism is different. The asymmetric hydrogenation of *O*- and *S*-heteroarenes is more likely similar to that for the asymmetric hydrogenation of alkenes. Up to now, a broad variety of chiral tetrahydrofurans, tetrahydrothiophenes, dihydrobenzofurans, and dihydrobenzothiophenes have been prepared via transition metal-catalyzed asymmetric hydrogenation.

Early in 2000, the first case of homogeneous asymmetric hydrogenation of two 2-substituted furans was reported by Studer and coworkers with a chiral

Rh/diphosphine catalyst, but much low (up to 24% ee) enantioselectivity was obtained [66]. Later in 2006, Albert and coworkers accomplished the asymmetric *cis*-hydrogenation of a 2,5-disubstituted furan with chiral Rh/diphosphine catalyst, leading to a chiral 2′,3′-dideoxynucleoside analogue with only 72% ee [67].

Pfaltz and coworkers investigated the asymmetric hydrogenation of several furans and benzofurans with Ir/*P*,*N*-ligand ((*S*)-**L19**) complexes; up to >99% ee values were achieved (Scheme 8.17) [68a]. Recently, the same group disclosed the asymmetric hydrogenation of different types of furans and benzofurans, including 2- and 3-substituted furans and benzofurans, and 2,4-disubstituted furans, with similar Ir/*P*,*N*-ligand (**L19a and L19c**) catalysts [68b]. Excellent enantioselectivities and high conversions could only be obtained with 3-substituted furans and 2- and 3-alkyl-substituted benzofurans.

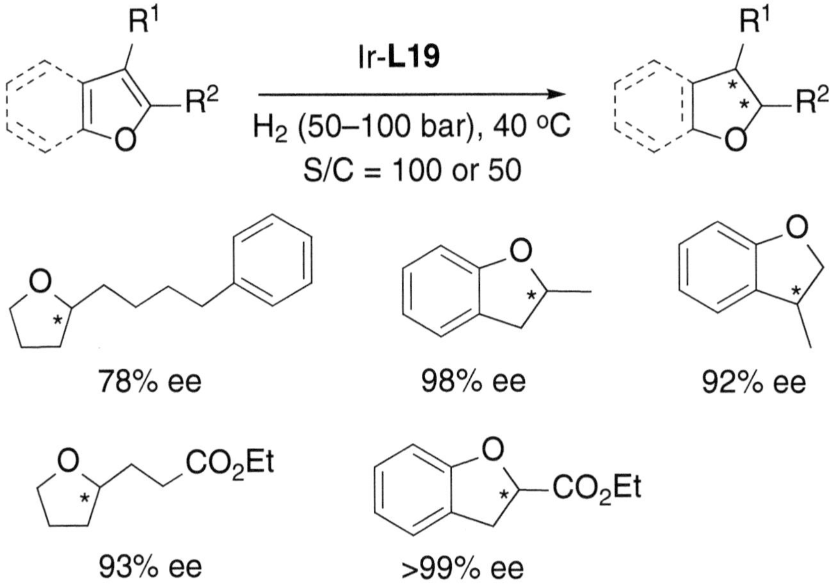

Scheme 8.17 Asymmetric hydrogenation of furans and benzofurans by Pfaltz and coworkers. Source: Based on Kaiser et al. [68a].

Ruthenium/NHC complexes were demonstrated to be highly efficient in the asymmetric hydrogenation of furans, benzofurans, thiophenes, and benzothiophenes by Glorius and coworkers (Scheme 8.18) [69]. $[Ru(cod)(2\text{-methylallyl})_2]$/SINpEt-$HBF_4$ ((*R*,*R*)-**L23**)/KO*t*Bu catalytic system was optimal for all substrates investigated. Good to excellent yields and enantioselectivities were obtained for 2-substituted benzofurans, 2-alkyl-5-aryl and 2-alkyl-4-arylfurans, 2- and 3-alkylbenzothiophenes, and 2-alkyl-5-electron-deficient arylthiophenes. This is the first example for the asymmetric hydrogenation of thiophenes and benzothiophenes.

Very recently, Fan and coworkers reported the first asymmetric hydrogenation of *in situ* generated isochromenylium intermediates by Cu/Ru tandem catalysis (Scheme 8.19) [70]. A broad range of *ortho*-(alkynyl)aryl ketones were converted to chiral 1*H*-isochromenes in high yields with good to excellent enantioselectivities.

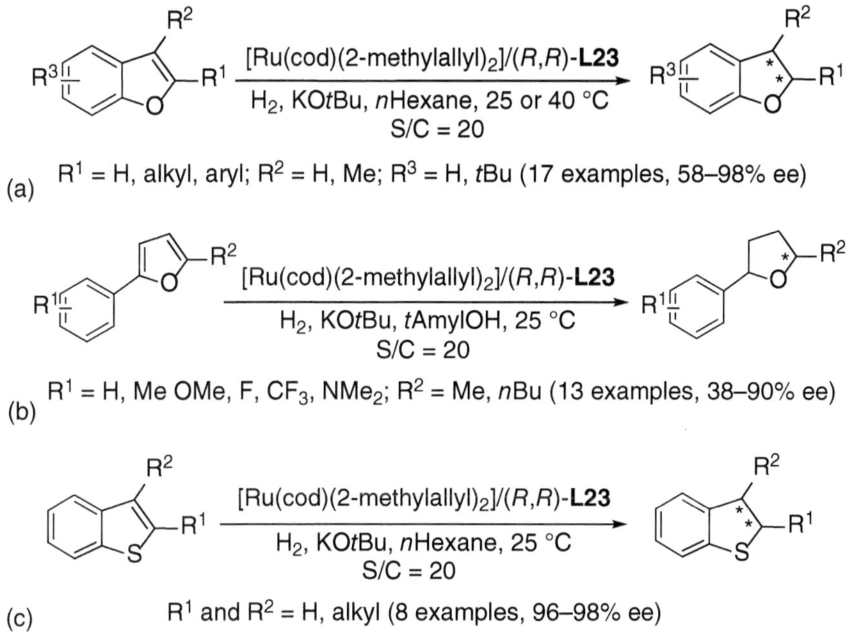

Scheme 8.18 Ru/NHC-catalyzed asymmetric hydrogenation of furans, benzofurans, thiophenes, and benzothiophenes by Glorius and coworkers. Source: (a) Modified from Ortega et al. [69a]. (b) Modified from Wysocki et al. [69b]. (c) Modified from Urban et al. [69c].

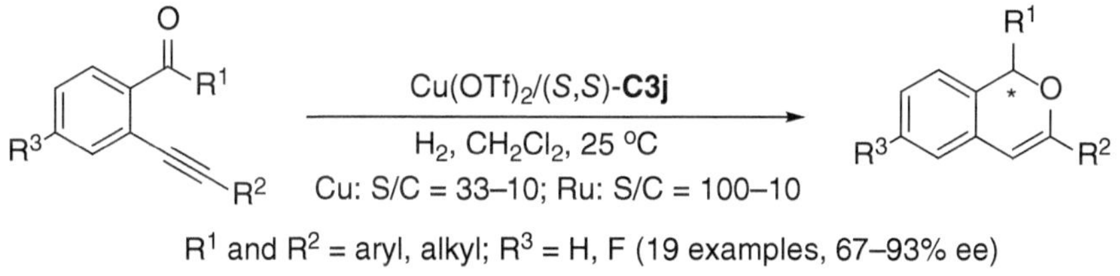

Scheme 8.19 Asymmetric hydrogenation of *in situ* generated isochromenylium intermediates with Cu/Ru tandem catalysis reported by Fan and coworkers. Source: Based on Miao et al. [70].

The mechanistic studies reveal that two catalytic cycles promoted by Cu and Ru catalysts, respectively, are involved. $Cu(OTf)_2$ catalyzes the cyclization reaction to generate isochromenylium intermediate, and Ru/diamine complex (*S*,*S*)-**C3j** further catalyzes the reduction of isochromenylium intermediate to afford the desired products. The chiral induction for the 1*H*-isochromenes products was rationalized by DFT calculations.

8.10 Summary and Conclusions

This chapter documents the homogeneous asymmetric hydrogenation of five- and six-membered single or benzofused heteroaromatic compounds, including quinolines, isoquinolines, quinoxalines, pyridines, pyrazines, indoles, pyrroles, *N*-heteroarenes with multiheterocycles, and *O*- and *S*-heteroarenes. Good to excellent activities and enantioselectivities were obtained for most examples summarized in this chapter. From the viewpoint of methodology development, several effective strategies for catalyst and substrate activation have been demonstrated to be successful and mature. However, considering the rich and diverse chemical compositions and structures of heteroarenes, further development of highly efficient and enantioselective catalytic systems and novel strategies for preparing useful chiral heterocycles (such as pharmaceuticals, materials, ligands, and catalysts) via asymmetric hydrogenation is still highly desirable. Meanwhile, computational chemistry becomes more important, not only in the design and selecting of suitable heteroarene substrates but also in the deep understanding of mechanism for different types of substrates with varied catalytic systems. We believe that there will be more breakthroughs made in this research field in the near future.

Acknowledgments

We are grateful for the financial support from the National Natural Science Foundation of China (Nos. 21790332 and 21521002) and CAS (QYZDJSSW-SLH023) for financial support.

References

1 (a) Wang, D.-S., Chen, Q.-A., Lu, S.-M., and Zhou, Y.-G. (2012). *Chem. Rev.* 112: 2557–2590. (b) Wiesenfeldt, M.P., Nairoukh, Z., Dalton, T., and Glorius, F. (2019). *Angew. Chem. Int. Ed.* 58: 10460–10470.

2 (a) Zhou, Y.-G. (2007). *Acc. Chem. Res.* 40: 1357–1366. (b) Kuwano, R. (2008). *Heterocycles* 76: 909–922. (c) He, Y.-M., Song, F.-T., and Fan, Q.-H. (2014). *Top. Curr. Chem.* 343: 145–190. (d) Zhao, D., Candish, L., Paul, D., and Glorius, F. (2016). *ACS Catal.* 6: 5978–5988. (e) Rueping, M., Sugiono, E., and Schoepke, F.R. (2010). *Synlett* 6: 852–865. (f) Meng, W., Feng, X., and Du, H. (2018). *Acc. Chem. Res.* 51: 191–201.

3 Balakrishna, B., Núñez-Rico, J.L., and Vidal-Ferran, A. (2015). *Eur. J. Org. Chem.*: 5293–5303.

4 Gualandi, A. and Savoia, D. (2016). *RSC Adv.* 6: 18419–18451.

5 Muthukrishnan, I., Sridharan, V., and Menéndez, J.C. (2019). *Chem. Rev.* 119: 5057–5191.

6 Wang, W.-B., Lu, S.-M., Yang, P.-Y. et al. (2003). *J. Am. Chem. Soc.* 125: 10536–10537.

7 Wang, D.-W., Wang, X.-B., Wang, D.-S. et al. (2009). *J. Org. Chem.* 74: 2780–2787.

8 (a) Xu, L., Lam, K.H., Ji, J. et al. (2005). *Chem. Commun.*: 1390–1392. (b) Tang, W.-J., Tan, J., Xu, L.-J. et al. (2010). *Adv. Synth. Catal.* 352: 1055–1062.

9 Tang, W., Sun, Y., Xu, L. et al. (2010). *Org. Biomol. Chem.* 8: 3464–3471.

10 Tadaoka, H., Cartigny, D., Nagano, T. et al. (2009). *Chem. Eur. J.* 15: 9990–9994.

11 (a) Lam, K.-H., Xu, L., Feng, L. et al. (2005). *Adv. Synth. Catal.* 347: 1755–1758. (b) Tang, W.-J., Zhu, S.-F., Xu, L.-J. et al. (2007). *Chem. Commun.*: 613–615. (c) Reetz, M.T. and Li, X. (2006). *Chem. Commun.*: 2159–2160. (d) Mršić, N., Lefort, L., Boogers, J.A.F. et al. (2008). *Adv. Synth. Catal.* 350: 1081–1089.

12 (a) Lu, S.-M., Han, X.-W., and Zhou, Y.-G. (2004). *Adv. Synth. Catal.* 346: 909–912. (b) Lu, S.-M. and Bolm, C. (2008). *Adv. Synth. Catal.* 350: 1101–1105.

13 Wen, J., Tan, R., Liu, S. et al. (2016). *Chem. Sci.* 7: 3047–3051.

14 (a) Wang, T., Zhuo, L.-G., Li, Z. et al. (2011). *J. Am. Chem. Soc.* 133: 9878–9891. (b) Zhou, H., Li, Z., Wang, Z. et al. (2008). *Angew. Chem. Int. Ed.* 47: 8464–8467. (c) Ding, Z.-Y., Wang, T., He, Y.-M. et al. (2013). *Adv. Synth. Catal.* 355: 3727–3735. (d) Yang, Z., Chen, F., He, Y.-M. et al. (2014). *Catal. Sci. Technol.* 4: 2887–2890. (e) Wang, Z.-J., Zhou, H.-F., Wang, T.-L. et al. (2009). *Green Chem.* 11: 767–769. (f) Li, Z.-W., Wang, T.-L., He, Y.-M. et al. (2008). *Org. Lett.* 10: 5265–5268.

15 Chen, Q.-A., Gao, K., Duan, Y. et al. (2012). *J. Am. Chem. Soc.* 134: 2442–2448.

16 Cai, X.-F., Guo, R.-N., Chen, M.-W. et al. (2014). *Chem. Eur. J.* 20: 7245–7248.

17 Cai, X.-F., Huang, W.-X., Chen, Z.-P., and Zhou, Y.-G. (2014). *Chem. Commun.* 50: 9588–9590.

18 (a) Zhang, Z. and Du, H. (2015). *Org. Lett.* 17: 2816–2819. (b) Zhang, Z. and Du, H. (2015). *Org. Lett.* 17: 6266–6269.

19 Murata, S., Sugimoto, T., and Matsuura, S. (1987). *Heterocycles* 26: 763–766.

20 Bianchini, C., Barbaro, P., Scapacci, G. et al. (1998). *Organometallics* 17: 3308–3310.

21 Tang, W., Xu, L., Fan, Q.-H. et al. (2009). *Angew. Chem. Int. Ed.* 48: 9135–9138.

22 Mršić, N., Jerphagnon, T., Minnaard, A.J. et al. (2009). *Adv. Synth. Catal.* 351: 2549–2552.

23 (a) Cartigny, D., Nagano, T., Ayad, T. et al. (2010). *Adv. Synth. Catal.* 352: 1886–1891. (b) Cartigny, D., Berhal, F., Nagano, T. et al. (2012). *J. Org. Chem.* 77: 4544–4556. (c) Nagano, T., Iimuro, A., Schwenk, R. et al. (2012). *Chem. Eur. J.* 18: 11578–11592.

24 Arai, N., Saruwatari, Y., Isobe, K., and Ohkuma, T. (2013). *Adv. Synth. Catal.* 355: 2769–2774.

25 Qin, J., Chen, F., Ding, Z. et al. (2011). *Org. Lett.* 13: 6568–6571.

26 Chen, Q.-A., Wang, D.-S., Zhou, Y.-G. et al. (2011). *J. Am. Chem. Soc.* 133: 6126–6129.

27 Fleischer, S., Zhou, S., Werkmeister, S. et al. (2013). *Chem. Eur. J.* 19: 4997–5003.

28 (a) Zhang, Z. and Du, H. (2015). *Angew. Chem. Int. Ed.* 54: 623–626. (b) Li, S., Meng, W., and Du, H. (2017). *Org. Lett.* 19: 2604–2606.

29 (a) Lu, S.-M., Wang, Y.-Q., Han, X.-W., and Zhou, Y.-G. (2006). *Angew. Chem. Int. Ed.* 45: 2260–2263. (b) Ye, Z.-S., Guo, R.-N., Cai, X.-F. et al. (2013). *Angew. Chem. Int. Ed.* 52: 3685–3689.

30 (a) Iimuro, A., Yamaji, K., Kandula, S. et al. (2013). *Angew. Chem. Int. Ed.* 52: 2046–2050. (b) Kita, Y., Yamaji, K., Higashida, K. et al. (2015). *Chem. Eur. J.* 21: 1915–1927.

31 Chen, M.-W., Ji, Y., Wang, J. et al. (2017). *Org. Lett.* 19: 4988–4991.

32 (a) Shi, L., Ye, Z.-S., Cao, L.-L. et al. (2012). *Angew. Chem. Int. Ed.* 51: 8286–8289. (b) Guo, R.-N., Cai, X.-F., Shi, L. et al. (2013). *Chem. Commun.* 49: 8537–8539.

33 Yu, Z., Jin, W., and Jiang, Q. (2012). *Angew. Chem. Int. Ed.* 51: 6060–6072.

34 Legault, C.Y. and Charette, A.B. (2005). *J. Am. Chem. Soc.* 127: 8966–8967.

35 Ye, Z.-S., Chen, M.-W., Chen, Q.-A. et al. (2012). *Angew. Chem. Int. Ed.* 51: 10181–10184.

36 Chang, M., Huang, Y., Liu, S. et al. (2014). *Angew. Chem. Int. Ed.* 53: 12761–12764.

37 Huang, W.-X., Yu, C.-B., Ji, Y. et al. (2016). *ACS Catal.* 6: 2368–2371.

38 (a) Qu, B., Mangunuru, H.P.R., Wei, X. et al. (2016). *Org. Lett.* 18: 4920–4923. (b) Qu, B., Mangunuru, H.P.R., Tcyrulnikov, S. et al. (2018). *Org. Lett.* 20: 1333–1337.

39 Renom-Carrasco, M., Gajewski, P., Pignataro, L. et al. (2016). *Chem. Eur. J.* 22: 9528–9532.

40 (a) Kita, Y., Iimuro, A., Hida, S., and Mashima, K. (2014). *Chem. Lett.* 43: 284–286. (b) Iimuro, A., Higashida, K., Kita, Y., and Mashima, K. (2016). *Adv. Synth. Catal.* 358: 1929–1933.

41 Chen, M.-W., Ye, Z.-S., Chen, Z.-P. et al. (2015). *Org. Chem. Front.* 2: 586–589.

42 Huang, W.-X., Liu, L.-J., Wu, B. et al. (2016). *Org. Lett.* 18: 3082–3085.

43 Higashida, K., Nagae, H., and Mashima, K. (2016). *Adv. Synth. Catal.* 358: 3949–3954.

44 Kuwano, R., Sato, K., Kurokawa, T. et al. (2000). *J. Am. Chem. Soc.* 122: 7614–7615.

45 (a) Kuwano, R., Kaneda, K., Ito, T. et al. (2004). *Org. Lett.* 6: 2213–2215. (b) Kuwano, R. and Kashiwabara, M. (2006). *Org. Lett.* 8: 2653–2655. (c) Kuwano, R., Kashiwabara, M., Ohsumi, M., and Kusano, H. (2008). *J. Am. Chem. Soc.* 130: 808–809.

46 Baeza, A. and Pfaltz, A. (2010). *Chem. Eur. J.* 16: 2036–2039.

47 Wang, D.-S., Chen, Q.-A., Li, W. et al. (2010). *J. Am. Chem. Soc.* 132: 8909–8911.

48 (a) Duan, Y., Li, L., Chen, M.-W. et al. (2014). *J. Am. Chem. Soc.* 136: 7688–7700. (b) Duan, Y., Chen, M.-W., Ye, Z.-S. et al. (2011). *Chem. Eur. J.* 17: 7193–7197.

49 Wang, D.-S., Ye, Z.-S., Chen, Q.-A. et al. (2011). *J. Am. Chem. Soc.* 133: 8866–8869.

50 Wen, J., Fan, X., Tan, R. et al. (2018). *Org. Lett.* 20: 2143–2147.

51 Yang, Z., Chen, F., He, Y. et al. (2016). *Angew. Chem. Int. Ed.* 55: 13863–13866.

52 Touge, T. and Arai, T. (2016). *J. Am. Chem. Soc.* 138: 11299–11305.

53 Chen, Z.-P. and Zhou, Y.-G. (2016). *Synthesis* 48: 1769–1781.

54 (a) Ortega, N., Tang, D.-T.D., Urban, S. et al. (2013). *Angew. Chem. Int. Ed.* 52: 9500–9503. (b) Schlepphorst, C., Wiesenfeldt, M.P., and Glorius, F. (2018). *Chem. Eur. J.* 24: 356–359.

55 Makida, Y., Saita, M., Kuramoto, T. et al. (2016). *Angew. Chem. Int. Ed.* 55: 11859–11862.

56 (a) Huang, W.-X., Yu, C.-B., Shi, L., and Zhou, Y.-G. (2014). *Org. Lett.* 16: 3324–3327. (b) Hu, S.-B., Chen, Z.-P., Song, B. et al. (2017). *Adv. Synth. Catal.* 359: 2762–2767. (c) Hu, S.-B., Zhai, X.-Y., Shen, H.-Q., and Zhou, Y.-G. (2018). *Adv. Synth. Catal.* 360: 1334–1339.

57 Wang, T., Chen, F., Qin, J. et al. (2013). *Angew. Chem. Int. Ed.* 52: 7172–7176.

58 (a) Zhang, J., Chen, F., He, Y.-M., and Fan, Q.-H. (2015). *Angew. Chem. Int. Ed.* 54: 4622–4625. (b) Ma, W., Chen, F., Liu, Y. et al. (2016). *Org. Lett.* 18: 2730–2733.

59 (a) Ma, W., Zhang, J., Xu, C. et al. (2016). *Angew. Chem. Int. Ed.* 55: 12891–12894. (b) Li, B., Xu, C., He, Y.-M. et al. (2018). *Chin. J. Chem.* 36: 1169–1173.

60 Kuwano, R., Kameyama, N., and Ikeda, R. (2011). *J. Am. Chem. Soc.* 133: 7312–7315.

61 Kuwano, R., Hashiguchi, Y., Ikeda, R., and Ishizuka, K. (2015). *Angew. Chem. Int. Ed.* 54: 2393–2396.

62 (a) Feng, G.-S., Chen, M.-W., Shi, L., and Zhou, Y.-G. (2018). *Angew. Chem. Int. Ed.* 57: 5853–5857. (b) Feng, G.-S., Shi, L., Meng, F.-J. et al. (2018). *Org. Lett.* 20: 6415–6419.

63 Chen, Z.-P., Chen, M.-W., Shi, L. et al. (2015). *Chem. Sci.* 6: 3415–3419.

64 Kita, Y., Higashida, K., Yamaji, K. et al. (2015). *Chem. Commun.* 51: 4380–4382.

65 Yang, Z., Chen, F., Zhang, S. et al. (2017). *Org. Lett.* 19: 1458–1461.

66 Studer, M., Wedemeyer-Exl, C., Spindler, F., and Blaser, H.-U. (2000). *Monatshefte für Chemie* 131: 1335–1343.

67 Feiertag, P., Albert, M., Nettekoven, U., and Spindler, F. (2006). *Org. Lett.* 8: 4133–4135.

68 (a) Kaiser, S., Smidt, S.P., and Pfaltz, A. (2006). *Angew. Chem. Int. Ed.* 45: 5194–5197. (b) Pauli, L., Tannert, R., Scheil, R., and Pfaltz, A. (2015). *Chem. Eur. J.* 21: 1482–1487.

69 (a) Ortega, N., Urban, S., Beiring, B., and Glorius, F. (2012). *Angew. Chem. Int. Ed.* 51: 1710–1713. (b) Wysocki, J., Ortega, N., and Glorius, F. (2014). *Angew. Chem. Int. Ed.* 53: 8751–8755. (c) Urban, S., Beiring, B., Ortega, N. et al. (2012). *J. Am. Chem. Soc.* 134: 15241–15244.

70 Miao, T., Tian, Z.-Y., He, Y.-M. et al. (2017). *Angew. Chem. Int. Ed.* 56: 4135–4139.

9

Asymmetric (Transfer) Hydrogenation of Imines

Itziar Peñafiel[1], Juan Mangas-Sánchez[2,3] and Carmen Claver[4]

[1]*Future Biomanufacturing Research Hub, University of Manchester, Manchester Institute of Biotechnology, 131 Princess Street, Manchester M1 7DN, UK*
[2]*Institute of Chemical Synthesis and Homogeneous Catalysis, CSIC-University of Zaragoza, Pedro Cerbuna 12, 50009 Zaragoza, Spain*
[3]*ARAID Foundation, Zaragoza, Spain*
[4]*Universidad Rovira i Virgili, Eurecat, Centre Tecnològic de Catalunya, C/Marcel·lí Domingo, 43007 Tarragona, Spain*

9.1 Asymmetric Hydrogenation of Imines

The use of chiral metal catalysts for the asymmetric hydrogenation of C=N double bonds constitutes one of the most largely applied methods for the preparation of chiral amines [1]. During the past decade, the development of robust and efficient catalytic systems has provided high activities and enantioselectivities through the overcoming of problems such as E/Z isomeric mixture generation and catalyst deactivation by the resultant amines [2, 3]. Thus, nowadays, the chiral metal-catalyzed asymmetric hydrogenation of imines supposes an effective methodology for the generation of chiral amines [4, 5]. The general equation for the asymmetric hydrogenation of imines is represented in Scheme 9.1.

In this section, the most relevant contributions are classified according to the metal. In the case of iridium, the different ligands involved in the catalytic system have been studied in the following subsections.

9.1.1 Iridium Catalysts

During the past three decades, the development of chiral ligands and chiral iridium catalysts has effectively provided chiral amines with excellent enantioselectivities under mild conditions involving a broad substrate scope. Thus, iridium cationic complexes and iridium neutral complexes have turned up into the most largely employed metal catalysts for the asymmetric hydrogenation of imines [6, 7].

9.1.1.1 (P,P) Ligands

Among the large variety of ligands containing two different phosphorus functionalities employed for the generation of iridium complexes [8–10], the impressive

Asymmetric Hydrogenation and Transfer Hydrogenation, First Edition.
Edited by Virginie Ratovelomanana-Vidal and Phannarath Phansavath.

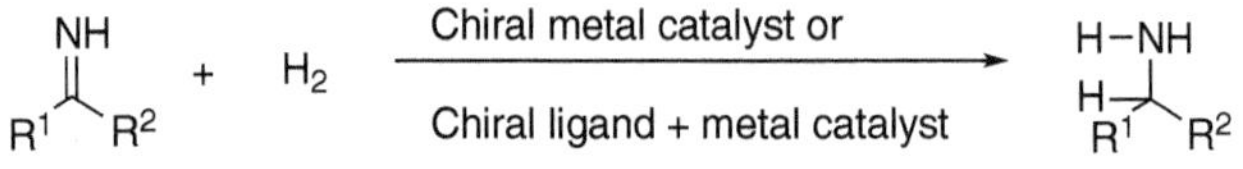

Scheme 9.1 General equation for the asymmetric hydrogenation of imines.

turnover number (TON) (superior to 1,000,000) provided by diphosphine ligand (*R*,*S*)-f-Xyliphos (Figure 9.1) and [Ir(μ-Cl)(cod)]$_2$ at large-scale reduction of imines is remarkable [11].

Additionally, [Ir(μ-Cl)(cod)]$_2$ has also been combined with ferrocene derivatives (*S*,*S*)-f-Binaphane and (*R*,*R*)-f-Binaphane (Scheme 9.2) providing enantioselectivities up to 99% for a range of chloride salts of NH imines and for less-explored unfunctionalized N—H imines (**1**, Scheme 9.2) [12, 13]. And *N*-substituted diarylmethanimines (**3**, Scheme 9.2) have been successfully hydrogenated (TON up to 4000 and ee 99%) using (*R*,*R*)-f-SpiroPhos [14].

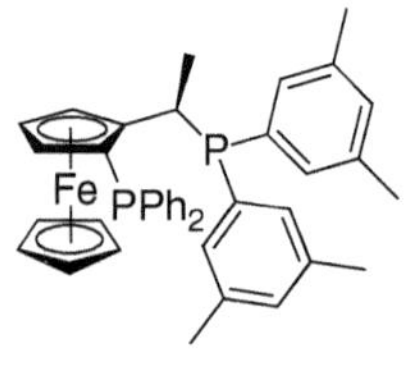

(*R*,*S*)-f-Xtliphos

Figure 9.1 Diphosphine ligand widely employed in industry.

Recently, the use of *N*-methylated ZhaoPhos as ligand (**5**, Scheme 9.3) has been successfully applied in the Ir-catalyzed asymmetric hydrogenation of cyclic sulfamidate imines (**6**) (up to 99% yield, 99% ee). The process can be run at gram scale exhibiting synthetic utility [15].

9.1.1.2 (P,N) Ligands

A significant number of successful attempts in the use of iridium complexes containing chiral oxazoline-based (P,N) ligands for the asymmetric hydrogenation of aryl methyl *N*-arylimines have been reported [16–20], and the important impact that the stereogenic center of the oxazoline unit has on the enantioselectivity of the reaction has been studied [21].

Recently, Verdaguer and coworkers have reported the employment of the stable cationic Ir(III)H catalyst (**8**) (Scheme 9.4) containing a P-stereogenic MaxPHOX ligand for the asymmetric hydrogenation of aryl imines with full conversion and in high enantiomeric excess at low temperature and using atmospheric H$_2$ pressure [22]. Variations on the chirality of the ligand allowed the asymmetric hydrogenation of methyl and alkyl imines [23].

A series of alkyl aryl *N*-phenyl imines (**11**) have been reduced at ambient hydrogen pressure with enantioselectivities up to 97% using a spirobiindane-based (**SIPHOX**) as ligand [24] (Scheme 9.5). Also, cationic spinPHOX L1 iridium complex phosphine–oxazoline (Scheme 9.5) has afforded this reduction, and a variation, spinPHOX L2 (Scheme 9.5), managed full conversions for the hydrogenation of aryl methyl *N*-benzyl imines (**11**) with enantioselectivities higher than 88% at hydrogen pressure from 1 to 5 atm [25].

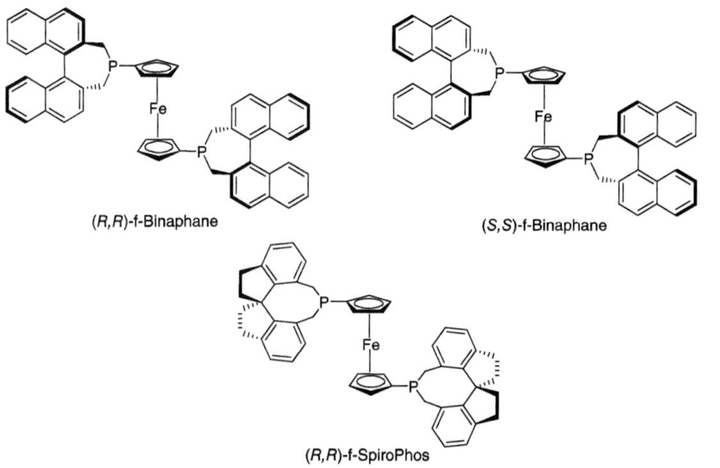

(*R*,*R*)-f-Binaphane

(*S*,*S*)-f-Binaphane

(*R*,*R*)-f-SpiroPhos

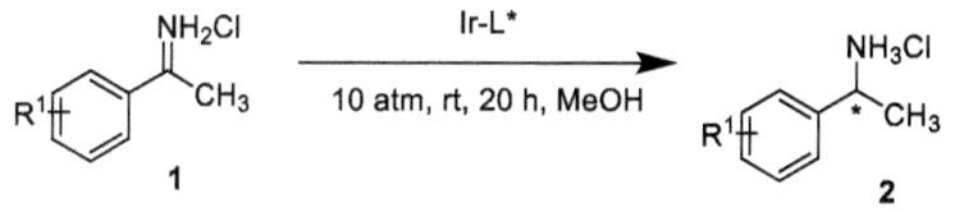

1 → **2**

2a (R^1 = 2-CH_3) ((*S*,*S*)-f-Binaphane) 95% ee, 92% yield
2b (R^1 = 4-CH_3) ((*S*,*S*)-f-Binaphane) 95% ee, 95% yield
2c (R^1 = H) ((*S*,*S*)-f-Binaphane) 93% ee, 93% yield
2d (R^1 = 4-MeO) ((*S*,*S*)-f-Binaphane) 93% ee, 95% yield
2e (R^1 = 4-F) ((*S*,*S*)-f-Binaphane) 92% ee, 95% yield
2f (R^1 = 4-Cl) ((*S*,*S*)-f-Binaphane) 94% ee, 95% yield
2g (R^1 = 4-Br) ((*S*,*S*)-f-Binaphane) 93% ee, 94% yield
2h (R^1 = 4-CF_3) ((*S*,*S*)-f-Binaphane) 93% ee, 93% yield
2i (R^1 = 3-CH_3) ((*S*,*S*)-f-Binaphane) 92% ee, 95% yield
2j (R^1 = 3-MeO) ((*S*,*S*)-f-Binaphane) 94% ee, 94% yield
2k (R^1 = 3-Cl) ((*S*,*S*)-f-Binaphane) 92% ee, 92% yield
2l (R^1 = 3-Br) ((*S*,*S*)-f-Binaphane) 91% ee, 93% yield

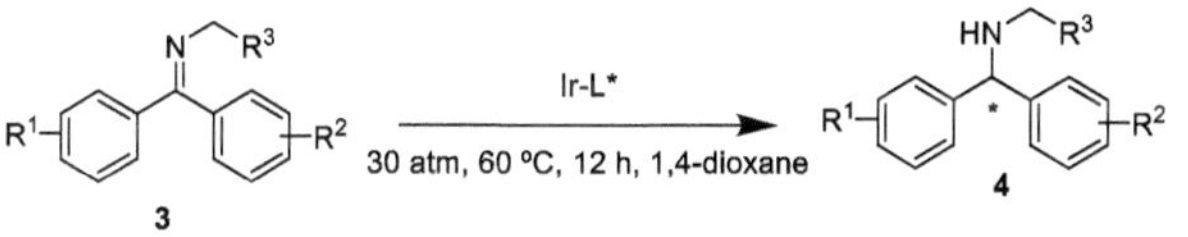

3 → **4**

4a (R^1 = 2-Me, R^2 = H, R^3 = $COOCH_3$) ((*R*,*R*)-f-SpiroPhos) 98% ee, 99% yield
4b (R^1 = 2-Me, R^2 = 3-OMe, R^3 = $COOCH_3$) ((*R*,*R*)-f-SpiroPhos) 94% ee, 97% yield
4c (R^1 = 2-Me, R^2 = 3-Me, R^3 = $COOCH_3$) ((*R*,*R*)-f-SpiroPhos) 96% ee, 97% yield
4d (R^1 = 2-Me, R^2 = 4-Me, R^3 = $COOCH_3$) ((*R*,*R*)-f-SpiroPhos) 97% ee, 99% yield
4e (R^1 = 2-OMe, R^2 = H, R^3 = $COOCH_3$) ((*R*,*R*)-f-SpiroPhos) 95% ee, 96% yield
4f (R^1 = 2-Cl, R^2 = 3-Me, R^3 = $COOCH_3$) ((*R*,*R*)-f-SpiroPhos) 97% ee, 97% yield
4g (R^1 = 2-Cl, R^2 = 4-F, R^3 = $COOCH_3$) ((*R*,*R*)-f-SpiroPhos) 97% ee, 98% yield
4h (R^1 = 2-Br, R^2 = H, R^3 = $COOCH_3$) ((*R*,*R*)-f-SpiroPhos) 96% ee, 98% yield
4i (R^1 = 2-F, R^2 = 3-Me, R^3 = $COOCH_3$) ((*R*,*R*)-f-SpiroPhos) 94% ee, 99% yield
4j (R^1 = 2-F, R^2 = 4-F, R^3 = $COOCH_3$) ((*R*,*R*)-f-SpiroPhos) 94% ee, 98% yield
4k (R^1 = 2-Me, R^2 = H, R^3 = COO^tBu) ((*R*,*R*)-f-SpiroPhos) 97% ee, 98% yield
4l (R^1 = 2-Me, R^2 = H, R^3 = Et) ((*R*,*R*)-f-SpiroPhos) 99% ee, 98% yield
4m (R^1 = H, R^2 = 3-Me, R^3 = $COOCH_3$) ((*R*,*R*)-f-SpiroPhos) 97% ee, 95% yield
4n (R^1 = H, R^2 = 4-F, R^3 = $COOCH_3$) ((*R*,*R*)-f-SpiroPhos) 87% ee, 96% yield

Scheme 9.2 Overview of selected (P,P) ligands and results of their use in iridium-mediated asymmetric hydrogenation of imines. Source: Refs [12] and [13, 14].

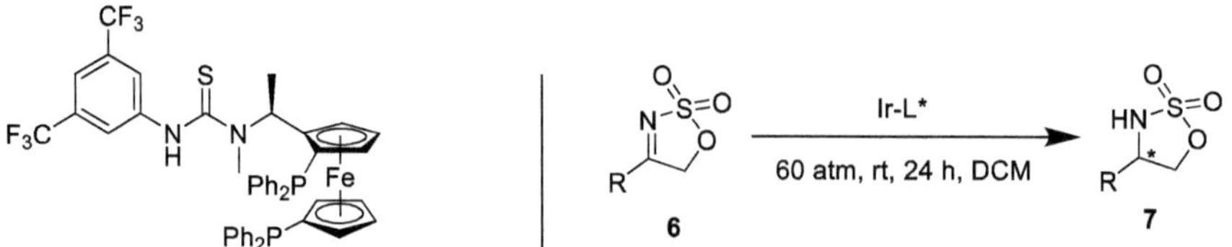

5

7a (R = Ph) 97% ee, 99% yield
7b (R = 4-MeOPh) 98% ee, 98% yield
7c (R = 2-MePh) 92% ee, 98% yield
7d (R = 4-MePh) 99% ee, 99% yield
7e (R = 3-MePh) 98% ee, 99% yield
7f (R = 2-FPh) 95% ee, 94% yield
7g (R = 4-FPh) 98% ee, 97% yield
7h (R = 3,5-diFPh) 96% ee, 98% yield
7i (R = 4-ClPh) 98% ee, 99% yield
7j (R = 3-ClPh) 98% ee, 96% yield
7k (R = 4-BrPh) 98% ee, 95% yield
7l (R = 3-BrPh) 98% ee, 96% yield
7m (R = 4-CO_2MePh) 98% ee, 96% yield
7n (R = naphthyl) 97% ee, 98% yield
7o (R = thiophenyl) 96% ee, 97% yield
7p (R = Bn) 73% ee, 98% yield

Scheme 9.3 Summary of results generated by N-methylated ZhaoPhos for the iridium-mediated asymmetric hydrogenation of imines.

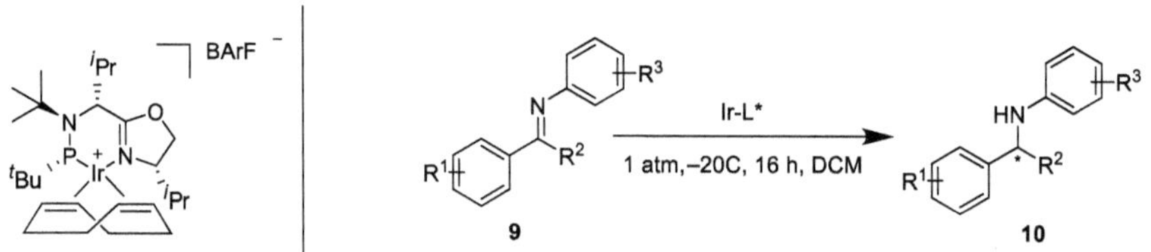

Ir-MaxPHOX complex 8

10a (R^1 = H, R^2 = Me, R^3 = H) 96% ee, 100% yield
10b (R^1 = 4-MeO, R^2 = Me, R^3 = H) 95% ee, 100% yield
10c (R^1 = 3-MeO, R^2 = Me, R^3 = H) 92% ee, 100% yield
10d (R^1 = 4-Cl, R^2 = Me, R^3 = H) 94% ee, 100% yield
10e (R^1 = H, R^2 = Me, R^3 = 4-OMe) 93% ee, 100% yield
10f (R^1 = H, R^2 = Me, R^3 = 3,5-diMe) 90% ee, 100% yield
10g (R^1 = H, R^2 = Me, R^3 = 4-Me) 95% ee, 100% yield

Scheme 9.4 Summary of results generated by complex **8** for the iridium-mediated asymmetric hydrogenation of imines.

Among different iridium P,N-ligands employed for the asymmetric hydrogenation of imines, the use of a ferrocene-based phosphine having a pyridine ring (**13**) constitutes an interesting alternative to the use of an oxazoline ring, allowing full conversions and good enantioselectivities (84–99% ee) (Scheme 9.6) to a range of aryl methyl and aryl alkyl *N*-aryl imines (**1**) under mild conditions (1 MPa) [26].

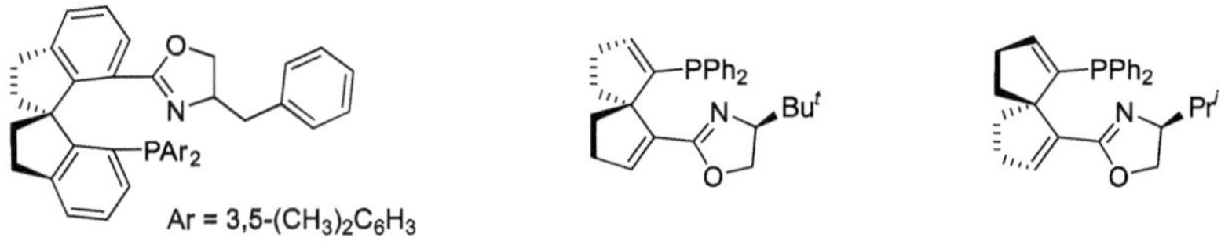

SIPHOX spinPHOX L1 spinPHOX L2

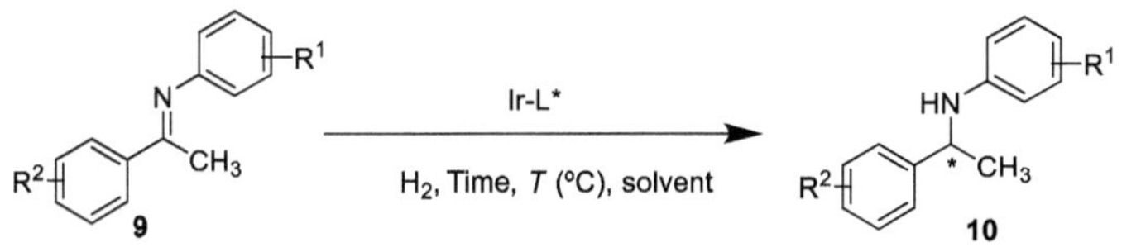

10a (R^1 = R^2 = H) (**SIPHOX**) 99% ee, 99% conv.
(**spinPHOX L1**) 91% ee, conv 92%
10c (R^1 = 4-OMe, R^2 = H) (**spinPHOX L1**) 90% ee, conv 95%
10d (R^1 = H, R^2 = 4-OMe (**SIPOX**) 94% ee, conv > 99.5%
(**spinPHOX L1**) 94% ee, conv 99%
10e (R^1 = H, R^2 = 4-Cl) (**SIPOX**) 90% ee, conv > 99.5%
(**spinPHOX L1**) 93.5% ee, conv 99%
10f (R^1 = H, R^2 = 4-Br) (**SIPOX**) 91% ee, conv > 99.5%
(**spinPHOX L1**) 95% ee, conv 99%
10g (R^1 = H, R^2 = 3-Cl) (**SHPOX**) 93% ee, conv > 99.5%
(**spinPHOX L1**) 93% ee, conv 99%
10h (R^1 = H, R^2 = 3-Br) (**SIPOX**) 92% ee, conv > 99.5%
(**spinPHOX L1**) 93% ee, conv 99%
10i (R^1 = 4-Cl, R^2 = H) (**SIPHOX**) 97% ee, conv > 99.5%
10j (R^1 = 4-Br, R^2 = H) (**SIPHOX**) 96% ee, conv > 99.5%
10k (R^1 = 3-Br, R^2 = H) (**SIPHOX**) 94% ee, conv > 99.5%

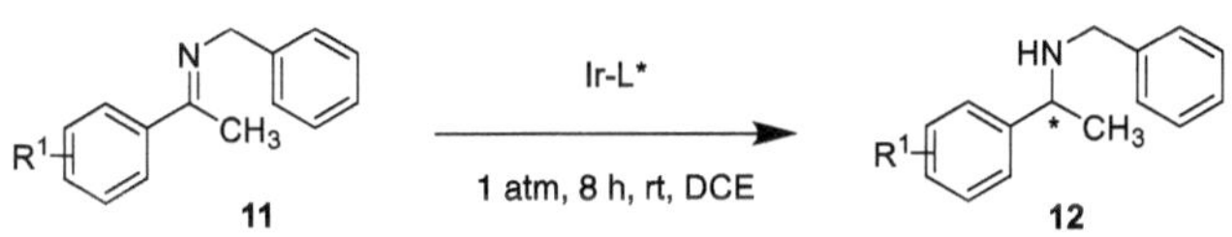

12a (R^1 = 4-MeO) (**spinPHOX L2**) 90% ee, conv. 99%
12b (R^1 = H) (**spinPHOX L2**) 91% ee, conv 99%
12c (R^1 = 4-Me) (**spinPHOX L2**) 89% ee, conv 99%
12d (R^1 = 4-Cl) (**spinPHOX L2**) 991% ee, conv 67%
12e (R^1 = 4-F) (**spinPHOX L2**) 88% ee, conv 99%
12f (R^1 = 4-Br) (**spinPHOX L2**) 89% ee, conv 90%

Scheme 9.5 Overview of selected phosphine–oxazoline ligands and results of their use in iridium-mediated asymmetric hydrogenation of acyclic imines.

Additionally, the use of diamine ligands such as *N*-tosyl-1,2-diphenylethane-1,2-diamine (TsDPEN), in combination with [Cp*Ir(III)] complex, and a bulky chiral phosphate counterion has mediated the hydrogenation of alkyl and dialkyl *N*-aryl imines with high enantioselectivities (92–98% ee and 92–95% ee, respectively) [27, 28].

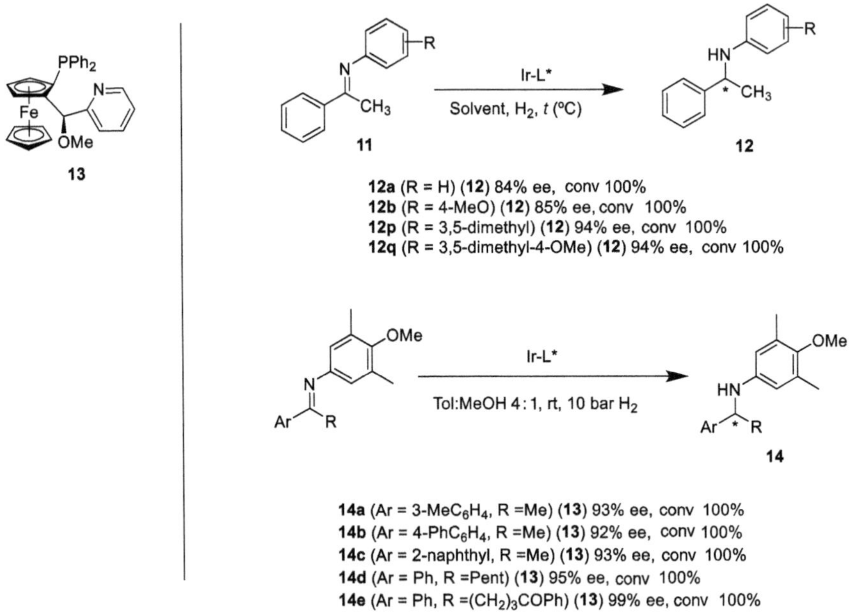

Scheme 9.6 Ferrocene-based (P,N) ligands employed in combination with iridium precursors and results of their use in asymmetric hydrogenation of imines.

9.1.1.3 P-Monodentate Ligands

Broadly, in metal-catalyzed asymmetric hydrogenation, bidentate ligands have proved superior effectiveness over monodentate ones due to the chelating effect; nevertheless, there are some cases where monodentate ligands offer a competent alternative leading to the corresponding chiral amine with high enantioselectivity. Here, it is important to stress the accessibility, air stability, low cost, and high modularity of monodentate phosphoramidite ligands. The reduction of aryl methyl and aryl alkyl *N*-aryl imines (**11** and **16**, respectively, Scheme 9.7) results in full conversion and high enantioselectivities (72–94% ee) [30, 31], employing the (S)-PipPhos ligand (Scheme 9.7).

Moreover, excellent conversions and enantioselectivities from 82% to 92% ee have been reported for the asymmetric hydrogenation of chloride salts of imines (**18**) using monodentate phosphoramidite ligand **15** (Scheme 9.7) [31].

Additionally, studies such as the one reported by Norbby and coworkers [32] are necessary for the development of improved chiral catalysts using a mechanism-guided analysis of candidate substrates and ligands to get a virtual screening of substrate/catalyst.

Undoubtedly, iridium is the most largely employed metal for the asymmetric hydrogenation of imines. The combination of other metals with monodentate (mostly phosphoramidites) or bidentate ligands (ferrocenyl and atropoisomeric diphosphines, phosphine–oxazoline, and diphosphine ligands) represents a promising alternative to iridium catalysts [33–35]. Some of them are summarized in the following paragraph.

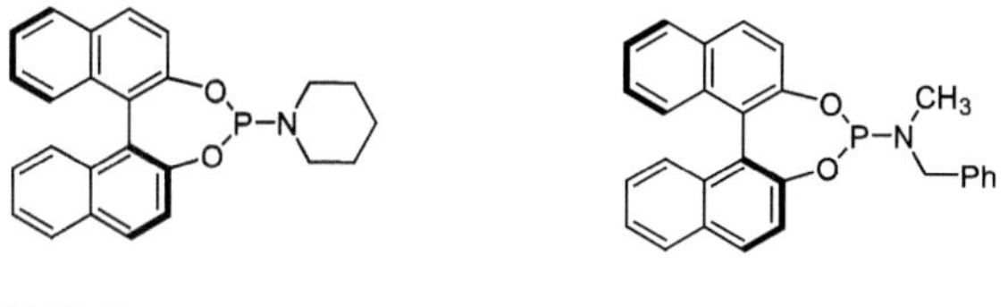

(*s*)-PipPhos **15**

12a (R = P ((**S**)-**PipPhos**) 87% ee, conv 100%
12l (Ar = 2-MeO-C_6H_4) ((**S**)-**PipPhos**) 97% ee, conv 100%
12m (Ar = 3,5-$(CH_3)_2$-C_6H_4) ((**S**)-**PipPhos**) 99% ee, conv 100%
12n (Ar =3,4,5-$(MeO)_3$-C_6H_4) ((**S**)-**PipPhos**) 99% ee, conv 100%

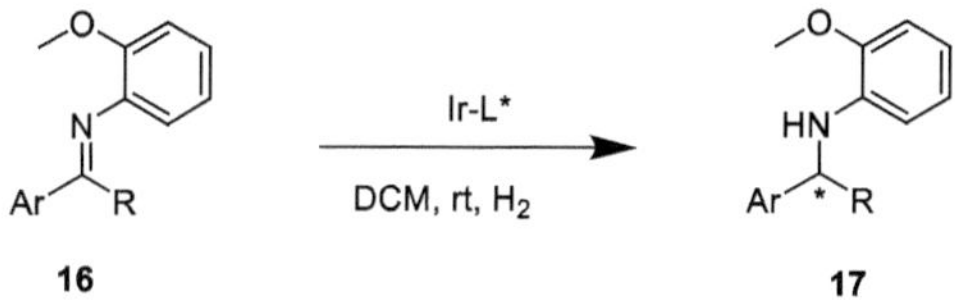

17a (Ar =naphthyl, R = Me) ((**S**)-**PipPhos**) 99% ee, conv 100%
17a (Ar =4-MeC_6H_4I, R = Me) ((**S**)-**PipPhos**) 98% ee, conv 100%
17a (Ar =4-F, R = Me) ((**S**)-**PipPhos**) 97% ee, conv 100%
17a (Ar = PhI, R = Pr) ((**S**)-**PipPhos**) 97% ee, conv100%

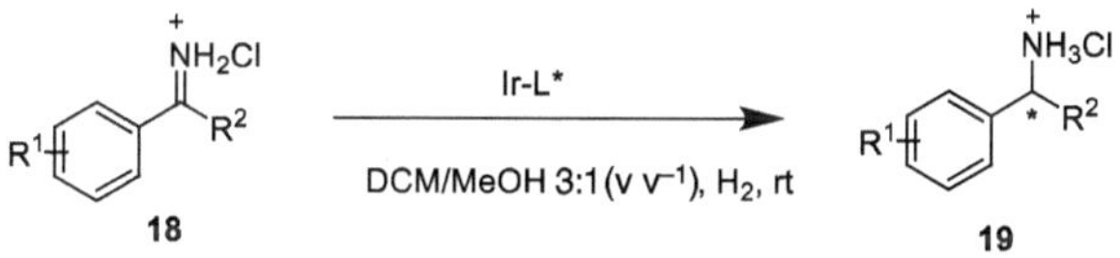

19m (R^1 = 2-Cl, R^2 = Ph) (**15**) 87% ee, conv 94%
19n (R^1 = 2-Br, R^2 = Ph) (**15**) 91% ee, conv 96%
19o (R^1 = 2-Me, R^2 = Ph) (**15**) 82% ee, conv 82%
19p (R^1 = 2-Me, R^2 = 4-OMePh) (**15**) 91% ee, conv 94%
19q (R^1 = 2-Cl-3-CF_3, R^2 = 4-MePh) (**15**) 92% ee, conv 93%

Scheme 9.7 Selection of (P) monodentate ligands employed in combination with iridium precursors and results of their use in asymmetric hydrogenation of imines. Source: Based on Hou et al. [29–31].

9.1.2 Rhodium and Palladium Catalysts

There are different reports of chiral rhodium catalysts with phosphine- or amine-derivative ligands providing high activity in the asymmetric hydrogenation of imines such as hydrazones and *N*-(diphenylphosphinyl)imines [28, 36, 37].

Also, the combination of rhodium precursor [(Rh(cod)$_2$)][BF$_4$] with chiral diphosphines such as (*S*,*S*,*R*,*R*)-TangPhos, (*R*,*S*)-DuanPhos, *S*-binapine, and (S)-C$_3$-TunePhos has provided the hydrogenation of a range of *N*-tosylimines with enantioselectivities of 80–99% [38].

The combination of a rhodium precursor with the methylated ZhaoPhos ligand (**5**) has led to the asymmetric hydrogenation of unprotected NH imines (**18**, Scheme 9.8), generating the corresponding chiral products with high yield and enantioselectivities (up to 97% and 95%, respectively) [39]. Moreover, the anion-bonding interaction between the chloride ion of the substrate and thiourea motif of the ligand has been proved essential for getting excellent enantioselectivity in this asymmetric hydrogenation [40].

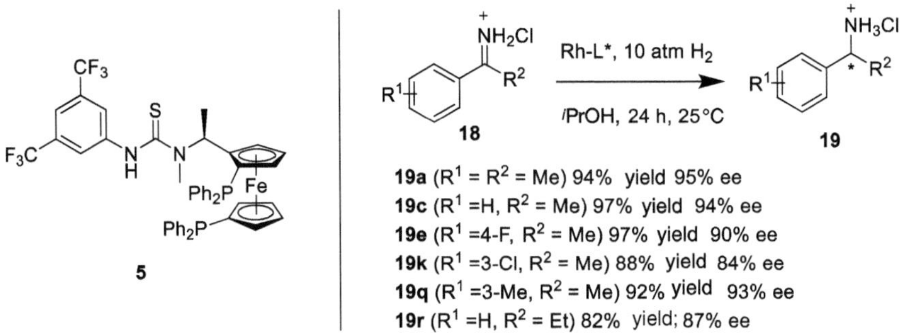

Scheme 9.8 Biphosphine–thiourea ligand **5** in Rh-catalyzed asymmetric hydrogenation of imines and the corresponding results.

There are few examples of Pd-catalyzed asymmetric hydrogenation of imines, and in most of them high hydrogen pressure is required [41–43]. However, an inexpensive Pd source such as palladium acetate was found to be a suitable catalyst precursor for the homogeneous asymmetric hydrogenation of sterically hindered *N*-tosylimines under 1 atm hydrogen pressure. The combination of this Pd salt with the chiral diphosphine ligand (*R*,*R*)-QuinoxP (**20**) provided the corresponding chiral amines (**22**) with excellent enantioselectivities (up to 99.9% ee) as well as high yields (Scheme 9.9) [44]. Additionally, the reaction can be conducted on a gram scale and applied to the synthesis of useful chiral products and N-ligand. Pd(OAc)$_2$ can also be used in the reduction of a wide range of acyclic and cyclic *N*-sulfonylimines under ambient pressure of hydrogen when it is combinated with Zn(OTf)$_2$ and the chiral biphosphorous ligand (R)-MeO-BIPHEP [(R)-(+)-(6,6-dimethoxybiphenyl-2,2-diyl)bis(diphenylphosphine)] [45].

9.2 Asymmetric Transfer Hydrogenation of Imines

Although one of the most explored reactions for the efficient preparation of chiral amines in high enantiopurity is the asymmetric reduction of imines using molecular hydrogen, alcohol is highly preferred as the hydrogen source due to several

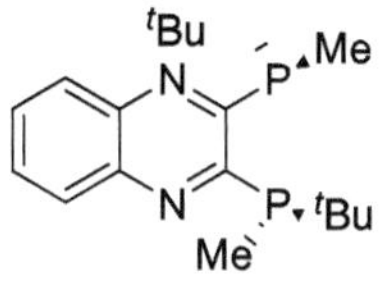

20

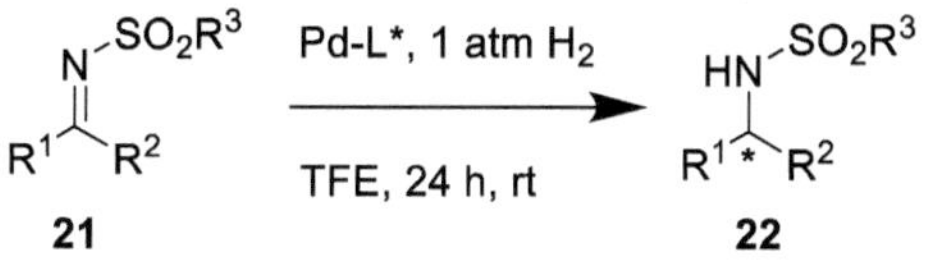

22a (R^1 = ^{t}Bu, R^2 = Ph, R^3 = 4-MePh) 99.9% ee, 99% yield
22b (R^1 = ^{t}Bu, R^2 = Ph, R^3 = 4-MeOPh) 99% ee, 96% yield
22c (R^1 = ^{t}Bu, R^2 = Ph, R^3 = 4-FPh) 96% ee, 95% yield
22d (R^1 = ^{t}Bu, R^2 = Ph, R^3 = Ph) 99.9% ee, 99% yield
22e (R^1 = ^{t}Bu, R^2 = Ph, R^3 = Me) 99.1% ee, 98% yield
22f (R^1 = ^{t}Am, R^2 = Ph, R^3 = H) 99.9% ee, 98% yield
22g (R^1 = Ad, R^2 = Ph, R^3 = H) 99% ee, 97% yield
22h (R^1 = ^{t}Bu, R^2 = 3-MePh, R^3 = H) 99.3% ee, 96% yield
22i (R^1 = ^{t}Bu, R^2 = 4-MePh, R^3 = H) 99.7% ee, 96% yield
22j (R^1 = ^{t}Bu, R^2 = 3-MeOPh, R^3 = H) 99.6% ee, 98% yield
22k (R^1 = ^{t}Bu, R^2 = 4-MeOPh, R^3 = H) 99.8% ee, 97% yield
22l (R^1 = ^{t}Bu, R^2 = 4-PhO, R^3 = H) 99.9% ee, 99% yield
22m (R^1 = ^{t}Bu, R^2 = 3-ClPh, R^3 = H) 99.8% ee, 96% yield
22n (R^1 = ^{t}Bu, R^2 = 4-ClPh, R^3 = H) 99.9% ee, 97% yield
22o (R^1 = ^{t}Bu, R^2 = 3-FPh, R^3 = H) 99.9% ee, 98% yield
22p (R^1 = ^{t}Bu, R^2 = 4-FPh, R^3 = H) 99.7% ee, 96% yield
22q (R^1 = ^{t}Bu, R^2 = 3,5-diMePh, R^3 = H) 99.7% ee, 98% yield
22r (R^1 = ^{t}Bu, R^2 = 3,5-di^{t}BuPh, R^3 = H) 99.8% ee, 96% yield

Scheme 9.9 (*R*,*R*)-QuinoxP ligand (**20**) employed in Pd-catalyzed asymmetric hydrogenation of acyclic imines and the corresponding results.

factors such as safety, sustainability, possible industrial application, and simple manipulation. More specifically, 2-propanol has been used almost exclusively, especially with ruthenium complexes, and also with iridium, rhodium, and iron complexes. In this section, the most relevant examples will be summarized.

9.2.1 Ruthenium Catalysts

Ruthenium-*N*-tosyl-1,2-diphenylethane-1,2-diamine (Ru-TsDPEN) was the first catalyst employed for the asymmetric transfer hydrogenation (ATH) of imines (Figure 9.2) [46]. The reaction took place using the formic acid/triethylamine azeotrope as the hydrogen source. Several examples have followed this first attempt via the development of a large variety of chiral N-sulfonylated 1,2-diamines as ligands for the ATH of ketones and imines. This prolificacy has generated a large number of articles and reviews [2, 47–49].

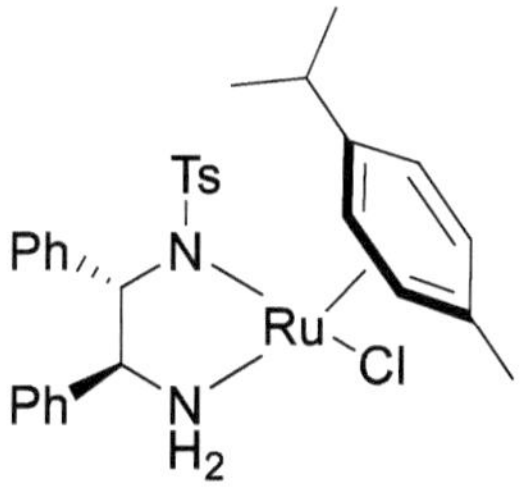

[Ru-TsDPEN]

Figure 9.2 Noyori's Ru catalyst.

A simple exchange of *p*-cymene substituent for mesitylene in Noyori's Ru catalyst allowed the generation of chiral trifluoromethyl-functionalized amines in high yields and enantioselectivities (93–99% ee) under mild conditions, using water as solvent, dimethylformamide as cosolvent, and sodium formate as hydrogen source [50].

The attempts done for the application of these ruthenium catalysts in greener reaction conditions are remarkable. Although no reaction was observed for acyclic amines, the water-soluble sulfonated ligand **23** (Scheme 9.10) showed high activity in the ATH of a range of cyclic amines (**25**, **27**, **29**) [51]. Also in a greener approach, different immobilization options have been described for catalyst recycling. Most of the immobilized systems are silica gel grafted and show an impressive recyclability, high conversions, and excellent enantioselectivities [52, 53]. Unfortunately, the majority of them gave much slower reactions than their homogeneous counterpart.

Polymeric functionalization has also been employed for the immobilization of *N*-sulfonyl-1,2-diphenylethane-1,2-diamine-type ligand **24** which has shown promising results for the reduction of either acyclic or cyclic imines (**25**, **27**, **29**) [54]. In this case, the reaction takes place using the azeotrope formic acid/triethylamine as hydrogen source in an organic solvent. The simple introduction of sulfonated groups in the polymeric ligand allows the performance of the reaction in water.

Other examples of Ru complexes providing high activity and enantioselectivity embrace the use of hydroxy/amine, pybox, or monodentate phosphite ligands. For example, the combination of ligand **31** and dichloro(*p*-cymene)ruthenium(II) reduced (R)-*N*-(*tert*-butanesulfinyl)ketimines followed by the removal of the sulfinyl group under mild acidic conditions with enantioselectivities up to 99% ee [55]. The use of pybox and monodentate phosphite ligands for the generation of ruthenium complex **32** has resulted in a highly efficient enantioselective catalytic system for the reduction of *N*-aryl imines derived from acetophenones (**11**, Scheme 9.11). The use of both hydrogen gas and isopropanol as hydrogen sources is completely tolerated by these catalytic systems [56].

9.2.2 Iridium and Rhodium Catalysts

Few examples of iridium and rhodium complexes have been described for the ATH of imines [49, 50, 57]. Some of them were originally designed for the transfer hydrogenation of carbonyl groups and have resulted in active catalysts for imines as substrates too. In this context, higher activities have been provided by the systems bearing rhodium as metal [42].

Those with *N*-tosyl-1,2-diphenylethane diamine (TsDPEN) or *N*-tosyl-1,2-diaminocyclohexane (TsDAC) as ligands provided ee values higher than 90% under similar reaction conditions than the ones having Ruthenium as metal center. For example, low loadings of both catalytic systems **36** and **37** provided high enantioselectivities and activities at low catalyst loadings for the ATH of cyclic imines (**25**, **27**) [58] (Scheme 9.12).

9.2.3 Iron Catalysts

Fe catalysts are used for the ATH of both ketones and imines with low catalyst loadings of 0.2 and 1 mol%, respectively [59–61]. It is remarkable that

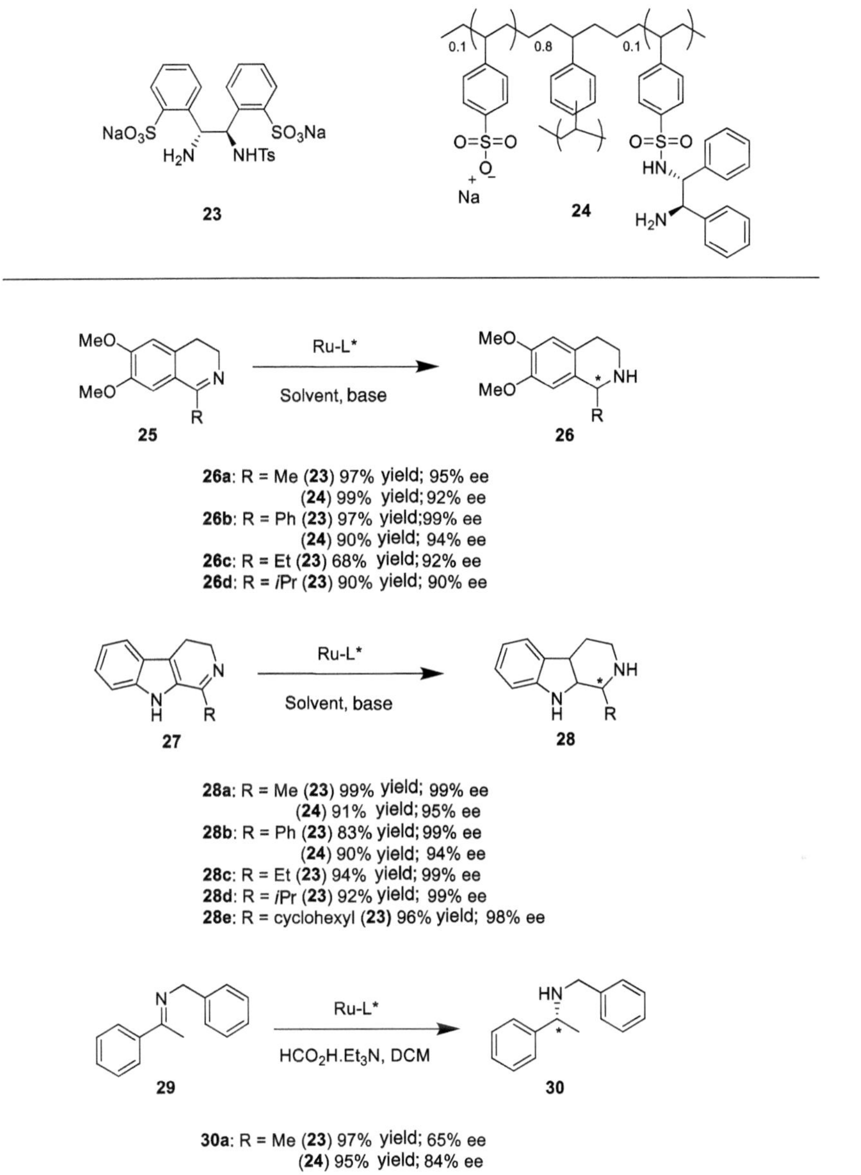

Scheme 9.10 Water-soluble and immobilized ligand/catalysts and the corresponding final products.

the first iron-catalyzed ATH of imines described was carried out using the iron(III)-based catalyst (**38**, Scheme 9.13) under mild reaction conditions. A series of *N*-(diphenylphosphinyl) ketimines (**40**) were reduced with high yields (85–95%) and excellent enantioselectivities (95–99%) [62]. Higher conversions have been achieved for these substrates using catalyst **39** in the presence of isopropanol as

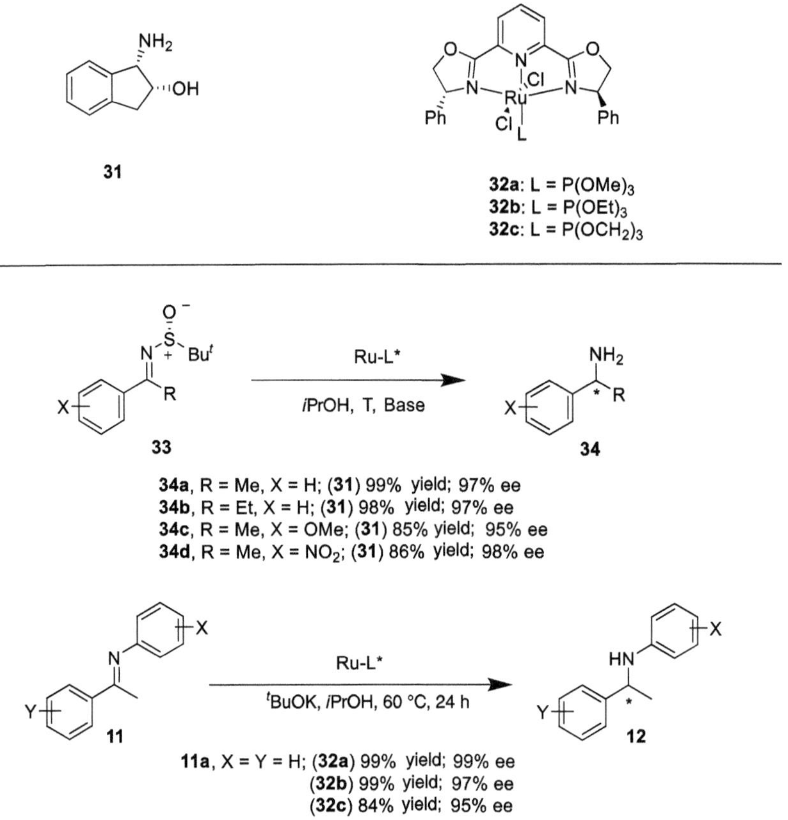

Scheme 9.11 Other ligands and Ru complexes and results of their use in imine ATH.

solvent and hydrogen source [63]. This catalyst has also afforded the asymmetric hydrogenation of imines activated with an *N*-(diphenylphosphinoyl) group (**42**).

9.3 New Approaches

9.3.1 Metal Free

The advent of frustrated Lewis pair (FLP) chemistry opened the door to metal-free asymmetric hydrogenations [64]. Since then, several efforts have been done in this direction, mechanistic studies have been carried out [65], and multiple chiral FLP catalysts have been developed for the asymmetric hydrogenation of imines [66–71]. Unfortunately, moderate enantioselectivities (30–85%) have been reached in the majority of cases.

Catalyst **44** (Scheme 9.14), which has been prepared via *in situ* hydroboration of a chiral internal diene with Piers' borane $HB(C_6F_5)_2$, has shown the highest activity and selectivity achieved to date for imine hydrogenation promoted by a chiral borane catalyst [72]. Yields in the range 83–99% with 83–95% ee were afforded for acyclic

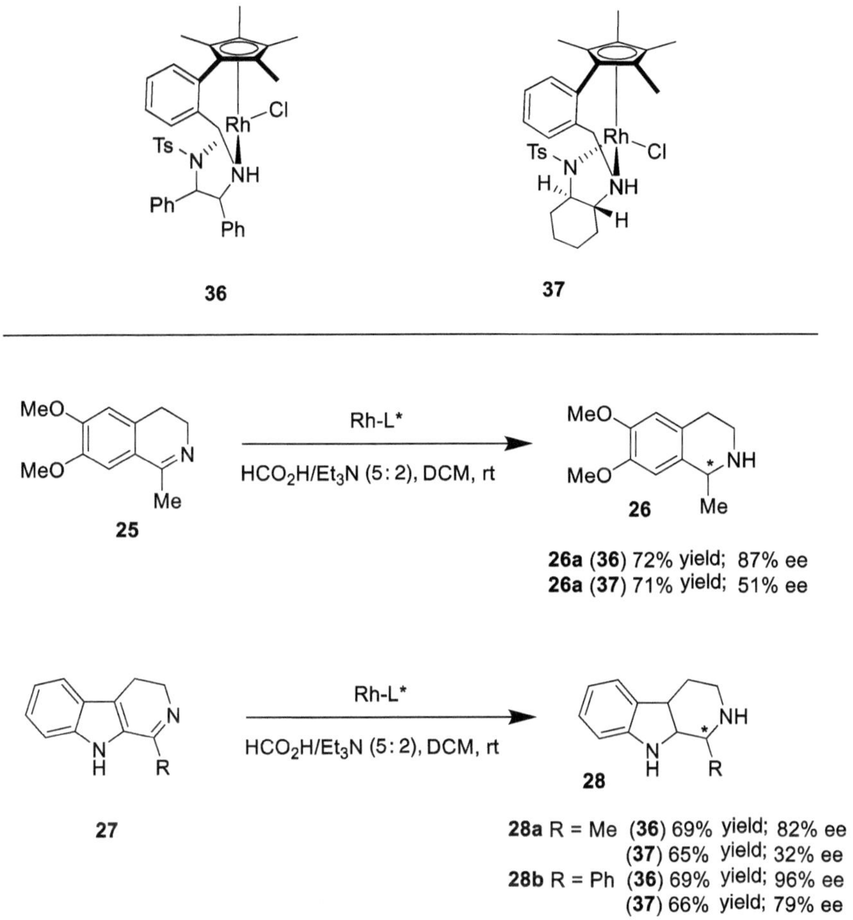

Scheme 9.12 Rh complexes and results of their use in imine ATH.

imines (**11**, Scheme 9.14) containing both electron-donating and -withdrawing substituents. The chiral boron Lewis acid generated tolerates the presence of a variety of substituents including halogen atoms and heterocyclic rings at different positions, providing a large number of chiral amines at low temperature and high H_2 pressure, showing a great potential utility.

9.3.2 Biocatalytic Imine Reduction

In recent years, enzymatic strategies have emerged as a real alternative to other catalytic approaches for the preparation of chemicals in various industries. Enzymes can operate under mild conditions and often display exquisite selectivity; therefore, they can offer a significant advantage to other synthetic approaches. Due to advances in bioinformatics and molecular biology techniques, the biocatalytic toolbox for the

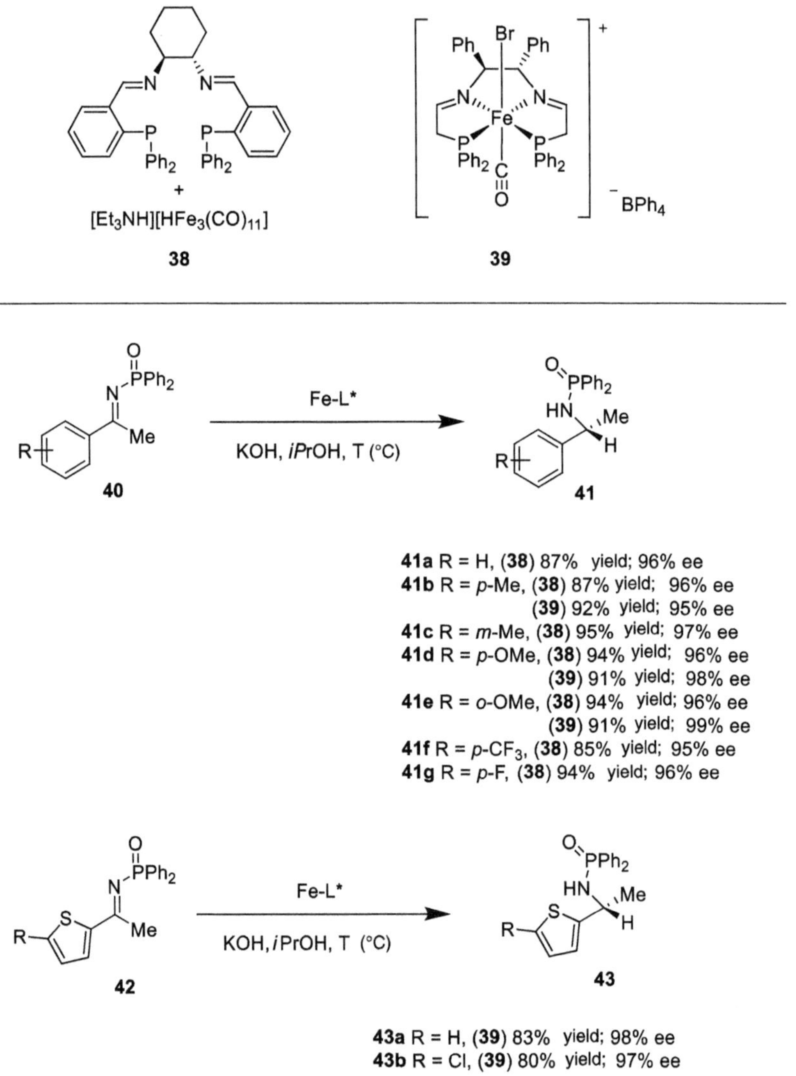

Scheme 9.13 Iron complexes and results for imine ATH.

preparation of chiral amines has been greatly expanded in the past decade. To classical approaches such as lipases [73], transaminases (TAs) [74], or amine oxidases [75], in recent years, processes involving the asymmetric reduction of prochiral imines have emerged as an alternative and expanded the portfolio of enzymes available for chiral amine synthesis. In this chapter, different enzymatic approaches involving artificial metalloenzymes and imine reductases (IREDs) for the reduction of imines are discussed.

9.3.2.1 Artificial Metalloenzymes

Enzymes display high catalytic performances in exquisite selectivity, and they are able to operate under mild reaction conditions. Even though tremendous effort

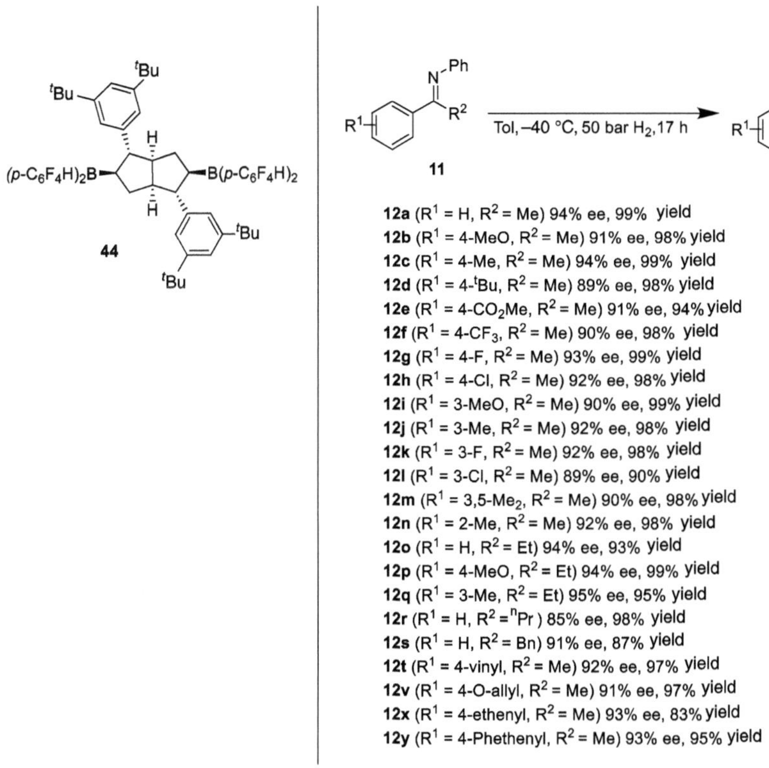

Scheme 9.14 Asymmetric hydrogenation of imines mediated by frustrated Lewis pair (FLP).

has recently been made in the directed evolution of enzymes to perform unnatural reactions [76], the plethora of transformations metal catalysis currently offers is yet to be matched. In this context, artificial metalloenzymes are synthetic biocatalysts that result from the integration of a metal catalyst within a protein scaffold in an attempt to get the best of both "catalytic worlds." Revolutionary work by Whitesides exploited the high affinity of the protein avidin for biotin, to anchor a metal catalyst to the cofactor and incorporate it into the protein cavity [77]. This pioneering work inspired different research groups, and in particular, the group of Ward has done extensive research in the field developing a series of biotinylated cofactors to carry out various reactions such as Suzuki coupling, allylic alkylation, olefin metathesis, sulfoxidation, olefin hydrogenation, and ATH reactions [78].

Using their previous experience in the ATH of ketones using metalloenzymes [79], Durrenberg et al. initially screened a series of d^5 and d^6 piano stool Ir and Ru complexes containing the biotinylated aminosulfonamide ligand (biot-*p*-L) using streptavidin as protein scaffold and sodium formate as hydride source for the synthesis of salsolidine (Scheme 9.15, **entries 1–5**) [80]. The complex η^5-C_5Me_5Ir(biot-*p*-L) showed the best activity obtaining (R)-salsolidine **46a** in full conversion and 57% ee. The artificial IRED (ATHase) was then engineered in order to both improve and invert the stereochemistry. A site-saturation library was generated in S112 position as

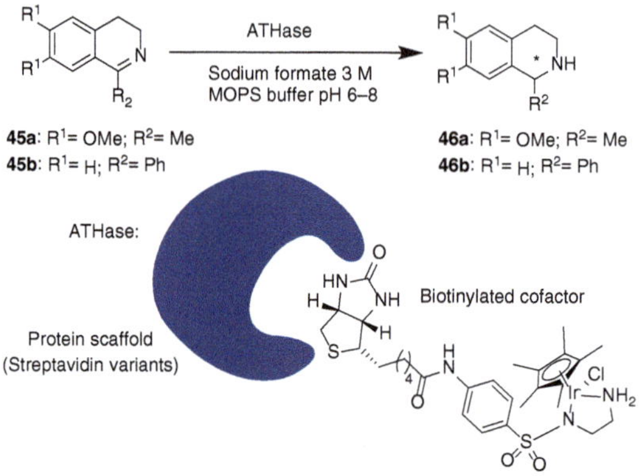

Entry	Substrate	Streptavidin variant	Conversion (%)	ee(%)
1	**45a**	No protein	>99	Racemic
2	**45a**	Wild-type	>99	57 (*R*)
3	**45a**	S112G	>99	60 (*R*)
4	**45a**	S112K	>99	78 (*S*)
5	**45a**	S112A	>99	96 (*R*)
6	**45b**	No protein	17	Racemic
7	**45b**	K121A	99	54
8	**45b**	N118P/K121A	96	59
9	**45b**	S112R/K121A	36	−27
10	**45b**	S112A/N118P/K121A	>99	86
11	**45b**	S112A/N118P/K121A/S122M	>99	92
12	**45b**	S112R/N118P/K121A/S122M	90	−63
13	**45b**	S112R/N118P/K121A/S122M/L124Y	99	−78

Scheme 9.15 Selected results for the reduction of cyclic imines **45a,b** using ATHases based on biotin–streptavidin technology.

it has been previously observed to play a major role in stereochemical control [80]. In this manner, they found an S112A variant to obtain (R)-**46a** in 96% ee and an S112K variant to afford (S)-**46a** in 78% ee. Further enzyme optimization was performed to improve the catalytic efficiency [81]. The authors envisaged that increasing the hydrophobicity in the active site would lead to an increase in the local concentration of the substrate. Crystallographic data showed 10 polar residues located in the proximity of the piano stool moiety; therefore, they gradually introduced hydrophobic substitutions on those sites. They could obtain a set on single, double, and triple mutants with higher k_{cat} values and catalytic efficiencies despite the increase in K_M. The best variant (D67V/S112A/K121L) showed a 40-fold increase in k_{cat} over the wild-type ATHase and a twofold increase compared to the free cofactor in solution.

Recently, Hestericová et al. further improved the catalytic performance of this ATHase via directed evolution [82]. Starting from the K121A variant, they targeted a series of residues within a 10-Å distance from the biotinylated metal moiety generating saturation libraries at those positions selecting the best variant in each round based on both conversion and enantiomeric excess using 1-phenyl dihydroisoquinoline **45b** as model substrate (Scheme 9.15, **entries 6–13**). After several rounds of mutagenesis, the authors obtained a series of enantiocomplementary variants to obtain both enantiomers of the amine **46b** in excellent conversions and enantiomeric excess.

Enzymes can be easily combined to create multistep processes without the need of tedious and time-consuming purification steps. In this manner, a collaboration between the groups of Ward and Turner led to the construction of an in vitro cascade by combining different oxidases with ATHase for the preparation of chiral amines from racemic amines [83]. An engineered monoamine oxidase from *Aspergillus niger* (MAO-N), which presents high selectivity toward the oxidation of (S)-amines to the corresponding imines, was combined with the ATHase. In this process, the (S)-enantiomer is oxidized by MAO-N to form the imine that is reduced by a (R)-selective ATHase; hence, the (R)-amine accumulates in the reaction mixture. The addition of catalase was necessary to remove the hydrogen peroxide produced in the oxidation step as the ATHase was inhibited. In this manner, they could perform the deracemization of tetrahydroisoquinolines and pyrrolidines obtaining the (R)-products in high enantiomeric excess.

9.3.2.2 Imine Reductases (IREDs)

IREDs are NADPH-dependent oxidoreductases that catalyze the asymmetric reduction of prochiral imines to the corresponding amines. Nearly a decade ago, Mitsukura et al. [84–86]. reported the imine-reducing activity of two NADPH-dependent oxidoreductases from *Streptomyces* sp. through an extensive screening of different microorganisms. Several academic and industrial groups have reported extensive research on these enzymes ever since [87–89], and some of those processes have been already intensified for their industrial application [89]. This chapter gives an overview of these biocatalysts focusing on their application in the asymmetric synthesis of amines.

Imine reduction in Nature is widespread, and IREDs are involved in the biosynthesis of metabolites including folate, siderophores, and antibiotics [90–95]. However, the use of IREDs in synthetic chemistry is quite recent. In a series of publications, Mitsukura et al. reported the discovery and characterization of two stereocomplementary IREDs from *Streptomyces* sp. (GF3546 and GF3587) with activity toward 2-methylpyrroline **47** [84–86] (Scheme 9.16). These studies established a starting point to other groups to investigate the synthetic potential of these two enzymes as well as discover new IREDs.

Pioneer work developed by the group of Turner showed the synthetic potential of these enzymes by firstly studying the asymmetric reduction of a series of cyclic imines catalyzed by the (S)-selective IRED from *Streptomyces* sp. GF3546 [96]. This enzyme showed a broad substrate scope from five-, six-, and seven-membered

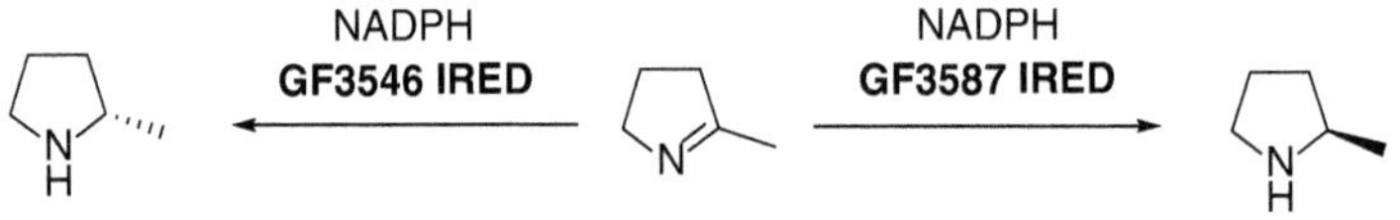

(S)-48 47 (R)-48

Scheme 9.16 Asymmetric reduction of 2-methylpyrroline by two stereocomplementary imine reductases from *Streptomyces* spp.

rings; benzo-fused heterocycles such as dihydroisoquinolines or β-carbolines, and iminium ions. In addition, the authors performed preparative-scale reactions using whole cells to highlight the industrial potential of this family of enzymes. Subsequently, a similar study was done for the (R)-selective IRED from *Streptomyces* sp. GF3587 [97]. The catalytic efficiency of this enzyme was found to be orders of magnitude higher than any other IRED reported to that date, and it was employed to prepare the alkaloid (R)-coniine in 90% yield and >99% ee on a gram scale. Using the structure of the (R)-IRED from *Streptomyces kanamyceticus* [98], a homology model was also constructed to give insight into potential residues for mutagenesis. Due to the importance of the products that can be accessed with this class of enzymes, other groups gained interest, and several IREDs have now been characterized expanding the biocatalytic C=N reduction landscape to a wide range of cyclic imines. By sequence similarity, Scheller et al. characterized three new enantiocomplementary IREDs from *Streptosporangium roseum* DSM 43021, *Streptomyces turgidiscabies*, and *Paenibacillus elgii* contained in that collection [99]. The groups of Turner and Grogan reported a new IRED from *Amycolatopsis orientalis* (*Ao*IRED) that catalyzes the reduction of prochiral imines and iminium ions displaying unique stereoselective properties [100]. Remarkably, the enzyme presents a switch in stereoselectivity toward methyl tetrahydroisoquinoline, depending on the state of the enzyme. Wetzl et al. performed successful bioreductions of a series of cyclic imines using a selection of bacterial IREDs including preparative-scale reactions with a substrate-feeding strategy to reduce substrate inhibition [101]. One of the main issues enzymes present is that they are normally active in aqueous media where most organic compounds present low solubility. The stability of different IREDs in organic solvents was studied by Maugeri et al. [102]. They reported the application of IREDs in microaqueous systems by testing IREDs from *Streptomyces* strains and an IRED from *P. elgii* B69 against different model substrates showing that these biocatalysts are also active in nonconventional media.

More recently, efforts have been made in order to expand the synthetic scope beyond simple cyclic imines. For instance, in a collaboration between Pfizer and the group of Turner, IREDs were applied for the asymmetric synthesis of dibenz[*c*,*e*]azepines showing that IREDs can also display selectivity toward different conformers in biaryl systems [103]. Employing a biocatalytic retrosynthetic approach, the authors designed a route involving a TA step coupled with a Suzuki–Miyaura cross-coupling to obtain the imine precursors. Using enantiocomplementary IREDs they could access a set of dibenz[*c*,*e*]azepines in high enantiomeric excess. Li and coworkers described the asymmetric synthesis of a set of substituted indolines using enantiocomplementary IREDs from *Paenibacillus lactis* in good conversions and stereoselectivities [104, 105]. In a series of works, the group of Gröger recently reported the IRED-catalyzed asymmetric synthesis of sulfur- and oxygen-containing heterocyles. Firstly, they described the bioreduction of 3-thiazolines and 2*H*-1,4-benzothiazines employing a collection of IREDs provided by Roche [106]. For the preparation of 3-thiazolines, they performed an Asinger-type reaction that was afterward carried out in a two-step sequential cascade manner [107] (Scheme 9.17a). They have also described the IRED-mediated reduction of 2*H*-1,4-benzoxazines employing a pH-based colorimetric screen for the quick identification of the best-performing biocatalysts within their IRED collection [106]. They found enantiocomplementary enzymes, and they could obtain the corresponding target chiral heterocycles in moderate to excellent conversions and good to excellent enantiomeric excess in this manner. Yao et al. used their in-house enzyme collection generated by gene mining to access the morphinan skeleton, which is present in different pharmaceuticals [108]. Furthermore, Qu and collaborators used a set of IREDs to prepare 1-aryl-substituted tetrahydroisoquinolines [109]. Through homology modeling, they identified a residue (W191) that was shown to play a major role in substrate accommodation. Mutation to smaller residues such as L and A led to variants with significantly lower K_M – suggestive of a more effective binding – which ultimately resulted in variants with increased catalytic efficiency. Finally, Nestl and coworkers developed an IRED-catalyzed process to access piperazines by combining dicarbonyls and diamines in which, after condensation, the *R*-selective IRED from *Myxococcus stipitatus* catalyzed a double reduction to obtain the final piperazines **55** [110] (Scheme 9.17b). In

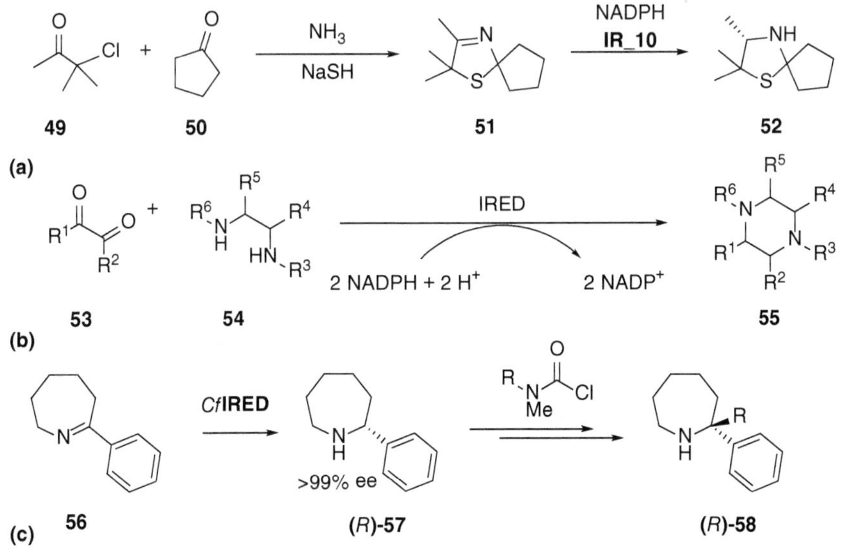

Scheme 9.17 Use of imine reductases for the asymmetric synthesis of different families of heterocycles.

a recent work, Zawodny et al. reported the chemoenzymatic synthesis of 2-aryl azepanes by combining an IRED bioreduction step with an organolithium-mediated functionalization [111]. Starting from different ketoamines, the reduction of the *in situ* formed seven-membered cyclic imines was studied with a series of novel IREDs from both genomic and metagenomic libraries. In this manner, they found enantiocomplementary IREDs to access both enantiomers. Subsequently, the authors used these scaffolds to prepare optically active heterocyclic α-tertiary amines by stereospecific aryl and alkyl migration of the corresponding lithiated ureas in excellent yields and enantiomeric ratios (Scheme 9.17c).

IREDs have also been applied into cascade processes to generate structural and stereochemical complexity in sequential steps without the isolation of the intermediates. For instance, Heath et al. developed a deracemization process in which amine oxidases and IREDs were combined to access a set of optically active two-substituted piperidines (S)-**60a–h** and pyrrolidines (S)-**59a,b** [112] (Scheme 9.18). France et al. reported a three-step cascade where carboxylic acid reductases (CARs), TAs, and IREDs were combined to prepare chiral piperidines from ketoacids exploiting the exquisite chemoselectivity enzymes often display [112]. In a first step, the carboxylic acid is reduced by a CAR from *Mycobacterium marinum* (mCAR) to the aldehyde that is then aminated by a commercial TA (ATA-113). After spontaneous cyclization, the corresponding cyclic imine is reduced by enantiocomplementary IREDs

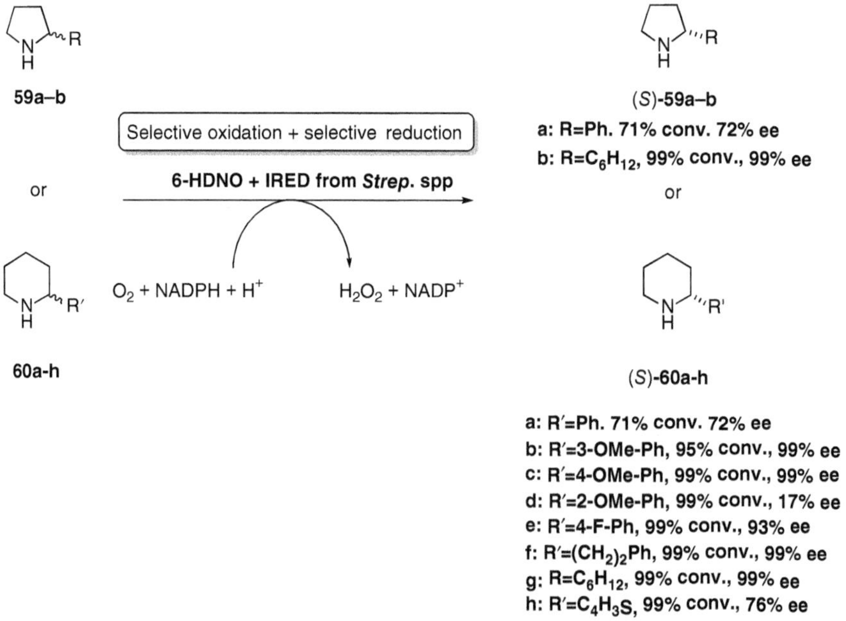

Scheme 9.18 Deracemization of a series of pyrrolidines **59a,b** and piperidines **60a–h** by combining an *R*-selective amine oxidase (6-HDNO) and an *S*-selective IRED from *Streptomyces* spp.

from *Streptomyces* sp. to obtain a variety of chiral-substituted piperidines and pyrrolidines. Starting from diamines, Nestl and coworkers reported the combination of a putrescine oxidase (PuO) variant and an IRED to synthesize substituted pyrrolidines from diamines [113]. The diamine was firstly oxidized to the aldehyde by a variant of the PuO from *Rhodococcus erythropolis* that spontaneously cyclized to form the cyclic imine. The corresponding optically active cyclic amines were subsequently obtained upon IRED-mediated reduction employing the *R*-selective IRED from *S. roseum* (*R*-IRED-*Sr*) and a hydrogenase for cofactor recycling.

9.4 Summary and Conclusions

Catalytic hydrogenation of imines, mediated by noble transition metals under high pressure of molecular hydrogen as one of the cleanest reducing agents, has been extensively studied and is an industrially applied process. Iridium is the most employed metal along with a variety of ligands. Nowadays, in response to the increasing social pressure toward a sustainable society, more environmentally friendly and less costly and harmful methodologies have been developed. Especially ruthenium complexes have allowed the efficient ATH of imines at low pressures using mostly alcohol as the hydrogen source, and new metal-free methodologies have shown promising results in the area. In conclusion, the preparation of chiral amines is possible through versatile, environmentally friendly, and economic procedures involving simple manipulation and cheap reagents.

Advances in molecular biology, bioinformatics, and enzyme discovery and greater speed in gene synthesis and sequencing have resulted in the field of biocatalysis to cover a wider range of transformations in the past decade. Particularly, for the asymmetric reduction of imines, different strategies described in this chapter have been recently developed, and due to further research in enzyme discovery and directed evolution, biocatalytic imine reduction is envisaged to expand its substrate scope in the next future and provide a greener alternative to traditional methodologies.

References

1 Seo, C.S.G. and Morris, R.H. (2018). *Organometallics* 38: 47–65.
2 Fleury-Brégeot, N., de la Fuente, V., Castillón, S., and Claver, C. (2010). *ChemCatChem* 2: 1346–1371.
3 Xie, J.H., Zhu, S.F., and Zhou, Q.L. (2011). *Chem. Rev.* 111: 1713–1760.
4 Yu, Z., Jin, W., and Jiang, Q. (2012). *Angew. Chem. Int. Ed.* 51: 6060–6072.
5 Fabrello, A., Bachelier, A., Urrutigoïty, M., and Kalck, P. (2010). *Coord. Chem. Rev.* 254: 273–287.
6 Blaser, H. and Spindler, F. (2009). *Org. React.* 74: 1.
7 Tang, W., Lau, C., Wu, X., and Xiao, J. (2014). *Synlett* 25: 81–84.
8 Hopmann, K.H. and Bayer, A. (2014). *Coord. Chem. Rev.* 268: 59–82.

9 Guiu, E., Munoz, B., Castillon, S., and Claver, C. (2003). *Adv. Synth. Catal.* 345: 169–171.

10 Guiu, E., Aghmiz, M., Díaz, Y. et al. (2006). *Eur. J. Org. Chem.* 3: 627–633.

11 Blaser, H.U. (2002). *Adv. Synth. Catal.* 344: 17–31.

12 Xiao, D. and Zhang, X. (2001). *Angew. Chem. Int. Ed.* 40: 3425–3428.

13 Bin Yu, C., Wang, D.W., and Zhou, Y.G. (2009). *J. Org. Chem.* 74: 5633–5635.

14 Kong, D., Li, M., Zi, G. et al. (2016). *J. Org. Chem.* 81: 6640–6648.

15 Liu, Y., Huang, Y., Yi, Z. et al. (2019). *Adv. Synth. Catal.* 361: 1582–1586.

16 Schnider, P., Koch, G., Prétôt, R. et al. (1997). *Chem. - Eur. J.* 3: 887–892.

17 Trifonova, A., Diesen, J.S., and Andersson, P.G. (2006). *Chem. -Eur. J.* 12: 2318–2328.

18 Schrems, M.G. and Pfaltz, A. (2009). *Chem. Commun.* 41: 6210–6212.

19 Guiu, E., Claver, C., Benet-Buchholz, J., and Castillón, S. (2004). *Tetrahedron Asymmetry* 15: 3365–3373.

20 Baeza, A. and Pfaltz, A. (2010). *Chem. -Eur. J.* 16: 4003–4009.

21 Blanc, C., Agbossou-Niedercorn, F., and Nowogrocki, G. (2004). *Tetrahedron Asymmetry* 15: 2159–2163.

22 Salomó, E., Rojo, P., Hernández-Lladó, P. et al. (2018). *J. Org. Chem.* 83: 4618–4627.

23 Salomo, E., Gallen, A., Sciortino, G. et al. (2018). *J. Am. Chem. Soc.* 140: 16967–16970.

24 Zhu, S.-F., Xie, J.-B., Zhang, Y.-Z. et al. (2006). *J. Am. Chem. Soc.* 128: 12886–12891.

25 Han, Z., Wang, Z., Zhang, X., and Ding, K. (2009). *Angew. Chem. Int. Ed.* 48: 5345–5349.

26 Cheemala, M.N. and Knochel, P. (2007). *Org. Lett.* 9: 3089–3092.

27 Li, C., Wang, C., Villa-Marcos, B., and Xiao, J. (2008). *J. Am. Chem. Soc.* 130: 14450–14451.

28 Tang, W., Johnston, S., Li, C. et al. (2013). *Chem. -Eur. J.* 19: 14187–14193.

29 Hou, G., Tao, R., Sun, Y. et al. (2010). *J. Am. Chem. Soc.* 132: 2124–2125.

30 Mršič, N., Jerphagnon, T., Minnaard, A.J. et al. (2009). *Adv. Synth. Catal.* 351: 2549–2552.

31 Mršić, N., Panella, L., Ijpeij, E.G. et al. (2011). *ChemCatChem* 3 (7): 1139–1142.

32 Tutkowski, B., Kerdphon, S., Limé, E. et al. (2018). *ACS Catal.* 8: 615–623.

33 Lagaditis, P.O., Sues, P.E., Sonnenberg, J.F. et al. (2014). *J. Am. Chem. Soc.* 136: 1367–1380.

34 Amézquita-Valencia, M. and Cabrera, A. (2013). *J. Mol. Catal. A Chem.* 366: 17–21.

35 Li, B., Chen, J., Zhang, Z. et al. (2019). *Angew. Chem. Int. Ed.* 58: 7329–7334.

36 Yoshikawap, N., Tan, L., Christopher McWilliams, J. et al. (2010). *Org. Lett.* 12: 276–279.

37 Burk, M.J. and Feaster, J.E. (1992). *J. Am. Chem. Soc.* 114: 6266–6267.

38 Yang, Q., Shang, G., Gao, W. et al. (2006). *Angew. Chem. Int. Ed.* 45: 3832–3835.

39 Zhao, Q., Wen, J., Tan, R. et al. (2014). *Angew. Chem. Int. Ed.* 53: 8467–8470.

40 Li, P., Huang, Y., Hu, X. et al. (2017). *Org. Lett.* 19: 3855–3858.

41 Abe, H., Amii, H., and Uneyama, K. (2001). *Org. Lett.* 3: 313–315.
42 Li, L., Wu, J., Wang, F. et al. (2007). *Green Chem.* 9: 23–25.
43 Fan, D., Liu, Y., Jia, J. et al. (2019). *Org. Lett.* 21: 1042–1045.
44 Chen, J., Zhang, Z., Li, B. et al. (2018). *Nat. Commun.* 9: 5000.
45 Gao, Y., Yang, F., Pu, D. et al. (2018). *Eur. J. Org. Chem.* 2018: 6274–6279.
46 Uematsu, N., Fujii, A., Hashiguchi, S. et al. (1996). *J. Am. Chem. Soc.* 118: 4916–4917.
47 Matuška, O., Zápal, J., Hrdličková, R. et al. (2016). *React. Kinet. Mech. Catal.* 118: 215–222.
48 Václavík, J., Šot, P., Vilhanová, B. et al. (2013). *Molecules* 18: 6804–6828.
49 Pan, H.J., Zhang, Y., Shan, C. et al. (2016). *Angew. Chem. Int. Ed.* 55: 9615–9619.
50 Wu, M., Cheng, T., Ji, M., and Liu, G. (2015). *J. Org. Chem.* 80: 3708–3713.
51 Wu, J., Wang, F., Ma, Y. et al. (2006). *Chem. Commun.* 16: 1766–1768.
52 Huang, X. and Ying, J.Y. (2007). *Chem. Commun.* 18: 1825–1827.
53 Li, J., Zhang, Y., Han, D. et al. (2009). *J. Mol. Catal. A Chem.* 298: 31–35.
54 Haraguchi, N., Tsuru, K., Arakawa, Y., and Itsuno, S. (2009). *Org. Biomol. Chem.* 7: 69–75.
55 Guijarro, D., Pablo, Ó., and Yus, M. (2009). *Tetrahedron Lett.* 50: 5386–5388.
56 Menéndez-Pedregal, E., Vaquero, M., Lastra, E. et al. (2015). *Chem. -Eur. J.* 21: 549–553.
57 Wang, C., Li, C., Wu, X. et al. (2009). *Angew. Chem. Int. Ed.* 48: 6524–6528.
58 Matharu, D.S., Martins, J.E.D., and Wills, P.M. (2008). *Chem. Asian J.* 3: 1374–1383.
59 Sues, P.E., Demmans, K.Z., and Morris, R.H. (2014). *Dalton Trans.* 43: 7650–7667.
60 Zuo, W. and Morris, R.H. (2015). *Nat. Protoc.* 10: 241–257.
61 Pan, H.J., Ng, T.W., and Zhao, Y. (2016). *Org. Biomol. Chem.* 14: 5490–5493.
62 Zhou, S., Fleischer, S., Junge, K. et al. (2010). *Angew. Chem. Int. Ed.* 49: 8121–8125.
63 Zuo, W., Lough, A.J., Li, Y.F., and Morris, R.H. (2013). *Science (80)* 342: 1080–1083.
64 Chen, D. and Klankermayer, J. (2008). *Chem. Commun.* 18: 2130–2131.
65 Hermeke, J., Mewald, M., and Oestreich, M. (2013). *J. Am. Chem. Soc.* 135: 17537–17546.
66 Chen, D., Wang, Y., and Klankermayer, J. (2010). *Angew. Chem. Int. Ed.* 49: 9475–9478.
67 Ghattas, G., Chen, D., Pan, F., and Klankermayer, J. (2012). *Dalton Trans.* 41: 9026–9028.
68 Chen, D., Leich, V., Pan, F., and Klankermayer, J. (2012). *Chem. -Eur. J.* 18: 5184–5187.
69 Lindqvist, M., Borre, K., Axenov, K. et al. (2015). *J. Am. Chem. Soc.* 137: 4038–4041.
70 Meng, W., Feng, X., and Du, H. (2017). *Acc. Chem. Res.* 51: 191–201.

71 Liu, X., Liu, T., Meng, W., and Du, H. (2018). *Org. Biomol. Chem.* 16: 8686–8689.
72 Tu, X., Zeng, N., Li, R. et al. (2018). *Angew. Chem. Int. Ed.* 57: 15096–15100.
73 Busto, E., Gotor-Fernández, V., and Gotor, V. (2011). *Chem. Rev.* 111: 3998–4035.
74 Slabu, I., Galman, J.L., Lloyd, R.C., and Turner, N.J. (2017). *ACS Catal.* 7: 8263–8284.
75 Batista, V.F., Galman, J.L., Pinto, D.C. et al. (2018). *ACS Catal.* 8: 11889–11907.
76 Arnold, F.H. (2018). *Angew. Chem. Int. Ed.* 57: 4143–4148.
77 Wilson, M.E. and Whitesides, G.M. (1978). *J. Am. Chem. Soc.* 100: 306–307.
78 Schwizer, F., Okamoto, Y., Heinisch, T. et al. (2018). *Chem. Rev.* 118: 142–231.
79 Letondor, C., Pordea, A., Humbert, N. et al. (2006). *J. Am. Chem. Soc.* 128: 8320–8328.
80 Dürrenberger, M., Heinisch, T., Wilson, Y.M. et al. (2011). *Angew. Chem. Int. Ed.* 50: 3026–3029.
81 Schwizer, F., Köhler, V., Dürrenberger, M. et al. (2013). *ACS Catal.* 3: 1752–1755.
82 Hestericová, M., Heinisch, T., Alonso-Cotchico, L. et al. (2018). *Angew. Chem. Int. Ed.* 57: 1863–1868.
83 Köhler, V., Wilson, Y.M., Dürrenberger, M. et al. (2013). *Nat. Chem.* 5: 93–99.
84 Mitsukura, K., Suzuki, M., Tada, K. et al. (2010). *Org. Biomol. Chem.* 8: 4533–4535.
85 Mitsukura, K., Kuramoto, T., Yoshida, T. et al. (2013). *Appl. Microbiol. Biotechnol.* 97: 8079–8086.
86 Mitsukura, K., Suzuki, M., Shinoda, S. et al. (2011). *Biosci. Biotechnol. Biochem.* 75: 1778–1782.
87 Mangas-Sanchez, J., France, S.P., Montgomery, S.L. et al. (2017). *Curr. Opin. Chem. Biol.* 37: 19–25.
88 Grogan, G. and Turner, N.J. (2016). *Chem. -Eur. J.* 22: 1900–1907.
89 Lenz, M., Borlinghaus, N., Weinmann, L., and Nestl, B.M. (2017). *World J. Microbiol. Biotechnol.* 33: 199.
90 Bornadel, A., Bisagni, S., Pushpanath, A. et al. (2019). *Org. Process Res. Dev.* 23: 1262–1268.
91 Ye, S., Braña, A.F., González-Sabín, J. et al. (2018). *Front. Microbiol.* 9: 1–12.
92 Posner, B.A., Li, L., Bethell, R. et al. (1996). *Biochemistry* 35: 1653–1663.
93 Meneely, K.M. and Lamb, A.L. (2012). *Biochemistry* 51: 9002–9013.
94 Meneely, K.M., Ronnebaum, T.A., Riley, A.P. et al. (2016). *Biochemistry* 55: 5423–5433.
95 Winzer, T., Kern, M., King, A.J. et al. (2015). *Science (80)* 349: 309–312.
96 Leipold, F., Hussain, S., Ghislieri, D., and Turner, N.J. (2013). *ChemCatChem* 5: 3505–3508.
97 Hussain, S., Leipold, F., Man, H. et al. (2015). *ChemCatChem* 7: 579–583.
98 Rodriguez-Mata, M., Frank, A., Wells, E. et al. (2013). *ChemBioChem* 14: 1372–1379.

99 Scheller, P.N., Fademrecht, S., Hofelzer, S. et al. (2014). *ChemBioChem* 15: 2201–2204.

100 Aleku, G.A., Man, H., France, S.P. et al. (2016). *ACS Catal.* 6: 3380–3889.

101 Wetzl, D., Berrera, M., Sandon, N. et al. (2015). *ChemBioChem* 16: 1749–1756.

102 Maugeri, Z. and Rother, D. (2016). *Adv. Synth. Catal.* 358: 2745–2750.

103 France, S.P., Aleku, G.A., Sharma, M. et al. (2017). *Angew. Chem. Int. Ed.* 56: 15589–15593.

104 Li, H., Luan, Z.J., Zheng, G.W., and Xu, J.H. (2015). *Adv. Synth. Catal.* 357: 1692–1696.

105 Li, H., Zhang, G.-X., Li, L.-M. et al. (2016). *ChemCatChem* 8: 724–727.

106 Zumbrägel, N., Machui, P., Nonnhoff, J., and Gröger, H. (2019). *J. Org. Chem.* 84: 1440–1447.

107 Zumbrägel, N. and Gröger, H. (2019). *J. Biotechnol.* 291: 35–40.

108 Yao, P., Xu, Z., Yu, S. et al. (2019). *Adv. Synth. Catal.* 361: 556–561.

109 Zhu, J., Tan, H., Yang, L. et al. (2017). *ACS Catal.* 7: 7003–7007.

110 Borlinghaus, N., Gergel, S., and Nestl, B.M. (2018). *ACS Catal.* 8: 3727–3732.

111 Zawodny, W., Marshall, J.R., Finnigan, J.D. et al. (2018). *Chem. Soc.* https://doi.org/10.1021/jacs.8b11891.

112 France, S.P., Hussain, S., Hill, A.M. et al. (2016). *ACS Catal.* 6: 3753–3759.

113 Al-Shameri, A., Borlinghaus, N., Weinmann, L. et al. (2019). *Green Chem.* 21: 1396–1400.

114 Farrow, S.C., Hagel, J.M., Beaudoin, G.a.W. et al. (2015). *Nat. Chem. Biol.* 11: 728–732.

115 Zumbrägel, N., Merten, C., Huber, S.M., and Gröger, H. (2018). *Nat. Commun.* 9: 1–9.

116 Heath, R.S., Pontini, M., Hussain, S., and Turner, N.J. (2016). *ChemCatChem* 8: 117–120.

10

Asymmetric Hydrogenation in Continuous-Flow Conditions

Gergely Farkas, József Madarász and József Bakos

University of Pannonia, Department of Organic Chemistry, Egyetem u. 10, H-8200, Veszprém, Hungary

10.1 Introduction

Asymmetric catalytic hydrogenation reactions performed in a continuous-flow system open up new horizons in green chemistry. Besides the high catalytic activity, selectivity, and atom economy associated with the hydrogenation process, the application of continuous-flow operations could ensure reduced reaction time, increased efficiency, better control and reproducibility, rapid optimization, screening and analysis, and improved process safety [1]. As a result, the development of asymmetric catalytic hydrogenation processes under continuous-flow conditions can be considered as one of the most strategic field of innovation toward greener production [2]. According to the comment of Knowles and Noyori, "in this present era of green chemistry, where environmentally benign reactions are a "must", asymmetric hydrogenation with its high yields, no byproducts, and very low catalyst usage will always play an important role" [3].

Continuous-flow technology is widely used in the petrochemical and bulk chemicals industry, but it was only after the turn of the twenty-first century it became a hot topic in the field of fine chemical industry. The lack of dedicated equipment and the absence of the link between chemical engineering and synthetic organic chemistry are the main reasons for the slow uptake of the flow technologies by the academic community [4]. The breakthrough of flow chemistry is mainly due to the advantageous features of the flow mode over the batch mode. Most importantly, the reactants are continuously introduced, react within a smaller reaction space under well-controlled conditions, and then are removed from the reactor [5]. Therefore, the key reaction parameters such as the mixing, heat transfer, or residence time can precisely be controlled, leading to higher productivity and lower amount of waste. Furthermore, flow technologies have improved hydrogenation outcomes by increasing substrate–gas–catalyst interactions, reducing the amount of undesirable side products and enhancing selectivity [6]. The improved gas–liquid contact can be achieved via the use of in-line gas mixing or gas-permeable membranes. These techniques can rapidly saturate the solvent with hydrogen gas and thus improve the reaction rate [7]. Additionally, an inherent feature of continuous-flow hydrogenation

Asymmetric Hydrogenation and Transfer Hydrogenation, First Edition.
Edited by Virginie Ratovelomanana-Vidal and Phannarath Phansavath.

reactions is that it minimizes the potential hazards associated with batch hydrogenation reactions [8]. Thus, the flow technology conforms to the 12 principles of green chemistry emphasizing the safer use of the (high-pressure) flammable hydrogen gas or pyrophoric catalysts and the importance of the excellent heat transfer. From an industrial point of view, the features detailed above have exceptional relevance due to their potential to the development of cost-effective production. The values of continuous-flow technology shape the future of pharmaceutical [9] and fine chemical [10] industry [11, 12]. The combination of highly (enantio)selective catalytic systems with flow chemistry further enhances the impact of the technology [13, 14].

The present chapter focuses on the main directions in asymmetric catalytic hydrogenation and transfer hydrogenation reactions performed in a flow reactor. The content is divided into four major categories based on the nature of the catalyst applied in the continuous-flow system: (i) chirally modified metal surfaces, (ii) organometallic catalysts, (iii) organocatalysts, and (iv) chiral auxiliary-aided syntheses.

Asymmetric catalytic reactions including hydrogenation performed in flow mode have been reviewed in several excellent monographs earlier [13–16]. The present chapter focuses exclusively on the continuous-flow asymmetric hydrogenation catalyzed by transition metals or organocatalysts as a narrower concept. Although enzymes also proved to be very effective in continuous-flow asymmetric hydrogenation [17], these systems are not included in the present chapter.

10.2 Chirally Modified Metal Surfaces

The modification of (supported) metal catalyst by the addition of an adsorbing chiral molecule (Figure 10.1) is one of the simplest ways to generate catalytically active chiral metal surfaces [18]. The stability, activity, and selectivity of these catalysts make them appropriate candidates for the application in continuous-flow reactors. The majority of reports in this field focus on the hydrogenation of activated ketones (Figure 10.2) on Pt surfaces modified with chiral alkaloids.

The first report on the application of chirally modified heterogeneous catalysts in continuous-flow hydrogenation dates back in 1979, when Orito and coworkers applied cinchonidine (CD)-modified Pt catalyst for the hydrogenation of

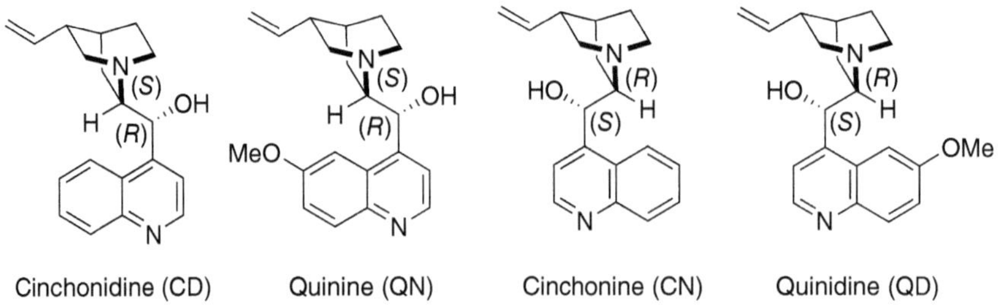

Figure 10.1 Alkaloids used for the chiral modification of metal surfaces.

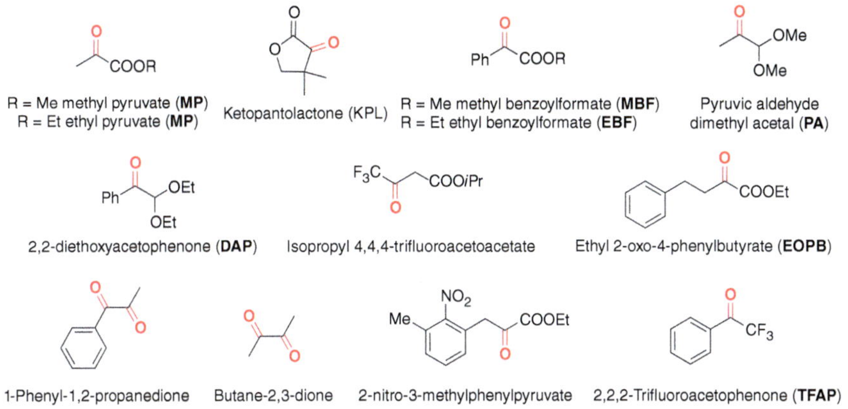

Figure 10.2 Activated ketones used as substrates in asymmetric hydrogenation reactions in continuous-flow systems.

α-ketoesters to obtain optically active products [19]. Due to the slow uptake of flow technology by the scientific community, the next asymmetric flow hydrogenation reaction was reported in 1991 by Ibbotson et al. [20]. As a preliminary study, a 6.3% Pt/silica catalyst modified with CD was synthesized and applied in a flow system for the asymmetric hydrogenation of methyl–pyruvate (MP). At a catalyst mass of 0.4 g and 12 ml h^{-1} flow rate using no solvent, the conversion remained low (~7%); however, the optical yields (up to ~80%) were approximately 20% higher than in the batch experiments. A considerable loss of the chiral modifying agent was detected only in the first 15 min and reduced to a minimum after the first hour, indicating that a certain amount of the CD was only physisorbed on the silica support.

Baiker and coworkers introduced an innovative concept in enantioselective flow hydrogenation applying continuous feeding of minute amounts of chiral modifier to the reactant system [21]. The potential of their methodology was illustrated in the hydrogenation of ketopantolactone (KPL) and ethyl pyruvate (EP) over 5% Pt/alumina at a 2800–3750 ppm CD concentration, relative to the substrate (Figure 10.3). Both KPL and EP could be hydrogenated with high optical purity (83.4% and 89.9% ee, respectively) at 40 bar H_2 pressure. Additionally, the hydrogenation of KPL for a longer period of time-on-stream revealed that after an

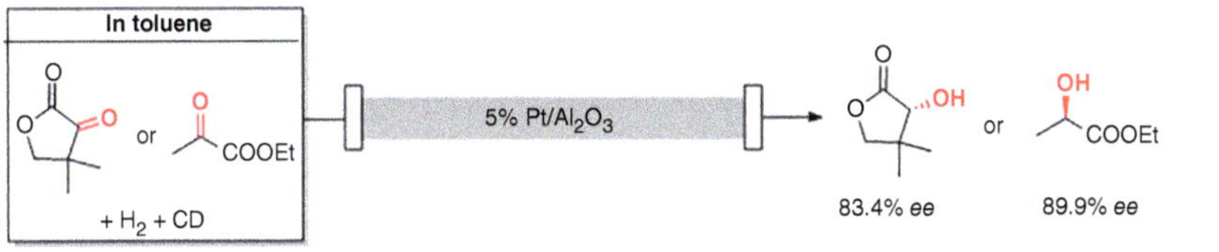

Catalyst: 1 g; solvent: toluene; substrate conc.: 0.078 M for KPL, 0.15 M for EP; CD/substrate molar ratio: 2800 ppm for KPL, 3750 ppm for EP; liquid flow rate: 5 mL min^{-1} for KPL and 2.5 mL min^{-1} for EP; H_2 pressure: 40 bar; T = 17 °C for KPL, 23 °C for EP.

Figure 10.3 Continuous feeding of the chiral modifier in asymmetric hydrogenation of EP and KPL.

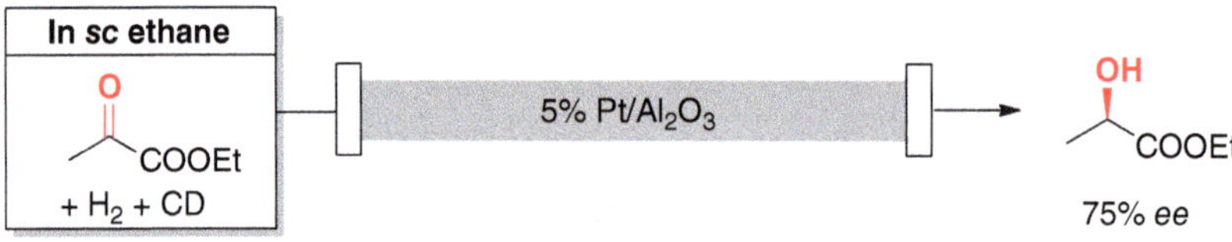

Catalyst: 100 mg 5% Pt/Al_2O_3 diluted with 900 mg Al_2O_3; solvent: *sc*-ethane; Solvent/Substrate/H_2: 500/1/10; Substrate/CD molar ratio: 2500/1; Substrate flow rate: 4.3 mmol min^{-1}; pressure: 100 bar; T = 303 K.

Figure 10.4 Asymmetric hydrogenation of EP in supercritical ethane. Source: Based on Wandeler et al. [22].

initial stabilization period, the ee only slowly decreased: the optical yield declined from 81.3% to 78.4% between two and six hour time-on-stream.

The same concept proved to be highly effective in the hydrogenation of isopropyl-4,4,4-trifluoroacetoacetate, where the chiral modifier and trifluoroacetic acid (TFA) as additive were continuously fed into the reactor containing the Pt/alumina catalyst [23]. The chiral fluorinated alcohol was synthesized with up to 90% ee and with 810 h^{-1} turnover frequency. Under optimized conditions the catalyst was used for 120 min with stable enantioselectivity; however, the decreasing conversion indicated the slow deactivation of the catalyst in the presence of TFA. The efficiency of Baiker's experimental setup was further increased when the continuous platinum-catalyzed hydrogenation of EP was performed in supercritical ethane (*sc*-ethane), as the first application of supercritical fluid in continuous asymmetric hydrogenation (Figure 10.4) [22]. The product ethyl lactate was synthesized with up to 75% ee and a turnover frequency (TOF) value of 15 s^{-1} at 70% conversion and ambient temperature, using the continuous dosing of CD. In a separate study the relation between the phase behavior of the supercritical system and the catalytic performance was thoroughly investigated [24].

The same research group also demonstrated that continuous-flow systems can be used to investigate the dynamics of chiral modification of Pt/alumina surfaces [25]. Simply changing the chiral modifier in the reactor feed for a second modifier that gives the opposite product enantiomer provides valuable experimental information on the relative adsorption strength of the modifiers. The catalytic experiments combined with an ATR-IR study clearly demonstrated the following adsorption abilities: CD>cinchonine (CN)>quinidine (QD). Moreover, in the continuous-flow fixed-bed reactor, EP was hydrogenated with ees up to 89% and with very high activities (TOF = 12 700 h^{-1}), at a modifier concentration of 307 ppm (relative to the substrate). As a preliminary study toward process intensification, Baiker and coworkers performed the asymmetric hydrogenation of EP over CD-modified Pt/Al_2O_3 catalyst in a trickle-bed reactor under reaction conditions optimized by experimental design [26]. The reaction setup afforded an average turnover frequency of 84 000 h^{-1} that corresponds to a production rate of 1.94 mol min^{-1} g_{Pt}^{-1}. Based on the comparison of these values to the literature results achieved in batch reactors, the authors concluded that under flow conditions the space–time yield is about one order of magnitude higher than in the batch system, indicating the

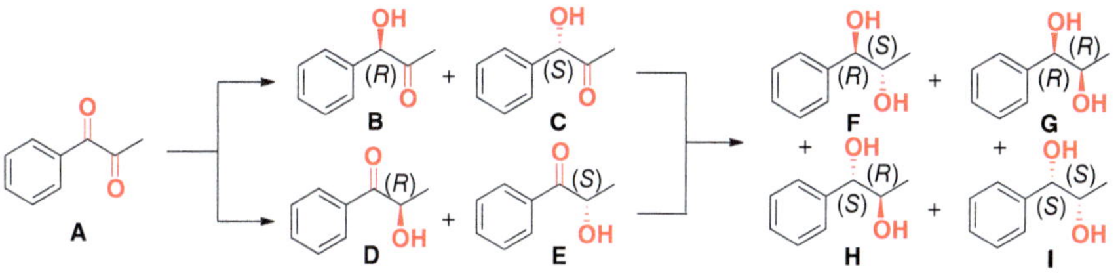

Figure 10.5 Asymmetric hydrogenation of 1-phenyl-1,2-propanedione on CD-modified knitted Pt/silica fiber catalyst. Source: Based on Toukoniitty et al. [28].

high potential of the flow process for the intensification of the catalytic reaction. However, the authors warn that in several cases (e.g. in the hydrogenation of substituted 2-pyrones) the hydrogenation processes are not suitable for the flow synthesis due to the counteracting effects of the modifier on the rate and the ee, thus limiting the beneficial use of the fixed-bed reactors.

In 2004, Garland and coworkers demonstrated the feasibility of the chiral fixed-bed reactors for the diverse synthesis of enantiomerically enriched chiral products [27]. The demodification of the solid catalyst and its subsequent remodification could be performed in the continuous-flow reactor, allowing the regeneration of the catalytic system using the same Pt/Al_2O_3 bed with multiple substrates and/or modifiers.

Toukoniitty and coworkers reported on the successful application of new knitted Pt/silica fiber catalyst modified with (−)-CD for the asymmetric hydrogenation of the challenging substrate 1-phenyl-1,2-propanedione (**A**) in a pressurized batch reactor and under continuous-flow conditions (Figure 10.5) [28]. The ee of compound **B** increased with increasing time-on-stream and after an initial period of c. 20 min in the flow reactor; the ee of compound **B** was the same as in the batch system (55%). The total yield of the diols (**F–I**) formed in the reaction increased with the reaction time in the batch reactor; however, it decreased with time (from 43% to 3%) in the flow system. After reaching the steady-state level, the ees of diol **F** formed in the slurry reactor (39%) and in the fixed-bed system (up to 35%) were commensurate. A decrease in the conversion of the starting material could be observed that indicated the deactivation of the catalyst.

Later, a mechanistic model for the catalyst deactivation was developed that assumed a first-order dependence on the reactant concentration [29]. In the hydrogenation of ethyl benzoylformate (EBF), the role of the bed diluent (Al_2O_3 vs. SiO_2) and the reaction medium (acetic acid vs. toluene) in determining the ee was investigated in order to study the adsorption of the chiral modifier CD [30]. It has been shown that in protic reaction media the adsorption of CD is negligible both on alumina and silica, but the steady-state ee can be obtained instantaneously as the modifier is adsorbed mainly on the metal and not on the support. Furthermore, maximally 93% ee could be obtained in the process.

A controversial phenomenon in the asymmetric hydrogenation catalyzed by chirally modified metal surfaces is the so-called ligand acceleration (LA) effect, i.e. the reactivity of chirally modified metals is higher compared to the unmodified metal

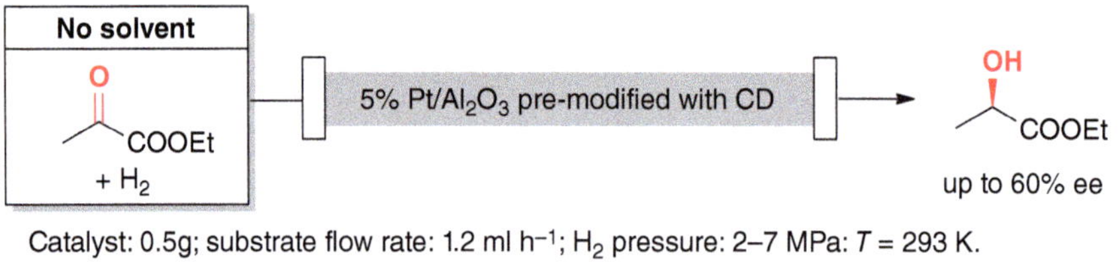

Figure 10.6 Asymmetric hydrogenation of neat EP on premodified CD-Pt/Al_2O_3 in flow. Source: Based on You et al. [33].

surfaces. Toukoniitty and Murzin applied the continuous-flow technique to investigate the origin of LA [31]. They found that in the hydrogenation of EP the effect is due to the CD's ability to prevent side reactions on Pt/Al_2O_3 surfaces that would lead to catalyst deactivation. Murzin and coworkers investigated the feasibility of the combination of continuous-flow synthesis of ethyl mandelate on a CD-modified Pt/Al_2O_3 catalyst and the simulated moving bed separation on a chiral stationary phase [32].

Li and coworkers performed the enantioselective hydrogenation of EP on 5 wt% Pt/Al_2O_3 premodified with CD in a fixed-bed reactor with or without a solvent (Figure 10.6) [33]. The catalyst (0.5 g) could be used with a complete conversion, and only a marginal drop of the activity could be observed during the 9–10-h-long continuous-flow operation at 1.2 ml h^{-1} substrate flow rate and at 2–7 MPa hydrogen pressure and ambient temperature. An enantioselectivity of up to 60% could be obtained, but it slowly decreased during the reaction.

The authors also demonstrated the beneficial use of supercritical ethane in the flow system, as well as the adsorption of CD on the Pt surface through a π-interaction. The asymmetric hydrogenation of ethyl-2-oxo-4-phenylbutyrate on CD-modified Pt/γ-Al_2O_3 catalyst was also performed in several solvents [34]. A stable conversion of up to 95% could be obtained in the nine-hours-long continuous process at 0.6 ml h^{-1} flow rate of the substrate and at 2.3–6 MPa H_2 pressure using 0.5 g of 5 wt% Pt/γ-Al_2O_3 for each experiment. It was also showed that the reduction of the ee with time-on-stream is considerably affected by the adsorption ability of the solvent used. The best ee (68%) could be obtained using toluene as a solvent.

Hutchings et al. performed gas-phase asymmetric hydrogenation of MP on Pt/γ-Al_2O_3 premodified with CD in a fixed-bed plug-flow reactor, as the first example of asymmetric hydrogenation with CD-modified supported metal in the gas phase (Figure 10.7) [35]. An enantioselectivity of 51% could be achieved in the gas phase at 2 bar hydrogen pressure and 40 °C in the presence of acetic acid additive. Furthermore, a series of experiments were conducted under similar conditions with Pt/SiO_2 and Pt/α-Al_2O_3 using MP, EP, and butane-2,3-dione as substrates. In each case, significant ee could be observed. Surprisingly, the premodified catalyst exhibited lower reaction rate compared to the unmodified catalyst, which is in sharp contrast with the rate enhancement observed for CD-modified catalysts in the liquid phase. The observed difference was attributed to the absence of the solvent.

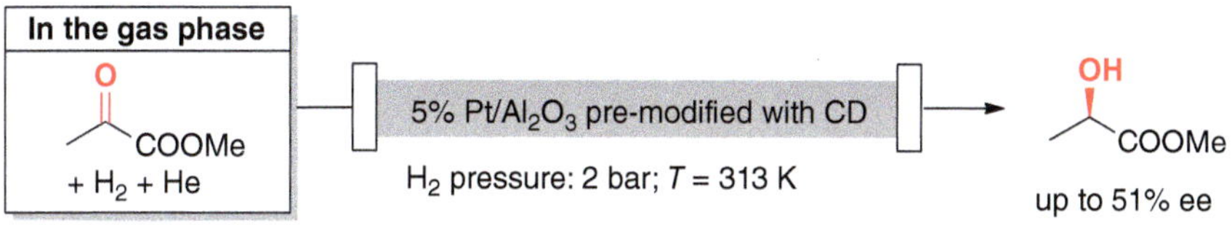

Figure 10.7 Asymmetric hydrogenation of MP in the gas phase using fixed-bed plug-flow reactor. Source: Based on von Arx et al. [35].

Later, their studies were extended to the liquid-phase asymmetric hydrogenation of alkyl pyruvate esters using premodified CD-Pt/γ-Al_2O_3 catalyst (0.1 g of CD/0.5 g of catalyst) under mild reaction conditions [36]. The continuous-flow system operating at 25 °C with 1.25 bar H_2 pressure produced (*R*)-ethyl lactate with a stable enantiomeric excess of >70% at 1 ml min^{-1} flow rate and 25 mmol l^{-1} substrate concentration. Again, in contrast to the previously observed rate enhancement in the presence of CD, the premodified catalyst proved to be less active than the unmodified one. This phenomenon was considered as the result of polymer formation on the surface of the catalyst under the catalytic conditions applied.

The continuous-flow technology has been successfully applied in the asymmetric hydrogenation of C=O double bonds, not only to increase the effectiveness of the process but also to elucidate the mechanism of the reaction. In this context, the work of Bartók, Fülöp, and Szőllősi largely contributed to the understanding of the mechanism of the Orito reaction [37, 38]. Their experimental results achieved using continuous microreactor system suggested that the rate enhancement of the hydrogenation is an intrinsic property of the modified surfaces [39–41]. However, the authors noted that the suppression of the catalyst deactivation by the addition of chiral modifiers as a possible explanation cannot be excluded. The asymmetric hydrogenation of several binary mixtures of activated ketones in the presence of chirally modified platinum catalysts gave outstanding enantioselectivities (up to 92% for methyl benzoylformate and pyruvic aldehyde dimethyl acetal) and provided valuable information on the reaction mechanism [42, 43]. Moreover, based on continuous-flow experiments fundamental mechanistic considerations could be disclosed concerning the role of intermediate complexes formed in the reaction [41, 44]. A previously unknown origin of the stereodifferentiation of the Orito reaction was detected by Garland and coworkers in the hydrogenation of EBF and EP on Pt/Al_2O_3, Pt/C, or powdered Pt catalysts [45]. After the enantioselective hydrogenation at 0 °C in the presence of a chiral modifier, the catalyst was washed at 0 °C so that the hydrogenation performed thereafter provided racemic product. However, when the cleaning was made at 50 °C, reproducible unmodified enantioselective hydrogenation (ees from −9% to −18%) reactions could be performed, and in several cases a reversal of the enantioselectivity with respect to the original enantioselective process could be observed. These phenomena were explained by the irreversible chiral modifications of the catalysts surface at higher cleaning temperature. As a continuation of this work, Bartók and coworkers presented a study on the hydrogenation of several activated ketones over cinchona alkaloid-modified chiral Pt catalysts using different solvents [46]. Based on the strong solvent and substrate

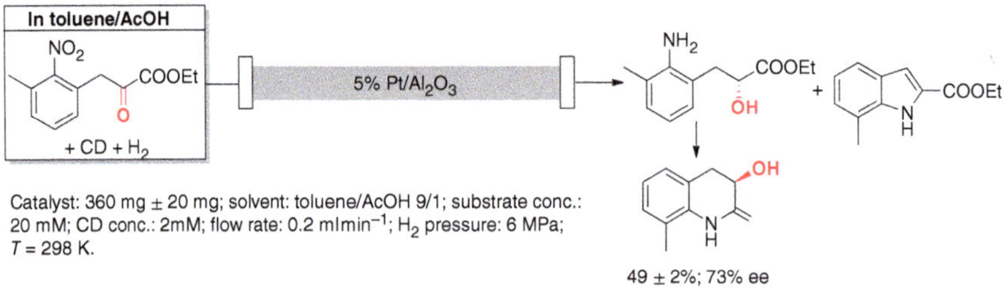

Figure 10.8 Pt-CD-catalyzed asymmetric cascade reaction in flow. Source: Based on Kovács et al. [47].

dependence it was established that the effect previously recognized by Garland et al. cannot be generalized to the enantioselective hydrogenation of activated ketones.

Fülöp and Szőllősi investigated the heterogeneous asymmetric cascade reaction of ethyl 2-nitro-3-methylphenylpyruvate over CD-modified platinum catalyst in a fixed-bed continuous-flow reactor (Figure 10.8) [47]. Although, the ees obtained in the flow reaction were commensurate to those achieved in the batch system, the selectivity of the final product quinolinone was significantly lower (up to 97 ± 1% vs. 49 ± 2%, respectively) in the flow process.

The continuous-flow hydrogenation on modified palladium surfaces was also studied by several workgroups. In these cases, however, a relatively high modifier/reactant molar ratio is necessary to maintain high enantioselectivity. In the hydrogenation of 4-methoxy-6-methyl-2-pyrone on CN-modified Pd/TiO_2, high enantio- (up to 84.5% ee) and chemoselectivity (100%) could be obtained only at 1% conversion [48]. Any efforts to achieve higher conversions led to a drastic fall of both the chemo- and enantioselectivity due to the hydrogenation of CN. The continuous-flow hydrogenation of α,β-unsaturated carboxylic acids on CD-modified Pd/Al_2O_3 resulted in slightly lower ees for aliphatic substrates but higher ees (up to 70%) in the case of α-phenylcinnamic acid compared to the batch reactions [49]. The authors also demonstrated that the benzylamine additive in the feed has a profound effect on the ee not only in the batch reactor but in the flow system too.

10.3 Well-defined Transition-metal Complexes

The asymmetric hydrogenation catalyzed by chiral transition-metal complexes is one of the most traditional and successful methods to produce optically active compounds. The enormous number of synthetic opportunities as well as the high productivity and selectivity available is clearly reflected in the 2001 Nobel Prize in chemistry. From an industrial perspective, however, the high cost of the metal and the chiral ligand as well as the difficulties connected to the separation of the toxic transition metal from the product raise serious concerns [50]. Therefore, the recycling of the valuable chiral catalyst and its separation from the product plays an important role in economical process design. Continuous-flow technology offers an ideal solution to the difficulties associated with recycling and efficient product separation. Furthermore, as the current approach to find new highly efficient chiral catalysts (chiral ligands) is still rather empirical, continuous-flow asymmetric hydrogenation emerges as a powerful tool for high-throughput screening of newly developed chiral catalysts.

10.3.1 Immobilized Systems

The high potential of homogeneous transition-metal complexes in synthetic organic chemistry has led to the conclusion that their immobilization on a support might broaden the scope of this approach in the industrial area. The beneficial use of immobilized chiral catalysts, however, requires the fulfillment of several strict criteria [51]: (i) simple catalyst preparation, (ii) better or comparable performance (activity and selectivity) of the immobilized system compared to the homogeneous catalyst, (iii) minimal leaching, (iv) mechanical, chemical, and thermal stability of the support, and (v) possibility of reuse without the loss of activity, which are crucial points to the utilization of an immobilized system. It is interesting to note that the immobilization of homogeneous catalysts is usually assumed to reduce the catalytic activity and/or the selectivity. The rationalization of the effect of the different supports on the catalyst performance, however, can largely contribute to improved catalytic properties compared to the homogeneous systems [52].

10.3.1.1 Covalently Anchored Ligands

Generally, the covalent attachment of the chiral ligand to the support represents an ideal solution to avoid leaching of the transition-metal complex due to the strong covalent bonding of the chiral ligand to the support. It is, however, important to mention that covalent immobilization techniques often require the functionalization of the chiral ligand. Such manipulations need the involvement of further synthetic steps, making the overall synthesis less environmentally benign and cost-effective. Furthermore, despite the strong covalent link between the ligand and the support, these tethered systems are prone to loss of the metal during the reaction. Consequently, the development of a simple and modular synthetic methodology for the preparation of covalently anchored stable transition-metal catalysts is highly desirable.

Van Leeuwen and Reek reported the synthesis of covalently anchored ruthenium catalysts and their application in the asymmetric transfer hydrogenation of acetophenone in batch and flow reactors [53]. A (1*R*,2*S*)-norephedrine-based ligand was immobilized on modified (methylated) silica as an inorganic support, and the addition of (*p*-cymene)-ruthenium(II) dimer provided the catalyst (Figure 10.9). A solution of 0.1 M acetophenone and 0.01 M *t*BuOK was passed through the catalyst bed (1 g containing 10–20 mg of Ru precursor). At an optimum flow rate of 700 μl h^{-1}, 89% ee and 90% conversion could be achieved. It was noted that the catalyst performed even better in the complete absence of the base (95% conv., 90% ee), which was needed only for the activation of the catalyst.

Additionally, increasing the substrate concentration to 0.8 M, the catalyst still provided the product with 29% conversion, and the ee was unchanged; space–time

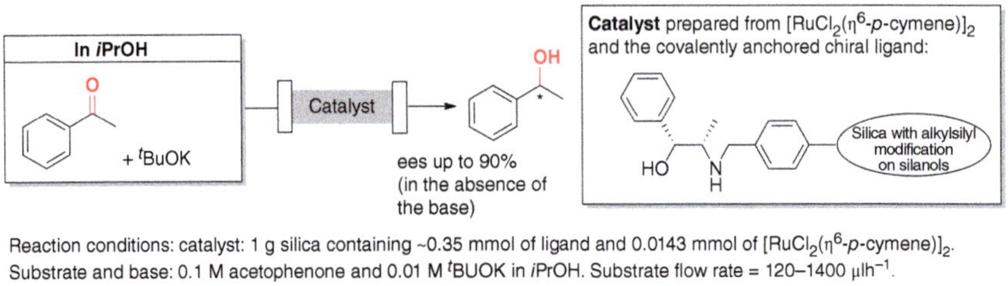

Figure 10.9 Asymmetric transfer hydrogenation by covalently anchored chiral ruthenium catalysts.

yields up to 39 g l^{-1} h^{-1} were obtained. The stability of the catalyst proved to be remarkable, as it was used for a week with no notable change neither in the activity nor in selectivity, and it was still active after three weeks of continuous use. These studies clearly evidence the high stability available by a covalently anchored transition-metal catalyst. The leaching of the ruthenium was negligible (1% during the 3–11-h interval). The remarkably high stability of the immobilized catalyst compared to that of the homogeneous system was attributed to the efficient site isolation on the support, i.e. the catalyst was prevented from the irreversible clustering, leading to catalytically inactive species. As an interesting addition, the immobilization strategies developed for the catalyst synthesis enabled fast and efficient catalyst screening. This was proved by the preparation of several other covalently anchored ligands that could also be used successfully in the transfer hydrogenation reaction.

10.3.1.2 Immobilization by the Augustine Method [54]

The immobilization of preformed transition-metal complexes by a methodology that does not involve the chiral ligand would have several advantages. At first, the activity and selectivity of the homogeneous system is assumed to be retained in this methodology. Second, the synthesis can be more modular, straightforward, and simple and does not need any additional functionalization steps. A possible way of the preparation of such immobilized systems is the attachment of the complex to the support through the metal. A convenient strategy developed by Augustine et al. uses heteropolyacid (HPA) as an anchoring agent between the support and the metal [54]. The immobilization procedure is simple as it permits the use of preformed transition-metal complexes and highly modular as there is no practical limitation on the transition-metal complexes and a number of supports can be used. Furthermore, under optimized conditions, the transition-metal catalysts preserve their activity and selectivity, and the immobilized systems exhibit outstanding reusability. Obviously, the major concern about immobilization by this technique is the sensitivity to solvent effects that could be responsible for the leaching of the catalyst. In this context, the nature of the HPA–metal and HPA–support interactions plays a crucial role in determining the catalytic properties. The attachment of the HPA to the support (alumina or silica) occurs via surface hydroxyl groups [55, 56], and the nature of HPA–metal interaction was found to be covalentlike [57].

Poliakoff and coworkers immobilized [Rh((*S*,*S*)-BDPP))(NBD)]$^+$ complex onto Al_2O_3 using phosphotungstic acid (PTA) as a linker [59]. The composite catalyst was then used in the continuous-flow asymmetric hydrogenation of dimethyl itaconate (DMIT) as a benchmark substrate in supercritical carbon dioxide. The optimum catalytic results (66% conversion and 63% ee) were obtained at 60 °C and 10 MPa pressure, and therefore a catalyst stability study was performed under these conditions. Only a minimal drop in the catalyst activity could be observed after an eight hours-long continuous operation, indicating the high stability of the catalyst under these conditions. Additionally, the leaching was also minimal, as <1 ppm total metal concentration has been determined in the product. The authors later extended their studies to the immobilization of several different types of chiral P,P-ligands and their catalytic screening in asymmetric hydrogenation of DMIT in continuous-flow *sc*CO_2 (Figure 10.10) [58]. Simple ditertiary phosphines (Skewphos (BDPP), Prophos, DIOP, and BINAP) and ferrocene-based chiral ligands (BoPhoz and ligands from the Josiphos, Walphos, Taniaphos, and Mandyphos series) were investigated as chiral selectors. It has been demonstrated that higher catalytic turnover (TON up to 4 600) could be achieved by the simple P,P-ligands compared to the ferrocenyl-based catalysts. However, the best ee (83%) was provided by the immobilized rhodium catalyst containing the ligand Josiphos 001. It has, however, clearly evidenced that the utilization of immobilized chiral Rh catalysts even under unusual conditions (in *sc*CO_2 at elevated temperature) in the continuous-flow

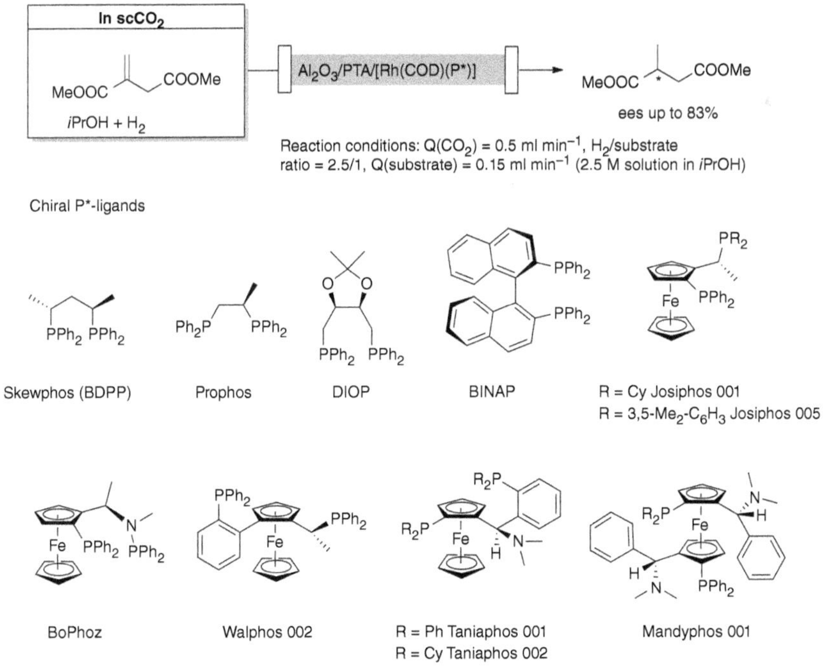

Figure 10.10 Asymmetric hydrogenation of dimethyl itaconate using immobilized Rh complexes of several chiral phosphine ligands. Source: Based on Stephenson et al. [58].

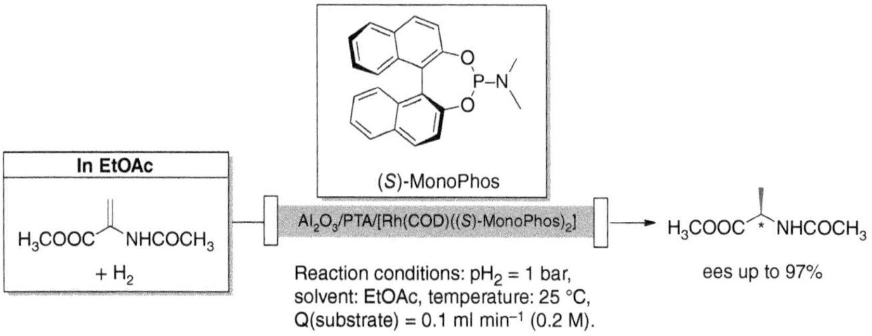

Figure 10.11 Asymmetric hydrogenation of methyl acetamidoacrylate using immobilized Rh–(*S*)-MonoPhos complex. Source: Based on Madarász et al. [60].

mode can be a powerful tool for the synthesis of enantiomerically enriched chiral products.

A thorough study on the reaction parameters affecting the asymmetric hydrogenation of DMIT in a trickle-bed reactor was disclosed by Simmons et al. It was shown that the initial reaction rate and the enantioselectivity of the reaction in the flow system can be improved by increasing the flow rate of the liquid phase. It was explained by the enhanced wetting efficiency and the improved H_2 mass transfer from the bulk liquid to the catalyst surface at higher rates. In the trickle-bed reactor, 99% conversion and up to 99.9% ee were obtained at a substrate/catalyst molar ratio of 223, at room temperature, and atmospheric pressure using 100 ml min^{-1} gas flow rate and 20 ml min^{-1} liquid flow rate. The reaction has been performed in the polar ethanol as solvent at a DMIT concentration of 0.3 M using supported [Rh((*R*,*R*)-Me-DuPhos)(COD)]BF_4 complex.

Bakos and coworkers immobilized [Rh(COD)((*S*)-MonoPhos)$_2$]BF_4 on commercially available and mesoporous alumina using PTA as a linker (Figure 10.11) [60]. The effects of the temperature, pressure, liquid flow rate, and substrate concentration were screened to gain optimum reaction parameters in the asymmetric hydrogenation of methyl acetamidoacrylate. Although the Rh–(*S*)-MonoPhos catalyst immobilized on the commercial Al_2O_3 with a surface area of 150 $m^2\ g^{-1}$ provided stable enantioselectivity (95–97%) for a 8.5-h-long continuous use, the conversion decreased from 95.5% to 74.4% under optimized reaction conditions (25 °C, 1 bar H_2 pressure, 0.1 ml min^{-1} flow rate, and 0.05 M substrate concentration) due to the leaching of the catalyst. In order to improve the catalyst stability, the support was changed for the mesoporous alumina with a surface area of 401 $m^2\ g^{-1}$. The product can be continuously obtained in >99% conversion and 96–97% ee for a total of 12 h at an increased substrate concentration of 0.2 M that corresponds to the production of 2.1 g (*R*)-*N*-acetylalanine during the reaction. The higher catalyst stability was attributed to the larger surface area, pore volume, and mesopore size of the latter support.

The same research group immobilized a chiral Rh complex based on the newly developed phosphine–phosphoramidite ligand UPPhos on mesoporous Al_2O_3 support (431 $m^2\ g^{-1}$) and used it in the asymmetric hydrogenation of

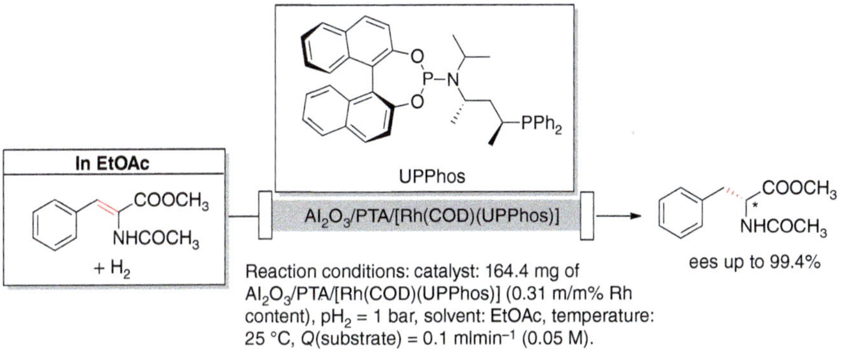

Figure 10.12 Asymmetric hydrogenation of (*Z*)-α-acetamidocinnamic acid methyl ester with immobilized Rh complex of the chiral ligand UPPhos. Source: Based on Balogh et al. [61].

(*Z*)-α-acetamidocinnamic acid methyl ester (Figure 10.12) [61]. Under optimized conditions (25 °C, 1 bar, 0.05 M substrate concentration, and 0.1 ml min^{-1} flow rate) using 164.4 mg [Rh(COD)(UPPhos)]/PTA/Al_2O_3 catalyst (~0.51 mg of Rh) and ethyl acetate as solvent, the system produced the product dimethyl methylsuccinate continuously with 99% conversion over 99% ee for six hours. Additionally, the hydrogenation was continued for an additional six hours when the conversion slightly decreased to 90% with an average of 97% ee.

In a separate study by Bakos et al., the use of phosphine–phosphite (POP-**1** and POP-**2**, Figure 10.13)-containing rhodium complexes was demonstrated in the asymmetric continuous-flow hydrogenation of DMIT and (*Z*)-α-acetamidocinnamic acid methyl ester (MAC) [62]. As phosphites are prone to hydrolysis and phosphines to oxidation, the work also aimed to evaluate the stability of the catalyst. Under optimized conditions, the [Rh(COD)(POP-**1**)]/PTA/Al_2O_3 catalysts based on commercially available alumina (150 m^2 g^{-1}) yielded ees up to 87% in the asymmetric hydrogenation of MAC and 98% when DMIT was used as substrate. Based on these initial results the catalyst [Rh(COD)(POP-**2**)]/PTA/Al_2O_3 with octahydrobinaphthyl moiety was applied in the hydrogenation of DMIT to produce the final product with a TON of 2678 and 95% ee, at a substrate concentration of 5 M. Furthermore, the same catalyst provided the hydrogenation product continuously for six hours with >96%

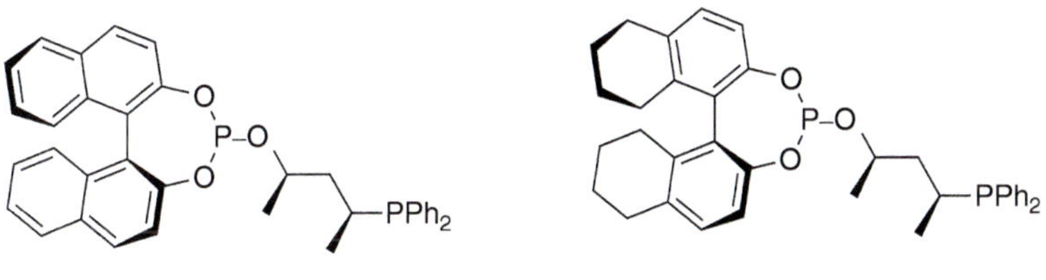

POP-1 POP-2

Figure 10.13 Phosphine–phosphite ligands used in Rh-catalyzed asymmetric hydrogenation under flow conditions.

conversion and 91–95% ee under optimized conditions (1 bar, 20 °C, 0.2 ml min^{-1} flow rate, and 0.05 M substrate concentration).

A nice demonstration of the scale-up of an asymmetric hydrogenation step in the synthesis of an active pharmaceutical ingredient (API) was reported by Poliakoff, Leitner, and Franciò [63]. The hydrogenation product, (*S*)-5-fluoropyrimidin-2-yl-ethanamine, is used in the synthesis of a JAK-2 kinase inhibitor already tested at the clinical level at AstraZeneca. Originally, two synthetic strategies have been developed for the synthesis of this key intermediate: an enzymatic transamination reaction and an Rh–(*S*,*S*)-ethyl-Duphos-catalyzed batch hydrogenation process. In both cases the final isolation is rather challenging, and extra derivatization steps are required. The authors therefore reported on translating the batch hydrogenation process to continuous-flow synthesis. In the first set of experiments, the flow process based on the [Rh(COD)((*S*,*S*)-EthylDuphos)]/PTA/Al_2O_3 catalyst was optimized in two different laboratories using different small-scale experimental setups to prove the robustness of the procedure. The catalyst itself was prepared in an on-line anchoring process, where the PTA/Al_2O_3 was loaded into the tubular reactor, dehydroxylated at 200 °C, and then an ethanolic solution of the chiral Rh complex was passed through it. Using a 0.6-M tetrahydrofuran (THF) solution of the substrate, an overall TON of approximately 5000 could be obtained within seven hours on stream, with complete conversion and ee>98%. According to these results, a space–time yield of 520 g l^{-1} h^{-1} has been achieved. The Rh contamination in the product stream was less than 1 ppm. The scale-up experiments were realized in a fully automated unit with two tubular reactors, each of 150 ml volume (Figure 10.14).

The product stream was directed to a gas chromatograph (GC); hence, the quality of the product could be analyzed continuously. If the product's quality did not reach the standard (90% conversion and 98% ee), the system automatically (i) change the

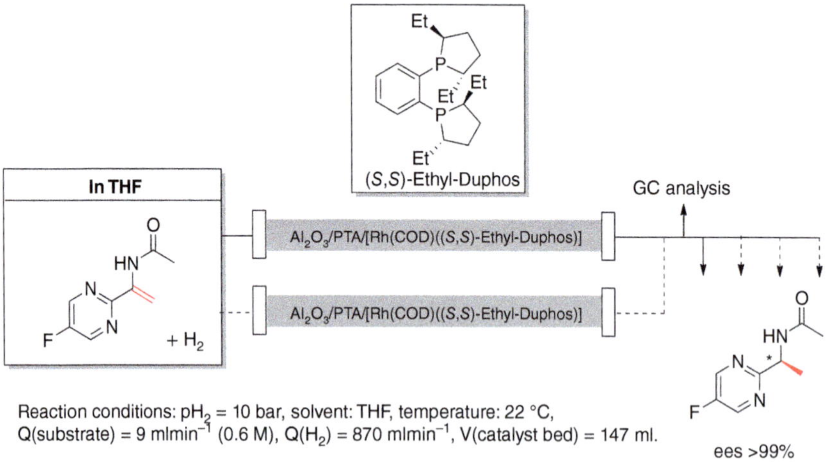

Figure 10.14 Schematic view of the automated continuous-flow unit equipped with two tubular reactors.

Table 10.1 Comparison of the original batch process and the continuous-flow system at 1 kg scale.

	Batch	Continuous-flow system
Number of runs	6	1
Volume of the reactor unit (ml)	660	150
Amount of catalyst (mmol)	10.1	0.68
ee (%)	>98.6	>98.6
Reaction time (h)	32[a)]	18
Process time (h)	64[a)]	18
H_2 pressure (bar)	4.75	10
Solvent (l)	4 (MeOH)	10 (THF)
Temperature (°C)	25	25
Estimated PMI[b)]	4.2	10
Rh contamination (ppm)[c)]	260	<1
Space–time yield ($g\,l^{-1}\,h^{-1}$)	47	399

a) Cumulative values for six batch reactions.
b) Process mass intensity (PMI) in the reaction step only.
c) In the crude product solution.

collecting vessel, (ii) decrease the flow rate, or (iii) direct the substrate stream to the other tubular reactor. After 18-h-long time-on-stream using THF as the solvent, the final product could be obtained with ≥98% purity and ≥99% ee. The workup procedure involves only the THF evaporation. The Rh content in the isolated product was again <1 ppm. The authors compared the efficiency of the flow process to that of the batch mode (Table 10.1). This simple case study unambiguously indicates the high potential of the flow technology in pharmaceutical industry.

Cole-Hamilton et al. studied the asymmetric hydrogenation of neat dibutyl itaconate (DBI) in a plug-flow reactor over [Rh(COD)((*R*,*R*)-MeDuphos)]/PTA/Al_2O_3 catalyst (Figure 10.15) [64]. After the careful investigation of the effects of hydrogen flow rate and pressure and the substrate flow rate on the activity and enantioselectivity, the reaction has been carried out under optimum conditions. At 0.2 mmol min^{-1} substrate flow rate and 9 mmol min^{-1} H_2 flow rate at 5 bar, the product dibutyl 2-methylsuccinate was achieved with 99% conversion in the first 47 h of time-on-stream that slowly decreased to 68% in the 83rd hour. An ee of 98% could be obtained in the first 23 h but later decreased to a value of 65%. The catalyst leaching was minimal after 23 h; less than 50 ppb Rh could be detected in the product. During this time, 41 mg of Rh–MeDuphos was used to prepare 278 mmol of (*S*)-dibutyl 2-methylsuccinate with no need for further purification.

10.3.1.3 Ionic Liquids as Matrices for Transition-metal Complex Catalysts

The combination of ionic liquid phases with transition-metal catalysts represents a convenient noncovalent immobilization strategy [66]. Continuous-flow

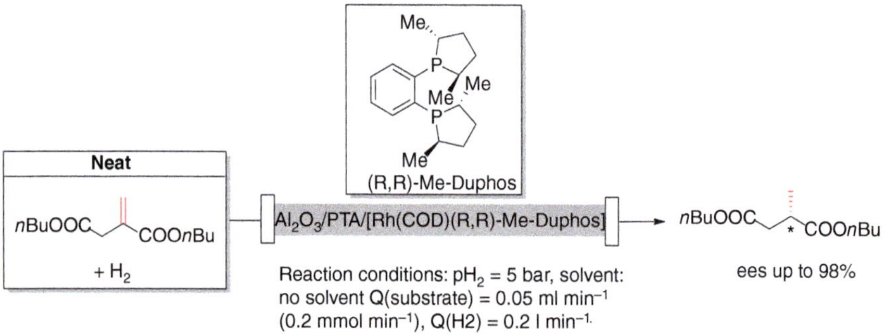

Figure 10.15 Asymmetric hydrogenation of dibutyl itaconate under continuous-flow solvent-free conditions. Source: Based on Duque et al. [64].

hydrogenation utilizing transition-metal catalysts in an IL phase, however, requires the use of gaseous substrates [67, 68] or the application of mobile phases that are non-IL-miscible [65, 69–72]. The first approach is strongly restricted by the limited number of suitable substrates with high volatility and catalysts with proper thermal stability. For the second methodology, the mobile phase must efficiently deliver the substrate to the IL phase and facilitate the separation of the products without the gradual dissolution of the ionic liquid phase. In this context, the utilization of *sc*CO_2 as the mobile phase, in combination with ionic liquid supports in a flow system, is a particularly attractive approach to develop highly stable and environmentally benign catalytic methodologies. The use of ionic liquids as catalyst matrices exhibits a number of inherent advantageous properties such as (i) the ease of preparation as the strategy allows the use of off-the-shelf catalysts, (ii) the fine-tuning of the catalytic system by simply varying the structure of the IL, and (iii) the contribution of the ionic liquid to the stabilization of the catalyst [73, 74]. The catalytic reaction takes place in a liquid-like environment that can be either the bulk ionic liquid phase (IL) or a small amount of ionic liquid supported on the surface of a solid material (supported ionic liquid phase, SILP). The continuous use of the IL/*sc*CO_2 method can be realized in a stirred tank reactor, while the SILP/*sc*CO_2 system can be operated in a tubular reactor containing the catalyst (Figure 10.16) [65]. Finally, it is also important to note that the use of supercritical CO_2 as the mobile phase requires extra operational cost associated with the compression, utilization, and possible recycling of carbon dioxide.

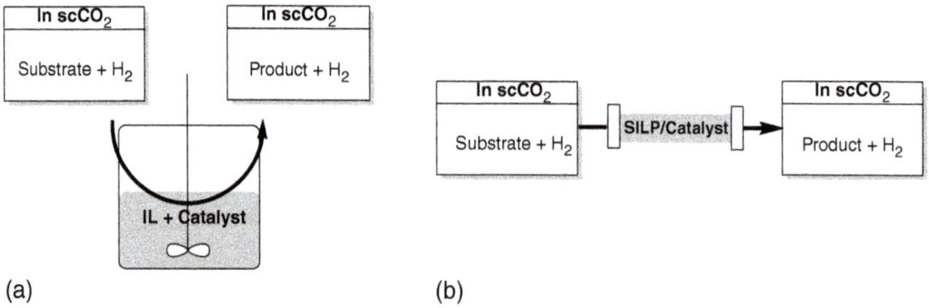

Figure 10.16 Schematic view of the continuous-flow systems based on IL/*sc*CO_2 (a) and SILP/*sc*CO_2 (b) Source: Based on Zhang et al. [65].

The first application of a continuous reactor for an enantioselective transformation using a SILP-based catalyst and *sc*CO_2 as the mobile phase was reported by Leitner and coworkers [69]. Dehydroxylated silica was chosen as support that was coated with [EMIM][NTf]$_2$ (EMIM = 1-ethyl-3-methylimidazolium) containing the commercially available (S_a,R_c)-1-naphthyl-QUINAPHOS-Rh complex as the precatalyst. The SILP/*sc*CO_2 biphasic system was used continuously in the asymmetric hydrogenation of DMIT for 65 h and provided a remarkable TON value of 115 000 with only a marginal drop of the conversion (>95%). In other words, 3 mg of the rhodium catalyst converted 50 g of substrate after 65 h with a space–time yield of 0.3 kg l^{-1} h^{-1}. Although, the ee started to decrease from 99% after approximately 10 h of time-on-stream, at the end of the process still high enantioselectivities, up to 70%, could be obtained. This might be due to the partial slow decomposition of the chiral catalyst to form nonselective catalytically active species. The same group later extended their studies to other ILs, catalysts, and supports in order to probe the interplay of these components under catalytic conditions [70]. It was found that the SILP catalyst based on a perfluoroalkyl–silica support is capable of maintaining very high activity (>99% conversion) and ee (>99%) for 10 h of continuous operation. Furthermore, as the presence of water was found to be responsible for the slow catalyst deactivation, the introduction of a water scavenger bed further enhanced the efficiency of the system (Figure 10.17). In this case >98% ee was achieved over 30 h on stream that corresponded to a TON value of 70 000. The chemically pure chiral product with high optical purity could be produced with space-time yields up to 0.7 kg l^{-1} h^{-1}.

Leitner and coworkers developed a continuous-flow system for the asymmetric hydrogenation of methyl 2-propionylacetate with Ru-BINAP immobilized in an imidazolium-based ionic liquid, and *sc*CO_2 was used as the mobile phase (IL/*sc*CO_2) [71]. It has been shown that an acidic additive has a profound effect on the catalyst activity and stability. By the addition of an acidic ionic liquid [HSO_3BBIm][BTA] maximum space–time yields of 180 g $l^{-1}h^{-1}$ were obtained with up to 82% enantioselectivity. The efficiency of continuous setup (Figure 10.18) was compared to

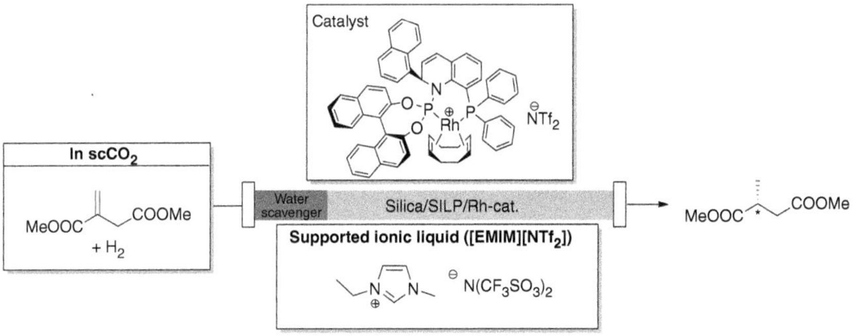

Reaction conditions: Catalyst: 0.8 g of supported Rh-catalyst (1.3 μmol) in [EMIM][NTf_2] on silica; water scanvenger: neat silica; temperature: 40 °C; pressure: 120 bar; $Q(CO_2)$ = 85 ml min^{-1}, $Q(H_2)$ = 10 ml min^{-1}, Q(substrate) = 0.78 g h^{-1}.

Figure 10.17 Asymmetric hydrogenation of dimethyl itaconate using SILP/*sc*CO_2 biphasic system.

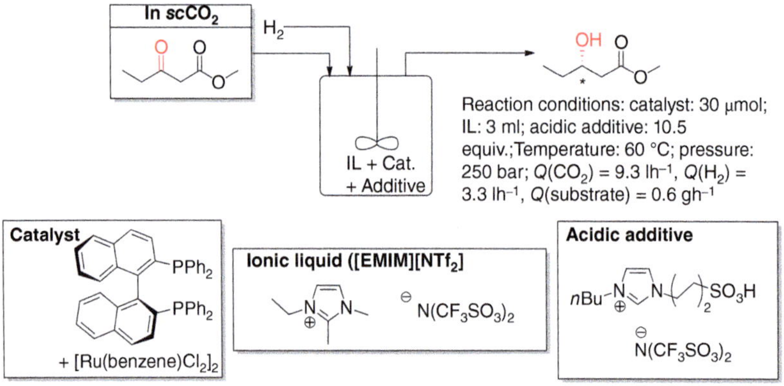

Figure 10.18 Continuous asymmetric hydrogenation of methyl 2-propionylacetate with IL/*sc*CO_2 system.

that of the batch system. It has been concluded that the flow process is remarkably more resource efficient and can be a promising alternative of the batch reactors. A major drawback arises from the reduced enantioselectivity of the process developed by the authors.

The utilization of the IL/*sc*CO_2 and the SILP/*sc*CO_2 systems was compared in the asymmetric hydrogenation of 1-(trifluoromethyl)-vinylacetate to 1-(trifluoromethyl)-ethylacetate, a useful chiral synthon for the production of pharmaceutically active compounds [65]. The SILP/*sc*CO_2 methodology using Rh–Xyl-QUINAPHOS as the catalyst afforded superior catalytic performance over the IL/*sc*CO_2 system. Very high stability and enantioselectivity could be obtained: the conversion remained above 90%, and the ee was around 80% in the first 130 h of continuous stream. After doubling the flow rates the conversion dropped to 70%, which is equivalent to a 1.5-fold increase in the TOF and the space–time yield. The continuous-flow system was operated for 233 h with 7.5 μmol of Rh catalyst, and the TON reached a value of 70 400.

Generally, the scope of the transition-metal/SILP/*sc*CO_2 systems under flow conditions is limited to substrates/products that have sufficient solubility in *sc*CO_2 [72]. Hence, compounds with highly functionalized structure or low volatility usually cannot be transformed efficiently using the SILP/*sc*CO_2 system. In order to overcome these difficulties, Leitner and Franciò et al. developed a methodology for the synthesis of (*S*)-5-fluoropyrimidin-2-yl-ethanamine, a key intermediate in the production of an API (see Section 10.3.1.2), with continuous asymmetric hydrogenation using modified *sc*CO_2 (*mod*CO_2). After the careful evaluation of the possible cosolvents, toluene was chosen as a suitable modifier. Again, the traces of water exerted a detrimental effect on the catalyst stability. Therefore, the introduction of a water scavenger predrying column and the utilization of a highly hydrophobic fluorous reverse-phase silica (FRP-SiO_2) as a support in combination with *mod*CO_2 led to a highly active, selective, and robust catalytic system. Under optimized condition, using the FRP-SiO_2/SILP/Rh–QUINAPHOS catalyst >99% conversion and ee were obtained during the 90-h-long time-on-stream, which corresponds to a total TON

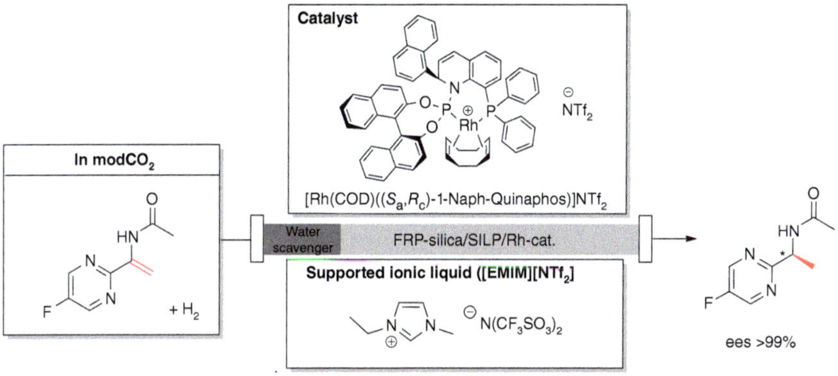

Reaction conditions: supported Rh-catalyst: 5.8 μmol; water scavenger: 3 Å molecular sieve; temperature: 50 °C; Pressure: 150 bar; Q(CO2) = 80 mlmin^{-1}, Q(H2) = 18 mlmin^{-1}, Q(substrate) = 0.037 mlmin^{-1} (0.18 M in toluene).

Figure 10.19 Asymmetric hydrogenation of *N*-(1-(5-fluoropyrimidin-2-yl)-vinyl)acetamide on a SILP-based catalyst.

value of 10 300 and a space–time yield of 24 g l^{-1} h^{-1} (Figure 10.19). An interesting addition from the authors is the comparison of their system to the catalyst based on Rh–DuPhos, immobilized by the Augustine method (see Section 10.3.1.2), both operating under flow conditions. The SILP-based catalyst proved to be less competitive than the latter approach due to its lower space–time yield (24 g l^{-1} h^{-1} vs. 399 g l^{-1} h^{-1}) and its higher operation cost associated with the higher pressure of CO_2.

Wasserscheid and coworkers developed a continuous gas-phase process for the asymmetric hydrogenation of methyl acetoacetate using a chiral SILP-based Ru catalyst [67]. The authors thoroughly investigated the role of the support materials and solvent additives in determining catalyst activity as well as the possible reasons for the catalyst deactivation. It has also been shown that the continuous cofeeding of gaseous methanol to the substrate has a profound effect on the activity. A stable conversion (~70%) was obtained after 35 h time-on-stream catalyst activation in a tubular plug-flow reactor and was maintained for 70 h. Enantioselectivities up to 80% could be obtained, and the total turnover number of the process reached a value of 2500. The studies were later extended to the gas-phase hydrogenation of the more challenging substrate MP with Ru complexes [68]. The continuous-flow mode ensured the thorough screening of the chiral ligand, the support material, the ionic liquid phase (anion and cation), the IL loading, and the acidic additive. Using the Ru-BINAP catalyst immobilized in 3-hydroxypropyl pyridinium bis(trifluoromethylsulfonyl)imide ([PrOHPyr][NTf_2]) on silica support, ees up to 30% could be obtained with stable yields of 80–84% for more than 50 h time-on-stream.

10.3.2 Homogeneous Systems

Homogeneous asymmetric hydrogenation has several remarkable advantages over the heterogeneous systems. First, it allows the development of a wide variety of soluble chiral transition-metal complexes by simply combining the corresponding transition-metal precursor with the chiral ligand. Very high activities and selectivities can be obtained even under mild reaction conditions. Furthermore, the intermediates of the catalytic cycle can relatively easily be investigated. One of the main concerns about homogeneous catalysis is the effective removal of the transition-metal contaminant from the reaction mixture that is strictly regulated by the pharmaceutical industry. Additionally, the cost of the expensive transition metal and the chiral ligand give further motivation to separate and recycle them in the catalytic reaction. The separation and recycling of homogeneous catalysts in continuous asymmetric hydrogenation was realized by membrane filtration or by reverse-flow adsorption technique.

Yamada et al. performed the cobalt-catalyzed asymmetric hydrogenation of various ketones using modified borohydride as the reducing agent [75]. The solution of the substrate containing the homogeneous catalyst and the solution of the modified borohydride were mixed in the microreactor, and the effluent was then directly poured into a vessel containing THF/AcOH to quench the reaction. At a 12-min residence time, up to 92% yield and 91% ee could be obtained for tetralone derivatives. Additionally, after the initial optimization, in a 9.75-h-long continuous experiment 3.1 g of 6-methoxytetralone was hydrogenated to produce the corresponding alcohol in 96% yield and 92% ee using 5 mol% optically active cobalt catalyst.

A tube-in-tube reactor was used in the diastereoselective hydrogenation of minimally functionalized olefins by Newton, Ley, and coworkers [76]. In their experimental setup, several Rh-, Ru-, and Ir-containing chiral metal complexes were screened. At a conversion of >99%, up to 76% ee could be obtained using 0.1 M substrate solution with 2.5 mol% catalyst at 15 bar H_2 pressure, 20 °C temperature, and 80 min reaction time. In order to further enhance the efficiency of the system, a second tube-in-tube reactor and a mixer chip were also introduced. Furthermore, a recirculating protocol was also developed that could increase the conversion from 49% to 75% at a catalyst concentration of 1 mol%.

A micromesh continuous reactor was presented by Gavriilidis and used in the rhodium-catalyzed asymmetric transfer hydrogenation of acetophenone in isopropanol [77, 78]. The micromesh system was able to efficiently remove the acetone by-product from the reaction mixture and thus prevent equilibrium reactions that can lower the enantioselectivity (Figure 10.20). The experimental setup provided superior performance over the batch system: at a substrate/catalyst molar ratio of 1000 and at 30 °C, the continuous system required 13 min to reach full conversion, in contrast to the batch reaction where 180 min reaction time was needed. In the mesh system, a final ee of 83%, and in the batch system 79% could be obtained.

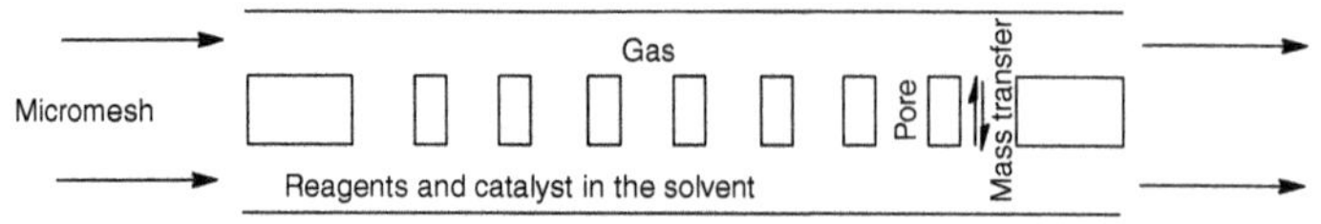

Figure 10.20 A schematic view of the micromesh interface.

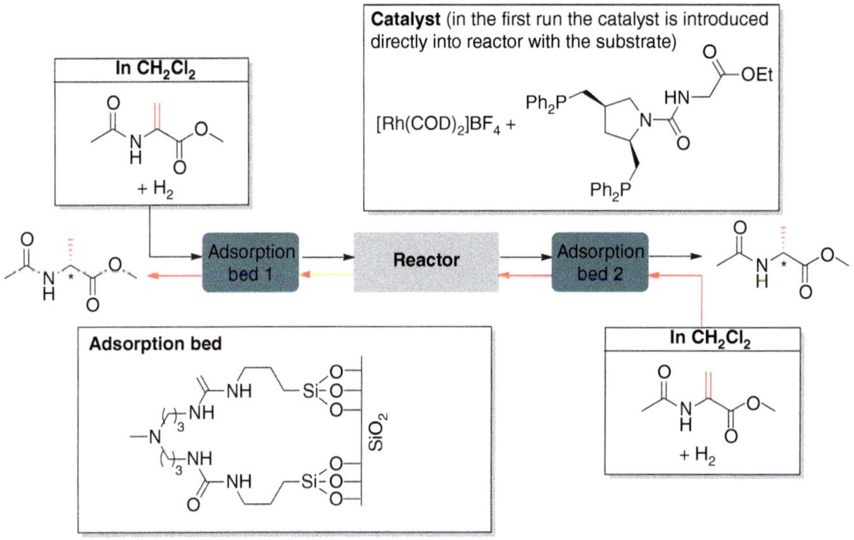

Figure 10.21 Reverse-flow setup for asymmetric hydrogenation developed by Reek and coworkers.

Reek and coworkers reported the development of a reverse-flow adsorption process based on a one-step adsorption as an efficient methodology for the recycling of homogeneous rhodium catalysts [79]. The basis of the method is the selective removal of the homogeneous catalysts from the reactor effluent by adsorption and the recycling of the catalyst to the reactor with its feed (Figure 10.21). The efficient adsorption of the catalyst was achieved using tailor-made binding motifs on the transition-metal complex (guest) and the adsorbent (host) and was based on hydrogen-bond and ionic interactions. In the asymmetric hydrogenation of methyl acetamidoacrylate 0.5 g of the host was used in each adsorption bed, and at a substrate/Rh molar ratio of 200, at room temperature, and 1 bar H_2 pressure 20% ee and 55.3% conversion (TOF = 138 h^{-1}) could be obtained in the first run. After eight consecutive cycles the conversion decreased to 26.2%, but the ee remained in the range of 18–16%.

A polysiloxane-bound homogeneous ruthenium catalyst (a chemzyme) was developed and applied in the asymmetric transfer hydrogenation of acetophenone in a continuously operated membrane reactor [80]. The polysiloxane moiety ensures the solubility and facilitates the ultrafiltration by the membrane. High ees (up to 94%) and space–time yields (578 g l^{-1} d^{-1}) could be achieved using isopropanol as a hydrogen donor and the solvent. The system was compared to the corresponding enzymatic systems. It was concluded that although the enzyme-catalyzed hydrogenation produces the product with higher enantioselectivities and activities, the solubility of the substrate may be low in the aqueous system, thus decreasing space–time yield. In contrast, the application of a chemzyme does not need any cofactor as a hydrogen mediator, and the space–time yields available can be higher by the optimization of the substrate and catalyst concentrations. A kinetic model of the transfer hydrogenation reaction has also been developed for the chemzyme

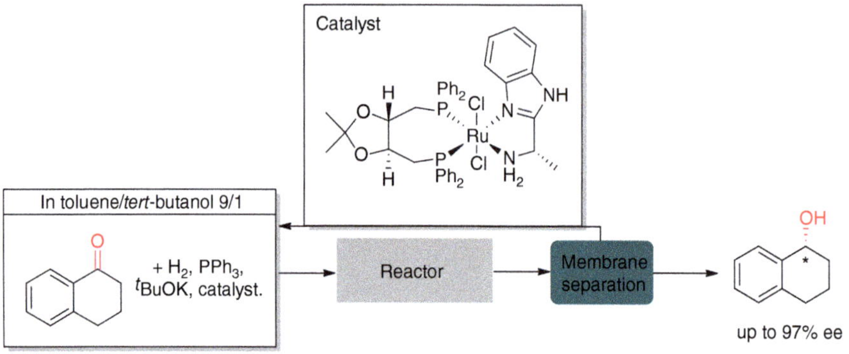

Figure 10.22 Continuous membrane separation of the catalyst in asymmetric hydrogenation of tetralone. Source: Based on O'Neal et al. [82].

reactor [81]. The reaction performed at the predicted conditions provided the product with 0.58 $kg\,l^{-1}\,d^{-1}$ space–time yield and 92.8% ee. These results are very close to the predicted values. Moreover, based on the model it has also been shown that the removal of acetone would result in a threefold increase in the total turnover number.

A continuous nanofiltration and recycling of a ruthenium diphosphine/diamine catalyst was developed by Jensen et al., and the system was applied in the asymmetric hydrogenation of tetralone in flow using H_2 as a hydrogen source (Figure 10.22) [82]. The continuous system was used for 24 h and provided the product with a TON value of c. 5000, and ees up to 93% could be obtained. Additionally, the catalyst leaching was minimal, and the product contained less than 250 ppb ruthenium. The 24-h continuous operation is equivalent to approximately 60 batch separation/recycle experiments.

Besides process intensification, the use of microreactors can be extremely fruitful in (i) fast catalyst screening, (ii) efficient optimization of reaction parameters, and (iii) the determination of reaction kinetics with as low reagent loading as possible. Bellefon and coworkers used different highly efficient microdevices and reactors for the thorough screening of a number of chiral rhodium–diphos systems in the asymmetric hydrogenation of prochiral substrates to investigate the nature of the gas–liquid reactions [83–85]. It has been clearly demonstrated that the catalytic tests can be accurately performed even with an extremely low catalyst loading. Furthermore, high activities and enantioselectivities could be obtained.

10.3.3 Self-supported Systems

In the self-supporting approach, the immobilization of chiral catalysts is performed by the self-assembly of the chiral multitopic ligands and transition-metal ions to form an organometallic polymer or network [86]. The resulting complex has low solubility in most organic solvents due to its polymeric nature and therefore can be used as an immobilized catalyst. It is also proposed that the system preserves the prosperous stereoelectronic features of the parent homogeneous catalyst and can

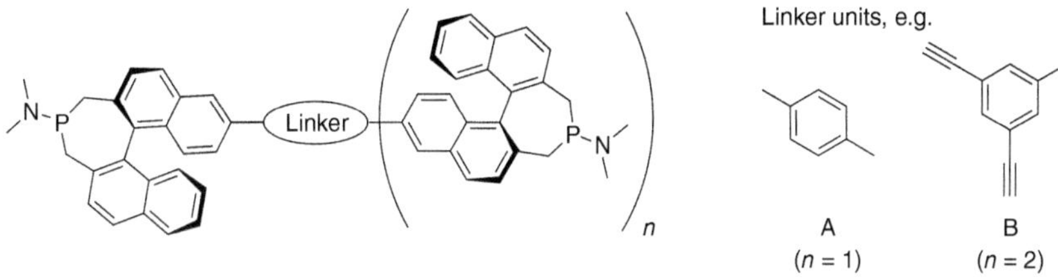

Figure 10.23 Examples of multitopic ligands for self-supported Rh catalysts developed by Ding et al.

effectively be used with high activity and selectivity similar to those of the homogeneous system in asymmetric catalytic hydrogenation.

Ding and coworkers developed a self-supported catalyst system for the asymmetric hydrogenation of α-dehydroamino acid methyl esters and 2-aryl-substituted enamides [87]. In their study, MonoPhos-based multitopic ligands were synthesized to produce the self-supported catalyst upon reaction with $[Rh(COD)_2]BF_4$ (Figure 10.23).

According to scanning electron microscopy, the catalyst is composed of micrometer-sized particles, and it has no crystallinity as evidenced by X-ray diffraction (XRD) measurements. It has been shown that approximately 37% of the Rh^I centers of the catalyst are available for acetamidocinnamic acid molecules in the hydrogenation process. Although the recycling of the catalyst in batch reactions indicated its deactivation, no decrease in the enantioselectivity (96–97%) occurred even after the 10th consecutive run in batch reactor. The slow deactivation was attributed to the catalyst decomposition during the recycle process, but the negative effect of air or moisture present in the system could not be ruled out. In order to enhance the catalyst stability the self-supported system was studied in the asymmetric hydrogenation of acetamidocrotonic acid methyl ester under flow conditions. The catalyst containing multitopic ligand **A** was mixed with packing material in order to avoid system blockage. Several inorganic materials were tested as $MgSO_4$, TiO_2, or activated carbon. The best performance was observed for the latest that could be explained with the improved dispersion of the self-supported catalyst in the reaction mixture in the presence of activated carbon. Under experimental conditions (catalyst: 60 mg (containing ligand **B**), activated carbon: 90 mg, substrate concentration: 1 g 200 ml^{-1} toluene, substrate flow rate: 0.05 ml min^{-1}, and H_2-flow rate: 3 ml min^{-1}), the hydrogenation product could be synthesized in >99% conversion and 97% ee for a total of 144 h. Furthermore, the tests clearly proved that the self-supporting process does not affect detrimentally the performance of the catalyst, and comparable or even better activities and enantioselectivities can be obtained as with the homogeneous analogues.

10.4 Organocatalysts

The combination of organocatalysis with continuous-flow technology offers an outstanding solution for the green synthesis of pharmaceutically important chiral

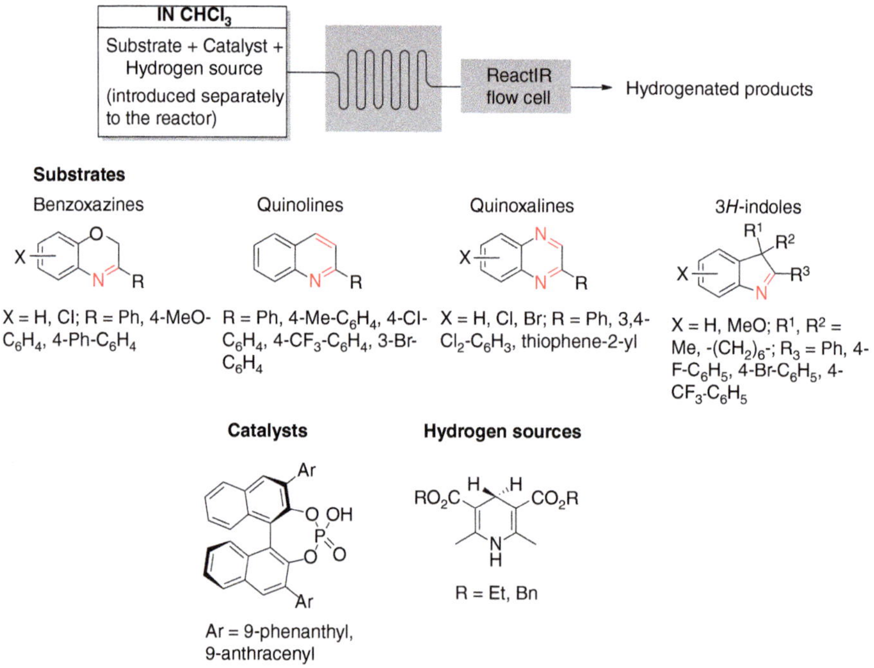

Figure 10.24 Organocatalytic asymmetric transfer hydrogenation of heteroaromatic compounds in flow. Source: Based on Rueping et al. [90].

ingredients [88, 89]. Although the asymmetric flow hydrogenation catalyzed by organocatalysts is not a general methodology of production, it is important to mention its high potential.

Rueping and coworkers reported on the first organocatalytic asymmetric transfer hydrogenation in a flow reactor (Figure 10.24) [90]. First, they examined the transfer hydrogenation of a benzoxazine-type substrate (X = H, R = Ph) in the presence of a Hantzsch dihydropyridine derivative (R = Et) as hydrogen donor. A binol-based chiral Brönsted acid (Ar = 9-phenanthryl) was used as catalyst. The reagents (substrate + the catalyst and the H-source) were introduced separately to the reactor. The glass microreactor was attached to a ReactIR flow cell which allowed the real-time analysis of the product mixture. In fact, both the substrate consumption and the product formation could be analyzed by the system, i.e. by the disappearance and the evolution of the corresponding IR signals. The real-time analysis ensured the fast optimization of the reaction parameters. Using 1.2 equiv of the hydrogen source and 2 mol% catalyst in $CHCl_3$ as solvent at 60 °C temperature, 0.1 ml min^{-1} flow rate, and 60 min residence time, the product could be obtained with 98% isolated yield and 98% ee in the hydrogenation of 3-phenyl–benzoxazine. Furthermore, the system was tested in the asymmetric hydrogenation of several 3-aryl benzoxazine derivatives, where ees up to 99% were obtained. The hydrogenation reactions of 2-aryl quinolines provided the product with up to 99% ees in the presence of only

0.5 mol% catalyst. In the hydrogenation of 2-phenyl quinoline the authors showed that transferring the reaction from flow to the batch system under the same reaction conditions results in a significant drop in the isolated yield of the product (96% vs. 67%). This effect was explained by the better heat transfer available in the flow system compared to the batch reactor. The asymmetric hydrogenation of more challenging substrates quinoxalines and 3*H*-indoles was also successfully performed to produce the corresponding tetrahydroquinoxalines and indolines with ees up to 94% and 99%, respectively.

Later, the same group presented an interesting study on a photocyclization–asymmetric hydrogenation cascade reaction in flow [91]. It was envisioned that 2-aminochalcones undergo photocyclization to form quinolines, and in the presence of a Binol-based Brönsted acid as catalyst and Hantzsch dihydropyridine as hydrogen source the corresponding enantioenriched tetrahydroquinolines can be produced in a subsequent hydrogenation step (Figure 10.25). The microfabricated glass reactor proved to be an efficient tool for the cascade reaction as it allowed better light penetration and uniform irradiation of the reaction mixture. High isolated yields (up to 88%) and enantioselectivities (up to 96%) were obtained starting from 2-aminochalcone derivatives under optimized conditions (2.4 equiv of Hantzsch dihydropyridine, 1 mol% catalyst, 0.03 M substrate concentration in $CHCl_3$, 55 °C, 0.1 ml min^{-1} flow rate, 60 min residence time, and $\lambda > 300$ nm). Again, a significant improvement of the product yield was observed when the reaction was conducted under flow conditions compared to the batch reaction.

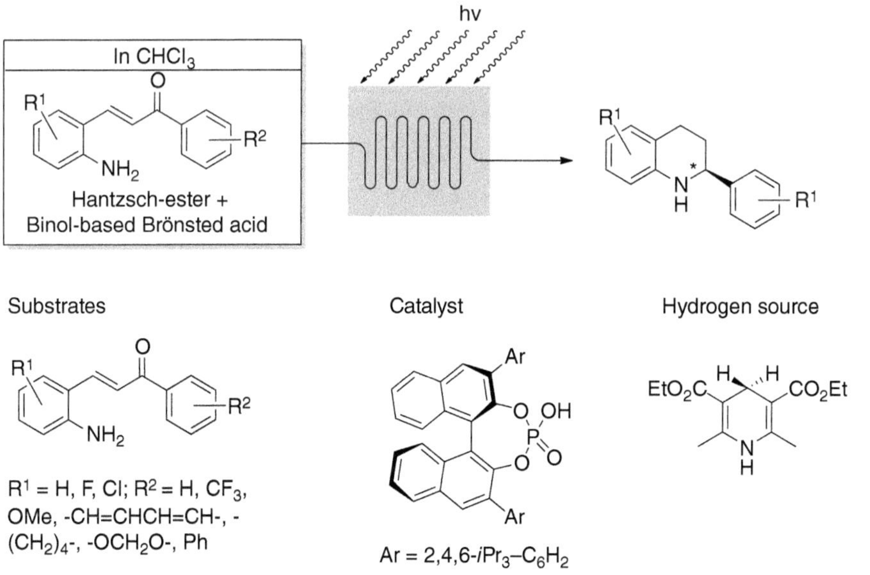

Figure 10.25 An organocatalytic photocyclization–transfer hydrogenation cascade in flow.

10.5 Chiral Auxiliary-controlled Asymmetric Hydrogenation in Flow

In view of the previous chapters, where highly active and selective catalytic systems were developed that used only a tiny amount of chiral catalyst to produce large amount of enantiomerically enriched product, the use of stoichiometric amount of chiral auxiliaries to transfer chiral information to the substrate seems to be wasteful. However, these strategies enable predictable, robust, and highly enantioselective chemical transformations or can be applied in cases where simple catalytic processes are not efficient. The drawbacks associated with the introduction, utilization, and removal of the auxiliary can be minimized using flow technology, i.e. by the integration of the chemical process into one multistep flow sequence [92].

Newman and coworkers reported the flow synthesis of β-chiral acid esters in a chiral auxiliary-mediated four-step reaction sequence [93]. The introduction of the chiral auxiliary (i), the asymmetric hydrogenation on Pd/C catalyst (ii), the removal of the auxiliary (iii), and its separation from the product (iv) enable the synthesis in one single telescoped process (Figure 10.26).

Additionally, the continuous separation of the chiral auxiliary (Oppolzer's sultam) allowed its recycling that enabled the use of formal substoichiometric amount of the chiral modifier. The acylation step (i) was realized in toluene under phase-transfer conditions using NaOH as a base. In the next step, the hydrogenation (ii) was performed on a Pd/C bed at 45 °C to produce the product with 97% yield and a diastereomeric ratio of 95/5. The chiral auxiliary was cleaved by methanolysis (3) using 7% NaOMe in methanol, resulting in the formation of the corresponding methyl ester. In the final step (iv), the chiral auxiliary (in the form of sodium salt) was directly extracted from the organic phase (org.) to the aqueous (aq.). The final ester products were isolated in 67–72% yields, and the diastereoselectivity of the hydrogenation step ranged from 92/8 to 98/2. In comparison to the batch process, the flow system presented similar to greatly improved yields and greatly reduced reaction times.

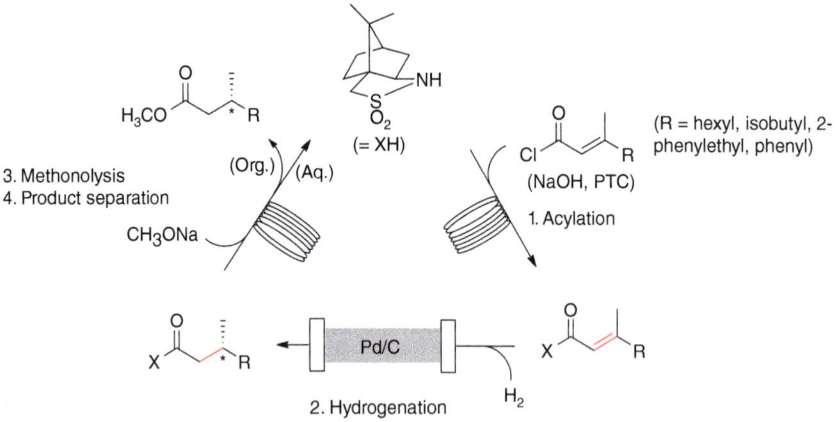

Figure 10.26 Schematic of the multistep flow reaction sequence.

Moreover, in the flow process the chiral auxiliary could be regenerated in 71–79% yield. This was utilized in a next set of experiments where the chiral auxiliary was continuously recycled. The system was continuously operating for 4.5 hours (~8 auxiliary recycles). The overall yields decreased only slightly, and the diastereomeric ratios of the products were the same as observed earlier.

10.6 Summary and Outlook

Continuous-flow technology provides a revolutionary change in the operational design of chemical processes both in research laboratory and in the fine chemical industry. The reduced reaction times, increased efficiency, excellent heat and mass transfer, better control and reproducibility, rapid optimization, screening and in-line analysis, and improved process safety largely contribute to the success of flow chemistry in the past two decades. In terms of green chemistry, continuous-flow technologies can make chemical transformations much more sustainable. Besides process intensification, the use of microreactors can be extremely fruitful in (i) fast catalyst screening, (ii) efficient optimization of reaction parameters, and (iii) in the determination of reaction kinetics with as low reagent loading as possible. Generally, the smaller amounts of solvents used in these processes, the reduced ecological footprint of chemical reactors, the safer realization of synthetic protocols, or the possibility of process integration into one reaction sequence are several of the many remarkable advantages of the flow technologies.

The wide variety of catalysts used for the asymmetric flow hydrogenation reactions performed clearly proves the high potential of the methodology not only in lab-scale processes but also in the industry. It is, however, important to note that the beneficial use of the technology in industrial scale requires the development of more active and stable catalysts. In this contribution, the synthesis of novel chiral ligands for organometallic catalysts still represents an essential part of chemical research.

Continuous-flow syntheses and organocatalytic processes are among the top emerging technologies [94]. The combination of asymmetric continuous-flow hydrogenation with organocatalysis might further enhance the ecological and economical value of the technology and can therefore be highlighted as a promising new direction in synthetic organic chemistry.

The asymmetric flow hydrogenation is not only "a spectacular strategy" but a real-world solution for the synthesis of biologically active synthons. Asymmetric hydrogenation in continuous-flow mode is of growing importance to a sustainable modern society, in which environmental protection is of increasing concern.

References

1 Gutmann, B., Cantillo, D., and Kappe, C.O. (2015). *Angew. Chem. Int. Ed.* 54: 6688–6728.
2 Baumann, M. and Baxendale, I.R. (2015). *Beilstein J. Org. Chem.* 11: 1194–1219.

3 Knowles, W.S. and Noyori, R. (2007). *Acc. Chem. Res.* 40: 1238–1239.
4 Glasnov, T. (2016). *Continuous-flow Chemistry in the Research Laboratory*. Switzerland: Springer International Publishing.
5 Irfan, M., Glasnov, T.N., and Kappe, C.O. (2011). *ChemSusChem* 4: 300–316.
6 Russel, C.C., Baker, J.R., Cossar, P.J., and McCluskey, A. (2017). *IntechOpen*: 269–288.
7 Cossar, P.J., Hizartzidis, L., Simone, I. et al. (2015). *Org. Biomol. Chem.* 13: 7119–7130.
8 Mallia, C.J. and Baxendale, I.R. (2016). *Org. Process Res. Dev.* 20: 327–360.
9 Porta, R., Benaglia, M., and Puglisi, A. (2016). *Org. Process Res. Dev.* 20: 2–25.
10 Masuda, K., Ichitsuka, T., Koumura, N. et al. (2018). *Tetrahedron* 74: 1705–1730.
11 Poechlauer, P., Colberg, J., Fisher, E. et al. (2013). *Org. Process Res. Dev.* 17: 1472–1478.
12 Hughes, D.L. (2018). *Org. Process Res. Dev.* 22: 13–20.
13 Mak, X.Y., Laurino, P., and Seeberger, P.H. (2009). *Beilstein J. Org. Chem.* 5 https://doi.org/10.3762/bjoc.5.19.
14 Zhao, D. and Ding, K. (2013). *ACS Catal.* 3: 928–944.
15 Rasheed, M., Elmore, S.C., and Wirth, T. (2011). *Catalytic Methods in Asymmetric Synthesis: Advanced Materials, Techniques, and Applications* (eds. M. Gruttadauria and F. Giacalone), 345–372. Hoboken, NJ: Wiley.
16 Ishitani, H., Saito, Y., and Kobayashi, S. (2016). *Top. Organomet. Chem.* 57: 213–248.
17 Thompson, M., Penafiel, I., Cosgrove, S.C., and Turner, N.J. (2019). *Org. Process Res. Dev.* 23: 9–18.
18 Mallat, T., Orglmeister, E., and Baiker, A. (2007). *Chem. Rev.* 107: 4863.
19 Orito, Y., Imai, S., Niwa, S., and Nguyen, G.-H. (1979). *J. Synth. Org. Chem. Jpn.* 37: 173–174.
20 Meheux, P.A., Ibbotson, A., and Wells, P.B. (1991). *J. Catal.* 128: 387–396.
21 Künzle, N., Hess, R., Mallat, T., and Baiker, A. (1999). *J. Catal.* 186: 239–241.
22 Wandeler, R., Künzle, N., Schneider, M.S. et al. (2001). *Chem. Commun.*: 673–674.
23 Künzle, N., Mallat, T., and Baiker, A. (2003). *Appl. Catal. A* 238: 251–257.
24 Wandeler, R., Künzle, N., Schneider, M.S. et al. (2000). *J. Catal.* 200: 377–388.
25 Meier, D.M., Ferri, D., Mallat, T., and Baiker, A. (2007). *J. Catal.* 248: 68–76.
26 Künzle, N., Solèr, J.-W., and Baiker, A. (2003). *Catal. Today* 79-80: 503–509.
27 Zhao, Y., Gao, F., Chen, and Garland, M. (2004). *J. Catal.* 221: 274–287.
28 Toukoniitty, E., Mäki-Arvela, P., Neyestanaki, A.K. et al. (2001). *Appl. Catal. A: Gen.* 216: 73–83.
29 Toukoniitty, E., Mäki-Arvela, P., Neyestanaki, A.K. et al. (2002). *Appl. Catal. A.* 235: 125–138.
30 Toukoniitty, E., Mäki-Arvela, P., Kumar, N. et al. (2004). *Catal. Lett.* 95: 179–183.
31 Toukoniitty, E. and Murzin, D.Y. (2006). *J. Catal.* 241: 96–102.
32 Martin, G., Pereira, C., Pettersson, F. et al. (2015). *Chem. Eng. Technol.* 38: 804–812.
33 You, X., Li, X., Xiang, S. et al. (2000). *Stud. Surf. Sci. Catal.* 130: 3375–3380.

34 Li, X. and Li, C. (2001). *Catal. Lett.* 77: 251–254.
35 von Arx, M., Dummer, N., Willock, D.J. et al. (2003). *Chem. Commun.*: 1926–1927.
36 Jenkins, R.J., McMorn, P., and Hutchings, G.J. (2005). *Catal. Lett.* 100: 255–258.
37 Szőllősi, G., Hermán, B., Fülöp, F., and Bartók, M. (2006). *React. Kinet. Catal. Lett.* 88: 391–398.
38 Mándity, I.M., Ötvös, S.B., Szőllősi, G., and Fülöp, F. (2016). *Chem. Rec.* 16: 1018–1033.
39 Szőllősi, G., Cserényi, S., Balázsik, K. et al. (2009). *J. Mol. Catal. A* 305: 155–160.
40 Szőllősi, G., Cserényi, S., Fülöp, F., and Bartók, M. (2008). *J. Catal.* 260: 245–253.
41 Szőllősi, G., Cserényi, S., Bucsi, I. et al. (2010). *Appl. Catal. A* 382: 263–271.
42 Szőllősi, G., Makra, Z., Fekete, M. et al. (2012). *Catal. Lett.* 142: 889–894.
43 Szőllősi, G., Makra, Z., Fülöp, F., and Bartók, M. (2011). *Catal. Lett.* 141: 1616–1620.
44 Szőllősi, G., Cserényi, S., and Bartók, M. (2010). *Catal. Lett.* 134: 264–269.
45 Gao, F., Chen, L., and Garland, M. (2006). *J. Catal.* 238: 402–411.
46 Cserényi, S., Szőllősi, G., Szőri, K. et al. (2010). *Catal. Commun.* 12: 14–19.
47 Kovács, L., Szőllősi, G., and Fülöp, F. (2015). *J. Flow. Chem.* 5: 210–215.
48 Künzle, N., Solèr, J.-W., Mallat, T., and Baiker, A. (2002). *J. Catal.* 210: 466–470.
49 Hermán, B., Szőllősi, G., Fülöp, F., and Bartók, M. (2007). *Appl. Catal. A: Gen.* 331: 39–43.
50 Hintermair, U., Franciò, G., and Leitner, W. (2011). *Chem. Commun.* 47: 3691–3701.
51 Vankelecom, I.F.J. and Jacobs, P.A. (2000). Catalyst immobilization on inorganic supports. In: *Chiral Catalyst Immobilization and Recycling* (eds. D.E. De Vos, I.F.J. Vankelecom and P.A. Jacobs), 19–42. Weinheim: Wiley-VCH.
52 Altava, B., Burguete, M.I., Garcia-Verdugo, E., and Luis, S.V. (2018). *Chem. Soc. Rev.* 47: 2722–2771.
53 Sandee, A.J., Petra, D.G.I., Reek, J.N.H. et al. (2001). *Chem. Eur. J.* 7: 1202–1208.
54 Augustine, R.L., Tanielyan, S.K., Anderson, S., and Yang, H. (1999). *Chem. Commun.*: 1257–1258.
55 Izumi, Y., Hasere, R., and Urabi, K. (1983). *J. Catal.* 84: 402–409.
56 Lefebvre, F. (1992). *J. Chem. Soc., Chem. Commun.*: 756–757.
57 Augustine, R.L., Tanielyan, S.K., Mahata, N. et al. (2003). *Appl. Catal. A: Gen.* 256: 69–76.
58 Stephenson, P., Kondor, B., Licence, P. et al. (2006). *Adv. Synth. Catal.* 348: 1605–1610.
59 Stephenson, P., Licence, P., Ross, S.K., and Poliakoff, M. (2004). *Green Chem.* 6: 521–523.
60 Madarász, J., Farkas, G., Balogh, S. et al. (2011). *J. Flow. Chem.* 2: 62–67.
61 Balogh, S., Farkas, G., Madarász, J. et al. (2012). *Green. Chem.* 14: 1146–1151.
62 Madarász, J., Nánási, B., Kovács, J. et al. (2018). *Monatsh. Chem.* 149: 19–25.
63 Amara, Z., Poliakoff, M., Duque, R. et al. (2016). *Org. Proc. Res. Dev.* 20: 1321–1327.

64 Duque, R., Pogorzelec, P.J., and Cole-Hamilton, D.J. (2013). *Angew. Chem. Int. Ed.* 52: 9805–9807.

65 Zhang, Z., Franciò, G., and Leitner, W. (2015). *ChemCatChem* 7: 1961–1965.

66 Wasserscheid, P. (2010). Transition metal catalysis in ionic liquids. In: *Handbook of Green Chemistry*, vol. 6 (eds. J. Dupont and L. Kollar), 65–91. Weinheim, Germany: Wiley-VCH Verlag GmbH & Co. ISBN: 978-3-52-762869-8.

67 Öchsner, E., Schneider, M.J., Meyer, C. et al. (2011). *Appl. Catal. A: Gen.* 399: 35–41.

68 Schneider, M.J., Haumann, M., and Wasserscheid, P. (2013). *J. Mol. Catal. A Chem.* 376: 103–110.

69 Hintermair, U., Höfener, T., Pullmann, T. et al. (2010). *ChemCatChem* 2: 150–154.

70 Hintermair, U., Franciò, G., and Leitner, W. (2013). *Chem. Eur. J.* 19: 4538–4547.

71 Theuerkauf, J., Franciò, G., and Leitner, W. (2013). *Adv. Synth. Catal.* 355: 209–219.

72 Geier, D., Schmitz, P., Walkowiak, J. et al. (2018). *ACS Catal.* 8: 3297–3303.

73 Burguete, M.I., García-Verdugo, E., and Luis, S.V. (2011). *Beilstein J. Org. Chem.* 7: 1347–1359.

74 Franciò, G., Hintermair, U., and Leitner, W. (2015). *Phil. Trans. R. Soc. A* 373: 20150005.

75 Hayashi, T., Kikuchi, S., Asano, Y. et al. (2012). *Org. Process Res. Dev.* 16: 1235–1240.

76 Newton, S., Ley, S.V., Arcé, E.C., and Grainger, D.M. (2012). *Adv. Synth. Catal.* 254: 1805–1812.

77 Sun, X. and Gavriilidis, A. (2008). *Org. Proc. Res. Dev.* 12: 1218–1222.

78 Sun, X. and Gavriilidis, A. (2009). *Chem. Eng. Technol.* 32: 1318–1325.

79 Marras, F., Leeuwen, P.W.N.M., and Reek, J.N.H. (2011). *Chem. Eur. J.* 17: 7460–7471.

80 Laue, S., Greiner, L., Wöltinger, J., and Liese, A. (2001). *Adv. Synth. Catal.* 343: 711–720.

81 Greiner, L., Laue, S., Liese, A., and Wandrey, C. (2006). *Chem. Eur. J.* 12: 1818–1823.

82 O'Neal, E.J., Lee, C.H., Brathwaite, J., and Jensen, K.F. (2015). *ACS Catal.* 5: 2615–2622.

83 de Bellefon, C., Pestre, N., Lamouille, T. et al. (2003). *Adv. Synth. Catal.* 345: 190–193.

84 Abdallah, R., Meille, V., Shaw, J. et al. (2004). *Chem. Commun.*: 372–373.

85 de Bellefon, C., Lamouille, T., Pestre, N. et al. (2005). *Catal. Today* 110: 179–187.

86 Guo, H. and Ding, Q. (2011). *Chimia* 65: 932–938.

87 Shi, L., Wang, X., Sandoval, C.A. et al. (2009). *Chem. Eur. J.* 15: 9855–9867.

88 Atodiresei, I., Vila, C., and Rueping, M. (2015). *ACS Catal.* 5: 1972–1985.

89 Puglysi, A., Benaglia, M., Porta, R., and Coccia, F. (2015). *Curr. Organocatal.* 2: 79–101.

90 Rueping, M., Bootwicha, T., and Sugiono, E. (2012). *Beilstein J. Org. Chem.* 8: 300–307.

91 Sugiono, E. and Rueping, M. (2013). *Beilstein J. Org. Chem.* 9: 2457–2462.
92 Britton, J. and Raston, C.L. (2017). *Chem. Soc. Rev.* 46: 1250–1271.
93 Sullivan, R.J. and Newman, S.G. (2018). *Chem. Sci.* 9: 2130–2134.
94 Gomollon-Bel, F. (2019). *Chem. Int.* 41: 12–17.

11

Organocatalytic Asymmetric Transfer Hydrogenation Reactions

Sayantani Das, Vijay N. Wakchaure and Benjamin List

Max-Planck-Institut für Kohlenforschung, Homogeneous Catalysis, Kaiser- Wilhelm-Platz 1, 45470, Mülheim an der Ruhr, Germany

11.1 Introduction

Hydrogenation is a fundamental chemical transformation in which molecular hydrogen (H_2) is used to "saturate" unsaturated organic molecules. Despite the seemingly simplistic nature of the reaction, hydrogen itself is normally not reactive toward organic molecules; therefore, the process typically requires promotion by transition-metal catalysts (e.g. Pd, Pt, Rh, Ni, and Ru). On the other hand, transfer hydrogenation refers to the addition of hydrogen to a molecule from a source other than molecular hydrogen itself. This strategy has become an attractive alternative to direct hydrogenations and has been applied in industrial processes. An example of a large-scale transfer hydrogenation is the "coal liquefaction" with hydrogen donor solvents such as tetralin [1]. The potential advantages of transfer hydrogenations over conventional hydrogenations include: (i) the systems do not require pressurized H_2 gas or special experimental setups, (ii) the hydrogen donors are often readily available and easy to handle, and (iii) the corresponding catalysts can be simple and readily accessible and handled under aerobic, noninert conditions [2].

The first known example of a transfer hydrogenation reaction dates back more than a century. In 1903, Knoevenagel [3] demonstrated that palladium black smoothly promotes the disproportionation of dimethyl 1,4-dihydroterephthalate to dimethyl terephthalate and *cis*-hexahydroterephthalate, in which a disproportionative hydrogen transfer between identical donor and acceptor units took place. In the 1960s [4], the pioneering work by Henbest, Mitchell, and coworkers showed that an iridium hydride complex could catalyze the hydrogenation of cyclohexanones and α,β-unsaturated ketones to alcohols using 2-propanol as a stoichiometric reductant. Since these seminal reports, catalytic transfer hydrogenations have been extensively studied and reported in the literature. Nowadays, the most prevalent hydrogen sources are hydrazine, dihydronaphthalene, 2-propanol [5], and formic acid [6] and are often used in the presence of metals such as Ru, Rh, Ir, and Sm or an organic molecule as the catalyst. Depending on the chosen catalyst, two methodologies of

Asymmetric Hydrogenation and Transfer Hydrogenation, First Edition.
Edited by Virginie Ratovelomanana-Vidal and Phannarath Phansavath.

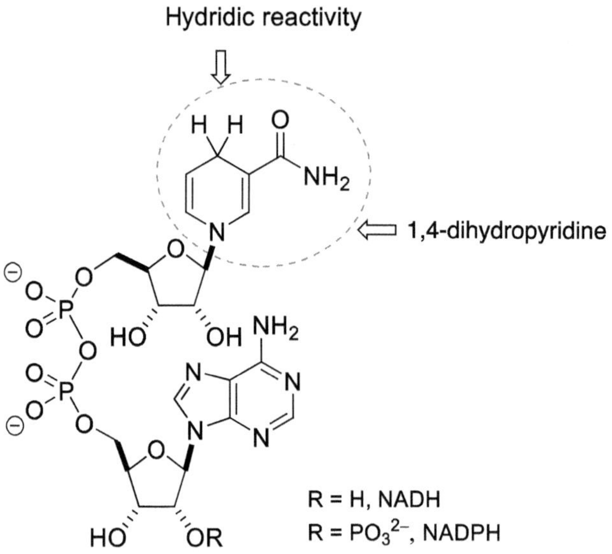

Figure 11.1 Structure of NADH and NADPH. Source: Based on Ref. [7].

asymmetric transfer hydrogenation (ATH) can be adopted: metal catalysis with chiral ligands and organocatalysis, often mimicking biological processes.

Nature performs ATH reactions in biological systems using NADH and NADPH (Figure 11.1) [7]. These cofactors contain a 1,4-dihydropyridine which is capable of delivering two hydrogen atoms to unsaturated C=C, C=N, and C=O bonds. Such cofactors have inspired chemists to develop organocatalytic ATH reactions [8] using 1,4-dihydropyridines, most importantly the so-called Hantzsch esters [9].

The first synthesis of 1,4-dihydropyridines was reported by Arthur Rudolf Hantzsch in 1881 [9], and the products were later named "Hantzsch esters". The synthesis involved the condensation between an aldehyde, two molecules of a β-ketoester, and ammonia to give a 1,4-dihydropyridine (Scheme 11.1).

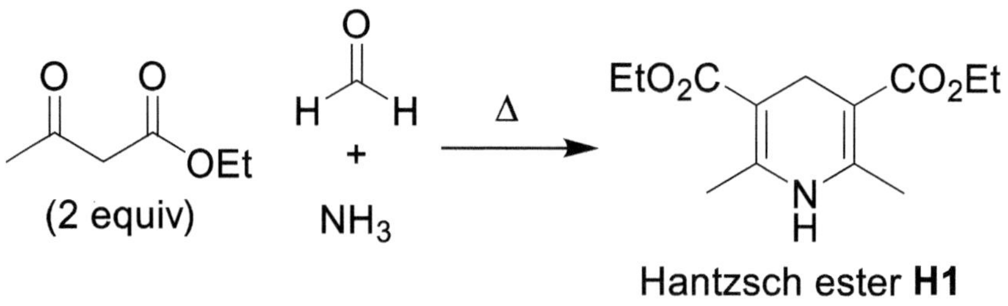

Scheme 11.1 The first synthesis of a Hantzsch ester.

One of the very early uses of the Hantzsch ester as a reductant was demonstrated by Westheimer and Mauzerall [10] in 1955 in the transfer hydrogenation of a carbonyl compound (Scheme 11.2). Lactic acid was produced via reduction of pyruvic acid with Hantzsch ester **H1** under neat conditions and at high temperature in 7%

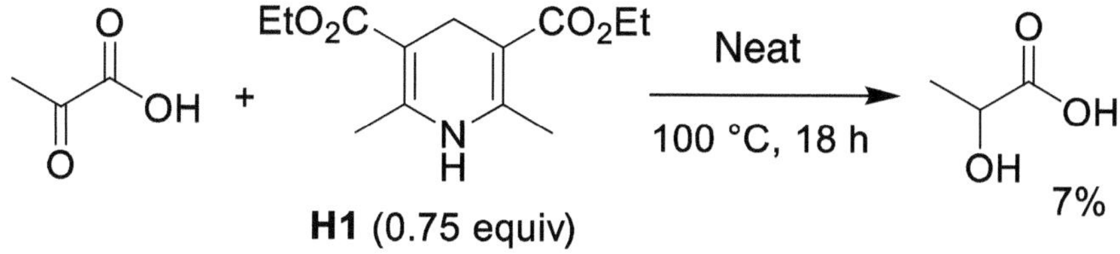

Scheme 11.2 Westheimer's reduction of pyruvic acid using Hantzsch ester (**H1**).

yield [11]. Since then, Hantzsch esters have emerged as powerful organic hydrogen sources in transfer hydrogenations [8].

In this chapter, we give an overview of the various types of organocatalytic ATH methods using organic hydrogen sources, with specific focus on the use of Hantzsch esters.

11.2 Reduction of C=C Double Bonds

The most frequently reported type of olefin modification is its reduction, i.e. the addition of a hydrogen molecule across the C=C bond. Intensive research on C=C bond reductions using metal catalysts and molecular hydrogen has been reported. More recently, C=C bond reductions via transfer hydrogenation in combination with organocatalysts have emerged as an alternative.

An early contribution to organocatalytic transfer hydrogenation reactions of alkenes was reported by List and coworkers in 2004 [12]. The conjugate reduction of α,β-unsaturated aldehydes in the presence of Hantzsch ester **H1** and a catalytic amount of ammonium salts such as dibenzylammonium trifluoroacetate at room temperature was developed (Scheme 11.3). This method is highly efficient, regio- and chemoselective, and is tolerant of a variety of functional groups that are typically sensitive to standard hydrogenation conditions.

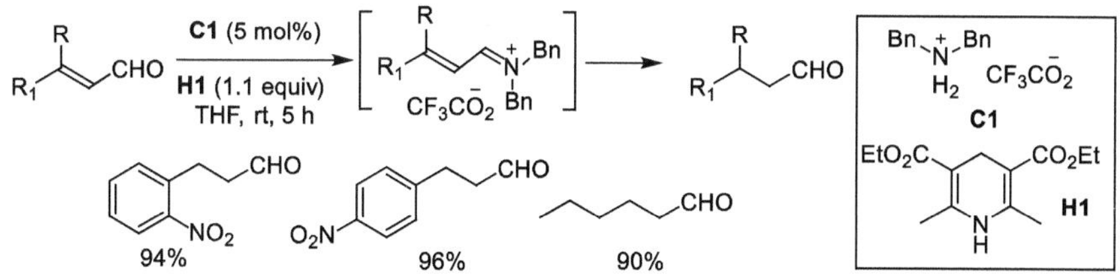

Scheme 11.3 The first organocatalytic transfer hydrogenation reaction of enals.

In this publication, the very first asymmetric version of this reaction using the hydrochloride salt of MacMillan-type imidazolidinone (**C2**) has also been described with good yield and enantioselectivity (Scheme 11.4).

Soon after, the List group advanced this methodology to the enantioselective reduction of several unsaturated aldehydes in very high yields and

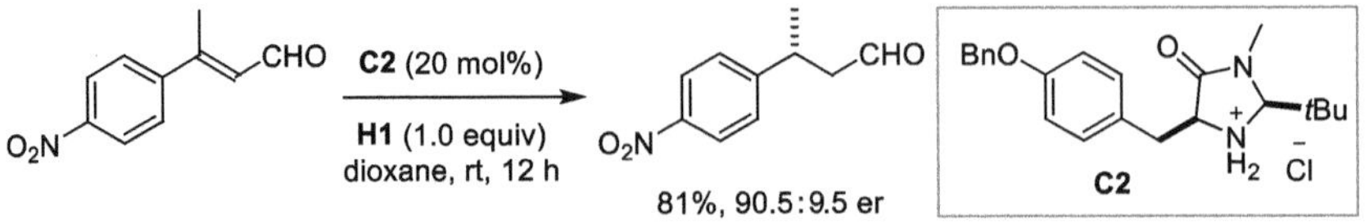

Scheme 11.4 The first organocatalytic ATH reaction of an enal.

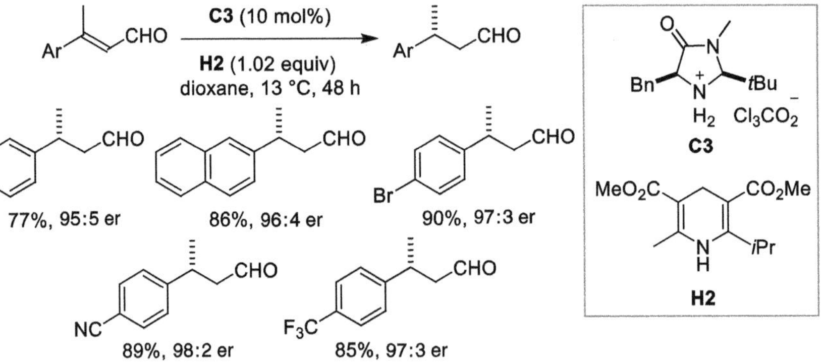

Scheme 11.5 The organocatalytic ATH reaction of α,β-unsaturated aldehydes. Source: Refs [13].

enantioselectivities using the trichloroacetate salt of an imidazolidinone (**C3**) as catalyst (Scheme 11.5) [13].

An interesting feature of this process is its *stereoconvergence*, i.e. starting from either the *Z*- or the *E*-isomer of the unsaturated aldehyde leads to the same enantiomer of the product. Therefore, the unsaturated aldehyde starting material of the reaction may be used as a mixture of *E*/*Z* isomers. It was proposed that the *E*- and *Z*-α,β-unsaturated iminium ion intermediates are in fast equilibrium via a dienamine species. The subsequent rate-limiting hydride transfer proceeds much faster with the *E*-isomer [$k(E) > k(Z)$], which predominantly affords the (*R*)-product (Figure 11.2).

Also, in 2005, MacMillan and coworkers [14] reported a similar study, where the monosubstituted catalyst **C4** could promote highly enantioselective reductions of trisubstituted aldehydes (irrespective of their geometric purity) at −30 °C (Scheme 11.6).

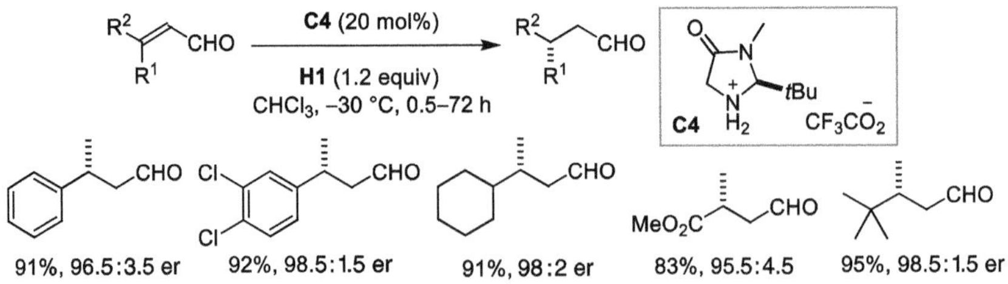

Scheme 11.6 The organocatalytic ATH reaction of α,β-unsaturated aldehydes.

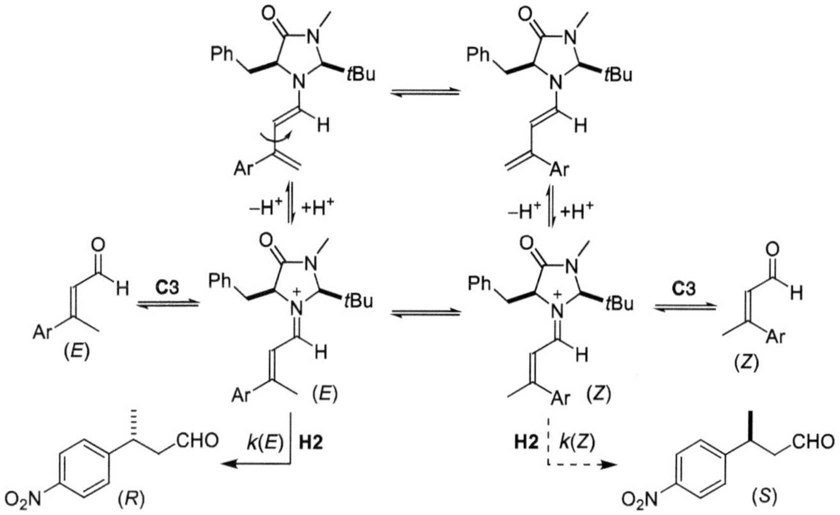

Figure 11.2 Proposed mechanism of asymmetric transfer hydrogenation catalyzed by the secondary amine salt, and the origin of enantioconvergence.

The same activation mode of iminium ion catalysis was later expanded to α,β-unsaturated ketones by List, MacMillan and coworkers [15]. The previously used catalysts for enal hydrogenation were found to be incompatible, due to the fact that ketones are sterically and electronically deactivated toward iminium formation in comparison to aldehydes. The furyl imidazolidinone catalyst (**C5**), which has been previously used for enantioselective Diels–Alder reactions with cyclic enones, promotes the highly enantioselective reduction of cyclic enones with several ring sizes using Hantzsch ester **H3** under mild conditions (Scheme 11.7).

In the meantime, the List group introduced the new concept "Asymmetric Counteranion Directed Catalysis (ACDC)" [16]. Intrigued by the observation of a strong counteranion effect on the yield and enantioselectivity of their transfer hydrogenation reaction [13], a new class of ammonium salt catalysts was designed, which consists of *achiral* ammonium ions and an enantiopure counteranion,

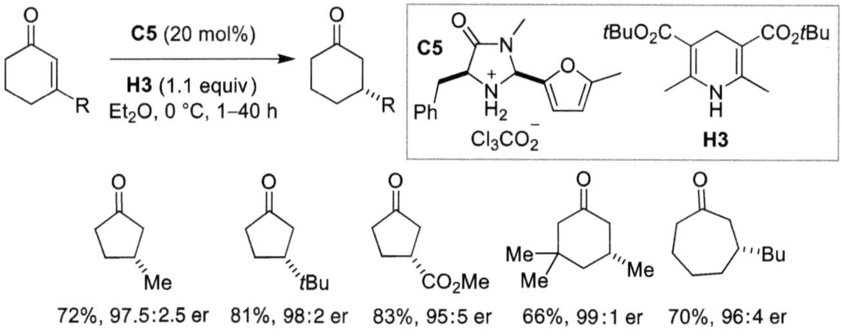

Scheme 11.7 The first organocatalytic ATH reaction of α,β-unsaturated ketones.

TRIP [17], a BINOL-derived phosphate anion. The chirality of the chiral phosphate anion was proposed to induce the enantioselectivity of the reaction via stereochemical communication within the iminium ion–phosphate pair. Under optimized conditions, morpholinium salt [16] **C6** could catalyze the enantioselective reduction of α,β-unsaturated aldehydes to furnish the corresponding saturated derivatives in up to >99:1 er, the highest yet reported for this class of substrates (Scheme 11.8). It was proposed that the reaction proceeds via an iminium ion with the chiral phosphate as a counteranion. There may be multiple interactions of the phosphate anion with the iminium ion including nonclassical CH· · ·O hydrogen bonding and an NH· · ·O hydrogen-bonding interaction with the Hantzsch ester. Such a compact ion–pair interaction in the TS could be responsible for the high enantioselectivity.

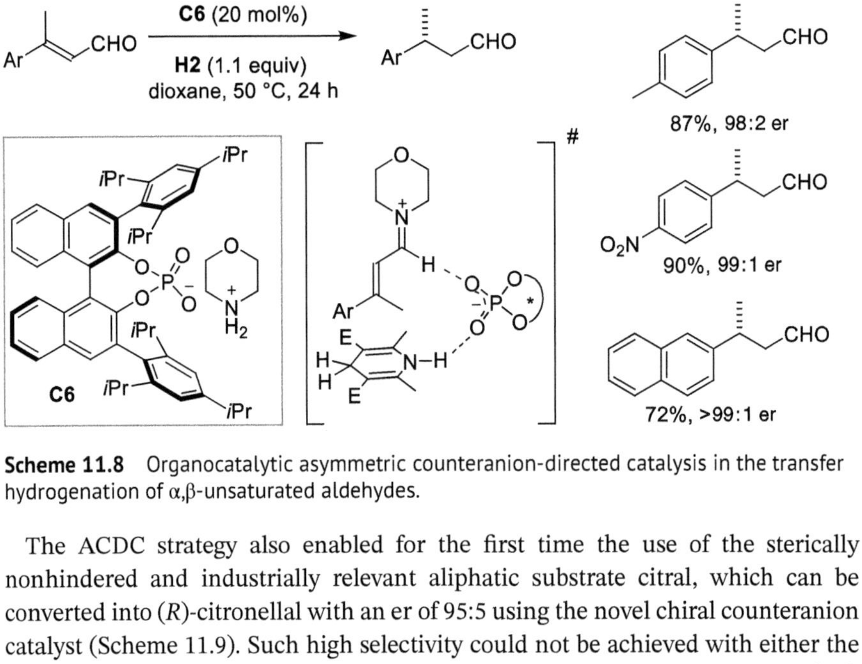

Scheme 11.8 Organocatalytic asymmetric counteranion-directed catalysis in the transfer hydrogenation of α,β-unsaturated aldehydes.

The ACDC strategy also enabled for the first time the use of the sterically nonhindered and industrially relevant aliphatic substrate citral, which can be converted into (*R*)-citronellal with an er of 95:5 using the novel chiral counteranion catalyst (Scheme 11.9). Such high selectivity could not be achieved with either the List group's previous system [13] or with that of MacMillan and coworkers [14]. Indeed, this had been the highest enantioselectivity so far reported for any catalytic asymmetric hydrogenation or transfer hydrogenation of citral.

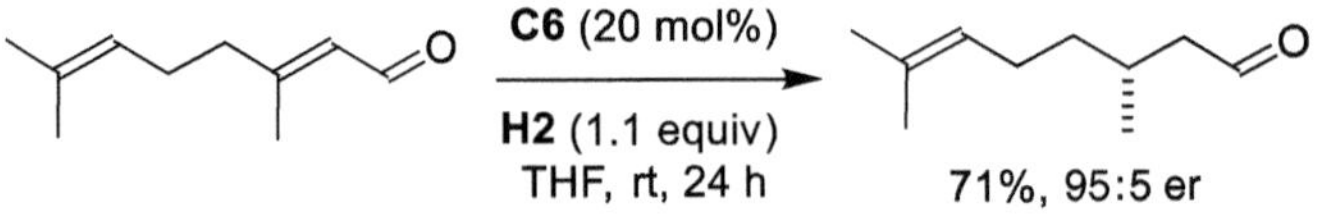

Scheme 11.9 Organocatalytic asymmetric transfer hydrogenation of (*E*)-citral.

The List group reasoned that the new strategy would be amenable to the activation of ketones using less bulky *primary* amine salts of TRIP, a phosphoric acid which had previously been introduced by the group as powerful catalyst for imine reductions and Pictet–Spengler cyclizations [18]. (*S*)-Valine ester-derived salt [15] **C7** proved to be the optimal catalyst capable of promoting highly selective reductions of a variety

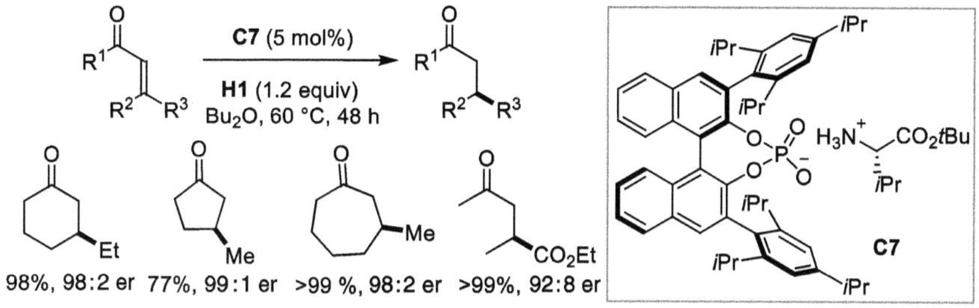

Scheme 11.10 Organocatalytic asymmetric counteranion-directed catalysis in transfer hydrogenation of α,β-unsaturated ketones.

of cyclic ketones (Scheme 11.10). Mechanistically, the reaction may again proceed via an iminium ion–phosphate pair, which is stabilized by hydrogen-bonding interactions. In addition to binding the iminium ion, the phosphate counteranion may also interact with the Hantzsch ester via an additional hydrogen bond. In that view, the activation mode would best be described as a combination of Brønsted acid catalysis and iminium ion catalysis.

In 2008, Kudo and coworkers [19] developed a supported catalyst for the ATH of enals. They demonstrated that a resin-supported N-terminal prolyl peptide with α,β-turn motif and hydrophobic polyleucine tether could effectively catalyze the ATH reaction of α,β-unsaturated aldehydes with Hantzsch esters in aqueous media (THF/H_2O) (Scheme 11.11). The products could be obtained in up to 76% yield and 98:2 enantiomeric ratio.

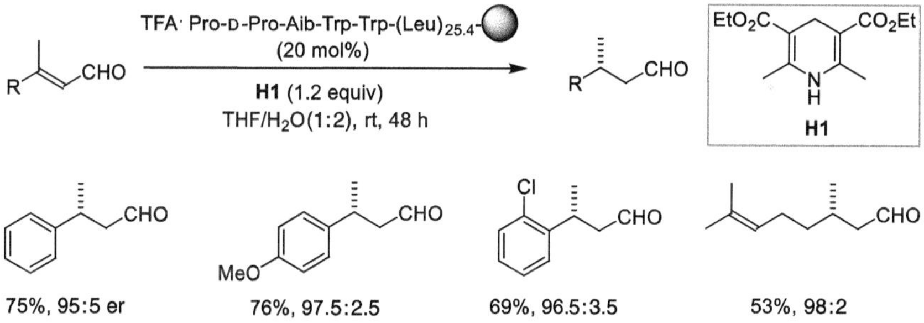

Scheme 11.11 Organocatalytic ATH in aqueous media using resin-supported peptide.

Chiral amines can also be used to reduce the enone functionality in complex molecules such as steroids [20]. Recently, Van Arman [21] and coworkers replaced the traditional Hantzsch ester with a Hantzsch amide and studied the non-enantioselective reduction of unsaturated linear ketones promoted by benzylammonium trifluoroacetate. The novelty of this process is that the unreacted Hantzsch amide and the bis-methylamidopyridine by-product are effectively removed by extraction, which is more challenging when using the more common Hantzsch diethyl ester.

Houk and coworkers [22] undertook a theoretical investigation on the ATH reaction of α,β-unsaturated ketones catalyzed by chiral imidazolidinone salts. Hybrid density functional theory (B3LYP/6-31G(d)) suggested that the *E*-iminium

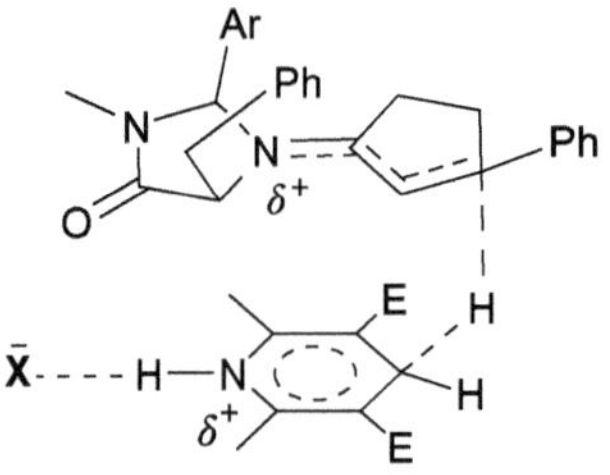

Figure 11.3 Proposed hydride attack transition state.

transition state is favored. Hydride transfer from the Hantzsch ester to the iminium is energetically favored from the less-hindered (bottom) face in an *anti*-fashion. The calculated energetic barrier of this transition state was 1.1 kcal mol^{-1} lower than its diastereomeric counterpart, consistent with the experimentally observed enantioselectivity (Figure 11.3).

List and coworkers also studied the ATH of β,β-disubstituted nitroolefins [23] and β-nitroacrylates [24] utilizing a chiral thiourea catalyst. A number of nitroolefins could be reduced in very high yields and enantiomeric excess (Scheme 11.12).

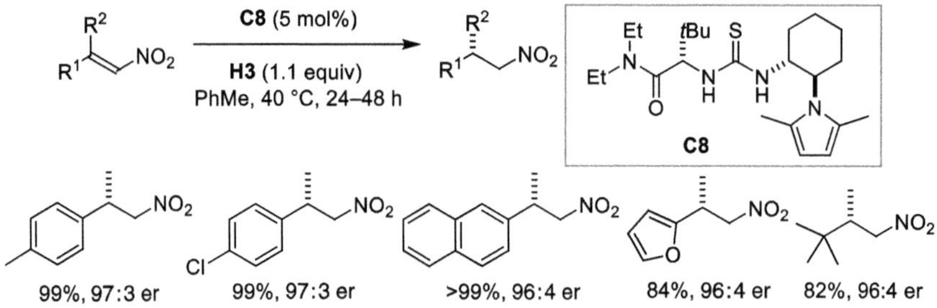

Scheme 11.12 The first example of thiourea catalysis in ATH of nitroolefins.

In case of β-nitroacrylates, the substrate reactivity was much higher (Scheme 11.13). The reactions could be performed at 0 °C instead of 40 °C, and the enantiomeric ratio of the resulting saturated β-nitroesters varied from 94.5:5.5 to 97.5:2.5. Also, the nitro group could be further reduced by hydrogen in the presence of palladium on activated charcoal to provide the corresponding β^2-amino acid with no loss of enantiopurity.

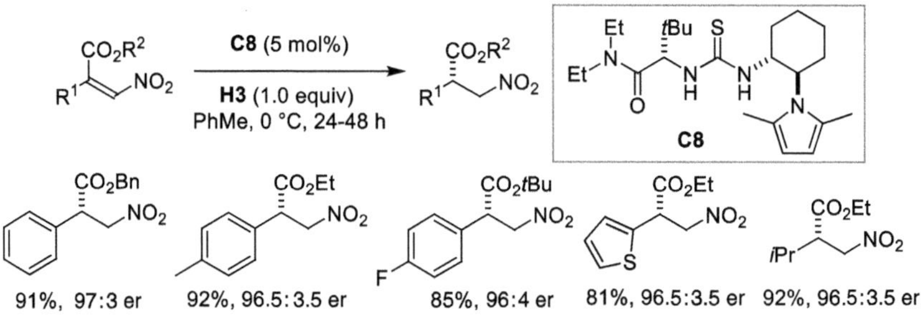

Scheme 11.13 The organocatalytic ATH reaction of β-nitroacrylates.

Later, Paradies and coworkers [25] used a bifunctional thiourea catalyst for the ATH reactions of nitroolefins with Hantzsch esters in high yields (76–99%) and good to moderate enantiomeric ratios (70:30–93.5:6.5). Their investigation showed that the Hantzsch dihydropyridine–NH bond significantly contributed to the efficiency and enantioselectivity of the transformation, suggesting H-bonding interactions with the hydroxy-substituted thiourea derivatives. This methodology was later extended to β-trifluoromethyl nitroalkenes [26] and β-aminonitroolefins [27]. The reduction products β-aminonitroalkanes may be transformed into a variety of useful substances relevant to medicinal chemistry.

Sun and coworkers [28] described an interesting organocatalytic transfer hydrogenation strategy for the asymmetric synthesis of 1,1-diarylethanes, an important class of compounds with broad medicinal and agricultural applications. Under mild conditions, a range of 1,1-diarylethanes, substituted with an *o*-hydroxyphenyl or indole unit, could be obtained with excellent yield and enantiomeric excess (Scheme 11.14). In some cases, triethylamine and TMSCl were added prior to product extraction to convert the products to their silylated derivatives in order to facilitate purification. The free hydroxyl group is required for the *in situ* generation of *o*-quinone methide-type intermediates where the fast hydride addition takes place on the side of the (*Z*)-*o*-quinone methide plane where the catalyst resides, giving rise to the observed stereoselectivity. They also extended the protocol to an asymmetric hydroarylation of 1,1-diarylalkenes with indoles for the synthesis of a range of highly enantioenriched 1,1,1-triarylethanes bearing acyclic all-carbon quaternary stereocenters.

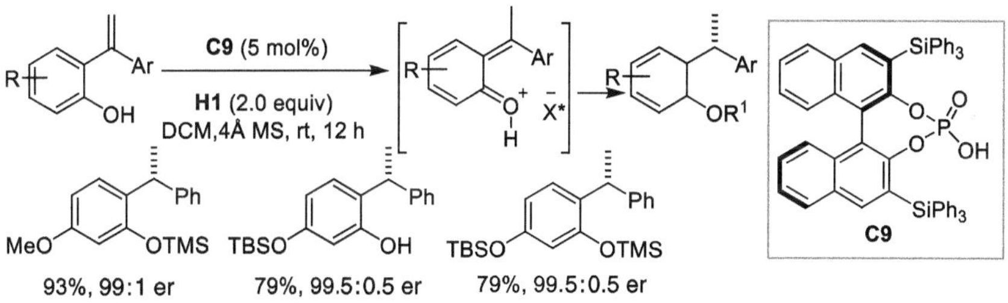

Scheme 11.14 Organocatalytic asymmetric synthesis of 1,1-diarylethanes by transfer hydrogenation.

11.3 Reduction of C=N Double Bonds

The first enantioselective transfer hydrogenation reaction of C=N double bonds with Hantzsch esters was claimed in 1989 by Singh and Batra [29]. α-Amino acid hydrochlorides or chiral acids mediated the reduction of imines using Hantzsch ester as the hydrogen source to afford amines with moderate enantioselectivity (up to 81.5:18.5 er). However, it is important to point out that no experimental details were revealed in this report. We have carefully reinvestigated these published experiments but have been unable to reproduce any of the reported enantioselectivities, either under catalytic or stoichiometric conditions. In all cases e.r.'s < 55:45 were obtained.

In 2005, a highly enantioselective transfer hydrogenation of imines using Hantzsch ester and chiral phosphoric acid (CPA) as catalyst to produce chiral amines under mild conditions was realized by List and coworkers and by Rueping and coworkers. The List group showed that 1 mol% of (*S*)-TRIP could catalyze the ATH of several PMP-protected (PMP: *p*-methoxylphenyl) imines using Hantzsch ester in toluene at 35 °C with high yields (80–98%) and enantioselectivities (90:10–96.5:3.5 er) (Scheme 11.15) [30]. The relatively less-reactive aliphatic isopropyl methyl-substituted imine could also be reduced with 80% yield and 95:5 er. Similar enantioselectivity (94:6 er) and yield (81%) were obtained in this first example of an enantioselective reductive amination with an organocatalyst. The catalyst loading could even be reduced to 0.1 mol% (J. Zhou and B. List, unpublished results) by performing the imine formation and reduction *in situ* in the presence of molecular sieves (MSs).

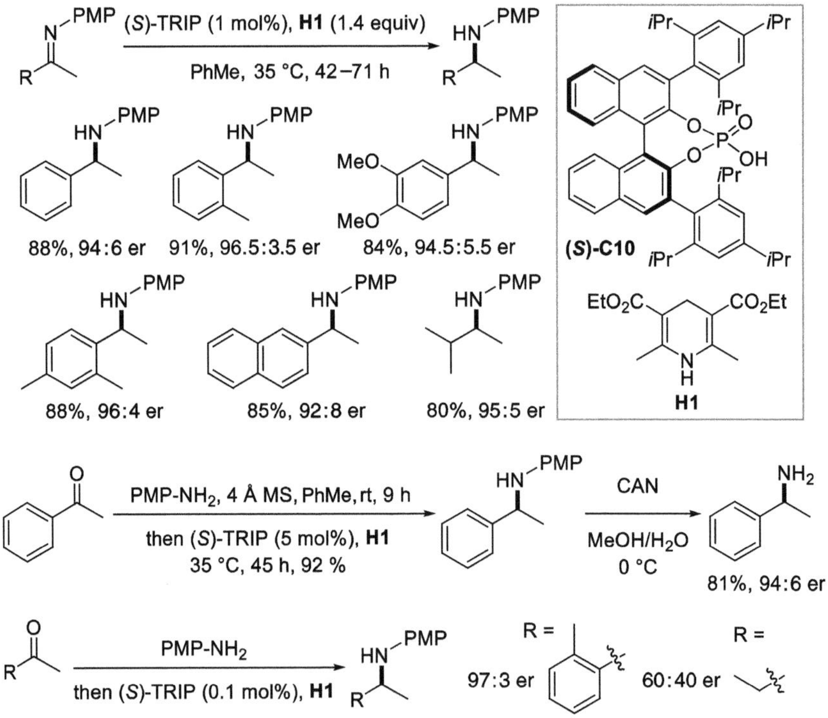

Scheme 11.15 CPA-catalyzed ATH of imines and first reductive amination of ketones. Source: Based on Ref. [30].

In the parallel studies from Rueping's group [31], the chiral Brønsted acid **C11** was identified as the optimal catalyst (Scheme 11.16). Compared to the List group protocol, it requires higher catalyst loading (20 mol%), higher temperature (60 °C), and longer reaction time (72 h) to give the desired amine product with lower yields (46%–91%) and enantioselectivities (84:16–92:8 er).

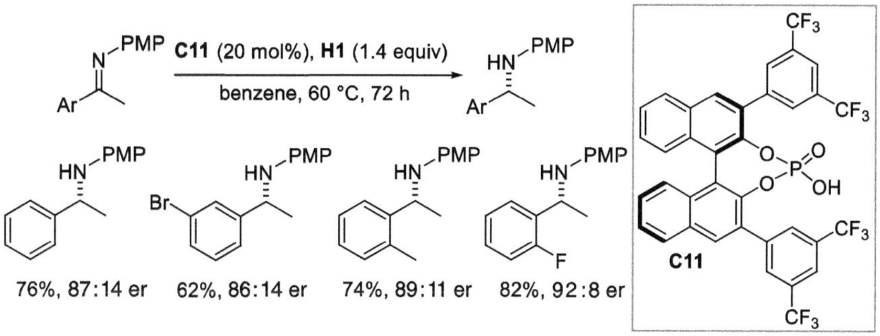

Scheme 11.16 CPA-catalyzed ATH of imines.

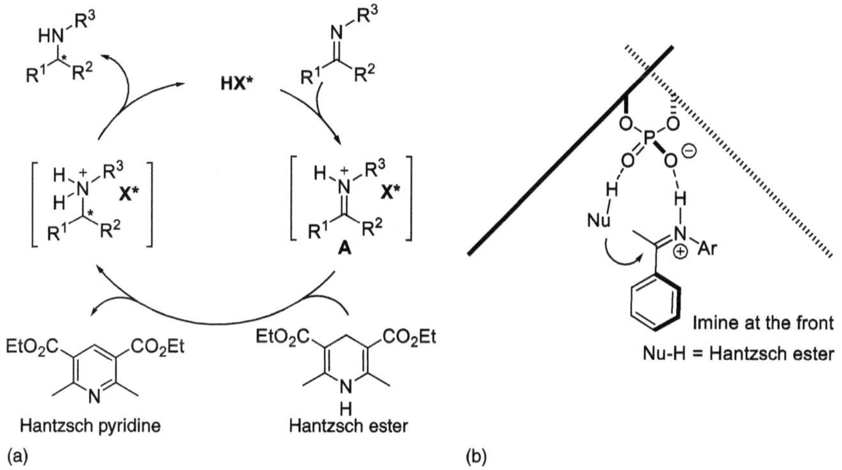

Figure 11.4 (a) The proposed catalytic cycle of CPA-catalyzed ATH of imines. (b) Stereochemical preference in CPA-catalyzed ATH of imines.

Both groups proposed a similar catalytic cycle, which is initiated by protonation of the imine from the chiral Brønsted acid catalyst (Figure 11.4a). The resulting iminium ion is paired, probably stabilized by hydrogen bonding, with a chiral enantiopure counteranion, and its reaction with the Hantzsch dihydropyridine (**H1**) could give an enantiomerically enriched amine and Hantzsch pyridine by-product. The concentration of the free acid HX* is presumably very small under the reaction conditions.

In 2009, Simón and Goodman [32] conducted a theoretical study toward elucidating the reaction mechanism and the origin of the observed enantioselectivity in the Hantzsch ester mediated, CPA-catalyzed ATH of imines. DFT calculations suggest a bifunctional activation mode of the CPA, in which the imine group is activated by the Brønsted acid which at the same time interacts with the Hantzsch ester. According to the calculations, the *Z*-imine involving transition state is lower in energy than the corresponding *E*-isomer (Figure 11.4b). The observed enantioselectivity was interpreted according to the "three-point contact model" [33], two hydrogen bonds with the phosphate, and the effect of steric hindrance.

Subsequently, MacMillan and coworkers reported an effective reductive amination of ketones using Hantzsch esters [34]. Phosphoric acid **C9**, with triphenylsilyl groups at the 3,3′-positions of the BINOL-derived backbone, was found to be optimal. Using 10 mol% of the catalyst and Hantzsch ester **H1** as the hydrogen source, a variety of aromatic and aliphatic ketones (*used in threefold excess*) were reductively aminated in good to high yields (49–92%) and with good to excellent enantioselectivities (90.5:9.5–98.5:1.5 er). In addition to *p*-anisidine (PMP–NH_2), heterocyclic amines could be used successfully with no loss in reaction efficiency or enantiocontrol. Notably, the efficient removal of water formed during the reaction seems to be important as the reaction rates and selectivities improved considerably upon using 5Å MS (Scheme 11.17).

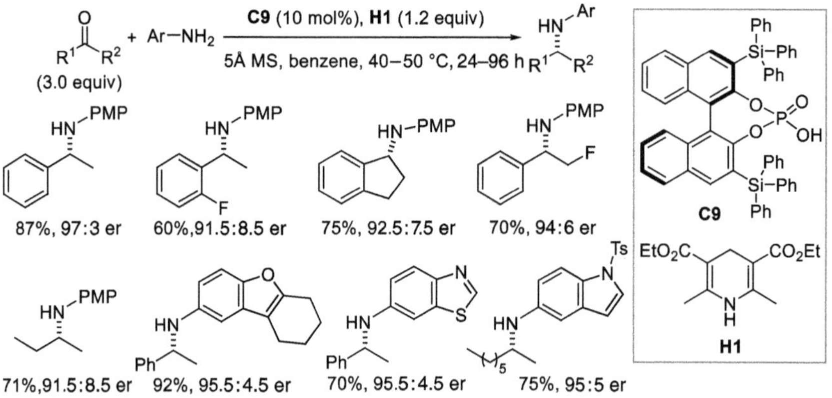

Scheme 11.17 Enantioselective reductive amination of ketones using Hantzsch ester.

In 2006, the List group discovered a unique CPA-catalyzed reductive amination of α-branched aldehydes via dynamic kinetic resolution (DKR) to produce the β-branched chiral amines (Scheme 11.18) [35]. A variety of racemic α-branched aldehydes were reductively aminated with aromatic amines, 5 mol% of (*R*)-TRIP,

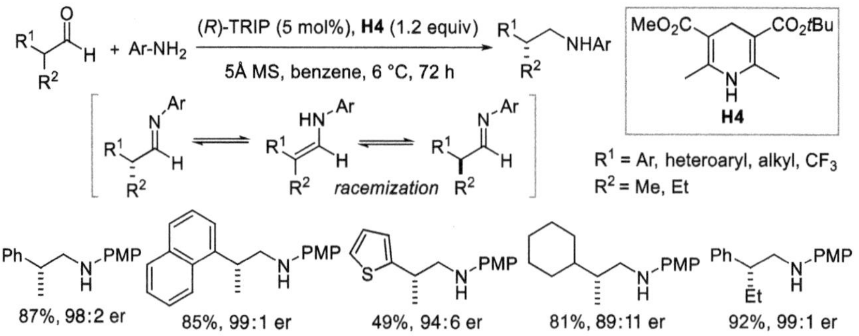

Scheme 11.18 CPA-catalyzed enantioselective reductive amination of α-branched aldehydes via DKR. Source: Based on Ref. [35].

and 1.2 equiv of modified Hantzsch ester **H4** in high yields (49–92%) and good to excellent enantioselectivities (90.5:9.5–99:1 er). Aliphatic aldehydes were also reductively aminated, however, with lower enantioselectivity. The key for the highly efficient enantiodifferentiating process via DKR is the fast equilibration between imine and enamine in the presence of the acid catalyst which enables a rapid racemization of the aldehyde substrate. The origin of the enantioselectivity of the reaction has been studied computationally by the Himo group [36].

In the same year, the Rueping group [37] found that benzoxazines, benzothiazines, and benzoxazinones could also be reduced under transfer hydrogenation conditions (Scheme 11.19). The best results are obtained with CPA **C12** and Hantzsch ester **H1** in $CHCl_3$ at ambient temperature. Remarkably, in some cases the catalyst loading could be lowered to 0.1 mol% without any significant decrease in yields or enantioselectivities. Further, the Rueping group [38] expanded the scope of the ATH reaction to benzodiazepinones and 3H-indoles and obtained the corresponding enantiomerically enriched saturated heterocycles.

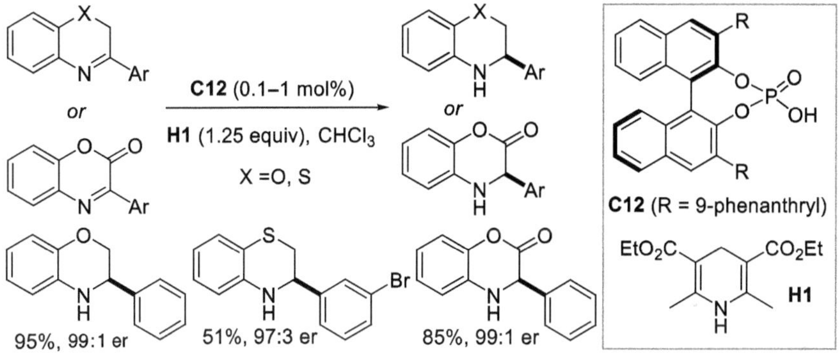

Scheme 11.19 CPA-catalyzed ATH of benzoxazines, benzothiazines and benzoxazinones with Hantzsch ester.

In the following years, several other groups further expanded these methodologies to CPA-catalyzed, Hantzsch ester-mediated reductions of α-imino esters [39], β,γ-alkynyl α-imino esters [40], enamides [41], racemic 2,3-dihydrobenzo[*b*][1,4]diazepines via DKR [42], *ortho*-hydroxyaryl alkyl N—H imines [43], and *N*-aryl *ortho*-hydroxybenzophenone imines [44] as well as *ortho*-hydroxybenzophenone N—H imines [45] to afford the corresponding amines in good to high yields and good to excellent enantioselectivities (Figure 11.5).

In 2010, the List group reported a CPA-catalyzed diastereo- and enantioselective reductive amination of racemic α-branched ketones via DKR (Scheme 11.20) [46]. An important feature of this method is its tolerance of a variety of different substituents while maintaining high yields and good to excellent diastereo- and enantioselectivity. Simple alkyl-substituted substrates are particularly reactive, requiring only 1 mol% catalyst loading, while sterically more-demanding substrates, as well as aromatic substrates, require slightly higher catalyst loadings. α,β-Unsaturated, α-branched ketones could be converted into the desired product

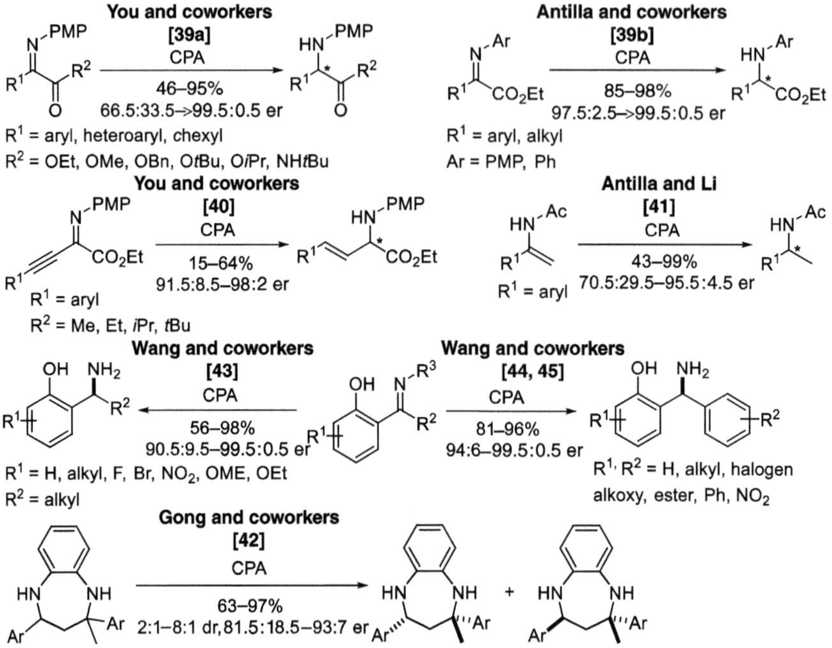

Figure 11.5 CPA-catalyzed ATH of variety of imines with Hantzsch ester.

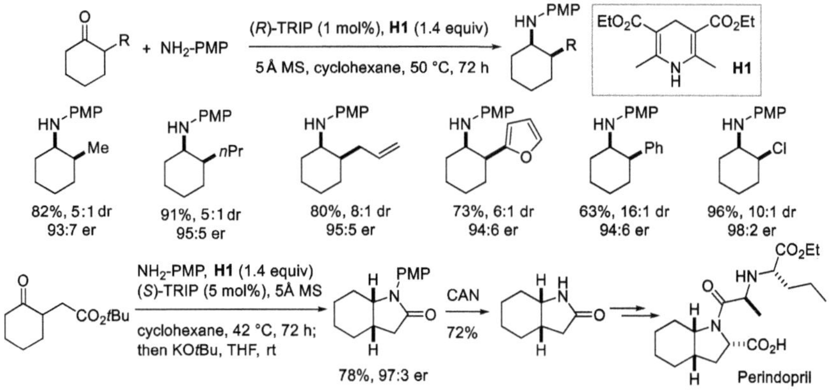

Scheme 11.20 CPA-catalyzed enantioselective reductive amination of α-branched ketones via DKR. Source: Based on Ref. [46].

in 54% yield and excellent selectivity (14:1 dr and 97:3 er). Based on this process, a key intermediate in the synthesis of Perindopril, a long-acting ACE inhibitor, was prepared with high enantioselectivity.

Typically, the reaction partners in organocatalytic ATH reactions of carbonyl compounds are aromatic amines. However, the removal of the installed aryl group, if desired, from the resulting amine usually requires relatively harsh conditions. In 2010, List and coworkers used benzylamine as the amine component in enantioselective reductive amination of ketones (Scheme 11.21) [47]. The method is noteworthy because the benzyl was introduced as the easily removable protecting

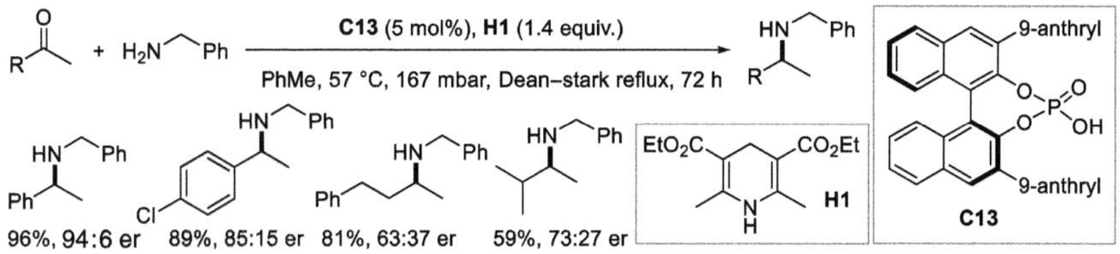

Scheme 11.21 CPA-catalyzed enantioselective reductive amination of ketones with benzyl amine. Source: Based on Ref. [47].

group. A Dean–Stark trap under refluxing conditions at reduced pressure was used to remove the water generated during the reaction. Aromatic ketones were reductively aminated in better enantioselectivity than aliphatic ones. Notable advantages of the method include the product purification via removal of Hantzsch pyridine by hydrolysis and the successful recycling of the catalyst without loss of the enantioselectivity. In the same year, Kumar et al. reported an alternative chiral acid in the enantioselective reductive amination of ketones with aryl amines [48].

In 2012, Zhou and coworkers [49] developed an interesting asymmetric reduction of racemic cyclic amide-derived hemiaminals with a Hantzsch ester (Scheme 11.22). In the presence of (*R*)-VAPOL-derived CPA **C14**, racemic 3-hydroxyisoindolinones generate an acyl-iminium ion intermediate by dehydration which can be efficiently reduced using Hantzsch ester **H3** to give chiral 3-substituted isoindolinones in modest yield (38–71%) and good to high enantioselectivity (80.5:19.5–97.5:2.5 er). An N-methylated substrate, lacking the proton required for iminium generation, did not undergo the reaction indicating the importance of the free N—H group. Later, they developed an asymmetric hydrogenation of imines and heteroaromatics using CPAs in combination with catalytic amounts of 9,10-dihydrophenanthridine (DHPD). DHPD was *in situ* regenerated from its oxidation product with hydrogen gas using a homogeneous ruthenium catalyst [49].

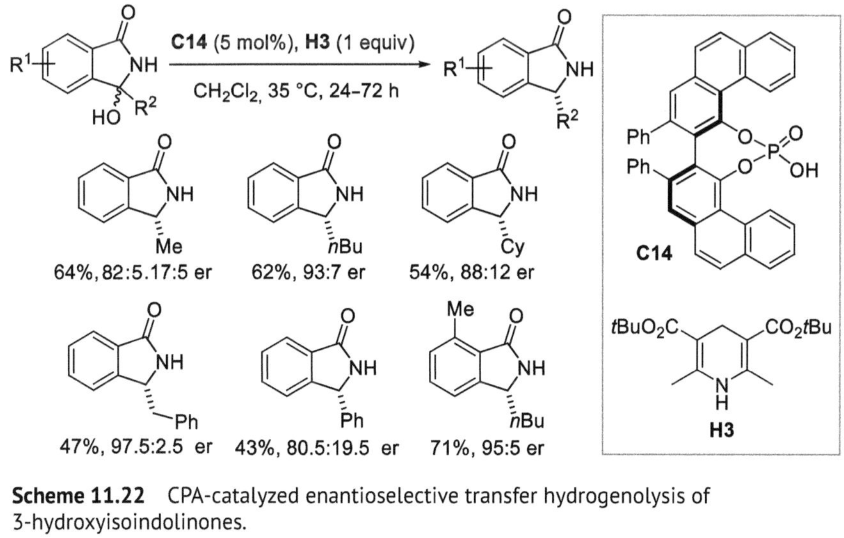

Scheme 11.22 CPA-catalyzed enantioselective transfer hydrogenolysis of 3-hydroxyisoindolinones.

In 2013, Toste and coworkers [50] discovered a single-operation deracemization of 3*H* indolines and tetrahydroquinolines. Phase separation enabled an asymmetric redox approach, in which a phosphoric acid catalyst and both oxidant and reductant were used together (Scheme 11.23). The solubility properties of the Hantzsch ester were altered to eliminate reagent quenching and thus to improve enantioenrichment of the isolated product. In a three-phase system, hexane/ether (or toluene)/water mixture, an oxopiperidinium salt was used to oxidize the cyclic amines and the resulting imines were subsequently reduced in the presence of 10 mol% of (*S*)-TRIP with modified Hantzsch ester **H5**. Various 3*H* indoline substrates were deracemized in excellent yields and enantioselectivties, while for tetrahydroquinolines, three or more equivalent of oxidant and reductant were required due to further oxidation of the intermediate dihydroquinoline to quinoline under the reaction conditions.

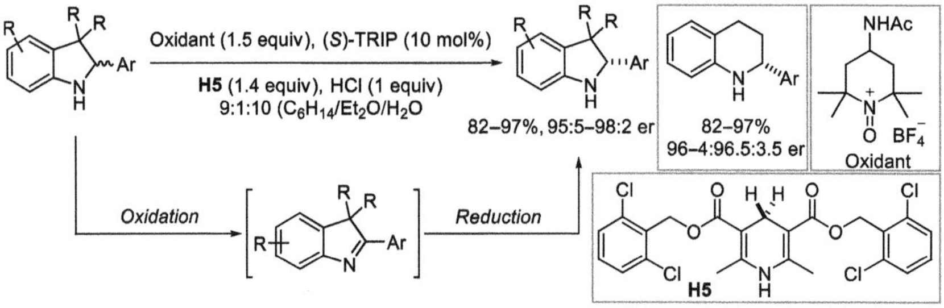

Scheme 11.23 Phase separation-enabled single-operation deracemization of 3*H*-indolines and tetrahydroquinolines.

In the same year, Akiyama and coworkers [51] developed the efficient oxidative kinetic resolution of 2-substituted indoline derivatives based on a CPA-catalyzed ATH from indoline to imine. By employing the imine as a hydrogen acceptor, several 2-substituted and 2,3-disubstituted indolines were resolved in high yields and excellent enantioselectivities; at the same time the chiral amines were also isolated in a nearly enantiopure form (Scheme 11.24). In continuation of the methodology, the same group recently reported a CPA-catalyzed oxidative kinetic resolution of cyclic secondary amines, including tetrahydroquinolines, using an aromatic imine as a hydrogen acceptor [52].

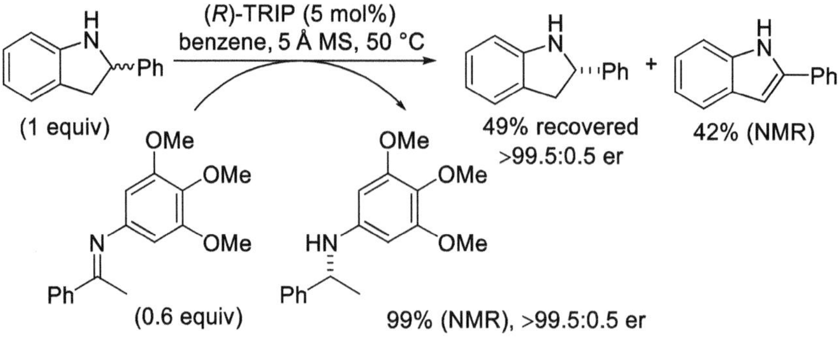

Scheme 11.24 CPA-catalyzed oxidative kinetic resolution of indolines.

In 2014, Tan and coworkers [53] reported a highly efficient strategy for the kinetic resolution of axially chiral BINAM derivatives based on an asymmetric reductive amination of aldehydes (Scheme 11.25). Bulky protecting groups in one of the amines of the racemic BINAM starting material were used to increase the rotation energy barrier and effective steric interaction, wherein Hantzsch ester **H1** or **H6** was used as the hydrogen source in combination with 10 mol% of CPA (**C15**). Functionalized aldehydes as well as many bulky protecting groups in one of the amines (sulfonyl, benzyl, 2-naphthyl methyl, Fmoc, amide, and thioamide) were well tolerated. The BINAM skeleton of the substrates could also be varied at the 6-position or 6,6′-positions (R^2 = R^3 = Br or TMS) with good yields and good to excellent enantioselectivities.

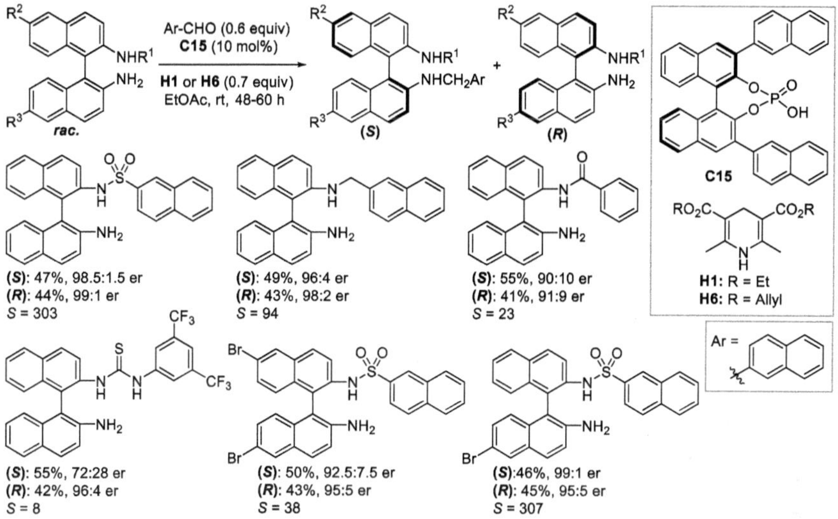

Scheme 11.25 CPA-catalyzed kinetic resolution of axially chiral BINAM derivatives. The selectivity factor was calculated as $S = \ln[(1-C)(1-\text{ee}\ (\mathbf{1}))]/\ln[(1-C)(1+\text{ee}\ (\mathbf{1}))]$, $C = \text{ee}(\mathbf{1})/(\text{ee}(\mathbf{3})+\text{ee}(\mathbf{1}))$. ee = enantiomeric excess.

In 2015, List and coworkers developed the ATH of *N*-alkyl imines with Hantzsch ester (Scheme 11.26) [54]. In this transformation, a more acidic chiral disulfonimide (DSI) derived from BINOL was used as catalyst to obtain high enantioselectivities. However, poor yields were initially obtained, presumably due to catalyst deactivation through salt formation with the highly basic amine product. The product inhibition pathway was successfully eliminated by *in situ* protection of the amine product with di-*tert*-butyl dicarbonate (Boc_2O). With the optimized condition, a variety of *N*-methyl imines were efficiently reduced with 5 mol% of (*R*)-DSI (**C16**) to afford the corresponding Boc-protected *N*-methyl amines with high yields and enantioselectivities. An additional strength of the process was demonstrated by utilizing substituents containing a variety of *N*-alkyl groups to provide the corresponding products in high yields and enantioselectivities. However, the lower enantioselectivities observed with aliphatic, C-substituted substrates illustrate a limitation of the process. The synthetic utility of this methodology

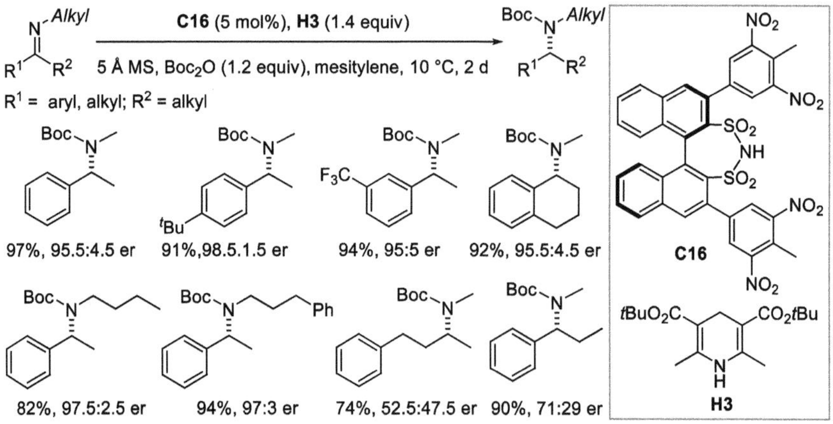

Scheme 11.26 Disulfonimide-catalyzed enantioselective transfer hydrogenation of *N*-alkyl imines with Hantzsch esters. Source: Based on Ref. [54].

was applied in the synthesis of several pharmaceutically relevant compounds, including (*S*)-Rivastigmine, NPS *R*-568·HCl, and (*R*)-Fendiline. Recently, Mazuela and coworkers [55] extended this methodology to *N*-alkyl aryl imino esters and obtained high yield and good to excellent enantioselectivities.

After the successful development of an *N*-alkyl imine reduction, the List group extended this concept to include the equally attractive but even more challenging N—H imines. However, investigation of the N—H imine reduction using Hantzsch ester **H3**, in the presence of 5 mol% of (*R*)-DSI **C17**, instead of enantioenriched primary amines, led to a remarkably direct and efficient entry to diastereo- and enantioenriched C_2-symmetric secondary amines (Scheme 11.27) [56]. Under optimized conditions, a variety of N—H imines efficiently underwent the reductive condensation to afford the corresponding C_2-symmetric secondary amines in good yields and with outstanding diastereo- and enantioselectivity. The authors proposed a catalytic cycle which is initiated by the protonation of the N—H imine by the chiral DSI catalyst. The resulting iminium ion pair **A** undergoes reduction with Hantzsch ester to give enantiomerically enriched primary amine salt **B**. Subsequently, this ammonium salt undergoes a rapid transimination with additional imine substrate to produce aminal **C** and, after liberation of ammonia, secondary iminium ion pair **D**. Finally, a subsequent reduction of intermediate **D** affords diastereo- and enantioenriched secondary amine product. A kinetic resolution in the second reduction step further contributes to the outstanding enantioselectivities observed.

In 2016, Akiyama and coworkers [57] developed a new strategy for the oxidative kinetic resolution of racemic indolines using a salicylaldehyde derivative as a pre-resolving agent. Employing CPA-catalyst **C18**, the *in situ* generated iminium intermediate was hydrogenated by the same enantiomer of indoline to afford another enantiomer of indoline by a self-redox mechanism. Products were obtained in good yields and with excellent enantioselectivities (Scheme 11.28).

In 2015, Akiyama and coworkers [58] reported a CPA-catalyzed intramolecular reductive amination under ATH condition and applied this process in the highly enantioselective synthesis of fused piperidine and pyrrolidine derivatives with

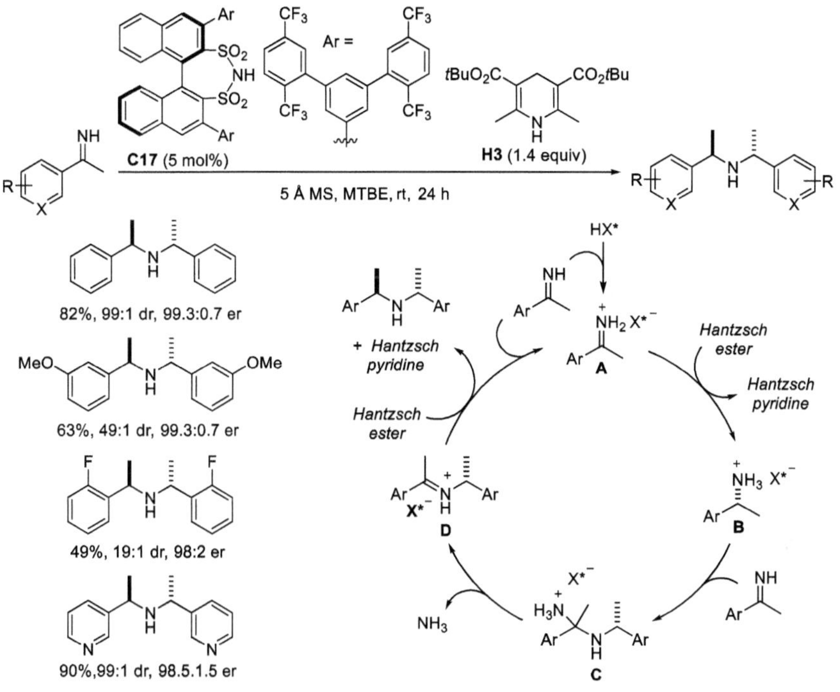

Scheme 11.27 Disulfonimide-catalyzed asymmetric reductive condensation of N–H imines and proposed catalytic cycle. Source: Modified from Ref. [56].

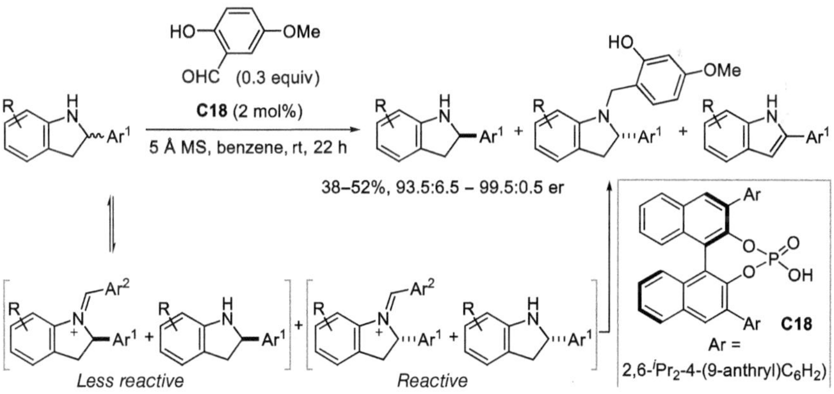

Scheme 11.28 Kinetic resolution of racemic indolines based on a self-redox reaction.

all-carbon stereogenic center. In 2016, Bousquet, Pélinski, and coworkers [59] developed an interesting strategy for ATH reactions of benzoxazines, where the reducing agent, the Hanztsch ester, was generated *in situ* in the presence of a catalytic amount of CPA.

Besides Hantzsch esters, benzothiazolines can also be employed as hydrogen source in ATH reactions. In 2009, Akiyama and coworkers [60] introduced benzothiazolines as a highly efficient hydrogen source for the CPA-catalyzed ATH reaction of imines to produce chiral amines in high yield and enantioselectivities

(Scheme 11.29). The best results were obtained with 2 mol% of (*R*)-TRIP in combination with 2-naphthyl-substituted benzothiazoline **BTZ1**. The group also showed that *in situ* generated benzothiazoline from 2-naphthalenecarbaldehyde and 2-aminophenol could accomplish the reduction with similar yield and enantioselectivity.

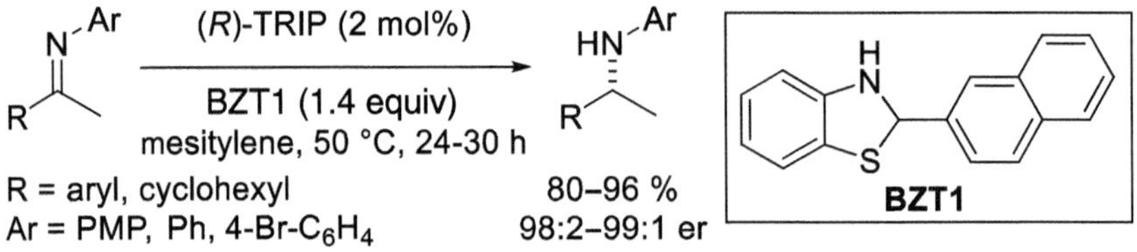

Scheme 11.29 The first enantioselective transfer hydrogenation of imines with BZT.

Further utilization of benzothiazolines as the hydrogen source in the CPA-catalyzed ATH reaction was explored by the same group in the reduction of α-imino esters [61], trifluoromethyl imines [62], difluoromethylimines [63], and in the reductive amination of trifluoromethyl ketones [62] and aliphatic ketones [64] with aryl amines in high yields and enantioselectivities. Akiyama and coworkers also demonstrated that the transfer hydrogenation reaction could be applied for the deuteration of imines [65]. In 2013, Jia and coworkers [66] also utilized a benzothiazoline in combination with CPA for the hydrogenolysis of racemic 3-aryl-3-hydroxyisoindolin-1-ones. In 2014, Akiyama and coworkers [67] further explored ATH of imines derived from propiophenone derivatives and systematically compared the performance of benzothiazoline vs. Hantzsch ester as reductants in the presence of (*R*)-TRIP. In comparison to Hantzsch esters, the benzothiazoline was found to give significantly higher enantioselectivities with this substrate class (Scheme 11.30).

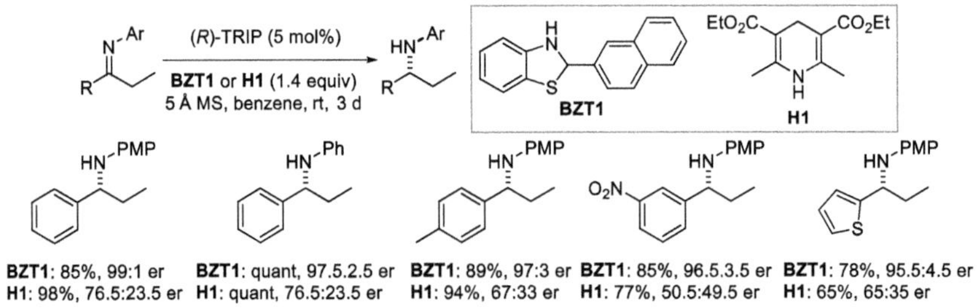

Scheme 11.30 Comparison of **BZT1** vs. **H1** as reductants in the imine reduction.

Akiyama and coworkers [68] also demonstrated that indoline could be used as hydrogen source in CPA-catalyzed ATH reactions of aryl imines. A variety of aryl imines were reduced in the presence of 5 mol% of (*R*)-TRIP and 2.5 equiv of 2-aryl indoline in good to excellent yields and enantioselectivities (Scheme 11.31). Further, utilizing similar reaction conditions, the reductive amination of aliphatic ketones with aryl amines was performed in high enantioselectivities. The authors

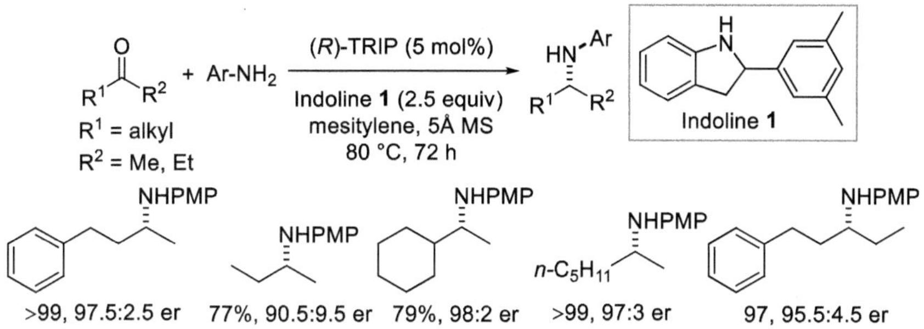

Scheme 11.31 Reductive amination of aliphatic ketones with indoline.

also showed that 2-aryl indoline could be regenerated by the reduction of 2-aryl indole after the ATH.

In the same year, Shimizu and coworkers [69] reported the enantioselective reduction of iminocyclobutenones using a CPA as catalyst and benzothiazoline as the reductant. The obtained chiral aminocyclobutenones when heated with different bases produced both *trans*- and *cis*-β-lactams in high yields and enantioselectivities. Later, Guo and coworkers [70] developed a CPA-catalyzed ATH of α-keto imines using benzothiazoline as the hydrogen source to produce chiral α-amino ketones in high yields and enantioselectivities. The chiral α-amino ketones could also be prepared by reductive amination of 1,2-diketones in high yields and with excellent regio- and enantioselectivities.

In an extension of the work reported earlier by the Gong group [42] for the synthesis of tetrahydrobenzodiazepines via phosphoric acid-catalyzed DKR of dihydrobenzodiazepines using Hantzsch ester, the Akiyama group [71] in 2016 used the combination of phosphoric acid **C19** and benzothiazoline reductant. Tetrahydrobenzodiazepine products bearing two tertiary centers were produced in excellent diastereo- and enantioselectivities (Scheme 11.32).

Recently, the Akiyama group [72] reported a CPA-catalyzed chemoselective ATH of trifluoromethyl alkynyl imines using benzothiazolines as the reductant, to give α-trifluoromethyl propargylamines in high yields and with excellent enantioselectivities. Subsequently, the Peng [73] group independently reported the ATH of fluorinated alkynyl imines. The synthetic utility of the method was demonstrated in the synthesis of chiral biologically active 2-(trifluoromethyl)-1,2-dihydroquinoline.

11.4 Cascade Reactions

A cascade reaction could be defined as a chemical transformation where several new carbon–carbon bonds and stereogenic centers are created in a highly controlled fashion from readily available starting materials in one pot. Over time, cascade reactions have been utilized as a powerful synthetic approach to build complex molecules. The first applications of ATH with Hantzsch esters as reductants in cascade processes were reported independently by List and coworkers [74] and by MacMillan and

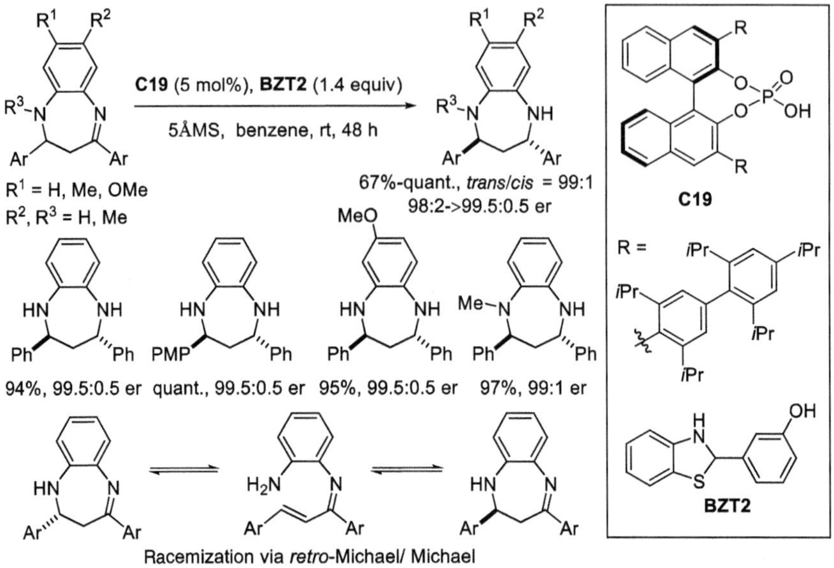

Scheme 11.32 CPA-catalyzed DKR of dihydrobenzodiazepines and racemization pathway.

coworkers [75] at the same time in 2005. The List group had used an imidazolidinone organocatalyst for a tandem conjugate reduction–Michael cyclization of enal enones yielding five- and six-membered carbacycles in high yields and enantiomeric excess (Scheme 11.33). Here the *in situ* generated enamine intermediate was trapped by the enone functionality of the molecule to yield the desired product. MacMillan used the same strategy, employing α,β-unsaturated aldehydes with Hantzsch esters and using electrophilic chlorine or fluorine sources to obtain chiral aldehydes with two stereogenic centers with excellent enantio- and diastereoselectivities (Scheme 11.34). By judiciously selecting the amine catalyst, the configuration of these two centers can be controlled. It was reported that a combined iminium–enamine dual catalytic cycle enabled a transfer hydrogenation first via iminium catalytic cycle, followed by an electrophilic halogenation via enamine catalytic cycle.

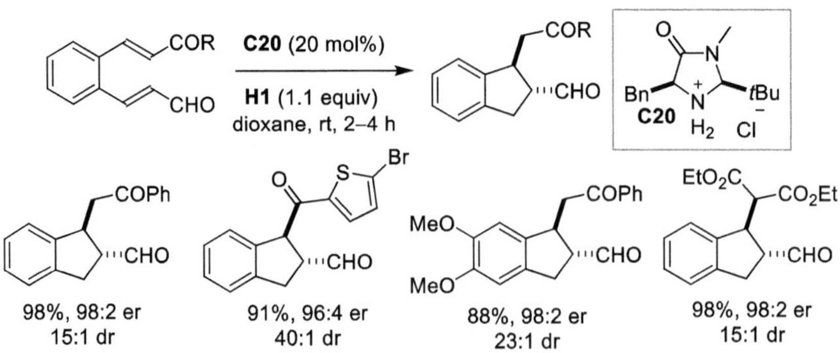

Scheme 11.33 Organocatalytic asymmetric reductive Michael cyclization.

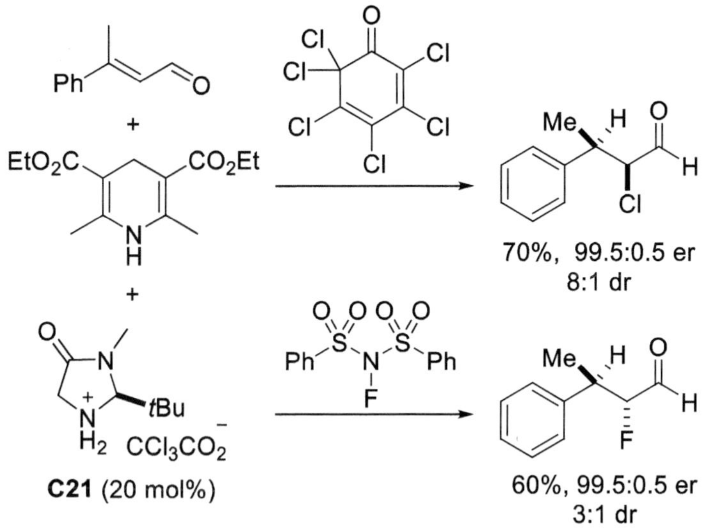

Scheme 11.34 Chiral amine-catalyzed domino reductive chlorination and fluorination.

Meanwhile, Brønsted acid catalysis has been introduced by List and coworker [76] in a cascade sequence where the acid catalyst is compatible with amine substrates and products. They have shown the concept of combining both enamine and iminium catalysis with asymmetric Brønsted acid catalysis to obtain chiral *cis*-3-substituted cyclohexylamines with excellent enantiomeric excess (Scheme 11.35). The cascade route proceeded via an aldolization–dehydration conjugate reduction–reductive amination that is catalyzed by a chiral Brønsted acid and accelerated by the achiral amine substrate.

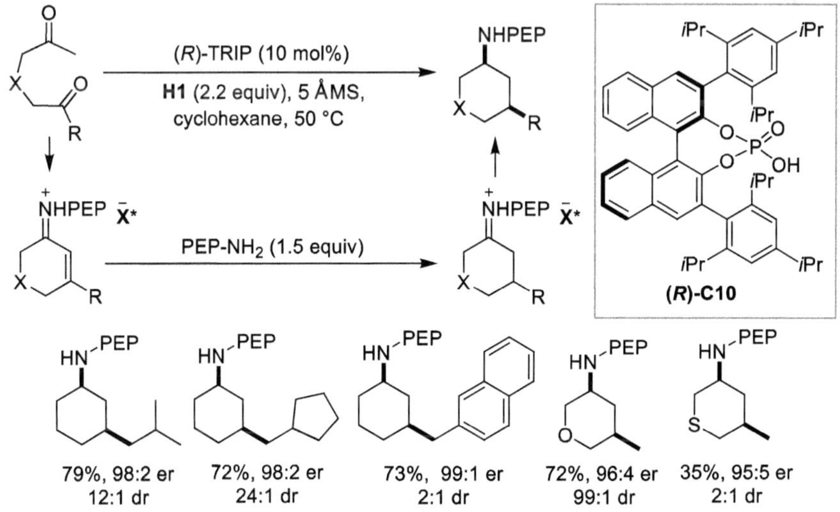

Scheme 11.35 Organocatalytic asymmetric reaction cascade to substituted cyclohexylamines.

Later, Rueping and coworkers [77] have demonstrated cascade sequences involving ATH reactions with Hantzsch esters to obtain tetrahydropyridines and azadecalinones with high enantiomeric excess via Brønsted acid catalysis from readily available enamines and α,β-unsaturated ketones (Scheme 11.36). This process is a great example in which the same Brønsted acid catalyzes six different reactions in a domino fashion with high level of enantiocontrol.

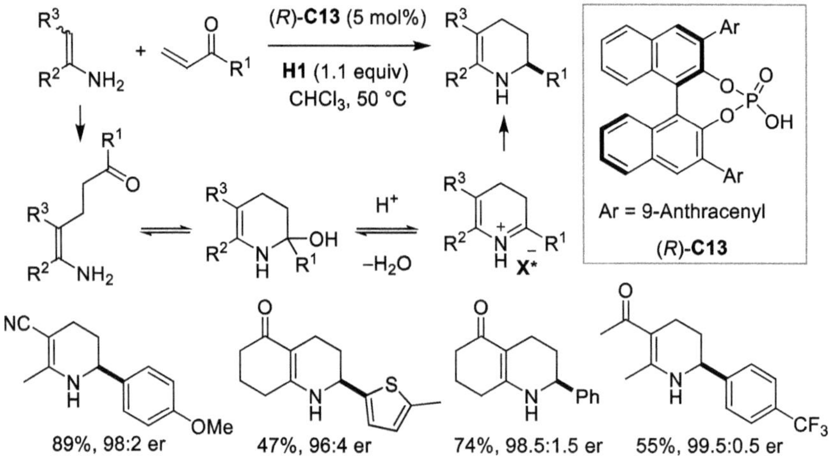

Scheme 11.36 Organocatalytic asymmetric reaction cascade to tetrahydropyridines.

Gong and coworkers [78] have used a combination of an achiral Lewis acid and a chiral Brønsted acid in a cascade method to obtain tetrahydroquinoline derivatives with three contiguous stereogenic centers in high stereoselectivity (>20:1 dr, 98.5:1.5 er) (Scheme 11.37). After these initial accounts, several cascade routes were reported to build complex molecules, including tetrahydroquinoline

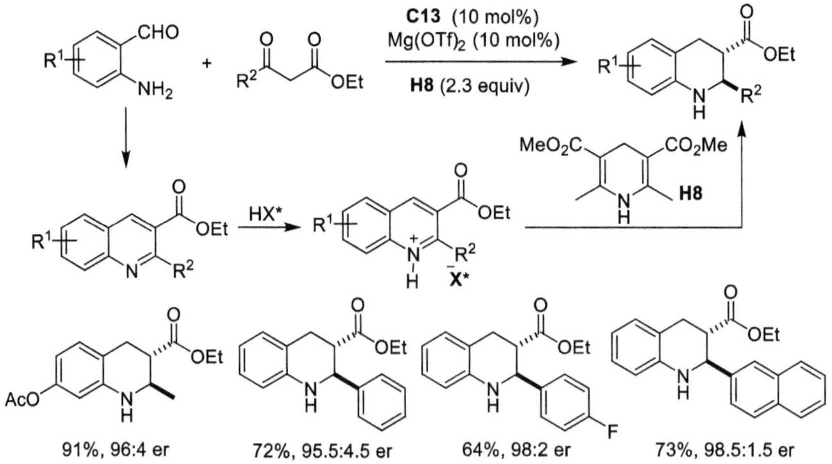

Scheme 11.37 Organocatalytic asymmetric reaction cascade to tetrahydropyridines.

[79], benzofused quinolizidines and indolizidines [80], cyclohexane carbaldehydes [81], spirotetrahydroquinolines [82], two-substituted tetrahydroquinolines [83], pyrrolidines, hexahydropyrrolizines, and octahydroindolizines [84], and tetrahydroquinoxalines and dihydroquinoxalinones [85].

You and Wang [86] developed a hydrogenative dearomatization of pyridines and asymmetric aza-Friedel–Crafts sequence using pyrroles as the nucleophile to obtain chiral piperidine scaffolds (Scheme 11.38). The first step is the protonation of the pyridine by a chiral phosphoric acid catalyst to generate a pyridinium salt, which is then reduced by 1,4-hydride transfer from the Hantzsch ester generating dihydropyridine intermediate. Then this intermediate upon protonation generates the iminium ion in the presence of phosphoric acid, and subsequently the iminium is trapped by a nucleophile in a aza-Friedel–Crafts manner.

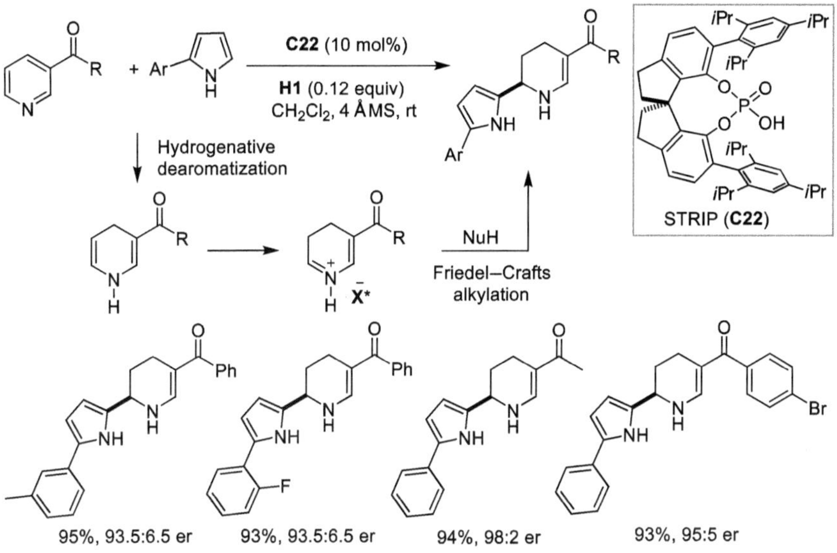

Scheme 11.38 Organocatalytic asymmetric domino hydrogenative dearomatization/aza-Friedel–Crafts reaction of pyridines.

Shi and coworkers [87] achieved an organocatalytic asymmetric arylative dearomatization of indoles through a tandem process involving 2,3-disubstituted indoles and quinone imine ketals (Scheme 11.39). First, the indole is arylated in the presence of a Brønsted acid and quinone imine ketals, and then the Hantzsch ester is added and an ATH reaction takes place. Enantiomerically pure indole derivatives that bear an all-carbon quaternary stereogenic center can be generated in high yields and excellent stereoselectivities (all dr >95:5, up to 99.5:0.5 er).

In 2015, Luo and coworkers [88] developed highly efficient Michael addition–protonation reactions of malononitriles to α-substituted vinyl ketones using chiral primary–tertiary diamine derived from L-phenylalanine as an organocatalyst in the presence of triflic acid. This work inspired them to develop a domino process which started with the reduction of a 2-benzylidenemalononitrile by a Hantzsch ester. They envisioned that Hantzsch esters may serve as a chemoselective hydride source

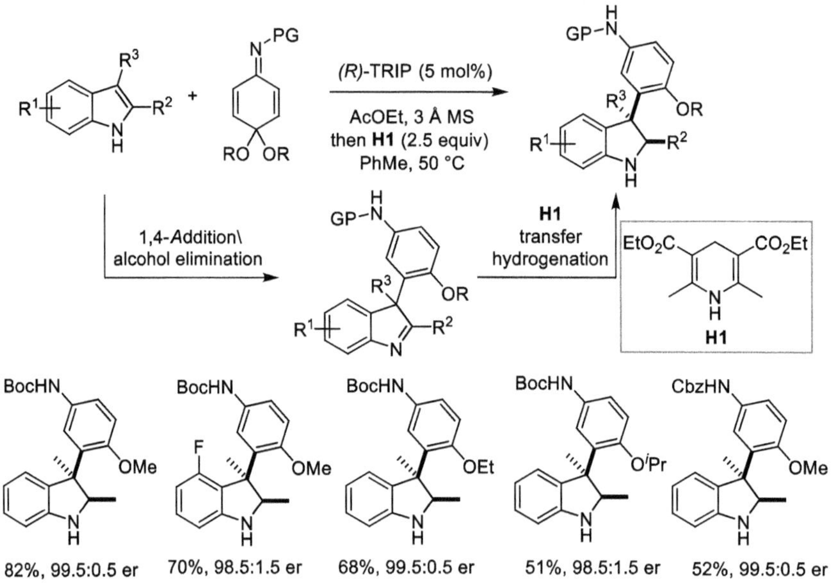

Scheme 11.39 Organocatalytic asymmetric arylative dearomatization of 2,3-disubstituted indoles.

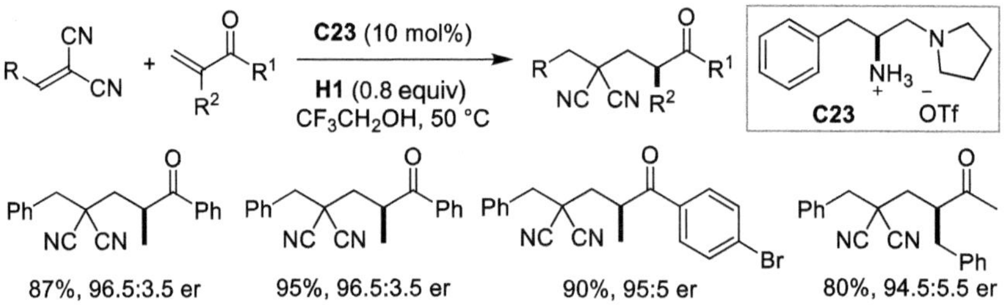

Scheme 11.40 Reductive Michael addition of malononitrile to α-substituted vinyl ketone.

to reduce 2-benzylidenemalononitrile and its derivatives, which would facilitate a one-pot enantioselective tandem reduction–Michael addition–protonation reaction with the same catalytic system (Scheme 11.40). Later, using the same concept, they developed a domino reduction–Michael addition protonation reaction between α-substituted vinyl ketones and alkylidene Meldrum acid derivatives [89].

List and Michrowska [90], building on their reductive Michael cyclizations, utilized a cascade strategy in the highly enantio- and diastereoselective total synthesis of ricciocarpin A where the asymmetric reductive Michael addition followed by Tishchenko reaction was the key feature of the design (Scheme 11.41). In the presence of catalyst **C20** and Hantzsch ester **H9**, the substrate was subjected to reductive Michael conditions to give a ketoaldehyde intermediate, which epimerizes in the presence of Samarium triisopropoxide followed by the Tishchenko reaction which occurred solely on the *trans*-ketoaldehyde to give the desired natural product. Several ricciocarpin analogues were then synthesized using this cascade sequence.

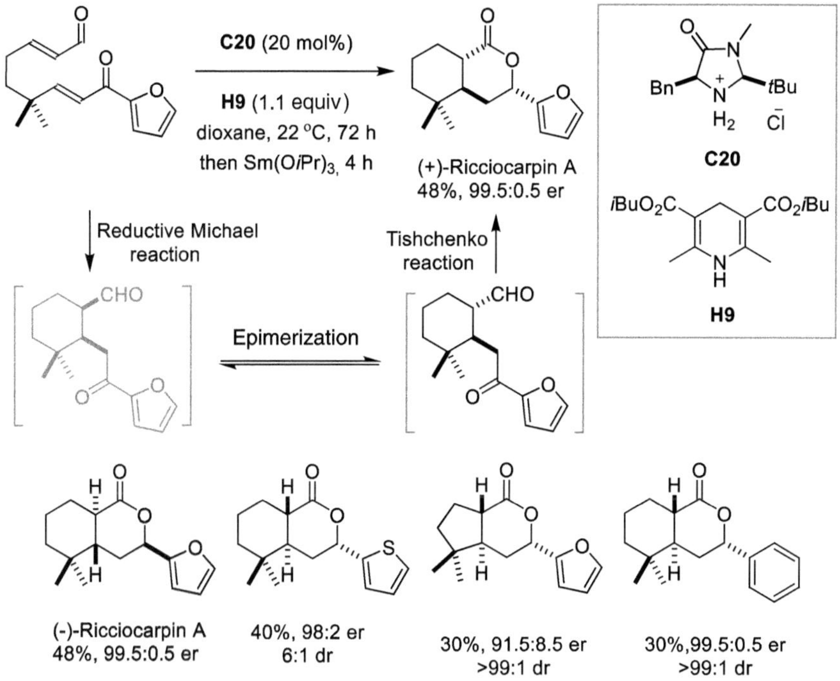

Scheme 11.41 Total synthesis of (+)-Ricciocarpin A and analogues via reductive Michael–Tishchenko cascade.

11.5 Dearomatization

Aromatic compounds are fundamental molecules in organic chemistry and are typically highly stable. As a consequence of this stability, high temperature and pressure are normally required to disrupt the aromaticity. One elegant strategy to build value-added products from arenes is their hydrogenative dearomatization. In principle, every aromatic compound can undergo such a transformation, but, due to the reduced resonance stabilization energy of heteroaromatics, such as furans, pyrroles, or pyridines, they are more susceptible to dearomative transformations than arenes. Importantly, the products of such reactions are useful in the pharmaceutical and agrochemical industries.

Rueping et al. [91] first showed that CPAs in combination with Hantzsch esters can be used to reduce quinolines to tetrahydroquinolines (Scheme 11.42). It was observed that nonpolar solvents were essential to obtain a high asymmetric induction. The proposed mechanism for this transformation consists of a cascade of reactions. First is the protonation of quinoline promoted by the Brønsted acid to generate an iminium, followed by a 1,4-addition of hydride to furnish an enamine. In the second cycle, protonation of the enamine to an iminium ion occurs, which is then subjected to another hydride transfer to yield the enantiomerically enriched tetrahydroquinoline and pyridine by-product.

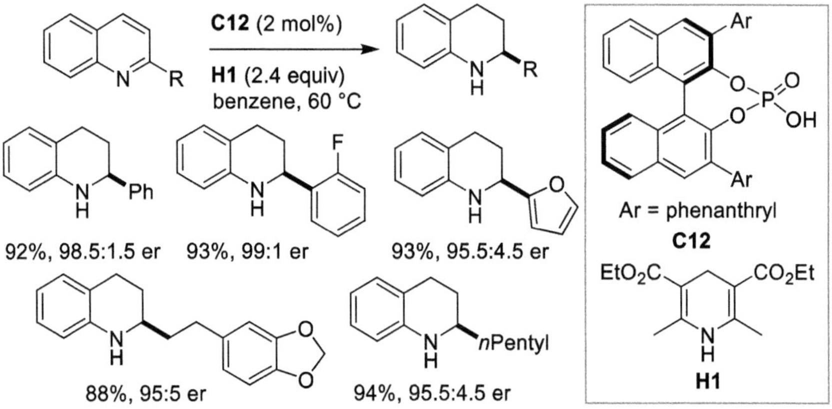

Scheme 11.42 Organocatalytic asymmetric transfer hydrogenation of quinolines.

After this work, many groups have used this strategy to achieve full or partial enantioselective hydrogenation of several heteroaromatic systems: trisubstituted pyridines [93], 2,3-disubstituted quinolines [94, 95], 3-substituted quinolones [95], 2- and 2,9-substituted 1,10-phenanthrolines [96], quinoxalines and the nonaromatic quinoxalinones [97], 6-fluoro-2-methylquinoline and 6,7-difluoro-3-methyl-3,4-dihydro-2*H*-benzo[1,4]oxazine [98], 4-substituted quinolines [99], 2,3-disubstituted nitroquinolines [100], 3-(trifluoromethyl) quinolones [101], and 3-trifluoromethyl thioquinolines [102] in high yields and enantiomeric excess. In 2014, Zhou and coworkers [103] envisioned that quinolin-3-amines can also be reduced using the combination of CPA and Hantzsch ester. They achieved the highest enantioselectivity using a mixed solvent system of 1,4-dioxane and CH_2Cl_2 with a ratio of 2:1 and (*S*)-TRIP as catalyst. In the same year, Pelinski and coworkers [104] demonstrated the partial hydrogenation of lactone-fused quinolones to give aza-podophyllotoxin derivatives using BINOL-based CPA (**C9**). Another development on the quinoline skeleton was the transfer hydrogenation of 1,2-dihydroquinoline to give 2,2,4-trimethyltetrahydroquinolines via the formation of highly reactive and unstable aza-*o*-xylylene (AOX)-type intermediates (Scheme 11.43) [92].

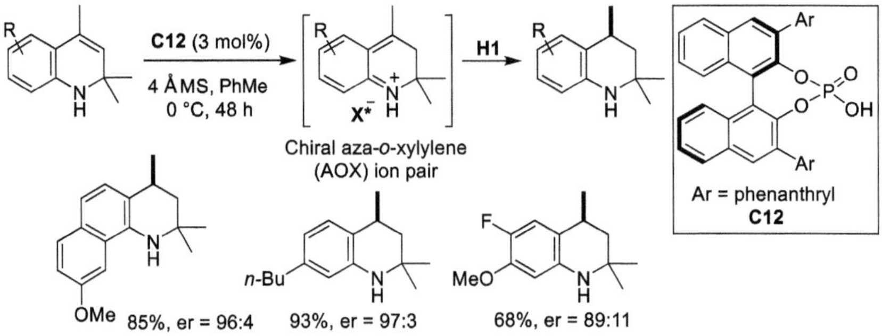

Scheme 11.43 Asymmetric transfer hydrogenation of 1,2-dihydroquinoline through the formation of AOX species. Source: Based on Ref. [92].

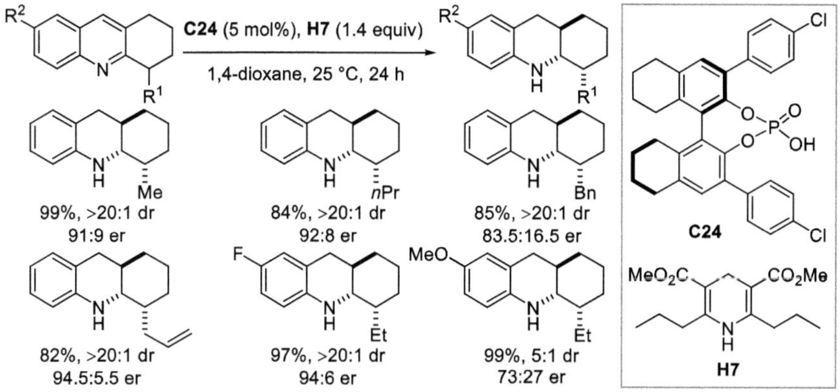

Scheme 11.44 Enantioselective synthesis of tetrahydroquinoline derivatives via DKR of 2,3-disubstituted quinolines.

Zhou and coworkers [105] developed an ATH process for the reduction of 2,3-disubstituted quinolines via DKR, which leads to the highly selective formation of chiral tricyclic amines with three contiguous stereogenic centers (Scheme 11.44). Various substituted quinolones were reduced using modified Hantzsch ester **H7** in combination with CPA (**C24**) in high yields and diastereo- and enantioselectivities.

Miller and Shugrue [106] developed a new type of CPA scaffold containing phosphothreonine peptides, which lacks C_2 symmetry in contrast to the privileged BINOL-based phosphoric acids. They used this new catalyst motif for enantioselective transfer hydrogenations of previously underexplored 8-aminoquinolines (Scheme 11.45). The mechanism of the reaction was studied by ^{1}H NMR, which revealed that the intermolecular interactions (through H-bonding) between the catalyst and substrates and also a rigid β-turn in the catalyst are responsible for the asymmetric induction.

Shi et al. [107] explored the transfer hydrogenation of isoquinoline using a chiral anion metathesis strategy to obtain protected 1-substituted 1,2-dihydroisoquinolines

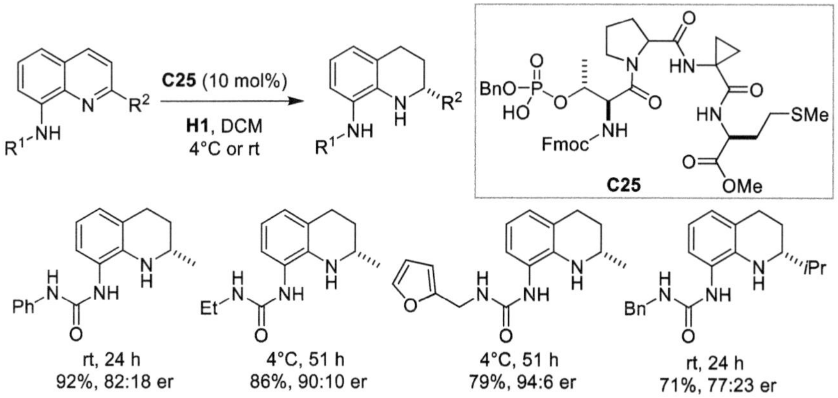

Scheme 11.45 Asymmetric transfer hydrogenation of 8-aminoquinolines.

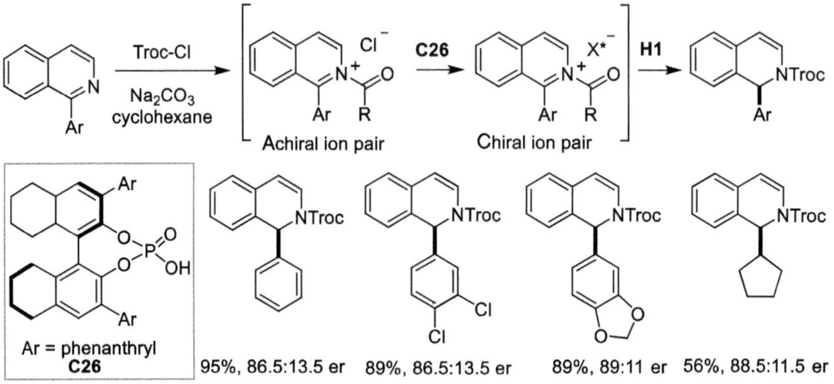

Scheme 11.46 Asymmetric transfer hydrogenation of 2-substituted isoquinolines.

with high yield and up to 89.5:10.5 enantiomeric ratio (Scheme 11.46). In the presence of Hantzsch ester and trichloroethylchloroformate and using an H_8-BINOL-derived CPA catalyst (**C26**), *N*-acylisoquinolinium chloride salt was generated which then undergoes anion metathesis to form a chiral contact ion pair which aids the enantioselectivity.

Between 2014 and 2016, developments have been made toward new CPAs for the ATH of heteroatomic systems (Figure 11.6). In 2014, Marinetti and coworkers [108] developed phosphoric acids with planar chiral paracyclophane scaffolds (**C27**) to promote the organocatalytic ATH of α-arylquinolines with up to 90% enantiomeric excess. Then, a new approach was introduced by Voituriez and Guinchard [109] based on the use of thiophosphonic acids possessing both a chiral backbone and a chiral phosphorus functionality. They synthesized P^V thiophostone derivatives of tri-*O*-benzyl-D-glucal and tested the ATH of 2-phenylquinoline. Shi and coworkers [110] developed a new type of phosphoric acid bearing a 5,5′-bitetralone scaffold which proved to be a highly effective catalyst in the ATH of 2-phenylquinoline and the Friedel–Crafts reaction of 2,2,2-trifluoroacetophenone. The X-ray structure of the catalyst showed that the torsion angle between the

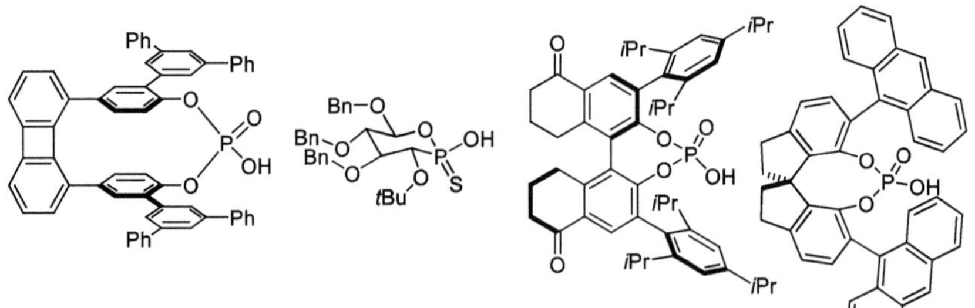

C27
Betzer, Marinetti and coworkers [108]

C28
Voituriez, Guinchard and coworkers [109]

C29
Shi and coworkers [110]

C30
Shi and coworkers [111]

Figure 11.6 Catalysts developed for the ATH of heteroaromatics. Source: Based on Ref. [108].

two phenyl planes of tetralone in **C29** is 62.00°, which is larger than that in TRIP (49.78–51.29°) [111], indicating a larger cavity which could be beneficial for other transformations. In the ATH of 2-phenylquinoline it was shown to be slightly more active than either TRIP or its H_8-analogue, and there was full conversion in less than 12 h at room temperature using only 0.2 mol% of the catalyst. The product was obtained in 99:1 enantiomeric ratio. Shi and coworkers [112] developed 1,1′-spirobiindane-7,7′-diol-derived CPAs as the catalyst for the ATH of heterocyclic compounds. In the ATH of 2-arylquinolines, tetrahydroquinolines were obtained in up to 93% yield with 1 mol% of the catalyst. 1,4 Benzoxazines could be reduced in very high yields and enantiomeric ratio varying between 95.5:4.5 and 99.5:0.5.

11.6 Conclusions

In this chapter, we have briefly summarized the utility of Hantzsch esters and related organic compounds in organocatalytic reductions of C=C and C=N bonds. The use of Hantzsch esters has occasionally been criticized for its lack of atom economy. While this criticism is obviously justified, an overall judgment of the best method for any given asymmetric reduction often depends on various factors. These may include the scale, reagent and catalyst costs, and safety issues. More importantly, if a given problem simply cannot be solved with hydrogen gas as the much more atom economic reagent, Hantzsch ester methods may find their utility, even on a technical scale.

References

1 Sato, Y., Inaba, A., Uemasu, I., and Kushiyama, S. (1986). *Fuel Process. Technol.* 14: 67–78.
2 (a) Gladiali, S. and Alberico, E. (2006). *Chem. Soc. Rev.* 35: 226–236. (b) Wang, D. and Astruc, D. (2015). *Chem. Rev.* 115: 6621–6686.
3 (a) Knoevenagel, E. and Bergdolt, B. (1903). *Ber. Dtsch. Chem. Ges.* 36: 2857–2860. (b) Wieland, H. (1912). *Ber. Dtsch. Chem. Ges.* 45: 484–493.
4 (a) Haddad, Y.M.Y., Henbest, H.B., Husbands, J., and Mitchell, T.R.B. (1964). *Proc. Chem. Soc., London*: 361. (b) Trocha-Grimshaw, M.J. and Henbest, H.B. (1968). *Chem. Commun.*: 1035–1036. (c) McPartlin, M. and Mason, R. (1967). *Chem. Zentralbl.* 138: 1.
5 Muñiz, K. (2005). *Angew. Chem. Int. Ed.* 44: 6622–6627.
6 Brieger, G. and Nestrick, T.J. (1974). *Chem. Rev.* 74: 567–580.
7 Nakamura, M., Bhatnagar, A., and Sadoshima, J. (2012). *Circ. Res.* 111: 604–610.
8 For selected reviews on organocatalytic ATH, see:(a)Ouellet, S.G., Walji, A.M., and Macmillan, D.W.C. (2007). *Acc. Chem. Res.* 40: 1327–1339. (b) Kampen, D., Reisinger, C.M., and List, B. (2010). *Top. Curr. Chem.* 291: 395–456. (c) Rueping, M., Dufour, J., and Schoepke, F.R. (2011). *Green Chem.* 13: 1084–1105. (d) de Vries, J.G. and Mrsic, N. (2011). *Catal. Sci. Technol.* 1:

727–735. (e) Shi, F. and Gong, L.-Z. (2012). *Angew. Chem., Int. Ed.* 51: 11423–11425. (f) Zheng, C. and You, S.-L. (2012). *Chem. Soc. Rev.* 41: 2498–2518. (g) McSkimming, A. and Colbran, S.B. (2013). *Chem. Soc. Rev.* 42: 5439–5488. (h) Bhadury, P.S. and Sun, Z. (2014). *Curr. Org. Chem.* 18: 127–150. (i) Foubelo, F. and Yus, M. (2015). *Chem. Rec.* 15: 907–924. j) Faisca Phillips, A.M. and Pombeiro, A.J.L. (2017). *Org. Biomol. Chem.* 15: 2307–2340.

9 (a) Hantzsch, A. (1881). *Ber. Dtsch. Chem. Ges.* 14: 1637–1638. (b) Hantzsch, A. (1882). *Justus Liebigs Ann. Chem.* 215: 1–82.

10 (a) Mauzerall, D. and Westheimer, F.H. (1955). *J. Am. Chem. Soc.* 77: 2261–2264. (b) Abeles, R. and Westheimer, F.H. (1958). *J. Am. Chem. Soc.* 80: 5459–5450.

11 Although first use of Hantzsch ester as a reducing agent in the reduction of maleic anhydride to succinic anhydride was demonstrated by Mumm and Diederichsen in 1939, they proposed the wrong structure for the Hantzsch ester as 1,2-dihydropyridine rather than a 1,4-dihydropyridine, see:Mumm, O. and Diederichsen, J. (1939). *Justus Liebigs Ann. Chem.* 538: 195–236.

12 Yang, J.W., Hechavarria Fonseca, M.T., and List, B. (2004). *Angew. Chem., Int. Ed.* 43: 6660–6662.

13 Yang, J.W., Hechavarria Fonseca, M.T., Vignola, N., and List, B. (2005). *Angew. Chem., Int. Ed.* 44: 108–110.

14 Ouellet, S.G., Tuttle, J.B., and MacMillan, D.W.C. (2005). *J. Am. Chem. Soc.* 127: 32–33.

15 (a) Tuttle, J.B., Ouellet, S.G., and MacMillan, D.W.C. (2006). *J. Am. Chem. Soc.* 128: 12662–12663. (b) Martin, N.J.A. and List, B. (2006). *J. Am. Chem. Soc.* 128: 13368–13369.

16 Mayer, S. and List, B. (2006). *Angew. Chem., Int. Ed.* 45: 4193–4195.

17 (a) Akiyama, T., Itoh, J., Yokota, K., and Fuchibe, K. (2004). *Angew. Chem. Int. Ed. Engl.* 43: 1566–1568. (b) Uraguchi, D. and Terada, M. (2004). *J. Am. Chem. Soc.* 126: 5356–5357.

18 Seayad, J., Seayad, A.M., and List, B. (2006). *J. Am. Chem. Soc.* 128: 1086–1087.

19 Akagawa, K., Akabane, H., Sakamoto, S., and Kudo, K. (2008). *Org. Lett.* 10: 2035–2037.

20 Ramachary, D.B., Sakthidevi, R., and Reddy, P.S. (2013). *RSC Adv.* 3: 13497–13506.

21 Van Arman, S.A., Zimmet, A.J., and Murray, I.E. (2016). *J. Org. Chem.* 81: 3528–3532.

22 Gutierrez, O., Iafe, R.G., and Houk, K.N. (2009). *Org. Lett.* 11: 4298–4301.

23 Martin, N.J.A., Ozores, L., and List, B. (2007). *J. Am. Chem. Soc.* 129: 8976–8977.

24 Martin, N.J.A., Cheng, X., and List, B. (2008). *J. Am. Chem. Soc.* 130: 13862–13863.

25 Schneider, J.F., Lauber, M.B., Muhr, V. et al. (2011). *Org. Biomol. Chem.* 9: 4323–4327.

26 (a) Martinelli, E., Vicini, A.C., Mancinelli, M. et al. (2015). *Chem. Commun.* 51: 658–660. (b) Massolo, E., Benaglia, M., Orlandi, M. et al. (2015). *Chem. Eur. J.* 21: 3589–3595.

27 Ferraro, A., Bernardi, L., and Fochi, M. (2016). *Adv. Synth. Catal.* 358: 1561–1565.

28 Wang, Z., Ai, F., Wang, Z. et al. (2015). *J. Am. Chem. Soc.* 137: 383–389.

29 Singh, S. and Batra, U.K. (1989). *Indian J. Chem., Sect. B* 28: 1–2.

30 Hoffmann, S., Seayad, A.M., and List, B. (2005). *Angew. Chem., Int. Ed.* 44: 7424–7427.

31 Rueping, M., Sugiono, E., Azap, C. et al. (2005). *Org. Lett.* 7: 3781–3783.

32 (a) Simon, L. and Goodman, J.M. (2008). *J. Am. Chem. Soc.* 130: 8741–8747. (b) Reid, J.P., Simon, L., and Goodman, J.M. (2016). *Acc. Chem. Res.* 49: 1029–1041.

33 Easson, L.H. and Stedman, E. (1933). *Biochem. J.* 27: 1257–1266.

34 Storer, R.I., Carrera, D.E., Ni, Y., and MacMillan, D.W.C. (2006). *J. Am. Chem. Soc.* 128: 84–86.

35 Hoffmann, S., Nicoletti, M., and List, B. (2006). *J. Am. Chem. Soc.* 128: 13074–13075.

36 Marcelli, T., Hammar, P., and Himo, F. (2009). *Adv. Synth. Catal.* 351: 525–529.

37 Rueping, M., Antonchik, A.P., and Theissmann, T. (2006). *Angew. Chem., Int. Ed.* 45: 6751–6755.

38 (a) Rueping, M., Merino, E., and Koenigs, R.M. (2010). *Adv. Synth. Catal.* 352: 2629–2634. (b) Rueping, M., Brinkmann, C., Antonchick, A.P., and Atodiresei, I. (2010). *Org. Lett.* 12: 4604–4607.

39 (a) Kang, Q., Zhao, Z.-A., and You, S.-L. (2007). *Adv. Synth. Catal.* 349: 1657–1660. (b) Li, G., Liang, Y., and Antilla, J.C. (2007). *J. Am. Chem. Soc.* 129: 5830–5831.

40 Kang, Q., Zhao, Z.-A., and You, S.-L. (2008). *Org. Lett.* 10: 2031–2034.

41 Li, G. and Antilla, J.C. (2009). *Org. Lett.* 11: 1075–1078.

42 Han, Z.-Y., Xiao, H., and Gong, L.-Z. (2009). *Bioorg. Med. Chem. Lett.* 19: 3729–3732.

43 Nguyen, T.B., Bousserouel, H., Wang, Q., and Gueritte, F. (2010). *Org. Lett.* 12: 4705–4707.

44 Nguyen, T.B., Bousserouel, H., Wang, Q., and Gueritte, F. (2011). *Adv. Synth. Catal.* 353: 257–262.

45 Nguyen, T.B., Wang, Q., and Guéritte, F. (2011). *Chem. Eur. J.* 17: 9576–9580.

46 Wakchaure, V.N., Zhou, J., Hoffmann, S., and List, B. (2010). *Angew. Chem., Int. Ed.* 49: 4612–4614.

47 Wakchaure, V.N., Nicoletti, M., Ratjen, L., and List, B. (2010). *Synlett*: 2708–2710.

48 Kumar, A., Sharma, S., and Maurya, R.A. (2010). *Adv. Synth. Catal.* 352: 2227–2232.

49 (a) Chen, M.-W., Chen, Q.-A., Duan, Y. et al. (2012). *Chem. Commun.* 48: 1698–1700. (b) Chen, Q.-A., Gao, K., Duan, Y. et al. (2012). *J. Am. Chem. Soc.* 134: 2442–2448.

50 Lackner, A.D., Samant, A.V., and Toste, F.D. (2013). *J. Am. Chem. Soc.* 135: 14090–14093.

51 Saito, K., Shibata, Y., Yamanaka, M., and Akiyama, T. (2013). *J. Am. Chem. Soc.* 135: 11740–11743.

52 Saito, K., Miyashita, H., and Akiyama, T. (2015). *Chem. Commun.* 51: 16648–16651.

53 Cheng, D.-J., Yan, L., Tian, S.-K. et al. (2014). *Angew. Chem., Int. Ed.* 53: 3684–3687.

54 Wakchaure, V.N., Kaib, P.S.J., Leutzsch, M., and List, B. (2015). *Angew. Chem., Int. Ed.* 54: 11852–11856.

55 Mazuela, J., Antonsson, T., Johansson, M.J. et al. (2017). *Org. Lett.* 19: 5541–5544.

56 Wakchaure, V.N. and List, B. (2016). *Angew. Chem., Int. Ed.* 55: 15775–15778.

57 Saito, K. and Akiyama, T. (2016). *Angew. Chem., Int. Ed.* 55: 3148–3152.

58 Mori, K., Miyake, A., and Akiyama, T. (2015). *Chem. Commun.* 51: 16107–16110.

59 Aillerie, A., Gosset, C., Dumont, C. et al. (2016). *RSC Adv.* 6: 54185–54188.

60 Zhu, C. and Akiyama, T. (2009). *Org. Lett.* 11: 4180–4183.

61 Zhu, C. and Akiyama, T. (2010). *Adv. Synth. Catal.* 352: 1846–1850.

62 Henseler, A., Kato, M., Mori, K., and Akiyama, T. (2011). *Angew. Chem., Int. Ed.* 50: 8180–8183.

63 Sakamoto, T., Horiguchi, K., Saito, K. et al. (2013). *Org. Chem.* 2: 943–946.

64 Saito, K. and Akiyama, T. (2012). *Chem. Commun.* 48: 4573–4575.

65 Sakamoto, T., Mori, K., and Akiyama, T. (2012). *Org. Lett.* 14: 3312–3315.

66 Zhou, J.-Q., Sheng, W.-J., Jia, J.-H. et al. (2013). *Tetrahedron Lett.* 54: 3082–3084.

67 Saito, K., Horiguchi, K., Shibata, Y. et al. (2014). *Chem. Eur. J.* 20: 7616–7620.

68 Saito, K., Miyashita, H., and Akiyama, T. (2014). *Org. Lett.* 16: 5312–5315.

69 Hachiya, I., Ito, A., and Shimizu, M. (2014). *Asian J. Org. Chem.* 3: 614–618.

70 Wen, W., Zeng, Y., Peng, L.-Y. et al. (2015). *Org. Lett.* 17: 3922–3925.

71 Horiguchi, K., Yamamoto, E., Saito, K. et al. (2016). *Chem. Eur. J.* 22: 8078–8083.

72 Miyagawa, M., Takashima, K., and Akiyama, T. (2018). *Synlett* 29: 1607–1610.

73 Chen, M.-W., Yang, Q., Deng, Z. et al. (2018). *J. Org. Chem.* 83: 8688–8694.

74 Yang, J.W., Fonseca, M.T.H., and List, B. (2005). *J. Am. Chem. Soc.* 127: 15036–15037.

75 Huang, Y., Walji, A.M., Larsen, C.H., and MacMillan, D.W.C. (2005). *J. Am. Chem. Soc.* 127: 15051–15053.

76 Zhou, J. and List, B. (2007). *J. Am. Chem. Soc.* 129: 7498–7499.

77 Rueping, M. and Antonchick, A.P. (2008). *Angew. Chem., Int. Ed.* 47: 5836–5838.

78 Ren, L., Lei, T., Ye, J.-X., and Gong, L.-Z. (2012). *Angew. Chem., Int. Ed.* 51: 771–774, S771/771-S771/763.

79 Enders, D., Liebich, J.X., and Raabe, G. (2010). *Chem. Eur. J.* 16: 9763–9766, S9763/9761-S9763/9721.

80 Rueping, M. and Hubener, L. (2011). *Synlett*: 1243–1246.

81 Rueping, M., Haack, K.L., Ieawsuwan, W. et al. (2011). *Chem. Commun.* 47: 3828–3830.

82 Wang, S.-G., Zhang, W., and You, S.-L. (2013). *Org. Lett.* 15: 1488–1491.

83 Liao, H.-H., Hsiao, C.-C., Sugiono, E., and Rueping, M. (2013). *Chem. Commun.* 49: 7953–7955.

84 Wang, Y.-G., Kumano, T., Kano, T., and Maruoka, K. (2009). *Org. Lett.* 11: 2027–2029.

85 Shi, F., Tan, W., Zhang, H.-H. et al. (2013). *Adv. Synth. Catal.* 355: 3715–3726.

86 Wang, S.-G. and You, S.-L. (2014). *Angew. Chem., Int. Ed.* 53: 2194–2197.

87 Zhang, Y.-C., Zhao, J.-J., Jiang, F. et al. (2014). *Angew. Chem., Int. Ed.* 53: 13912–13915.

88 Fu, N., Zhang, L., and Luo, S. (2015). *Org. Lett.* 17: 382–385.

89 Fu, N., Guo, Y., Zhang, L., and Luo, S. (2015). *Synthesis* 47: 2207–2216.

90 Michrowska, A. and List, B. (2009). *Nat. Chem.* 1: 225.

91 Rueping, M., Antonchick, A.P., and Theissmann, T. (2006). *Angew. Chem., Int. Ed.* 45: 3683–3686.

92 Li, G., Liu, H., Lv, G. et al. (2015). *Org. Lett.* 17: 4125–4127.

93 Rueping, M. and Antonchick, A.P. (2007). *Angew. Chem., Int. Ed.* 46: 4562–4565.

94 Guo, Q.-S., Du, D.-M., and Xu, J. (2008). *Angew. Chem., Int. Ed.* 47: 759–762.

95 Rueping, M., Theissmann, T., Raja, S., and Bats, J.W. (2008). *Adv. Synth. Catal.* 350: 1001–1006.

96 Metallinos, C., Barrett, F.B., and Xu, S. (2008). *Synlett*: 720–724.

97 Rueping, M., Tato, F., and Schoepke, F.R. (2010). *Chem. Eur. J.* 16: 2688–2691.

98 Rueping, M., Stoeckel, M., Sugiono, E., and Theissmann, T. (2010). *Tetrahedron* 66: 6565–6568.

99 Rueping, M., Theissmann, T., Stoeckel, M., and Antonchick, A.P. (2011). *Org. Biomol. Chem.* 9: 6844–6850.

100 Cai, X.-F., Chen, M.-W., Ye, Z.-S. et al. (2013). *Chem. -Asian J.* 8: 1381–1385.

101 Guo, R.-N., Chen, Z.-P., Cai, X.-F., and Zhou, Y.-G. (2014). *Synthesis* 46: 2751–2756.

102 Zhou, J., Zhang, Q.-F., Zhao, W.-H., and Jiang, G.-F. (2016). *Org. Biomol. Chem.* 14: 6937–6941.

103 Cai, X.-F., Guo, R.-N., Feng, G.-S. et al. (2014). *Org. Lett.* 16: 2680–2683.

104 Aillerie, A., Lemau de Talance, V., Moncomble, A. et al. (2014). *Org. Lett.* 16: 2982–2985.

105 Chen, M.-W., Cai, X.-F., Chen, Z.-P. et al. (2014). *Chem. Commun.* 50: 12526–12529.

106 Shugrue, C.R. and Miller, S.J. (2015). *Angew. Chem., Int. Ed.* 54: 11173–11176.

107 Shi, L., Ji, Y., Huang, W., and Zhou, Y. (2014). *Acta Chim. Sin.* 72: 820–824.

108 Isaac, K., Stemper, J., Servajean, V. et al. (2014). *J. Org. Chem.* 79: 9639–9646.

109 Ferry, A., Stemper, J., Marinetti, A. et al. (2014). *Eur. J. Org. Chem.* 2014: 188–193.

110 Wang, Y., Liu, W., Ren, W., and Shi, Y. (2015). *Org. Lett.* 17: 4976–4979.
111 Klussmann, M., Ratjen, L., Hoffmann, S. et al. (2010). *Synlett* 2010: 2189–2192.
112 Zhang, Y., Zhao, R., Bao, R.L.-Y., and Shi, L. (2015). *Eur. J. Org. Chem.* 2015: 3344–3351.

Index

Asymmetric Hydrogenation and Transfer Hydrogenation, First Edition.
Edited by Virginie Ratovelomanana-Vidal and Phannarath Phansavath.